Edward R. T. Tiekink, Julio Zukerman-Schpector (Eds.)
Multi-Component Crystals

Also of Interest

Highlights in Applied Mineralogy
Soraya Heuss-Aßbichler, Georg Amthauer, Melanie John (Eds.), 2017
ISBN 978-3-11-049122-7, e-ISBN (PDF) 978-3-11-049734-2,
e-ISBN (EPUB) 978-3-11-049508-9

Highlights in Mineralogical Crystallography
Thomas Armbruster, Rosa Micaela Danisi (Eds.), 2015
ISBN 978-3-11-041704-3, e-ISBN (PDF) 978-3-11-041710-4,
e-ISBN (EPUB) 978-3-11-041721-0

Phononic Crystals.
Artificial Crystals for Sonic, Acoustic, and Elastic Waves
Vincent Laude
ISBN 978-3-11-030265-3, e-ISBN (PDF) 978-3-11-030266-0,
e-ISBN (EPUB) 978-3-11-038791-9

Modern X-Ray Analysis on Single Crystals. A Practical Guide
Peter Luger, 2014
ISBN 978-3-11-030823-5, e-ISBN (PDF) 978-3-11-030828-0,
e-ISBN (EPUB) 978-3-11-037061-4

Symmetry. Through the Eyes of Old Masters
Emil Makovicky, 2016
ISBN 978-3-11-041705-0, e-ISBN (PDF) 978-3-11-041714-2,
e-ISBN (EPUB) 978-3-11-041719-7

Multi-Component Crystals

Synthesis, Concepts, Function

Edited by
Edward R. T. Tiekink, Julio Zukerman-Schpector

DE GRUYTER

Editors
Prof. Dr. Edward R. T. Tiekink
Sunway University
Faculty of Science & Technology
Centre for Crystalline Materials
No 5 Jalan Universiti
47500 Bandar Sunway
Malaysia
edward.tiekink@gmail.com

Prof. Dr. Julio Zukerman-Schpector
Federal University of São Carlos
Department of Chemistry
Laboratório de Cristalografia, Estereodinâmica e
Modelagem Molecular
C. P. 676
13565-905 São Carlos, SP
Brazil
julio@power.ufscar.br

ISBN 978-3-11-046365-1
e-ISBN (PDF) 978-3-11-046495-5
e-ISBN (EPUB) 978-3-11-046379-8
Set-ISBN 978-3-11-046496-2

Library of Congress Cataloging-in-Publication Data
A CIP catalog record for this book has been applied for at the Library of Congress.

Bibliographic information published by the Deutsche Nationalbibliothek
The Deutsche Nationalbibliothek lists this publication in the Deutsche Nationalbibliografie; detailed bibliographic data are available on the Internet at http://dnb.dnb.de.

Cover image: Nneirda/iStock/thinkstock
Typesetting: le-tex publishing services GmbH, Leipzig
Printing and binding: CPI books GmbH, Leck
♾ Printed on acid-free paper
Printed in Germany

www.degruyter.com

Preface

As the title of this volume suggests, this book covers topics in the general area of multi-component co-crystals, interpreted broadly to include crystals comprising at least two distinct components. The volume deals with two comprehensive themes; namely applications, as these are the primary motivations for studies in this area, and fundamentals, as these are ever evolving to higher levels of sophistication. Without dispute, the most prominent application of multi-component crystals is within the pharmaceutical industry. Here, an active pharmaceutical ingredient (API) is co-crystallised with another molecule, which is often drawn from a list of generally regarded as safe (GRAS) molecules to generate a new material that, in favourable circumstances, presents advantageous properties for the delivery of the co-crystal for therapeutic effect compared with the original API. Therefore, it is appropriate that the first chapter of this volume is devoted to the principles of co-crystal formation, particularly in relation to how this relates to the pharmaceutical industry. This theme is continued in Chapter 2, with a practical focus on how this technology can be exploited to provide new ways of delivering molecules with poor solubility for the treatment of tuberculosis. Chapter 3 links the first two and subsequent chapters, while also elaborating on general guidelines, e.g. by employing well-defined physical descriptors for intelligence-based multi-component crystal design, rather than "hit and miss" methods, of pharmaceutical co-crystals and those relevant to other applications. The subsequent chapters devoted to "applications" also serve to highlight the diverse impact of multi-component crystals upon materials science. Thus, co-crystal technology can be employed to moderate photochromism of the N-salicylideneaniline chromophore (Chapter 4). The focus of Chapter 5 is upon approaches based on prior knowledge of related systems, such as pyridyl derivatives, for the generation of co-crystals comprising quinoline or isomeric isoquinoline with a variety of coformers. Similarly, a bottom–up methodology is discussed in Chapter 6 to enable the formation of multi-component co-crystals of N-oxides.

Thus far, attention has largely been focussed upon ambient conditions; however, Chapter 7 moves towards discussing the impact upon multi-component crystals within conditions of low temperature and/or high pressure. In Chapter 8, the design of multi-component crystals is discussed, whereupon potentially reactive components are brought in close proximity to enable, for example, [2 + 2] cycloaddition reactions stimulated by UV light to occur to form molecules that prove almost intractable by conventional wet chemistry methods. In recognition of the reality that many co-crystals are generated from solution, Chapter 10 provides an overview of the phase diagrams, kinetics and thermodynamics of different modes, such as cooling, slow evaporation, of crystallising multi-component crystals from solution.

Although the emphasis thus far has been upon how co-crystals might be formed and their applications, the remaining chapters in this volume are devoted to ascer-

https://doi.org/10.1515/9783110464955-201

taining self-assembly patterns in multi-component crystals. To demonstrate, concepts relating to halogen bonding are discussed in Chapter 10, with particular reference to the co-crystal/salt continuum. In Chapter 11, stacking interactions and their energetics between benzene molecules, coordinated and non-coordinated, are evaluated with clear evidence provided for the particular stability of off-set interactions. In Chapter 12, halogen bonding is again featured, but this time its complementarity with hydrogen bonding in carbonyl and thiocarbonyl species is evaluated. Hydrogen bonding interactions between molecules containing diamide and pyridyl residues in co-crystals and metal complexes form the focus of Chapter 13. Finally, in Chapter 14, the influence of the main group element (lone pair)$\cdots\pi$(arene) interactions, operating in isolation from other supramolecular synthons, in solvates, the main group elements compound is surveyed with an emphasis on the supramolecular architectures they sustain.

Overall, a diverse range of topics ranging from applications to fundamentals are included in this volume to provide a state of art snapshot of key components within the burgeoning field of crystal engineering.

May, 2017

Edward R. T. Tiekink, Sunway University
Julio Zukerman–Schpector, Universidade Federal de São Carlos

Contents

List of contributors

Srinivasulu Aitipamula
Institute of Chemical and Engineering Sciences
1 Pesek Road, Jurong Island 627833, Singapore
srinivasulu_aitipamula@ices.a-star.edu.sg

Reginald B. H. Tan
National University of Singapore,
4, Engineering Drive 4, 117576 Singapore
Institute of Chemical and Engineering Sciences
1 Pesek Road, Jurong Island 627833, Singapore
reginald_tan@ices.a-star.edu.sg

Artem Surov
Russian Academy of Sciences
Institute of Solution Chemistry
Academicheskaya St., 1
1153045 Ivanovo, Russia
aos@isc-ras.ru

Alex Manin
Russian Academy of Sciences
Institute of Solution Chemistry
Academicheskaya St., 1
1153045 Ivanovo, Russia
anm@isc-ras.ru

German Perlovich
Russian Academy of Sciences
Institute of Solution Chemistry
Academicheskaya St., 1
1153045 Ivanovo, Russia
germanper@yandex.ru

Anna Krawczuk
Jagiellonian University in Krakow
Faculty of Chemistry
Gronostajowa 3
30-387 Krakow
Poland
krawczuk@chemia.uj.edu.pl

Marlena Gryl
Jagiellonian University in Krakow
Faculty of Chemistry
Gronostajowa 3
30-387 Krakow
Poland
gryl@chemia.uj.edu.pl

Hidehiro Uekusa
Tokyo Institute of Technology
Department of Chemistry and Materials Science
12-1-H62, Ookayama 2, Meguro-ku
Tokyo 152-8551
Japan
uekusa@chem.titech.ac.jp

Kohei Johmoto
Tokyo Institute of Technology
Department of Chemistry and Materials Science
12-1-H62, Ookayama 2, Meguro-ku
Tokyo 152-8551
Japan
johmoto@chem.titech.ac.jp

Jubaraj B. Baruah
Indian Institute of Technology
Department of Chemistry
Guwahati 781 039
India
juba@iitg.ernet.in

Elena Boldyreva
Institute of Solid State Chemistry and
Mechanochemistry
Siberian Branch of Russian Academy of Sciences
Kutateladze st. 18
630128 Novosibirsk
Russia
eboldyreva@yahoo.com

https://doi.org/10.1515/9783110464955-202

Leonard R. MacGillivray
University of Iowa
Department of Chemistry
Iowa City, Iowa 52358
USA
len-macgillivray@uiowa.edu

Kristin M. Hutchins
Texas Tech University
Department of Chemistry & Biochemistry
Lubbock, TX 79409
USA
kristin.hutchins@ttu.edu

Ryan H. Groeneman
University of Iowa
Department of Chemistry
Iowa City, IA 52242
USA
ryangroeneman19@webster.edu

Jelena Stojakovič
University of Iowa
Department of Chemistry
Iowa City, Iowa 52358
USA
jstojakovic@illumina.com

Tom Leyssens
Université Catholique de Louvain
Place L. Pasteur 1
1348 Louvain la Neuve
Belgium
tom.leyssens@uclouvain.be

Joop H. Ter Horst
Strathclyde University
99 George street, Glasgow G1 1RD
UK
Joop.terHorst@strath.ac.uk

Marc Fourmigue
Université de Rennes 1 & CNRS
Institut des Sciences Chimiques de Rennes (ISCR)
Campus de Beaulieu, 35042 Rennes
France
marc.fourmigue@univ-rennes1.fr

Snežana Zarić
University of Belgrade
Department of Chemistry
Studentski trg 12-16
11000 Belgrade
Serbia
szaric@chem.bg.ac.rs

Dušan Malenov
University of Belgrade
Department of Chemistry
Studentski trg 12-16
11000 Belgrade
Serbia
malenov@chem.bg.ac.rs

Ivana Antonijević
University of Belgrade
Institute of Chemistry, Technology and Metallurgy
Njegoševa 12
11001 Belgrade
Serbia
ivana@chem.bg.ac.rs

Orion B. Berryman
University of Montana
Chemistry Department
Missoula, MT 59812
USA
orion.berryman@umontana.edu

Daniel A. Decato
University of Montana
Chemistry Department
Missoula, MT 59812
USA
daniel.decato@umconnect.umt.edu

Edward R. T. Tiekink
Sunway University
School of Science & Technology
Research Centre for Crystalline Materials
No 5 Jalan Universiti
47500 Bandar Sunway
Malaysia
edwardt@sunway.edu.my

Julio Zukerman-Schpector
Federal University of São Carlos
Department of Chemistry
Laboratório de Cristalografia, Estereodinâmica e Modelagem Molecular
C.P. 676
13565-905 São Carlos, SP
Brazil
julio@power.ufscar.br

Srinivasulu Aitipamula and Reginald B. H. Tan

1 Pharmaceutical co-crystals: crystal engineering and applications

1.1 Introduction

1.1.1 Scope

The physicochemical properties of a drug depend on the properties of the active pharmaceutical ingredient (API) that it contains and its formulation type. A majority of existing and newly discovered APIs are known to pose severe challenges in drug development. These challenges are primarily associated with the poor aqueous solubility of APIs which is directly responsible for the poor bioavailability of the corresponding drugs in the patient. Therefore, pre-formulation of APIs requires a thorough knowledge of their solid state characteristics and physicochemical properties. Co-crystals, which are crystalline solids that contain more than one solid component, have received significant attention recently for modifying the physicochemical properties of APIs. Pharmaceutical co-crystals contain an API and a benign co-former generally regarded as a safe material for human consumption. Salts are thus far the most preferred solid forms for solubility improvement of APIs due to their greater affinity towards water molecules. However, the application of salt formation is confined to only APIs that are ionisable. In this regard, co-crystallisation has a potential advantage over salt formation as it can be applied even to non-ionisable APIs. Over the past decade, the co-crystallisation technique has been applied to a number of APIs that resulted in improved properties including solubility, dissolution rate, and stability. This chapter highlights the role of crystal engineering in the design of co-crystals, applications of co-crystals in drug development, and examples of representative co-crystal systems.

1.1.2 Different solid forms

The identification of a suitable solid form is the first and foremost step when developing a drug formulation. For a solid form to be suitable for development, it should have specific solid state characteristics. For example, the solid form should be stable under different humidity and temperature conditions, the solubility and dissolution rate should meet the requirements of the mode of action, and the solid form should possess good bioavailability, flowability and compressibility. A wide range of solid forms has been explored to select a suitable solid form for formulation development. These include amorphous forms, salts, polymorphs, hydrates and solvates [1]. While amorph-

https://doi.org/10.1515/9783110464955-001

ous forms are thermodynamically unstable and convert to crystalline forms, hydrates and solvates may lose water or solvent from the crystal lattice during processing or manufacturing. Polymorphs, which differ in their supramolecular assembly, are well known for their phase transformations to the most stable polymorphic form [2]. Salts, on the other hand, are by far the most common solids for drug development. Salt is formed when an acid reacts with a base, resulting in a molecular salt mediated by ionic interactions [3]. Solid solutions and eutectics have also been explored for some APIs. Applications of various solid forms of APIs in drug development are widely reviewed. Co-crystals constitute a subset of the broad category of multi-component crystals. Over the past decade, the technique of co-crystallisation has helped to broaden the landscape for solid form APIs and proved promising in modifying physicochemical properties. The definition of what constitutes a co-crystal has been the subject of continuous debate, yet there is no universally accepted definition. A generally accepted definition of co-crystals that has received a broad consensus is, "Co-crystals are solids that are crystalline single phase materials composed of two or more different molecular and/or ionic compounds generally in a stoichiometric ratio, and which are neither solvates nor simple salts" [4]. This definition distinguishes co-crystals from other solid forms, such as salts and hydrates, and infers that co-crystals are neutral molecular complexes containing more than one solid component.

1.1.3 History of co-crystals

The discovery of the first co-crystal dates back to 1844 when Wöhler realised that mixing a solution of benzoquinone and hydroquinone results in a crystalline substance that was later named quinhydrone [5]. The co-crystal was structurally characterised in 1958 and comprises zig-zag chains of benzoquinone and hydroquinone molecules mediated by O–H···O hydrogen bonds. Co-crystals have traditionally been termed molecular compounds, heteromolecular complexes and addition compounds in the older literature. The idea that two molecules can be combined to form a co-crystal was pursued by many researchers before the term "co-crystal" was coined. For example, co-crystallisation of drug molecules has been demonstrated as early as 1937 by Hoffman–La-Roche (1937), who made a co-crystal of pyrithyldione (a sedative and tranquiliser) and propyphenazone (a non-steroidal anti-inflammatory drug, NSAID) [6]. Subsequently, several API co-crystals were reported well before the formal beginning of the directed synthesis of pharmaceutical co-crystals. Some of the early examples include co-crystals containing antipyrene and salicylic acid, theophylline and phenobarbital, pyrimidine and barbituric acid, and sulfaproxyline and caffeine [7, 8].

The term co-crystal was first used, as indexed in SciFinder (accessed on 27 July 2016), by Lawton and Lopez in 1963 in a patent that indicated co-crystals (termed solid crystalline complexes) involving organic amines and bisphenol. These co-crystals were declared as useful photosensitive compositions that may be developed to produce a coloured image simply by the application of heat. However, the conceptual understanding of current co-crystal design strategies has been credited to Etter, whose seminal work on co-crystallisation experiments with some substituted urea molecules led to the formulation of hydrogen bonding rules in organic crystals [9]. Two of the three most prominent rules pertaining to the design of co-crystals are (1) all good proton donors and acceptors are used in hydrogen bonding and (2) the best proton donors and acceptors remaining after intramolecular hydrogen bond formation form intermolecular hydrogen bonds with one another. APIs are generally pre-disposed to multiple functional groups, and therefore, Etter's rules of hydrogen bonding serve as essential guiding principles for the design of API co-crystals. The concept of supramolecular synthon [10], or hydrogen bonded building units in crystal structures, by Desiraju, was guided by the vast structural information contained in the Cambridge Structural Database (CSD) [11], and has furthered co-crystal design as a directed synthetic approach based on crystal engineering strategies.

1.1.4 Role of pharmaceutical co-crystals in drug development

The internal structure of solids determines their properties. With respect to APIs, physicochemical properties such as stability, solubility, melting point, dissolution rate, and density. depend on the way molecules interact with each other in the solid state. Therefore, understanding the internal structure is a prerequisite in any attempt to modify the physicochemical properties of APIs. Solid oral dosage forms have evolved as the most preferred delivery systems for most therapeutic agents due to ease of handling, shelf-life, patient compliance and stability. Almost 80% of the currently marketed drugs are administered as oral dosage forms and the majority of these have low solubility and pose challenges to formulation development. Drug formulators often rely on techniques such as micronisation, solid dispersion, encapsulation, salt formation, and amorphous forms. for addressing solubility issues [12]. However, these techniques have inherent drawbacks with respect to manufacturing and may result in the reduced stability of the resulting formulation. In this regard, pharmaceutical co-crystals provide a novel and effective way of modifying physicochemical properties of APIs without changing their chemical nature, thereby maintaining their therapeutic activity. Pharmaceutical co-crystals are made with co-formers that are chosen from the list of Generally Recognised as Safe (GRAS) chemicals. GRAS chemicals are food additives considered safe for human consumption [13]. The list of GRAS chemicals constitutes a wide range of chemicals, which include aldehydes, alcohols, carboxylic acids, amides, and sweeteners. Therefore, the diversity of GRAS chemicals in terms of their structure and physicochemical properties provides an additional means of selecting a suitable co-former for a target change in the API. As opposed to salt formation, the co-crystallisation technique can be applied to both ionisable and non-ionisable drugs, and recent developments suggest that co-crystals of salts and their solvates/hydrates/polymorphs are becoming more frequent. Such co-crystals not only provide a fundamental understanding of co-crystallisation, but also serve as valuable solids for patent protection and intellectual property.

In summary, over the past decade, co-crystals have demonstrated significant promise in drug development. The physicochemical and pharmacokinetic properties of drug substances, such as solubility and dissolution rate, bioavailability, particle morphology and size, tableting and compaction, melting point, physical form, biochemical and hydration stability, and permeability have been modified by the co-crystallisation technique [14–16]. For example, it was recently shown that the solubility and dissolution rate of a non-ionisable anti-inflammatory drug, ethenzamide (EA), can be improved by co-crystallisation with multiple co-formers [17]. In addition, some of these co-crystals have also been shown to exist in multiple polymorphic forms, making them suitable candidates for studying structure/property correlations. The impact of co-crystallisation on various properties of APIs has been detailed with representative case studies in Section 1.3.

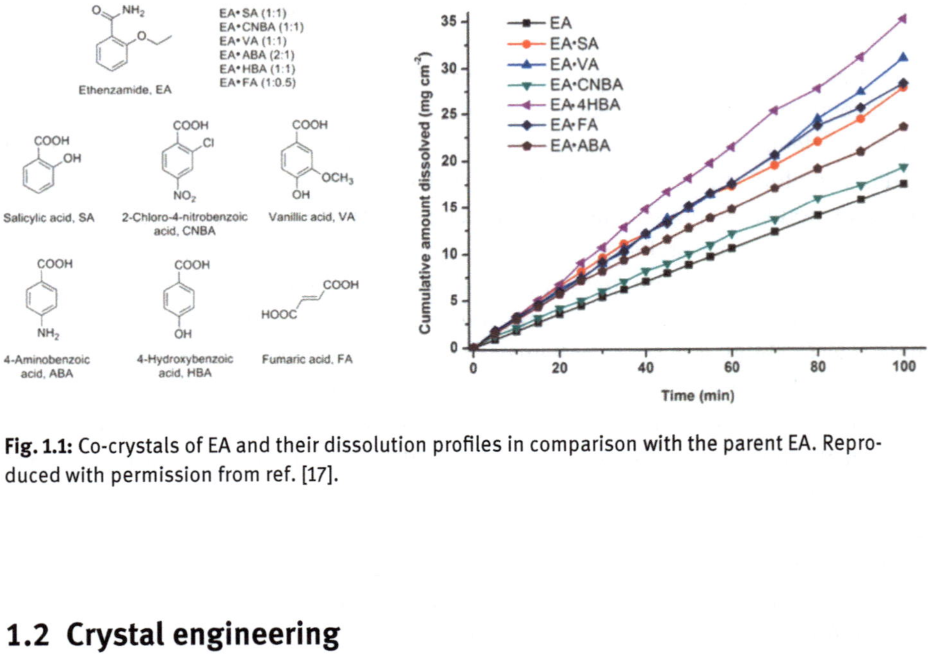

Fig. 1.1: Co-crystals of EA and their dissolution profiles in comparison with the parent EA. Reproduced with permission from ref. [17].

1.2 Crystal engineering

The term "crystal engineering" was first coined by Pepinsky in 1955 in the context of preparing metal-organic complexes to obtain advantageous properties [18]. However, Schmidt has been credited for formally implementing crystal engineering concepts for preparing crystals that undergo photochemical reactions, utilising the knowledge of intermolecular contact geometries of the known crystal structures [19]. Crystal engineering witnessed a rapid growth in the 1980s, especially in the fields of organic, metal-organic and inorganic solids. A modern definition for crystal engineering that has stood the test of time was provided by Desiraju in 1989 [20], who stated crystal engineering is "the understanding of intermolecular interactions in the context of crystal packing and in the utilisation of such understanding in the design of new solids with desired physical and chemical properties". Molecular crystals are viewed as an ensemble of molecules connected via intermolecular interactions such as hydrogen bonds, ionic or electrostatic bonds, metal-ligand bonds, secondary bonding, halogen bonding and van der Waals interactions. The resultant crystal structure is a manifestation of self-assembled molecules that provides detailed information on the way the molecules interact with adjacent molecules through the aforementioned intermolecular interactions. Therefore, the knowledge of intermolecular interactions, their nature and role in forming a specific crystal structure, is important in crystal engineering. Structural store houses such as the CSD and the inorganic crystal structure database (ICSD), provide a wide breadth of information pertaining to published crystal structures which can be used for statistical analysis of intermolecular interactions for the subsequent design of desired crystals. Self-assembly or molecular recognition in crystals is the result of intermolecular interactions involving functional groups present on the molecules. Recurring patterns of intermolecular interactions in a family of crystal structures are termed as supramolecular synthons, which were defined by Desiraju as "structural units within supermolecules which can be formed and/or assembled by known or conceivable synthetic operations involving intermolecular interactions" [10]. Supramolecular synthons are further classified as two basic types. The one that involves the same functional groups is called a supramolecular homosynthon and the one that involves different but complementary functional groups is called a supramolecular heterosynthon [21]. Co-crystallisation is the outcome of molecular recognition between two or more different molecules. Therefore, the concept of a supramolecular heterosynthon plays a vital role in the design of co-crystals.

Figure 1.2 demonstrates a crystal engineering strategy for the design of co-crystals of a platelet aggregation inhibition drug, triflusal (TFA) [22]. TFA poses multiple challenges in formulation development as it is practically insoluble in water, has a half-life of 0.5 h and rapidly converts to its deacetylated metabolite [23]. In addition, TFA is incompatible with all the excipients commonly used in chemical and pharmaceutical technology. Therefore, the marketed formulations do not contain any excipient. Co-crystals are possible alternatives to address stability issues of TFA. TFA has a carboxylic acid group that can form either acid-acid dimer synthon or acid-acid catemer synthon. If one were to make co-crystals of TFA, the first step would be to identify complementary functional groups that would form intermolecular interactions with the acid group. A survey of the crystal structures deposited in the CSD revealed that the carboxylic acid group forms strong heterosynthons with pyridine and amide groups. The high probability of occurrence of the acid–amide and acid–pyridine heterosynthons suggests that co-crystallisation of TFA with molecules that contain amide and pyridine functional groups could potentially yield co-crystals. Therefore, co-crystal screening of TFA was conducted with amide or pyridine functionalised compounds selected from the list of GRAS chemicals. Crystal structure analysis of the co-crystals revealed that all the co-crystals feature the expected heterosynthons. Some of the co-crystals have been subsequently proved stable at different conditions and claimed as promising candidates for developing stable TFA formulations. The following sections detail various types of intermolecular interactions that are being used for the crystal engineering of pharmaceutical co-crystals.

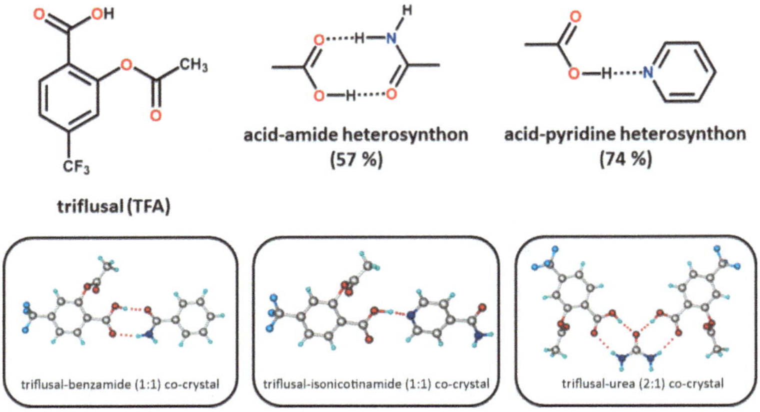

Fig. 1.2: A crystal engineering strategy for preparing co-crystals of triflusal (TFA) with molecules that contain amide and pyridine functional groups. Reproduced with permission from ref. [22].

1.2.1 Design of co-crystals based on conventional hydrogen bonds

Co-crystals are generally constructed from intermolecular interactions such as hydrogen bonding, halogen bonding and π···π interactions. The majority of the pharmaceutical co-crystals reported to date involve conventional hydrogen bonds of the type, N–H···O, N–H···N, O–H···O and O–H···N hydrogen bonds. This observation is most likely due to the fact that the functional groups such as carboxylic acid, amide, pyridine and hydroxyl are ubiquitous in APIs. In addition, these hydrogen bonds are relatively strong and directional, and thereby are prevalent in co-crystals. Some commonly used supramolecular synthons for co-crystal design are shown in Figure 1.3.

The use of strong hydrogen bonds in the design of co-crystals has been demonstrated widely. In general, heterosynthons are more energetically favourable and robust compared to the homosynthons, e.g. the acid-amide heterosynthons is favoured over both carboxylic acid and amide homosynthons. Similarly, hydroxyl-

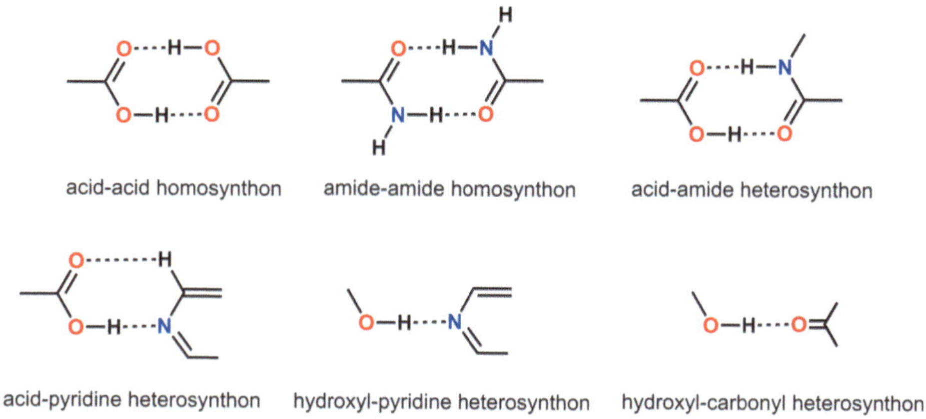

Fig. 1.3: Typical supramolecular synthons utilised for crystal engineering of pharmaceutical co-crystals.

pyridine and carboxylic acid-pyridine are strongly favoured compared to their respective homosynthons. As shown in Figure 1.2, triflusal (TFA) preferentially forms co-crystals with amide- and pyridine-containing molecules and all the co-crystals feature the expected acid-amide or acid-pyridine heterosynthons [22]. A number of co-crystals of APIs containing functional groups that can form strong hydrogen bonds have been reported. For example, the ability of an anti-epileptic and analgesic drug, carbamazepine, to form more than thirty diverse co-crystals has been credited to its ability to form strong hydrogen bonds involving its amide group [24]. The co-crystals of carbamazepine have been proved valuable for improving solubility, stability and dissolution rates. According to a recent report, a co-crystal of carbamazepine and saccharin was shown to be bioequivalent to carbamazepine form III in dogs, making the co-crystal a promising solid form for the development of carbamazepine formulations [25].

A recent case study that relies on strong hydrogen bonds for co-crystal formation involves a chemotherapy agent, tegafur (TG). TG has been used for the treatment of colorectal cancer. TG suffers from a severe drawback mainly due to non-uniform oral absorption, its short biological half-life due to rapid metabolism and its low solubility necessitating a high dose administration [26]. In order to discover a fast-dissolving and highly soluble solid form, a co-crystal screen was conducted. As shown in Figure 1.4, TG has a –CO–NH–CO– moiety and forms the imide-imide homosynthon in its parent crystal structure. Crystal structures that were retrieved from the CSD suggested that molecules that contain hydroxyl, pyridine and amide functional groups could potentially interrupt the imide-imide synthon to result in a co-crystal. Co-crystal screening with GRAS chemicals resulted in five co-crystals, with nicotinamide, isonicotinamide, 4-hydroxybenzoic acid, theophylline and pyrogallol, respectively [27]. All the co-crystals featured the expected supramolecular synthons. The evaluation of physico-

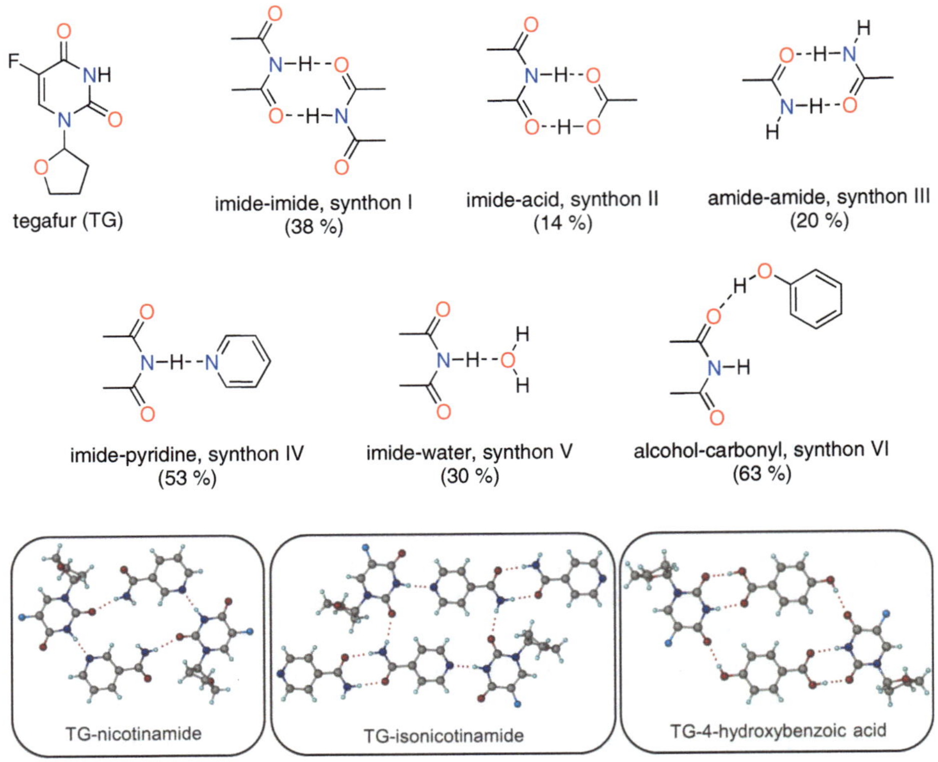

Fig. 1.4: Crystal engineering of TG co-crystals: plausible hydrogen bond synthons and some of the observed co-crystals. Reproduced with permission from ref. [27]. Copyright ©2014, The American Chemical Society.

chemical properties revealed that some of the co-crystals improved the solubility and dissolution rate of TG.

1.2.2 Design of co-crystals based on halogen bonds

The term halogen bond was introduced to describe any non-covalent interaction involving halogen atoms as electrophilic species. A more recent definition was formulated based on the consensus of an IUPAC task group that stated that "a halogen bond occurs when there is evidence of a net attractive interaction between an electrophilic region associated with a halogen atom in a molecular entity and a nucleophilic region in another, or the same, molecular entity" [28]. Over the years, the halogen bond has garnered significant interest not only from a structural chemistry point of view, but also from a crystal engineering perspective. The fact that the strength of halogen bonds is comparable with hydrogen bonds, i.e. ranging between weak (ca 5 kJ mol^{-1})

to strong (180 kJ mol^{-1} [29], and that halogen bonds exhibit high directionality, make them good design tools for the directed synthesis of supramolecules and higher-dimensional architectures. Accordingly, the use of halogen-bonds as a design element in crystal engineering has been known for more than 30 years. For example, Desiraju and co-workers used halogen-bond synthons to form a solvate between urotropin and CBr_4 [30]. Similarly, the synthesis of several one-dimensional (1D), 2D and 3D self-assembled halogen-bonded networks has been reported by several groups. Some typical halogen bonds that are frequently used in crystal engineering are shown in Figure 1.5.

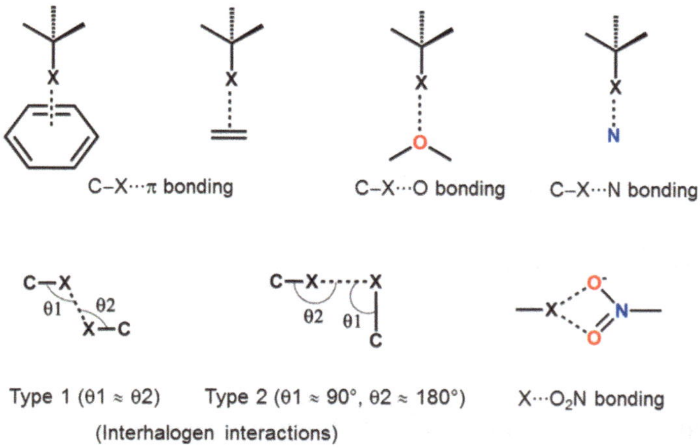

Fig. 1.5: Examples of typical supramolecular synthons mediated by halogen bonds.

The pioneering work of the research groups of Aakeroy, Nangia, Jones and Metrangolo and Resnati exemplified the use of halogen-bonds in the directed synthesis of co-crystal systems [31]. The design of higher-order multi-component crystals such as ternary co-crystals was also made possible by combining halogen bonds with hydrogen bonds. For example, a series of ternary co-crystals mediated by strong hydrogen bonded synthons such as acid-amide and amide-amide and X···O_2N halogen-bonds have been reported recently [32].

The abundance of halogen atoms in pharmaceutical compounds necessitates greater insight into halogen bonds for the development of crystal engineering strategies. Although there have been a number of co-crystals of neutral organic molecules that were formed through halogen bonding, a survey of the current literature on pharmaceutical co-crystals suggests that halogen bond mediated pharmaceutical co-crystals are quite scarce. A lone example that illustrates the key role of halogen bonding in driving the formation of a pharmaceutical co-crystal has been reported recently by the group of Metrangolo, Resnati and Terraneo [33]. Four co-crystals of an iodinated antimicrobial agent, 3-iodo-2-propynyl-*N*-butylcarbamate (IPBC) have been prepared and characterised. It was found that halogen bonding is the key interaction responsible for the self-assembly of the pharmaceutical co-crystals, wherein the halogen bond does not interfere with the hydrogen bonds present in the parent IPBC crystal structure (Figure 1.6). Some of the co-crystals have improved powder flow properties and thermal stability. The co-crystals further demonstrate the applications of the combined use of halogen and hydrogen bonds to obtain new co-crystals of halogenated APIs.

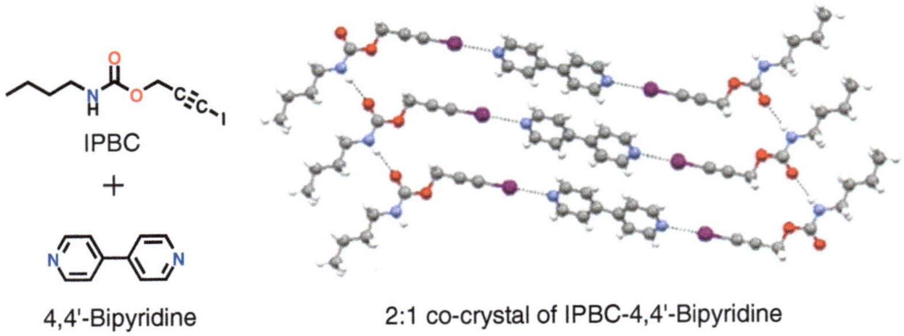

Fig. 1.6: A pharmaceutical co-crystal involving IPBC and 4,4′-bipyridine mediated by I···N halogen-bond. Reproduced with permission from ref. [33]. Copyright ©2013, The American Chemical Society.

1.2.3 Co-crystal design strategies for APIs that lack strong hydrogen bonding groups

Inter- and intramolecular interactions play a major role in dictating crystal structure. Therefore, understanding the nature of various intermolecular interactions, regardless of their being strong or weak, is a prerequisite in the design of crystalline materials. While the structure-directing role of strong hydrogen bonds has been well-established for crystal engineering applications, weaker interactions of the type π···π, C–H···π, O–H···π, N–H···π, C–H···X, etc. are also important for the self-assembly of organic molecules, and their geometrical preference and cooperative action can significantly contribute towards the conformation of organic molecules. Therefore, it is not surprising that studies concerning weak interactions are important and facilitate the design of co-crystals. For example, in the case of 4,4′-bis-hydroxyazobenzene, π···π and C–H···N interactions play a structure-directing role in the synthesis of ternary co-crystals [34].

The majority of the reported pharmaceutical co-crystals have been designed based on functional groups that can form strong hydrogen bonds with the co-formers. However, a sub-set of APIs, i.e. APIs devoid of functional groups that can form strong hydrogen bonds, has thus far been not been well explored for co-crystal design. Compared to the traditional way of designing co-crystals, the design of co-crystals for molecules that lack potential hydrogen bonding sites is rather difficult. For example, extensive co-crystal screening on the carbonyl-only containing molecules griseofulvin (GF), spironolactone and artemisinin each formed only a single co-crystal, except in the case of artemisinin, which formed two co-crystals [35–37](Figure 1.7). The difficulty in finding co-crystals for this type of API underscores the importance and need for more reliable co-crystal design strategies. The likelihood of co-crystal formation in the case of artemisinin has also been assessed using molecular descriptors which involve parameters such as molecular shape, size, polarity, and hydrogen bonding propensity [36]. A comparison of molecular descriptors of artemisinin and co-formers suggested that of the 75 co-formers chosen for the initial co-crystal screening, 33 would not have been expected to form co-crystal with artemisinin and could have been avoided in the initial co-crystal screen. Interestingly, despite favourable molecular descriptors for the rest of the co-formers and artemisinin, only two resulted in co-crystals. The limited success with the molecular descriptor approach suggests that a deeper understanding is required for formulating co-crystal design strategies for this class of API.

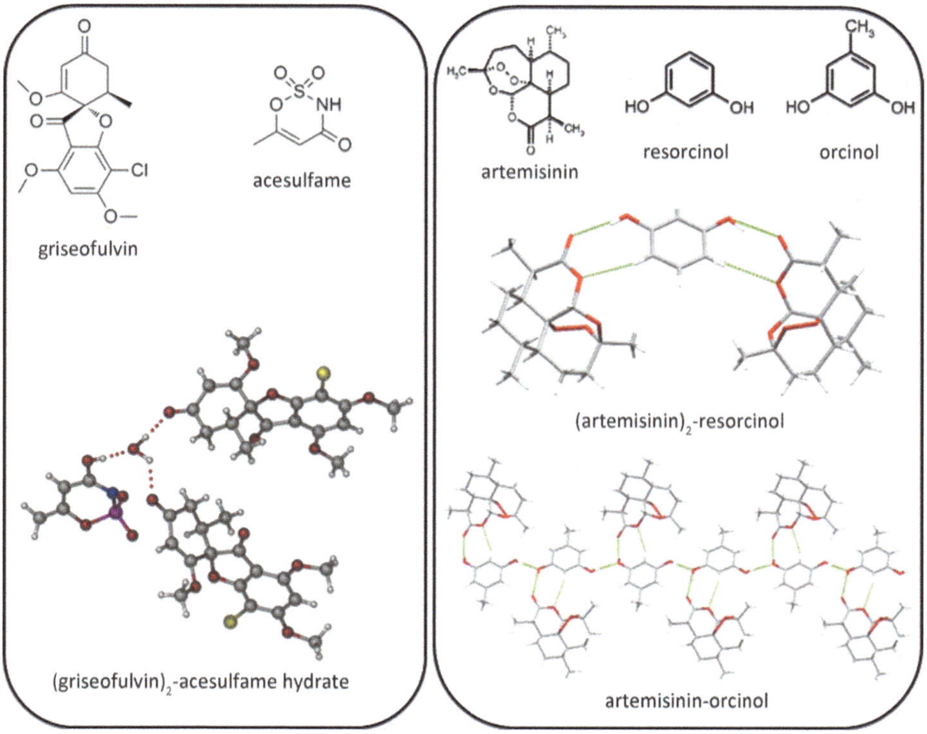

Fig. 1.7: Examples of co-crystals of molecules that lack strong hydrogen bonding groups, GF (left) and artemisinin (right). Reproduced from ref. [35] and ref. [36]. Copyright Royal Society of Chemistry, 2010.

1.2.4 Ionic co-crystals

The term 'ionic co-crystal' was coined by Braga and co-workers in 2010 [38], but the discovery of ionic co-crystals dates back to 1783 when de Romé de L'Isle observed a habit change of NaCl crystals when they were crystallised from an aqueous solution of urea [39]. Ionic co-crystals are generally formed by an organic molecule and an ionic salt (an alkali halide or organic cation halide) and are sustained by charge-assisted hydrogen bonds and/or coordination bonds (Figure 1.8). While an ionic co-crystal that contains alkali halides could be classified as a coordination polymer, an ionic co-crystal that contains an organic cation halide could be classified as a co-crystal of salts. Interest in ionic co-crystals has been growing in recent years and contemporary literature suggests that these co-crystals are also useful for the modification of physicochemical properties in APIs, such as dissolution rate, thermal stability, crystal habit and hygroscopicity. Compared to the conventional co-crystal design strategies for neutral organic molecules, the design of ionic co-crystals is rather difficult, and systematic studies that demonstrate crystal engineering of ionic co-crystals are seldom reported. With respect to the ionic co-crystals that contain alkali halides, a priori knowledge of the metal coordination patterns would help in the design strategy. However, the coordination number and geometry of alkaline and alkali earth metals are not easily predictable. Therefore, co-crystal formation for most of the reported ionic co-crystals has been first realised by trial and error methods. Most of the ionic co-crystals that contain organic cation halides are of hydrochloride salts. Chloride anions are known to be good hydrogen bonding acceptors and capable of forming intermolecular interactions with additional neutral molecules. Therefore, co-formers such as benzoic, fumaric, and succinic acids have been chosen for making salt co-crystals of hydrochloride salts of fluoxetine, ciprofloxacin, moxifloxacin, and saxagliptin [40]. The co-crystals proved effective in modulating physicochemical properties such as dissolution rate, solubility and stability.

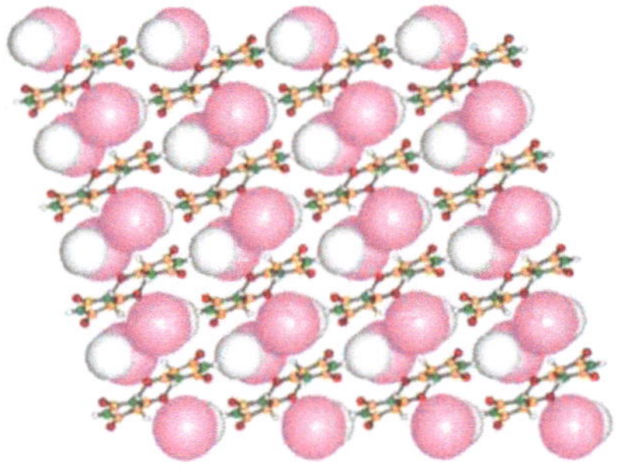

Fig. 1.8: An example of an ionic co-crystal: a co-crystal of benzoic acid (BA) and CsI showing the BA hydrogen bonded dimers within the piles of caesium and iodide ions (light-grey and pink spheres represent Cs^+ and I^-, respectively. Reproduced with permission from ref. [38]. Copyright Royal Society of Chemistry, 2010.

1.3 Applications of pharmaceutical co-crystals

The main motivation behind the development of pharmaceutical co-crystals is their ability to modify pharmaceutical properties of APIs. Pharmaceutical co-crystals have found diverse applications in many areas of drug development. These include, but are not limited to: improvement in solubility, dissolution rate, and permeability, which directly impact bioavailability, hygroscopicity, chemical and photo-stability. These in turn have a profound effect on storage stability, and mechanical properties that affect tabletability and processability. There have been attempts to correlate the resulting properties with the crystal structure of co-crystals and it has been found that co-formers that interact with drug molecules via heteromeric interactions (between co-formers and drug) greatly impact upon the physicochemical properties of APIs. In this section, the impact of co-crystallisation on some physicochemical properties of APIs is detailed with representative examples.

1.3.1 Enhancing bioavailability

The bioavailability of a drug depends on several factors including aqueous-solubility, dissolution rate, and drug permeability. For a Biopharmaceutics Classification System (BCS) class II drug, solubility behaviour is the key determinant of its oral bioavailability. Therefore, the impact of co-crystallisation on the solubility and dissolution rate of APIs will have potential implications on the bioavailability of the parent API.

The amount of drug dissolved in the solvent at the equilibrium state is defined as the solubility and the rate at which this equilibrium state is reached is the dissolution rate. Thus, solubility is a thermodynamic parameter while the dissolution rate is a kinetic phenomenon. Co-crystals offer potential advantages in terms of improving the solubility and dissolution rate of APIs. Babu and Nangia reviewed the solubility advantages offered by co-crystals and amorphous drugs and concluded that these species exhibit very similar solubility and dissolution rate profiles [16]. However, co-crystals, being in their crystalline state, score well on several fronts. For example, co-crystals that tend to be stable to humidity and storage are generally stable to drug processing, wet granulation, tableting, and compaction. The solubility and dissolution rate advantage of co-crystals has been demonstrated in co-crystals of several APIs. For example, Remenar et al. showed that the solubility of an anti-fungal drug can be improved by its L-malic acid and L-tartaric acid co-crystals [41]. Co-crystals have proved effective in improving the dissolution rate of APIs. For example, it was recently shown that the dissolution rate of the co-crystal hydrate of GF with acesulfame in pH 7.5 phosphate buffer at 37 °C is significantly faster than the parent GF [35]. As shown in Figure 1.9, the co-crystal reaches equilibrium within approximately 20 min. On the other hand, the dissolution rate of amorphous and crystalline GF is slower and reaches equilibrium only after 2 h. The striking similarity between the dissolution profiles of amorphous

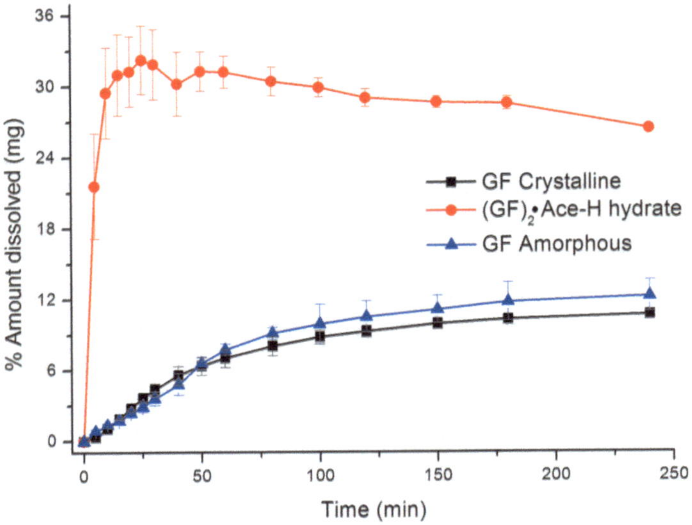

Fig. 1.9: Powder dissolution profile of GF and $(GF)_2 \cdot$ Ace–H hydrate in pH 7.5 buffer. Notice that the initial dissolution rate of $(GF)_2 \cdot$ Ace–H hydrate is significantly faster than that of the parent GF. Reprinted from ref. [35]. Copyright ©2012, American Chemical Society.

and crystalline GF is due to the rapid conversion of amorphous GF into crystalline GF in the dissolution medium.

Co-crystals have not only been used for improving the solubility and dissolution rate of APIs, but also to lower the solubility of the parent drug. For example, Nangia and co-workers successfully made co-crystals of sulfacetamide, an antibiotic used to treat conjunctivitis and other ocular ailments, and found that a co-crystal with caffeine showed lower solubility than the parent sulfacetamide [42]. Interestingly, the therapeutic efficacy of sulfacetamide in treating eye infections is severely limited by physiological constraints such as tear flow and reflex blinking which results in considerable drug loss. This necessitates frequent dosing which is often inconvenient to patients and amounts to excess drug loading. The lower solubility of the sulfacetamide-caffeine co-crystal, which is due to stronger hydrogen bonds and denser crystal packing in the co-crystal, could potentially address the poor residence time and faster elimination issues of sulfacetamide, in a fashion similar to extended release formulations.

The impact of co-crystallisation on the bioavailability has been demonstrated by in vivo bioavailability studies. For example, McNamara et al. prepared a co-crystal of a low solubility development candidate for neuopathic pain, 2-[4-(4-chloro-2-fluorophenoxy)phenyl]pyrimidine-4-carboxamide (I) with glutaric acid, and evaluated its physicochemical properties and bioavailability [43]. The co-crystal was stable at 40 °C/75% RH and 60 °C for 2 months. In addition, the high-aqueous solubility of glutaric acid contributes to the 18 times increase in the dissolution rate in water for the co-crystals. The bioavailability studies in beagle dogs confirmed that the area under the curve (AUC) values at both low dose (5 mg/kg) and high dose (50 mg/kg) for the co-crystal are much higher than for I. In another study, Childs et al. prepared a co-crystal involving danazol and vanillin, and compared its dissolution and bioavailability data with the formulated co-crystal. It has been shown that an aqueous suspension of the co-crystal that contained 1% of vitamin E-TPGS and 2% Klucel LR Pharm hydroxypropylcellulose (HPC) improved the bioavailability of the co-crystal over 10 times compared to the danazol polymorph. On the other hand, the pure co-crystal showed only a modest 1.7 times improvement in bioavailability (Figure 1.10) [44]. Similarly, the bioavailability of APIs such as metaxalone, carbamazepine, meloxicam, baicalein, sildenafil, danazol and a flavonoid quercetin have been improved by co-crystallisation with pharmaceutically acceptable co-formers.

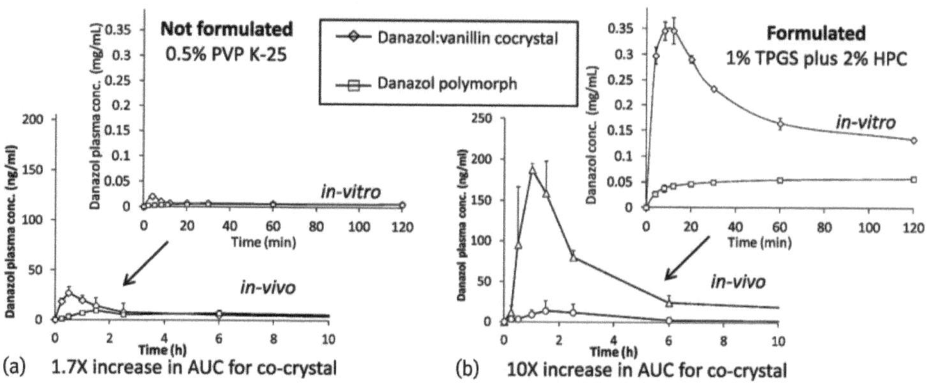

Fig. 1.10: In vitro dissolution data and in vivo plasma concentration for the danazol co-crystal and polymorph, shown for, the unformulated suspension (a) containing 0.5% PVP K-25 as a suspending agent and the formulated suspension (b) containing 1% TPGS and 2% HPC. Reproduced with permission from ref. [44]. Copyright ©2013, The American Chemical Society.

Permeability across biological membranes is a key factor that determines the absorption and distribution profiles of drugs. The flux or the amount of drug transported per unit time and per unit area of the membrane is a function of the permeability of the drug. The maximum drug absorption occurs when the drug has maximum permeability and maximum concentration at the site of absorption. The parallel artificial membrane permeability assay and Caco-2 cells are the most frequently used in vitro models to assess intestinal permeability. The presence of a co-former together with the parent API in a co-crystal leads to changes in the polarity of the relevant chemical entity thereby affecting the drug's permeability. Applications of co-crystals in modulating drug permeability have been reported only recently. For example, Desiraju and co-workers demonstrated the improvement in the permeability of a BCS class II, diuretic drug, hydrochlorothiazide (HCT) [45]. Five co-crystals with pharmaceutically acceptable co-formers, namely resorcinol (RES), 4-aminobenzoic acid (PABA), nicotinic acid (NIC), nicotinamide (NCT) and succinimide (SAM), were prepared and studied for their solubility and diffusion/permeability. It was found that most of the co-crystals showed enhanced solubility and flux/permeability. The solubility and permeability product best represents the impact of co-crystallisation on the pharmacokinetic properties of HCT. As shown in Figure 1.11, HCT-PABA co-crystals show the highest solubility-permeability product followed by the HCT-NIC co-crystal. The improved permeability of the co-crystal was attributed to the sulfonamide-amide heterosynthon formed between the API and the co-former. In contrast, the parent API in its polymorphs forms the sulfonamide-sulfonamide homosynthon.

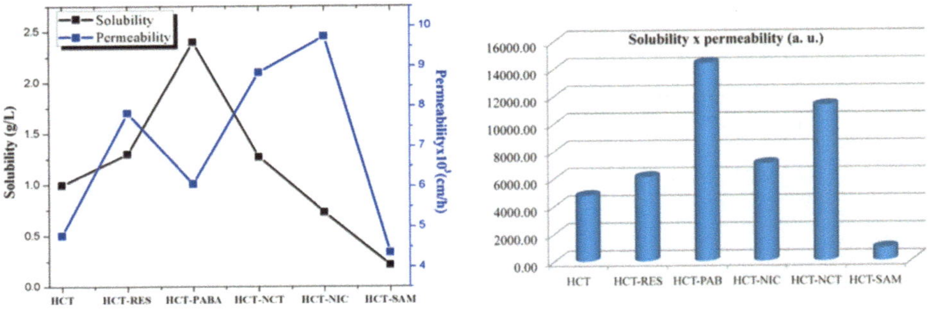

Fig. 1.11: Solubility and permeability interplay in co-crystals of HCT (left) and the product of solubility and permeability in HCT and its co-crystals (right). Reproduced with permission from ref. [45]. Copyright ©2015, The American Chemical Society.

A similar co-crystallisation strategy has been demonstrated to modulate the membrane permeability of an anti-neoplastic drug, 5-fluorouracil (5FU) [46]. 5FU is a BCS class III drug that shows good aqueous solubility but poor permeability. Approaches such as pro-drug and chemical enhancers have been evaluated to enhance the permeability of 5FU with different degrees of success. Chen and co-workers prepared three novel co-crystals of 5FU with the co-formers 3-hydroxybenzoic acid, PABA and cinnamic acid. A permeability study using a Franz diffusion cell with a silicone membrane revealed that the cumulative amount of drug permeated per unit area (Q_n) vs time indicated that the diffusion of the drug through membrane is linear and that the co-crystals showed a greater penetration rate than the pure drug (Figure 1.12). It was also found that under similar experimental conditions, physical mixtures of 5FU with 3-hydroxybenzoic acid and with cinnamic acid show similar cumulative permeated amounts per unit area to pure 5FU, while the physical mixture of 5FU and PABA show poorer diffusion. The better/improved permeability of 5FU when present in co-crystals compared to the physical mixture was rationalised based on the crystal structures of the co-crystals and packing efficiency. All three co-crystals are similar in their crystal structures in which the drug–drug intermolecular interactions in the parent 5FU are

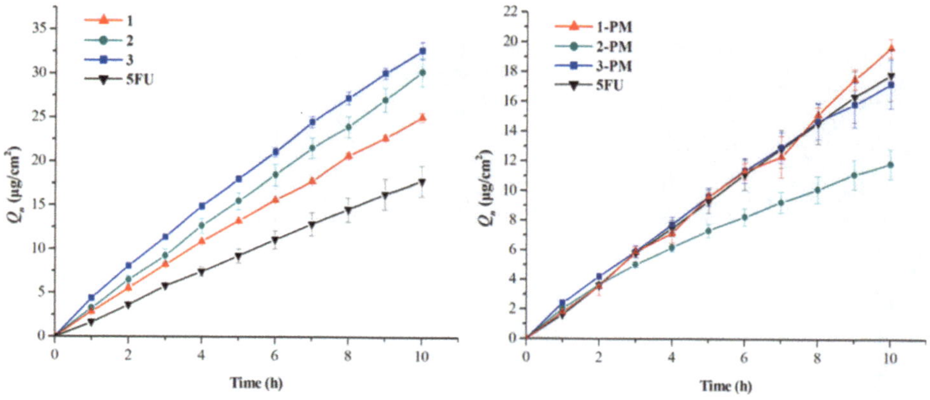

Fig. 1.12: Comparison of cumulative amount per unit area permeated (Q_n) of 5FU with the co-crystals (left) and physical mixture (right). **1**: 5FU-3-hydroxybenzoic acid, **2**: 5FU-PABA and **3**: 5FU-cinnamic acid. Reproduced with permission from ref. [46]. Copyright ©2016, The American Chemical Society.

replaced by drug-co-former interactions in the co-crystals with the calculated density of all three co-crystals being lower than the parent 5FU.

In addition to the case studies described above, co-crystals of a bronchodilator drug, theophylline, with isomeric aminobenzoic acids [47], an anti-HIV drug, acyclovir, with dicarboxylic acids [48] and a non-steroidal anti-inflammatory drug, indomethacin, with 2-hydroxy-4-methylpyridine, 2-methoxy-5-nitroaniline and saccharin have also been shown to modulate permeability of the parent drug [49].

1.3.2 Improving storage stability

The physical and chemical stability of a solid drug substance in the presence of atmospheric moisture is of concern to drug development as it has practical implications for processing, packaging, storage and formulation. In some instances, an anhydrous crystal form converts to a hydrate at high relative humidity (RH) conditions making it non-viable for further development. One such API that undergoes anhydrous to hydrate transformation is caffeine, which is a central nervous system stimulant and a smooth muscle relaxant. Caffeine is known to form a non-stoichiometric channel hydrate that contains 0.8 moles of water per mole of caffeine. Jones and co-workers prepared a series of caffeine co-crystals with aliphatic dicarboxylic acids and evaluated their stability with respect to atmospheric humidity [50]. Caffeine readily forms co-crystals with dicarboxylic acids typically in the 2 : 1 and 1 : 1 stoichiometries. While some of the co-crystals dissociate into crystalline starting components at high RH conditions, the caffeine-oxalic acid co-crystal was stable at all RHs and for all time points, and demonstrated superior stability to anhydrous caffeine (Figure 1.13).

• 2:1 Caffeine/oxalic acid • 2:1 Caffeine/malonic acid

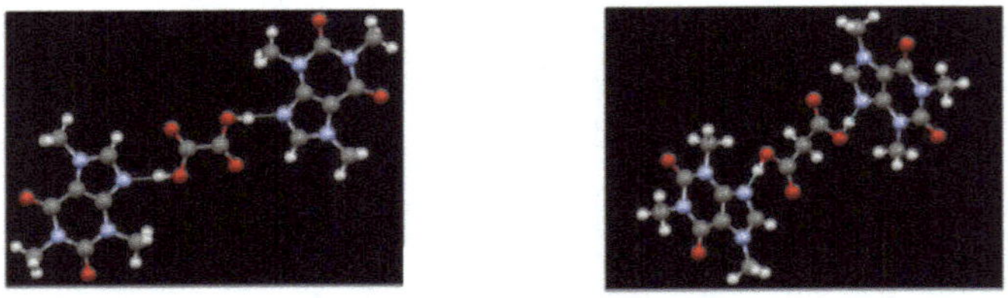

2:1 Caffeine/oxalic acid

RH	1 day	3 days	7 days	7 weeks
0%	✓	✓	✓	✓
43%	✓	✓	✓	✓
75%	✓	✓	✓	✓
98%	✓	✓	✓	✓

2:1 Caffeine/malonic acid

RH	1 day	3 days	7 days	7 weeks
0%	✓	✓	✓	✓
43%	✓	✓	✓	✓
75%	✓	✓	✓	✓
98%	✓	✓	✓	✗

Fig. 1.13: Hydrogen bonding and stability of caffeine co-crystals. Reproduced with permission from ref. [50]. Copyright ©2005, The American Chemical Society.

In vivo biochemical transformation of andrographolide, a diterpenoid lactone bioactive agent used in traditional medicine in China, India and Southeast Asian countries for anti-viral, anti-inflammatory, anti-cancer and anti-malarial applications, limits its bioavailability (2.67%) and aqueous-solubility (46 mg L^{-1}) [51]. A solution to the decreased efficacy of andrographolide owing to its biotransformation to the sulfate derivative will enhance the utility of andrographolide in medicine. Based on the hydroxyl groups available on the andrographolide, Nangia and co-workers chose molecules that contain C=O, COOH and OH functional groups for co-crystal screening that could potentially form O–H···O hydrogen bonds in the co-crystals [52]. A total of five co-crystals with GRAS chemicals were found and structurally characterised. An in vitro synthetic method was developed to mimic the in vivo biotransformation conditions. In this method, andrographolide and its co-crystals were subjected to HSO_3^- (in the presence of HSO_4^- and an excess amount of Na_2SO_3) to generate the sulfated product by the nucleophilic attack of the HSO_3^- ion at the β position of the α,β-unsaturated lactone moiety of andrographolide (Figure 1.14). The reaction was monitored by means of ^{1}H and ^{13}C NMR spectroscopy. While andrographolide and most of the co-crystals were found to transform to the metabolite, no transformation was found in the case of the co-crystal with salicylic acid. The acidic nature of salicylic acid was reasoned to provide an inhibitory effect on the chemical transformation of andrographolide. The carboxylic acid of the salicylic acid in the co-crystal is able to titrate HSO_3^- to H_2SO_3 and sodium salicylate, which stops/inhibits the subsequent addition reaction. The co-crystal has also exhibited a dissolution rate that is three times faster and a drug release that was two times higher than that of pure andrographolide.

Exposure to light may induce certain changes in drug products and affect the efficacy of the drug product. The majority of APIs are white in appearance, meaning they do not absorb light in the visible region, but may absorb in the UV region as a con-

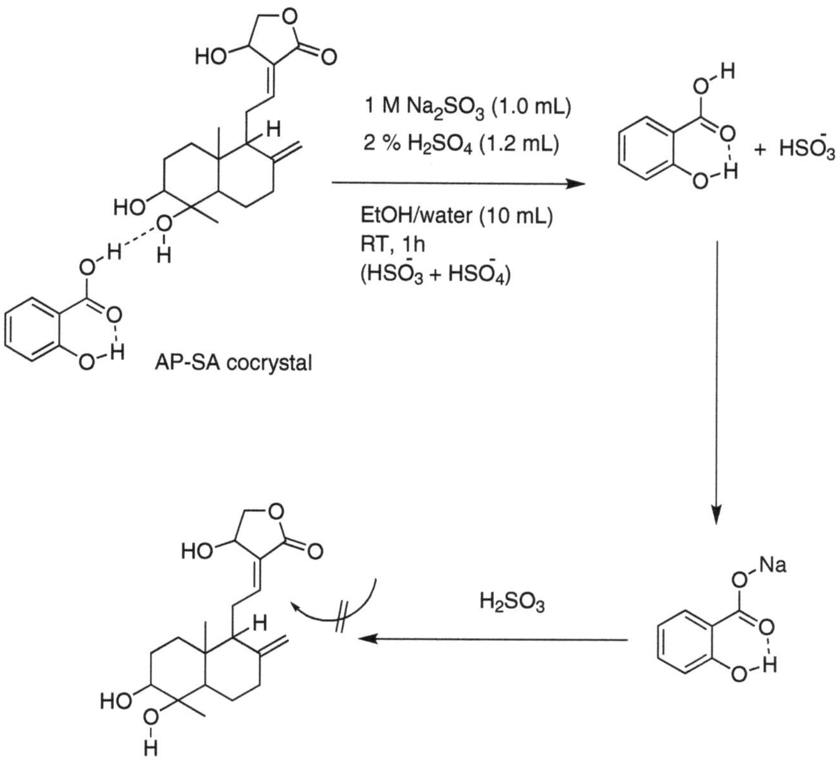

Fig. 1.14: A plausible mechanism for the inhibition of the reaction by the andrographolide-salicylic acid co-crystal. The sulfonate metabolite is not formed, as andrographolide is stable in the presence of salicylic acid because it quenches the HSO_3^-. Copyright ©2013 by John Wiley & Sons, Inc. Reprinted from ref. [52] by permission of John Wiley & Sons, Inc.

sequence of their chemical structure. The presence of aromatic residues and conjugated double bonds containing N, S, or O in the structure is usually associated with the ability of the molecule to absorb light. Some light-sensitive drugs are rapidly affected by the light and become discoloured due to the formation of degraded products. Therefore, knowledge of photochemical and photophysical properties of APIs is essential for handling, packaging, and labelling drug substances and drug products. A number of APIs are known to be photo-labile. For example, APIs such as chlorpromazine, carbamazepine, temozolamide, furosemide, nefidipine and nitrofurantoin undergo degradation upon irradiation by light. Attempts have been made to address the photo-stability issues of APIs by co-crystallisation with appropriate co-formers. For example, Vangala et al. showed co-crystals of an anti-bacterial drug, nitrofurantoin (NF), help to circumvent chemical degradation of NF under UV irradiation [53]. The improved photo-stability (Figure 1.15) of the co-crystals was attributed to restricted photo-isomerisation and the prevention of direct availability of incident photons to C=N of NF.

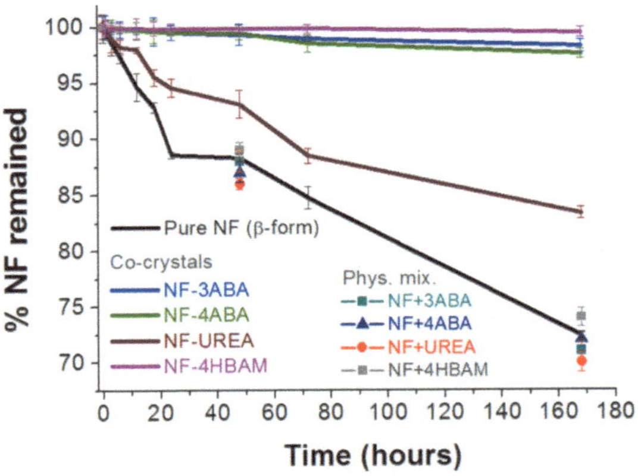

Fig. 1.15: UV irradiation of co-crystal and physical mixture samples involving NF for up to a week and the photo-degradation curves with error bars show the enhanced photo-stability of co-crystals with 3-aminobenzoic acid (3ABA), 4-aminobenzoic acid (4ABA) and 4-hydroxybenzamide (4HBAM). Reproduced with permission from ref. [53]. Copyright ©2012, The American Chemical Society.

In all co-crystals, NF forms hydrogen bond synthons with co-formers and such strong intermolecular interactions hinder photo-isomerisation.

A recent study by Nangia and co-workers has demonstrated applications of the co-crystallisation technique for the inhibition of discolouration of an anti-cancer drug, temozolamide (TMZ) [54]. TMZ is the only approved drug for the treatment of malignant brain tumours. The drug reacts with moisture and undergoes degradation to form a by-product, amino-imidazolecarboxamide. The degradation of TMZ is also associated with discoloration from white to tan-brown, which makes patients anxious and suspicious about drug efficacy. The stability issues of TMZ have been solved through a co-crystal formed with succinic acid. As shown in Figure 1.16, upon exposure to accelerated conditions (40 °C, 75% RH), the co-crystal did not show any discoloration or decomposition for up to 6 months. By contrast, TMZ degraded within a few weeks at ambient and accelerated conditions resulting in colour change from light pink to dark brown over the course of 5 weeks. The co-crystal was not only stable for over 6 months,

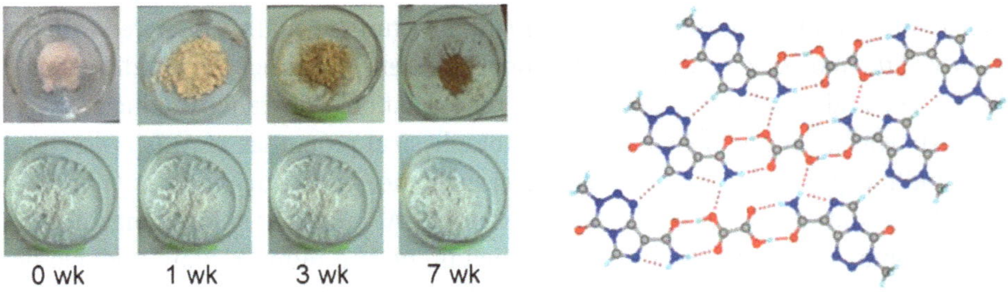

Fig. 1.16: Physical stability and colour comparison of pure TMZ (top panel) and TMZ–SA co-crystal (bottom panel) under accelerated ICH conditions of 40 °C and 75% RH. Pure TMZ showed darkening of colour from pink/light-tan to dark-brown over the course of 7 weeks, whereas the drug co-crystal remained white for a longer duration, i.e. 28 weeks (left). Amide–acid heterosynthon of N–H···O and O–H···O hydrogen bonds in the TMZ-succinic acid (2 : 1) co-crystal (right). Copyright ©2012 by John Wiley & Sons, Inc. Reprinted from ref. [54] by permission of John Wiley & Sons, Inc.

but was also found to be bio-equivalent to the reference drug, making it a promising alternative solid for the development of stable temozolamide drug formulations.

1.3.3 Desirable mechanical properties

Many materials, including drugs, undergo mechanical deformation upon the application of external force. In drug development, this is important as drugs undergo processes of milling, filling, tableting and compaction of powders. Therefore, understanding crystal packing in the context of mechanochemical properties is useful for modulating the mechanical properties of solid materials. In this regard, co-crystallisation techniques are promising for altering the crystal packing and mechanical properties of APIs. For example, Jones and co-workers applied co-crystallisation techniques to improve the tabletability of the widely used analgesic drug, paracetamol (PCA) [55]. PCA exists in three polymorphic forms and among these, two have been thoroughly studied. Metastable Form II, which forms flat hydrogen bonded layers in the crystal, is suitable for tableting. However, the thermodynamically stable Form I, with corrugated hydrogen bonded layers, has been developed as tablets with large amounts of binders to prevent chipping and disintegration. Co-crystal screening resulted in four co-crystals with the co-formers oxalic acid, theophylline, naphthalene and phenazine. As shown in Figure 1.17, Form I produces tablets with extensive chipping. On the other hand, all four co-crystals readily formed tablets. A rationale for the better tabletability of the co-crystals was based on the crystal structures of the co-crystals, as each co-crystal features a layered structure and presence of slip planes that help to enhance its tabletability.

Fig. 1.17: Tablets resulting from compression experiments involving PCA Form I and four co-crystals. Notice the poor shape and chipped tablets of PCA Form I. Copyright ©2009 by John Wiley & Sons, Inc. Reprinted from ref. [55] by permission of John Wiley & Sons, Inc.

Greater detail of the mechanical properties of molecular crystals can be deduced by nanoindentation, a technique used to extract elastic modulus (E) and hardness (H) of the material from load-displacement measurements. E and H are measures of the resistance offered by the materials to elastic and plastic deformations, respectively. High values of both E and H imply that the material is resistant to deformation and hence is brittle. Desiraju and co-workers have studied the mechanical properties of an antifungal drug, voriconazole (VOR) by nanoindentation [56]. The drug in its pure form is too soft for tableting and compacting. A number of salts and co-crystals were prepared and analysed for their mechanical properties. The stronger ionic interactions and lack of slip planes in the salts make them harder than VOR. On the other hand, the presence of a large number of slip planes and smaller number of interactions makes the co-crystals softer than VOR. The superior hardness of ionic salts compared to the VOR offer ease of handling and better tableting, making them suitable for development as new VOR formulations (Figure 1.18).

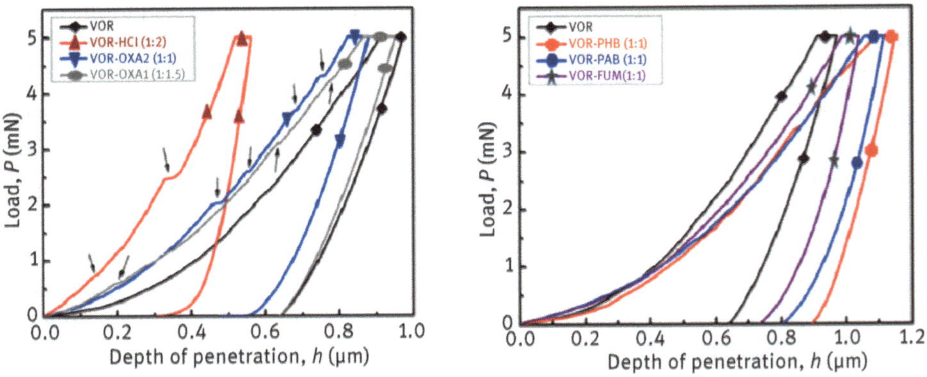

Fig. 1.18: Representative *P*–*h* curves obtained for VOR and its salts (left) and co-crystals (right). The *P*–*h* responses obtained for VOR crystals are smooth, indicating homogeneous plastic deformation. In contrast, all salts showed discrete displacement bursts, referred to as 'pop-ins', which indicates plastic deformation intermittently. The *P*–*h* curves obtained for the VOR co-crystals are smooth and a significant pile-up of indentation impressions was observed. Reproduced with permission from ref. [56]. Copyright ©2015, The American Chemical Society.

1.3.4 Energetic co-crystals

Energetic materials are compounds with high amounts of stored chemical energy, containing both oxidiser and fuel components that produce high temperature combustion products. The literature suggests that only a very small number of energetic materials have been developed as high explosives due to the need for using cost-effective production methods and meeting certain rigid performance targets [57]. Applications of co-crystallisation techniques in improving the properties of energetic materials have only been demonstrated recently. This is partly due to the fact that energetic materials are primarily defined by nitro groups and thus devoid of functional groups that form predictable inter-molecular interactions. Therefore, co-crystal design for the energetic materials should rely only on weaker interactions mediated by nitro groups and the size and shape mimicry of the components. Landenberger and Matzger have prepared 17 co-crystals of 2,4,6-trinitrotoluene (TNT) [57] with aromatic co-formers. Crystal structure analysis of the co-crystals revealed that donor-acceptor $\pi\cdots\pi$ interactions drive co-crystal formation. The strong polarising effect of nitro groups makes the TNT an electron-poor π-system. On the other hand, the chosen co-formers all possess relatively electron-rich aromatic systems. Therefore, donor-acceptor interaction between electron-poor TNT and electron-rich aromatic rings is the most favourable interaction. The evaluation of the properties to validate the viability of the co-crystal materials for energetic applications revealed that the co-crystals indeed increase density and thermal stability. However, the use of non-energetic compounds as co-formers led to the dilution of explosive power. It was reasoned that a co-crystal that contains two energetic materials would help to achieve the goal of improving the explosiveness of existing energetic materials. Such an energetic-energetic co-crystal has been found recently with 2,4,6,8,10,12-hexanitro-2,4,6,8,10,12-hexaazaisowurtzitane (CL-20) and TNT in a 1 : 1 molar ratio (Figure 1.19) [58]. CL-20 possesses favourable explosive characteristics, but its explosive applications are limited due to high production costs and high sensitivity. The co-crystal overcomes the problem of effective dilution of explosives by retaining the explosiveness of both CL-20 and TNT, and showed excellent impact sensitivity. Heating to 136 °C converts the co-crystal to liquid TNT and CL-20, and cooling the resultant mixture does not lead to the re-crystallisation of the

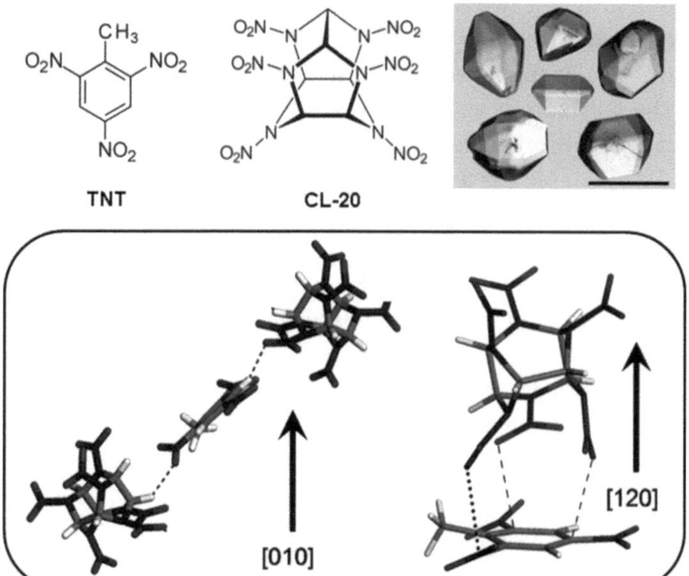

Fig. 1.19: Molecular components of the energetic-energetic co-crystal, TNT and CL-20, prismatic habit of the co-crystal, and interactions between TNT and CL-20. Copyright ©2011 by John Wiley & Sons, Inc. Reprinted from ref. [58] by permission of John Wiley & Sons, Inc.

co-crystal. This implies that an insensitive energetic material (co-crystal) that is not easily detonated by impact is transformed into a sensitive energetic material (CL-20) that is readily detonated by impact through heating. Therefore, the energetic-energetic co-crystal demonstrated the potential of co-crystallisation to produce effective explosive materials.

1.3.5 Drug–drug co-crystals

In general, pharmaceutical co-crystals constitute an API and one or more pharmaceutical acceptable co-formers. When the co-crystals are prepared with two or more APIs, the resulting co-crystals are termed drug–drug or multi-API co-crystals [59]. Thus, drug–drug co-crystals represent a sub-set of pharmaceutical co-crystals. This class of co-crystal has potential applications in developing combination drugs or fixed-dose combinations, which are defined as combinations of two or more APIs produced in a single tablet. Combination drug products are becoming increasingly popular because of advantages such as a greater therapeutic effect cf. component drugs administered separately, reduction of the number of prescriptions and administrative costs and an increase in patient compliance. The idea of using two or more APIs to form a co-crystal for therapeutic benefit is not new and has been implemented even before pharmaceutical co-crystals became popular (see Introduction to this chapter).

A drug–drug co-crystal involving two NSAID's, EA and gentisic acid, has been reported recently [59]. Interestingly, the co-crystal exists in three polymorphic forms. All the polymorphs were structurally characterised and their polymorphic phase transformations evaluated. The polymorphs show improved dissolution rates compared to the parent EA. Uekusa and co-workers recently reported a co-crystal of metformin (MET) with gliclazide (GLI) [60]. MET is a first line medication for type 2 diabetes, whereas GLI is a hypoglycaemic agent that is used for long-term treatment of diabetes mellitus. Both drugs pose formulation challenges in terms of their dissolution rate and hygroscopicity. A combination product involving MET and GLI has been suggested as a better medication for glycemic control. A 1 : 1 co-crystal was obtained by solid state grinding and solution crystallisation, and the crystal structure was determined. An evaluation of the physicochemical properties revealed that the dissolution rate of the drug–drug co-crystal was significantly greater than that of GLI itself (Figure 1.20). In the dynamic vapour sorption measurements, the co-crystal showed negligible sorption, on the other hand MET converts to its hydrate at higher RH conditions. Thus, the drug–drug co-crystal simultaneously addresses issues concerning the poor dissolution rate of GLI and hygroscopicity of MET. Drug–drug co-crystals are now frequently reported. For example, Evora et al. reported a co-crystal involving pyrazinamide and diflunisal [61]. Jiang et al. reported a series of drug–drug co-crystals of dapsone and showed that the co-crystals have better solubility and dissolution rates than dapsone [62]. Other recent examples of drug–drug co-crystals include lamivudine-

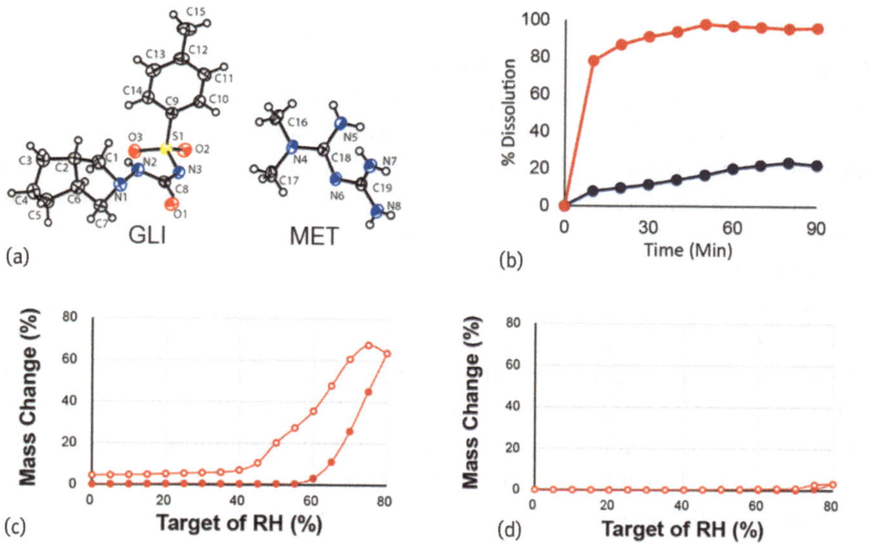

Fig. 1.20: A multi-drug co-crystal of MET and GLI, (a) asymmetric unit of the co-crystal, (b) dissolution profiles of GLI (blue) and MET-GLI co-crystal (red) in acidic medium (pH 1.2), (c) and (d) DVS charts of MET and MET-GLI respectively, showing negligible water sorption in the co-crystal. Reproduced with permission from ref. [60]. Copyright ©2016, The American Chemical Society.

zidovudine, curcumin-artemether, pioglitazone-imatinib, rosiglitazone-exenatide, sildenafil-aspirin, furosemide-pentoxifylline and hypochlorothiazide-pyrazinamide.

1.4 Polymorphism in co-crystals

Polymorphs refer to crystalline forms with distinct crystal structures of the same compound [2]. Polymorphism can exert a great influence on solid state properties of APIs, such as stability, solubility, bioavailability, and processability. Therefore, identifying novel polymorphs and establishing their stability profile are routinely conducted in pre-formulation stages of drug development. Akin to single component systems, a co-crystal can also exist in polymorphic forms [63]. In a recent exhaustive study on polymorphism in co-crystals, 114 co-crystals were identified from the literature [63]. A survey of these co-crystals revealed that it is not appropriate to distinguish co-crystals from single component crystals with respect to polymorphism and finding a polymorph of a co-crystal relies solely on the ability to find the right experimental conditions. Thus, with the recent interest in the development of pharmaceutical co-crystals, there is also an increase in the number of polymorphic co-crystals being reported. For example, it was shown that as many as four co-crystals of analgesic drug, EA, with gentisic acid, saccharin, 3,5-dinitrobenzoic acid and ethylmalonic acid exist in polymorphic forms [59, 64–66]. Some other recent examples of polymorphic co-crystals include celecoxib-δ-valerolactam, sulfacetamide-acetamide (dimorphs), orcinol-4,4′-bipyridine (tetramorphs), furosemide-nicotinamie (tetramorphs), carbamazepine-nicotinamide (dimorphs), gallic acid–succinimide (trimorphs). The impact of polymorphism on the performance of co-crystals has also been demonstrated in some case studies. For example, Goud and Nangia reported two polymorphs of a co-crystal involving a sulfonamide anti-biotic, sulfacetamide and acetamide [42]. The polymorphs were classified as synthon polymorphs based on the distinct hydrogen bond synthons in the crystal structures. An evaluation of the dissolution rates revealed that the polymorphs show differences in their rate of dissolution: metastable and stable polymorphs showing 1.6 and 1.3 times, respectively, faster dissolution rate than the parent sulfacetamide. In another study, Gonnade and co-workers showed that a 2 : 1 co-crystal of furosemide with 4,4′-bipyridine exists in two polymorphic forms that have distinct colours (Figure 1.21) even though both furosemide and 4,4′-bipyridine are colourless [67]. The colour difference between the two polymorphs was ascribed to the different π-stacking patterns between the constituent molecules. Density functional theory (DFT) calculations revealed the decrease of HOMO-LUMO gap from one polymorph to the other leading to the colour change from orange to light yellow.

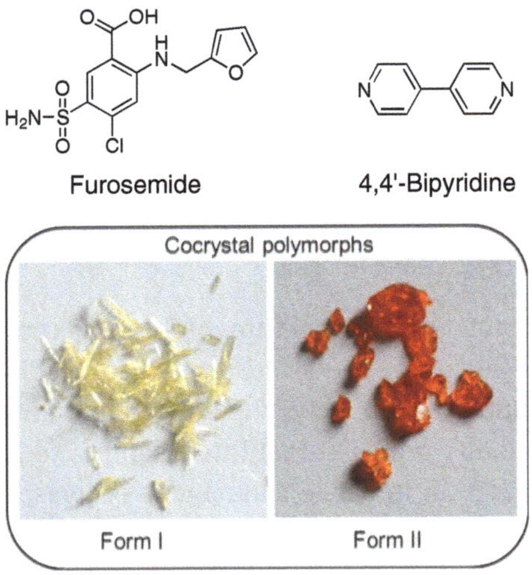

Fig. 1.21: Distinct colours of the polymorphs of the co-crystal furosemide-4,4′-bipyridine (2 : 1). Reproduced with permission from ref. [67]. Copyright ©2016, The American Chemical Society.

1.5 Co-crystals on the market

The US Food and Drug Administration (US-FDA) [68] and European Medicines Agency (EMA) [69], have released a draft guidance and reflection paper, respectively, on the subject of the regulatory classification of pharmaceutical co-crystals. While the

FDA classified pharmaceutical co-crystals as drug product intermediates and the co-formers used to make the co-crystals as excipients [68], EMA defined co-crystals as being solid state variants of APIs, aligning them with salts, polymorphs, hydrates and solvates [69]. The recognition by these regulatory bodies is expected to boost the development of co-crystals. US-FDA has recently approved a co-crystal hydrate that contains disodium valsartan and monosodium sacubitril [70]. The co-crystal was launched by Novartis with a trade name of Entresto for the treatment of chronic heart failure. The drug shows mortality benefits and has a patient tolerance similar to enalapril, another drug used for the treatment of heart failure. Notwithstanding this recent example, there are other marketed drug products that contain active ingredients in the form of co-crystals. For example, sodium valproate-valproic acid, caffeine citrate, escitalopram oxalate-oxalic acid, all fall under co-crystal classification. A number of recently discovered co-crystals are also showing promise for commercialisation. For example, a novel drug–drug co-crystal containing a NSAID, celecoxib, and an opioid drug, tramadol, which is being developed by Esteve R&D, showed promising results in Phase II clinical trials [71]. The results revealed excellent therapeutic benefits for the co-crystal at low doses of each active ingredient and has significantly superior analgesic efficacy compared to comparable doses of tramadol. Another promising co-crystal, currently in phase III clinical trials, is the co-crystal of an anti-diabetic drug ertugliflozin with 5-oxo-proline [72].

1.6 Conclusions

The field of pharmaceutical co-crystals has witnessed significant progress within the short span of a decade. A number of case studies have highlighted the role of co-crystal design strategies in the engineering of drug properties. Crystal engineering played a key role in the design of co-crystals that is complemented by the ever growing crystallographic data available from the CSD. Co-crystals address a broad range of issues in drug development that are associated with the physicochemical properties of active ingredients. Solubility, stability, permeability, mechanical, pharmacokinetic properties, etc. are effectively modulated by co-crystallisation approach. Recent US-FDA and EMA guidelines on regulatory classification of co-crystals have further validated the co-crystallisation approach for modifying the properties of active ingredients. Besides the recently approved co-crystal, Entresto, for the treatment of chronic heart failure, many co-crystals have shown promise in clinical trials and are expected to secure a place on the market. These developments provide further impetus for development of co-crystals and place pharmaceutical co-crystals in the elite group of solids alongside salts for addressing formulation issues in drug development.

Acknowledgement: We greatly acknowledge financial support from the Science and Engineering Research Council of A*STAR, Singapore.

Bibliography

[1] Byrn SR. Solid-State Chemistry of Drugs: Academic Press; 1982.
[2] Bernstein J. Polymorphism – A Perspective. Cryst. Growth Des. 2011, 11, 632–650.
[3] Wouters J, Quere L, Editors. Pharmaceutical Salts And Co-crystals. [In: RSC Drug Discovery Ser., 2012; 16]: Royal Society of Chemistry; 2012.
[4] Aitipamula S, Banerjee R, Bansal AK, et al. Polymorphs, Salts, and Cocrystals: What's in a Name? Cryst. Growth Des. 2012, 12, 2147–2152.
[5] Wöhler F. Untersuchungen über das Chinon. Annalen 1844, 51, 145–163.
[6] Lemmerer A, Bernstein J, Griesser UJ, et al. A Tale of Two Polymorphic Pharmaceuticals: Pyrithyldione and Propyphenazone and their 1937 Co-crystal Patent. Chem. – Eur. J. 2011, 17, 13445–13460, S13445/13441–S13445/13448.
[7] Ghosh M, Basak AK, Mazumdar SK, Sheldrick B. Structure and conformation of the 1 : 1 molecular complex sulfaproxyline-caffeine. Acta Crystallogr. C 1991, 47, 577–580.
[8] Singh TP, Vijayan M. Structural studies of analgesics and their interactions. II. The crystal structure of a 1 : 1 complex between antipyrine and salicyclic acid (salipyrine). Acta Crystallogr. Section B 1974, 30, 557–562.
[9] Etter MC. Encoding and decoding hydrogen-bond patterns of organic compounds. Acc. Chem. Res. 1990, 23, 120–126.
[10] Desiraju GR. Supramolecular synthons in crystal engineering – a new organic synthesis. Angew. Chem., Int. Edn. Engl. 1995, 34, 2311–2327.
[11] Allen FH. The Cambridge Structural Database: a quarter of a million crystal structures and rising. Acta Crystallogr. B Structural Science 2002, B58, 380–388.
[12] Morissette SL, Almarsson O, Peterson ML, et al. High-throughput crystallization: polymorphs, salts, co-crystals and solvates of pharmaceutical solids. Adv. Drug Delivery Rev. 2004, 56, 275–300.
[13] Remenar J, MacPhee M, Peterson ML, Morissette SL, Almarsson O, inventors; Transform Pharmaceuticals, Inc., USA, assignee. Novel conazole crystalline forms and related processes, pharmaceutical compositions and methods patent WO2003101392A2. 2003.
[14] Duggirala NK, Perry ML, Almarsson O, Zaworotko MJ. Pharmaceutical cocrystals: along the path to improved medicines. Chem. Commun. 2016, 52, 640–655.
[15] Bolla G, Nangia A. Pharmaceutical cocrystals: walking the talk. Chem. Commun. 2016, 52, 8342–8360.
[16] Babu NJ, Nangia A. Solubility Advantage of Amorphous Drugs and Pharmaceutical Cocrystals. Cryst. Growth Des. 2011, 11, 2662–2679.
[17] Aitipamula S, Wong ABH, Chow PS, Tan RBH. Pharmaceutical cocrystals of ethenzamide: structural, solubility and dissolution studies. CrystEngComm 2012, 14, 8515–8524.
[18] Pepinsky R. Crystal engineering: New concepts in Crystallography. Phys. Rev. 1955, 100, 971–971.
[19] Schmidt GMJ. Photodimerization in the solid state. Pure Appl. Chem. 1971, 27, 647–678.
[20] Desiraju GR. Crystal Engineering: The Design of Organic Solids. Amsterdam: Elsevier; 1989.
[21] Walsh RDB, Bradner MW, Fleischman S, et al. Crystal engineering of the composition of pharmaceutical phases. Chemical Communications (Cambridge, United Kingdom) 2003, 186–187.
[22] Aitipamula S, Mapp LK, Wong ABH, Chow PS, Tan RBH. Novel pharmaceutical cocrystals of triflusal: crystal engineering and physicochemical characterization. CrystEngComm 2015, 17, 9323–9335.
[23] Planavila A, Rodriguez-Calvo R, de Arriba AF, et al. Inhibition of cardiac hypertrophy by triflusal (4-trifluoromethyl derivative of salicylate) and its active metabolite. Mol. Pharmacol. 2006, 69, 1174–1181.

[24] Childs SL, Wood PA, Rodríguez-Hornedo N, Reddy LS, Hardcastle KI. Analysis of 50 Crystal Structures Containing Carbamazepine Using the Materials Module of Mercury CSD. Cryst. Growth Des. 2009, 9, 1869–1888.

[25] Hickey MB, Peterson ML, Scoppettuolo LA, et al. Performance comparison of a co-crystal of carbamazepine with marketed product. Eur. J. Pharm. Biopharm. 2007, 67, 112–119.

[26] Cooke JWB, Bright R, Coleman MJ, Jenkins KP. Process Research and Development of a Dihydropyrimidine Dehydrogenase Inactivator: Large-Scale Preparation of Eniluracil Using a Sonogashira Coupling. Org. Proc. Res. Develop. 2001, 5, 383–386.

[27] Aitipamula S, Chow PS, Tan RBH. Crystal Engineering of Tegafur Cocrystals: Structural Analysis and Physicochemical Properties. Cryst. Growth Des. 2014, 14, 6557–6569.

[28] Desiraju R, Ho PS, Kloo L, et al. Definition of the halogen bond (IUPAC Recommendations 2013). In: Pure and Applied Chemistry; 2013:1711.

[29] Metrangolo P, Resnati G, Pilati T, Liantonio R, Meyer F. Engineering functional materials by halogen bonding. J. Poly. Sci. A: Poly. Chem. 2007, 45, 1–15.

[30] Reddy DS, Craig DC, Rae AD, Desiraju GR. N···Br mediated diamondoid network in the crystalline complex carbon tetrabromide: hexamethylenetetramine. J. Chem. Soc., Chem. Commun. 1993, 1737–1739.

[31] Metrangolo P, Meyer F, Pilati T, Resnati G, Terraneo G. Halogen Bonding in Supramolecular Chemistry. Angew. Chem., Int. Edn 2008, 47, 6114–6127.

[32] Tothadi S, Sanphui P, Desiraju GR. Obtaining Synthon Modularity in Ternary Cocrystals with Hydrogen Bonds and Halogen Bonds. Cryst. Growth Des. 2014, 14, 5293–5302.

[33] Baldrighi M, Cavallo G, Chierotti MR, et al. Halogen Bonding and Pharmaceutical Cocrystals: The Case of a Widely Used Preservative. Mol. Pharm. 2013, 10, 1760–1772.

[34] Chakraborty S, Rajput L, Desiraju GR. Designing Ternary Co-crystals with Stacking Interactions and Weak Hydrogen Bonds. 4,4′-Bis-hydroxyazobenzene. Cryst. Growth Des. 2014, 14, 2571–2577.

[35] Aitipamula S, Vangala VR, Chow PS, Tan RBH. Cocrystal Hydrate of an Antifungal Drug, Griseofulvin, with Promising Physicochemical Properties. Cryst. Growth Des. 2012, 12, 5858–5863.

[36] Karki S, Friscic T, Fabian L, Jones W. New solid forms of artemisinin obtained through cocrystallisation. CrystEngComm 2010, 12, 4038–4041.

[37] Takata N, Takano R, Uekusa H, Hayashi Y, Terada K. A Spironolactone-Saccharin 1 : 1 Cocrystal Hemihydrate. Cryst. Growth Des. 2010, 10, 2116–2122.

[38] Braga D, Grepioni F, Maini L, Prosperi S, Gobetto R, Chierotti MR. From unexpected reactions to a new family of ionic co-crystals: the case of barbituric acid with alkali bromides and caesium iodide. Chem. Commun. 2010, 46, 7715–7717.

[39] De Romé de L'Isle JBL. Crystallographie. 2nd edn.; 1783.

[40] Childs SL, Chyall LJ, Dunlap JT, Smolenskaya VN, Stahly BC, Stahly GP. Crystal Engineering Approach To Forming Cocrystals of Amine Hydrochlorides with Organic Acids. Molecular Complexes of Fluoxetine Hydrochloride with Benzoic, Succinic, and Fumaric Acids. J. Am. Chem. Soc. 2004, 126, 13335–13342.

[41] Remenar JF, Morissette SL, Peterson ML, et al. Crystal engineering of novel cocrystals of a triazole drug with 1,4-dicarboxylic acids. J. Am. Chem. Soc. 2003, 125, 8456–8457.

[42] Rajesh Goud N, Khan RA, Nangia A. Modulating the solubility of sulfacetamide by means of cocrystals. CrystEngComm 2014, 16, 5859–5869.

[43] McNamara DP, Childs SL, Giordano J, et al. Use of a glutaric acid cocrystal to improve oral bioavailability of a low solubility API. Pharm. Res. 2006, 23, 1888–1897.

[44] Childs SL, Kandi P, Lingireddy SR. Formulation of a Danazol Cocrystal with Controlled Supersaturation Plays an Essential Role in Improving Bioavailability. Mole. Pharm. 2013, 10, 3112–3127.

[45] Sanphui P, Devi VK, Clara D, Malviya N, Ganguly S, Desiraju GR. Cocrystals of Hydrochlorothiazide: Solubility and Diffusion/Permeability Enhancements through Drug-Coformer Interactions. Mol. Pharm. 2015, 12, 1615–1622.
[46] Dai X-L, Li S, Chen J-M, Lu T-B. Improving the Membrane Permeability of 5-Fluorouracil via Cocrystallization. Cryst. Growth Des. 2016.
[47] Saikia B, Bora P, Khatioda R, Sarma B. Hydrogen bond synthons in the interplay of solubility and membrane permeability/diffusion in variable stoichiometry drug cocrystals. Cryst. Growth Des. 2015, 15, 5593–5603.
[48] Sarkar A, Rohani S. Cocrystals of acyclovir with promising physicochemical properties. J. Pharm. Sci. 2015, 104, 98–105.
[49] Ferretti V, Dalpiaz A, Bertolasi V, et al. Indomethacin co-crystals and their parent mixtures: does the intestinal barrier recognize them differently? Mol. Pharm. 2015, 12, 1501–1511.
[50] Trask AV, Motherwell WDS, Jones W. Pharmaceutical Cocrystallization: Engineering a Remedy for Caffeine Hydration. Cryst. Growth Des. 2005, 5, 1013–1021.
[51] Chen M, Xie C, Liu L. Solubility of Andrographolide in Various Solvents from (288.2 to 323.2) K. J. Chem. Eng. Data 2010, 55, 5297–5298.
[52] Suresh K, Goud NR, Nangia A. Andrographolide: Solving Chemical Instability and Poor Solubility by Means of Cocrystals. Chem. – Asian J. 2013, 8, 3032–3041.
[53] Vangala VR, Chow PS, Tan RBH. Co-Crystals and Co-Crystal Hydrates of the Antibiotic Nitrofurantoin: Structural Studies and Physicochemical Properties. Cryst. Growth Des. 2012, 12, 5925–5938.
[54] Babu NJ, Sanphui P, Nangia A. Crystal engineering of stable temozolomide cocrystals. Chem. Asian J. 2012, 7, 2274–2285.
[55] Karki S, Friscic T, Fabian L, Laity PR, Day GM, Jones W. Improving mechanical properties of crystalline solids by cocrystal formation: new compressible forms of paracetamol. Adv. Mater. (Weinheim, Fed. Repub. Ger.) 2009, 21, 3905–3909.
[56] Sanphui P, Mishra MK, Ramamurty U, Desiraju GR. Tuning Mechanical Properties of Pharmaceutical Crystals with Multicomponent Crystals: Voriconazole as a Case Study. Molecular Pharmaceutics 2015, 12, 889–897.
[57] Landenberger KB, Matzger AJ. Cocrystal Engineering of a Prototype Energetic Material: Supramolecular Chemistry of 2,4,6-Trinitrotoluene. Cryst. Growth Des. 2010, 10, 5341–5347.
[58] Bolton O, Matzger AJ. Improved stability and smart-material functionality realized in an energetic cocrystal. Angew. Chem. 2011, 50, 8960–8963.
[59] Aitipamula S, Chow PS, Tan RBH. Trimorphs of a pharmaceutical cocrystal involving two active pharmaceutical ingredients: potential relevance to combination drugs. CrystEngComm 2009, 11, 1823–1827.
[60] Putra OD, Furuishi T, Yonemochi E, Terada K, Uekusa H. Drug–Drug Multicomponent Crystals as an Effective Technique to Overcome Weaknesses in Parent Drugs. Cryst. Growth Des. 2016, 16, 3577–3581.
[61] Evora AOL, Castro RAE, Maria TMR, et al. Pyrazinamide-Diflunisal: A New Dual-Drug Co-Crystal. Cryst. Growth Des. 2011, 11, 4780–4788.
[62] Jiang L, Huang Y, Zhang Q, He H, Xu Y, Mei X. Preparation and Solid-State Characterization of Dapsone Drug-Drug Co-Crystals. Cryst. Growth Des. 2014, 14, 4562–4573.
[63] Aitipamula S, Chow PS, Tan RBH. Polymorphism in cocrystals. A review and assessment of its significance. Cryst. Eng. Comm. 2014, 16, 3451–3465.
[64] Aitipamula S, Chow PS, Tan RBH. Polymorphs and Solvates of a Cocrystal Involving an Analgesic Drug, Ethenzamide, and 3,5-Dinitrobenzoic Acid. Cryst. Growth Des. 2010, 10, 2229–2238.

[65] Aitipamula S, Chow PS, Tan RBH. Dimorphs of a 1 : 1 cocrystal of ethenzamide and saccharin: solid-state grinding methods result in metastable polymorph. CrystEngComm 2009, 11, 889–895.

[66] Aitipamula S, Wong ABH, Chow PS, Tan RBH. Polymorphism and phase transformations of a cocrystal of nicotinamide and pimelic acid. Cryst. Eng. Comm. 2012, 14, 8193–8198.

[67] Sangtani ESS, Thorat SH, Gawade RL, Jha KK, Munshi P, Gonnade RG. Furosemide Cocrystals with Pyridines: An Interesting Case of Color Cocrystal Polymorphism. Cryst. Growth Des. 2015, 15, 5858–5872.

[68] Guidance for Industry: Regulatory Classification of Pharmaceutical Co-crystals; Food and Drug Administration: Silver Spring, MD, December 2011. In.

[69] EMA Reflection Paper on the Use of Cocrystals and other Solid State Forms of Active Substances in Medicinal Products. European Medicines Agency. 2014.

[70] https://www.novartis.com/news/media-releases/novartis-new-heartfailure-medicine-lcz696-now-called-entrestotm-approved-fda (accessed 28 July, 2016).

[71] Ranzani LS, Aldea AF. US patent 2013/0109659 A1. 2013.

[72] Mascitti V, Thuma BA, Smith AC, et al. On the importance of synthetic organic chemistry in drug discovery: reflections on the discovery of antidiabetic agent ertugliflozin. Med. Chem. Comm. 2013, 4, 101–111.

German L. Perlovich, Artem O. Surov, and Alex N. Manin

2 Pharmaceutical multi-component crystals: improving the efficacy of anti-tuberculous agents

2.1 Introduction

Fluoroquinolones represent an important class of broad-spectrum anti-biotics that are widely used for the treatment of several types of bacterial infections and inflammation [1]. Ciprofloxacin (CIP), norfloxacin (NOR) and enrofloxacin (ENR) are second generation fluoroquinolone drugs (Figure 2.1) and show potent activity against Gram-negative and Gram-positive bacteria [2]. Patented by Bayer A.G., ciprofloxacin and norfloxacin drugs were the first members of the fluoroquinolone family approved for clinical use in the USA [3, 4]. These drugs are highly effective in treating bacterial infections of the lower respiratory tract and skin tissue [5, 6]. The proposed molecular mechanism of action entails the inhibition of cell division by acting on bacterial topoisomerase II (DNA gyrase) and the topoisomerase IV complex, which interferes with bacterial DNA replication, transcription, repair and recombination [7, 8].

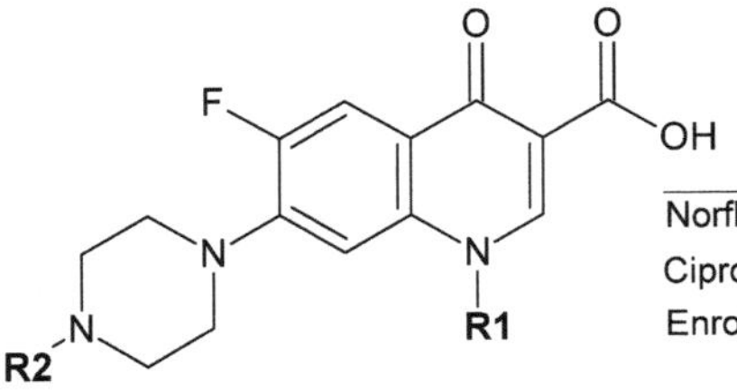

	R1	R2
Norfloxacin	ethyl	H
Ciprofloxacin	cyclopropyl	H
Enrofloxacin	cyclopropyl	ethyl

Fig. 2.1: Structural formulae of the key drugs discussed herein.

Despite the potent biological activity in vivo, fluoroquinolones suffer from low solubility at neutral pH [9] and poor permeability across biological membranes [10]. This makes them a class IV drug according to the Biopharmaceutics Classification System (BCS) [11]. This is a most challenging case for drug development as well as formulation design. However, it is widely accepted that formulation approaches similar to those for BCS class II drugs (improvement of solubility and dissolution behaviour) could be applied to BCS class IV drugs [12]. One of the common strategies to modify the aqueous solubility and/or dissolution rate of a drug is salt formation because of the favourable hydration of ionised species in water compared to neutral or zwitterionic forms. Indeed, a number of fluoroquinolones available on the market are formulated as hydrochloride salts [13]. However, there are some potential drawbacks of using chloride

https://doi.org/10.1515/9783110464955-002

as a salt former, including the high acidity in parenteral formulations, the risk of corrosion of industrial equipment, a decreased solubility in the stomach due to the common ion effect, etc. [14]. To overcome these problems, alternative multi-component crystalline forms of fluoroquinolones with different pharmaceutically acceptable organic counter-ions have been developed in recent years. For example, a number of ciprofloxacin and norfloxacin salts with citric, tartaric, malonic, benzoic acids [15], saccharine [16–18], succinic acid [19], diflunisal and indoprofen [20] have been prepared and characterised. In addition, a multi-component molecular complex comprising ciprofloxacin and norfloxacin ions in the solid state has been described by Vitorino et al. [21]. It has been also reported that salt formation affects the octanol-water partition coefficients of ciprofloxacin and norfloxacin, which allows the tuning of relevant pharmacological properties of the drugs by systematic counter-ion alteration [22]. Hu et al. [23] obtained molecular salts of norfloxacin with adipic acid, mucic acid, hydroxyl isomers of benzoic acid, naphthalene-1,5-disulfonic acid and naphthalene-2-sulfonic acid. Unfortunately, the properties of the enrofloxacin salts/co-crystals are not as promising as those for CIP and NOR. The study by Karanam et al. [24] should be mentioned in this context as it describes a wide screening of salts/co-crystals with dicarboxylic acids, benzoic acid derivatives, etc. It also describes crystal structures of the four novel salts of enrofloxacin with maleic, fumaric, succinic and oxalic acids and studies their thermochemical properties.

The goal of the present work is to both summarise and analyse literature crystallographic data for ciprofloxacin, norfloxacin and enrofloxacin with dicarboxylic acids. In addition, an attempt to systematise thermochemical data (melting points of individual components and their salts/co-crystals) has been undertaken in order to predict the directed design of two-component molecular crystals of the drugs. In the final part of the work, a comparison is made between the dissolution behaviours of different solid fluoroquinolone forms.

2.2 Crystal structure analysis

2.2.1 CSD analysis of fluoroquinolone multi-component crystals

Fluoroquinolones are a good example of drugs where the solubility challenge can be successfully overcome by utilising crystal engineering strategies. The literature analysis and CSD (Cambridge Structural Database) survey show that many efforts have been made to extend the range of fluoroquinolone solid forms through the formation of multi-component crystals with various co-formers. In particular, significant attention has been paid to ciprofloxacin, norfloxacin and enrofloxacin. According to the classification recently proposed by Grothe et al. [25], all the currently reported fluoroquinolone multi-component crystals of which there are 67 examples can be categorised into six classes, namely hydrates (solvates), salts, salt hydrates (solvates), salt co-crystals, salt co-crystal hydrates (solvates) and co-crystal solvates. Their relative distribution is shown in Figure 2.2.

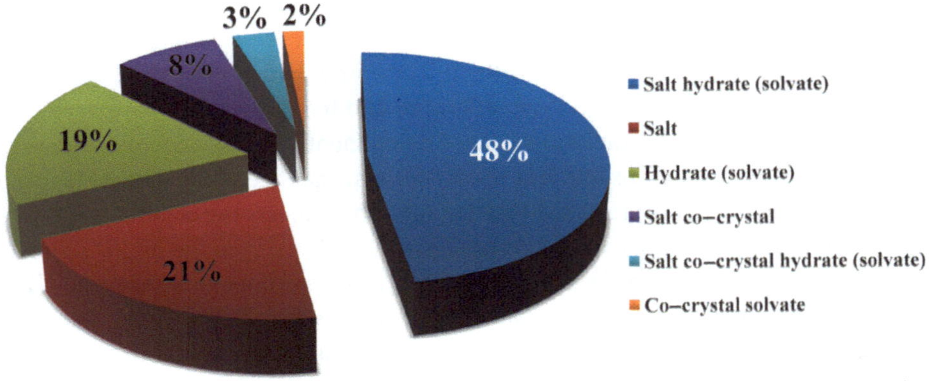

Fig. 2.2: Distribution of the 67 solid forms of multi-component crystals of fluoroquinolones.

It is evident from Figure 2.2 that approximately half of the fluoroquinolone multi-component crystals under ambient conditions are salt hydrates (solvates), i.e. ternary systems that contain one to four solvent molecules (mainly water) in the crystallographic asymmetric unit. This class of solid form also includes four-component structures such as the salt ciprofloxacin citrate tetrahydrofuran trihydrate (GUYWUU), i.e. with two different solvent molecules in the crystal. Simple hydrates (solvates) with various water stoichiometry comprise approximately 19% of solid forms. Approximately the same contribution is made by simple salts (21%), and approximately 8% of crystal structures are found to be salt co-crystals. Some complex systems such as salt co-crystal hydrates are also observed. However, co-crystal formation for the fluoroquinolones is limited due to their high basicity. In fact, the solvated co-crystal of norfloxacin with isonicotinamide (VETVUM) seems to be the only reported example. Another issue to consider is the polymorphism of the fluoroquinolone multi-component crystals. However, to date, this field does not seem to have been systematically explored and only one polymorphic salt is described in the literature, namely, dimorphic ciprofloxacin saccharinate. In addition, the X-ray crystallographic parameters of the 1 : 1 ciprofloxacin salt with maleic acid reported recently in independent studies [26, 27] are found to be considerably different, which indicates that the structures should be considered polymorphic. Such diversity of fluoroquinolone solid forms seems likely to be a consequence of the combination of the chemical features of the compounds (high basicity, several sites for hydrogen-bonding, broad aromatic system) and the distinct supramolecular structure of their crystalline complexes (the presence of hydrophobic and hydrophilic domains in the crystal structure that are accessible to solvent molecules).

Tab. 2.1: Results of the analysis of hydrogen-bonding motifs occurring between the different components in the crystal structures of fluoroquinolone salts with mono- and dicarboxylic acids.

		API – SF[a]	API – H_2O– SF	SF– SF	SF – H_2O– SF	H_2O– H_2O	API – API
anhydrous							
CIP + maleic (1 : 1) Form I	QUKHIP	✓					
CIP + maleic (1 : 1) Form II	QUKHIP01	✓					✓
CIP + TartAc (1 : 1)	ORUYOQ	✓		✓			
CIP + diflunisal (1 : 1)	KOFFAO	✓					
NOR + adipic (1 : 1)	LOQKOT	✓		✓			
NOR + phthalic (1 : 1)	HIGSAT	✓					
NOR + isophthalic (1 : 1)	HIGSIB	✓					
NOR + terephthalic (1 : 1.5)	UPAYOA	✓		✓			✓
ENR + maleic (1 : 1)	PILTAH	✓					
ENR + succinic (1 : 1)	PILSOU	✓		✓			
ENR + fumaric (1 : 1)	PILSUA	✓		✓			
ENR + oxalic (2 : 1)	KIXWEV	✓					
1(H_2O)							
CIP + fumaric + H_2O (1 : 1 : 1)	QUKHUB	✓			✓		
CIP + citric + H_2O (1 : 1 : 1)[b]	ORUYIK	✓		✓			
CIP + indoprofen + H_2O (1 : 1 : 1)	KOFFES	✓	✓				
NOR + maleic + H_2O (1 : 1 : 1)	VETWIB	✓	✓		✓		
NOR + succinic + H_2O (1 : 0.5 : 1)	VETWAT	✓			✓		
NOR + 3-hydroxybenzoic + H_2O (1 : 1 : 1)	LOQLAG	✓		✓	✓		✓
NOR + salycilic + H_2O (1 : 1 : 1)	LOQLOU	✓	✓		✓		
NOR + 4-hydroxybenzoic + H_2O (1 : 1 : 1)	LOQLUA	✓			✓		
NOR + benzoic + H_2O (1 : 1 : 1)	ORUYEG	✓			✓		
ENR + citric + H_2O (1 : 1 : 1)	GECCUO	✓		✓	✓		
2(H_2O)							
CIP + malonic + H_2O (1 : 1 : 2)	ORUZAD	✓	✓		✓	✓	
CIP + adipic + H_2O (1 : 0.5 : 1)	QUKHOV	✓	✓		✓	✓	
NOR + oxalic + $2H_2O$ (1 : 0.5 : 1)	ORUYUW	✓	✓		✓	✓	
NOR + malonic + $2H_2O$ (1 : 1 : 2)	VETWEX	✓	✓		✓	✓	
NOR + TartAc + $2H_2O$ (1 : 1 : 2)	ORUYAC	✓	✓	✓	✓		
3(H_2O)							
CIP + succinic + $3H_2O$ (1 : 1 : 3)	QABTOF	✓	✓	✓	✓	✓	
ENR + oxalic + $3H_2O$ (1 : 0.5 : 3)	PILSIO	✓			✓	✓	
CIP + fumaric + $3H_2O$ (2 : 1 : 3)	GUYWII	✓	✓		✓		
4(H_2O)							
CIP + succinic + $4H_2O$ (2 : 1 : 4)[c]	QABTUL	✓	✓		✓		

[a] API – Active Pharmaceutical Ingredient, and SF – Salt Former
[b] Hydrogen atoms of the water molecule were not provided
[c] Water molecules are disordered

The fluoroquinolones are known to have a great propensity for hydration in the solid state (Figure 2.2). Six hydrates with multiple stoichiometries are claimed to exist for norfloxacin, three hydrated structures are reported for ciprofloxacin while only one hexahydrate of enrofloxacin are described in the literature. As Figure 2.2 indicates, the same trend is preserved in the fluoroquinolones salts. However, in contrast to clathrate-like structures of hydrates of pure fluoroquinolones, in most salts, water molecules play an important role by connecting the different components of the crystal via a network of hydrogen bonds. Table 2.1 summarises the analysis of hydrogen bonds occurring between different components in the respective crystal structures with various mono- and dicarboxylic acids. The structures are arranged in the order of increasing number of water molecules in the asymmetric unit.

In anhydrous structures, hydrogen bonds can be formed between the API and salt former (API–SF), salt former ions (SF–SF) and API ions (API–API). Table 2.1 indicates that for anhydrous fluoroquinolone salts, (SF–SF) hydrogen bonds are observed in approximately 40% of structures, while (API–API) hydrogen bonds are found in only two crystals. In hydrated salts, water molecules occupy the most hydrophilic domains of the crystal. Consequently, in monohydrates, a water molecule is mainly disposed between SF ions forming water bridges. A good example of such a structure is cipro-

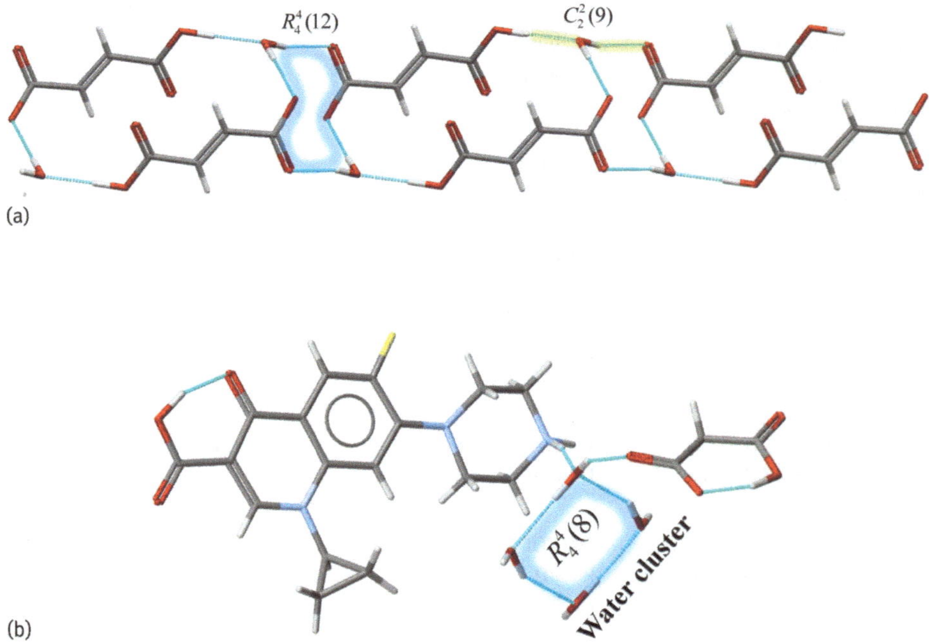

Fig. 2.3: (a) Hydrogen bonded ring and chain motifs occurring between fumarate ions and water molecules in the [CIP + Fum + H_2O] (1 : 1 : 1) co-crystal and (b) hydrogen bonded water cluster in the [CIP + malonic + H_2O] (1 : 1 : 2) co-crystal.

floxacin fumarate monohydrate (1 : 1 : 1), where fumarate ions are arranged into hydrogen-bonded chains and rings via water bridges (Figure 2.3a). In dihydrates, one of the water molecules is embedded between the API and SF replacing a hydrogen bond. In addition, different water molecules are connected to form closed-ring clusters (Figure 2.3b). A further increase in the number of water molecules in the crystal enlarges hydrogen bonded water clusters but does not change the hydrogen bonding pattern.

2.2.2 Analysis of supramolecular synthons

As the next step, crystal structures of fluoroquinolone salts with mono- and dicarboxylic acids were analysed in order to identify and classify the most probable and structure directing supramolecular synthons. Table 2.2 contains a schematic illustration of the observed synthons and their description in terms of Graph-Set notation [29, 30]. Figure 2.4 shows the distribution of the ring motifs formed by hydrogen bonds in the crystal structures of fluoroquinolone salts.

The distribution shows clearly that the $R_4^4(12)$ ring motif is the most frequent configuration of hydrogen bonds in the considered structures. There are three alternative versions of the implementation of this motif. Most of the structures contain "anhydrous" centro-symmetric tetrameric units, which include two APIs and two SF ions (Synthon 1, Table 2.2). Synthon 2 is formed by one API ion, two SF ions and one water molecule, whereas two water molecules participate in Synthon 3 formation. The next frequently observed ring motifs are $R_4^2(8)$ (Synthon 4) and $R_1^2(5)$ (Synthons 5, 6 and 7). It should be noted that all of the mentioned motifs can be written using the following

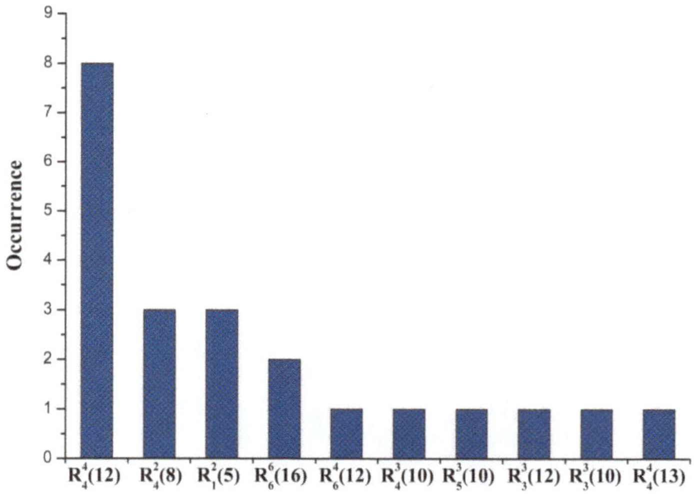

Fig. 2.4: The distribution of hydrogen bonded ring motifs in the crystal structures of fluoroquinolone salts.

Tab. 2.2: Illustration of the supramolecular synthons found in fluoroquinolone salts with mono- and dicarboxylic acids and their description in terms of Graph-Set notation.

Graph-set notation	Schematic illustration of synthon	Synthon type	CSD Refcode
$R^4_4(12)$	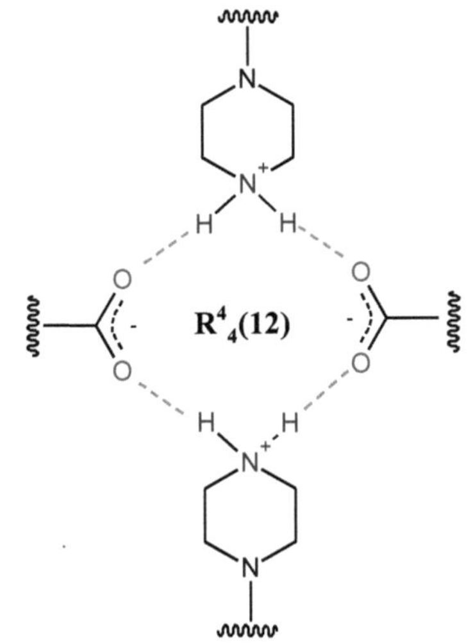	Synthon 1	ORUYIK ORUYIK KOFFES VETWAT LOQLUA ORUYEG
	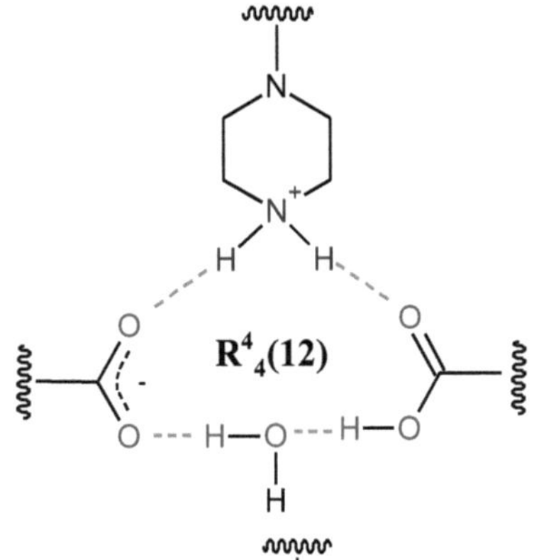	Synthon 2	QUKHUB
	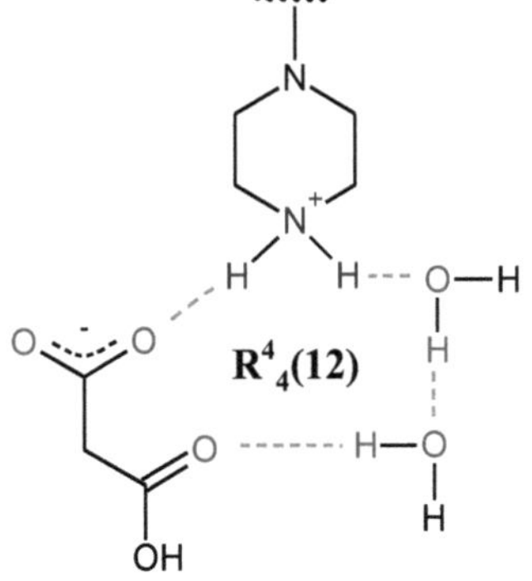	Synthon 3	VETWEX

Tab. 2.2 (continued).

Graph-set notation	Schematic illustration of synthon	Synthon type	CSD Refcode
$R_4^2(8)$	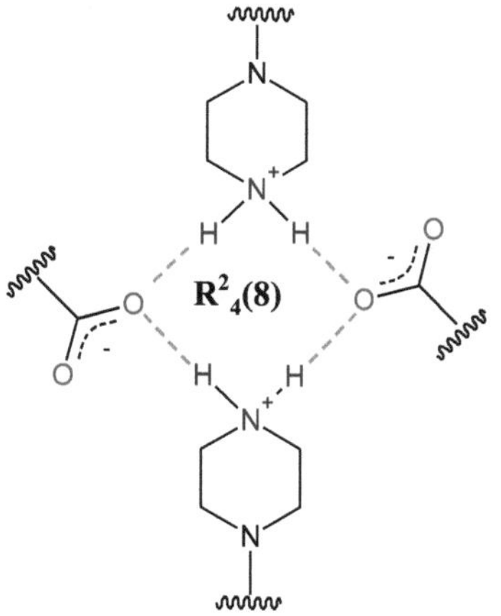	Synthon 4	QUKHIP QABTUL HIGSIB
$R_1^2(5)$	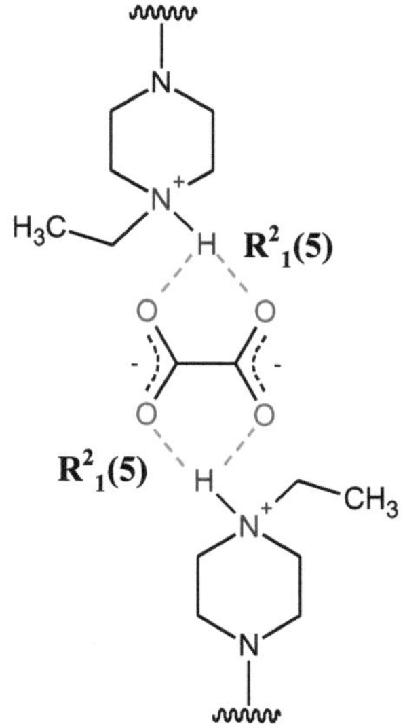	Synthon 5	PILSIO

Tab. 2.2 (continued).

Graph-set notation	Schematic illustration of synthon	Synthon type	CSD Refcode
	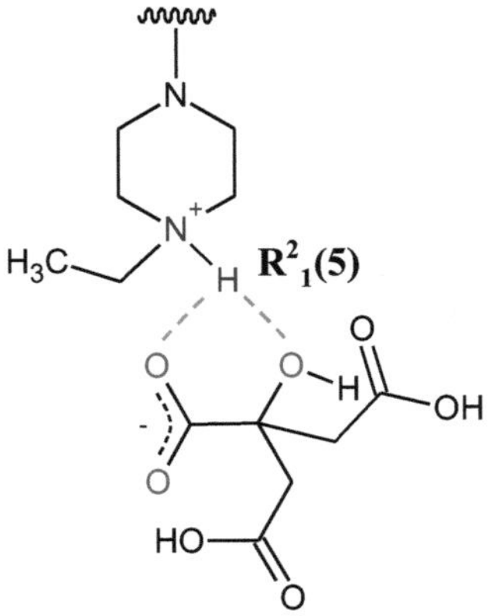	Synthon 6	GECCUO
	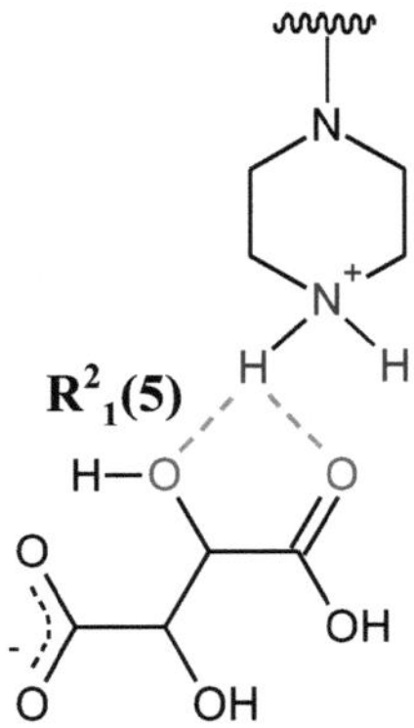	Synthon 7	ORUYAC
$R^3_4(10)$	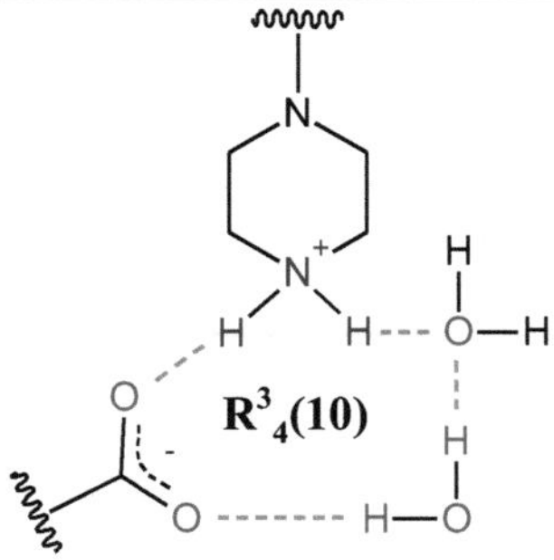	Synthon 8	ORUZAD

Tab. 2.2 (continued).

Graph-set notation	Schematic illustration of synthon	Synthon type	CSD Refcode
$R_6^6(16)$	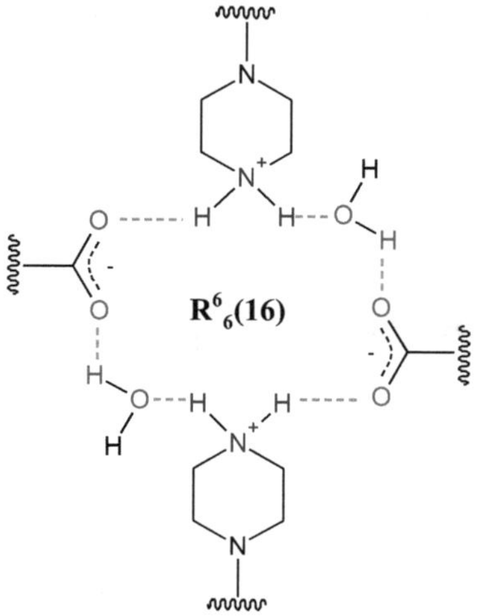	Synthon 9	ORUYUW LOQLOU
$R_6^4(12)$	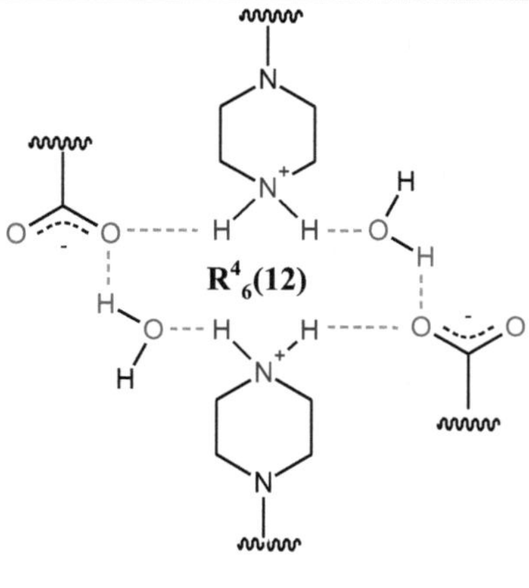	Synthon 10	QUKHOV

Tab. 2.2 (continued).

Graph-set notation	Schematic illustration of synthon	Synthon type	CSD Refcode
$R_3^3(12)$	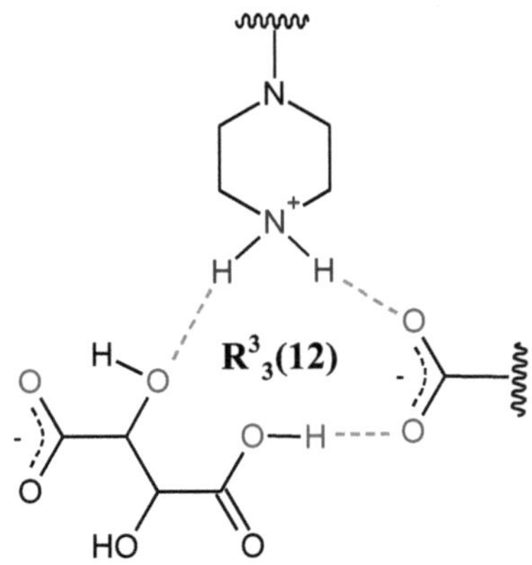	Synthon 11	ORUYOQ
$R_5^3(10)$	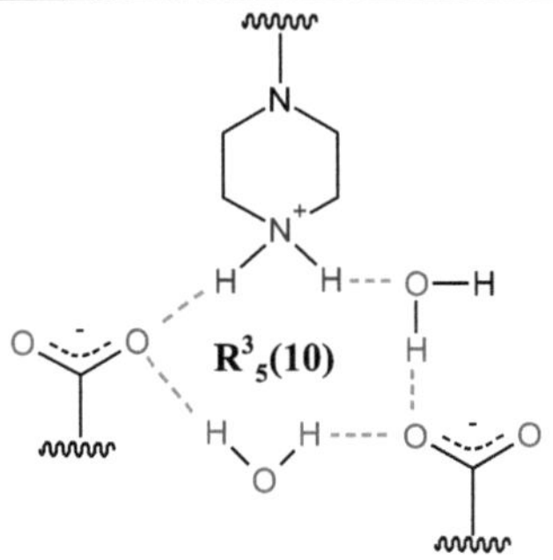	Synthon 12	QABTUL
$R_4^4(13)$	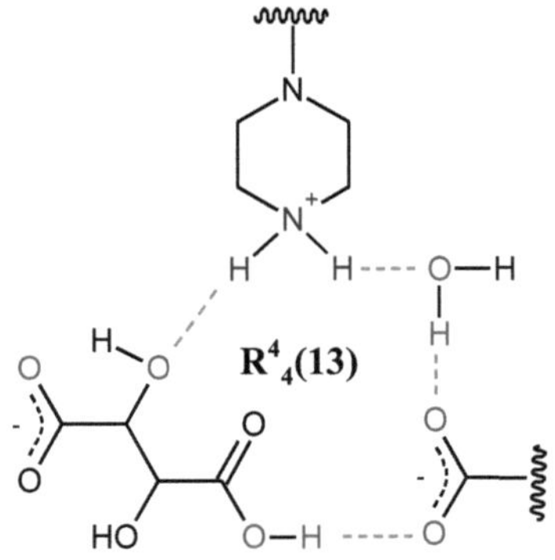	Synthon 13	ORUYAC

Tab. 2.2 (continued).

Graph-set notation	Schematic illustration of synthon	Synthon type	CSD Refcode
$R_3^3(10)$	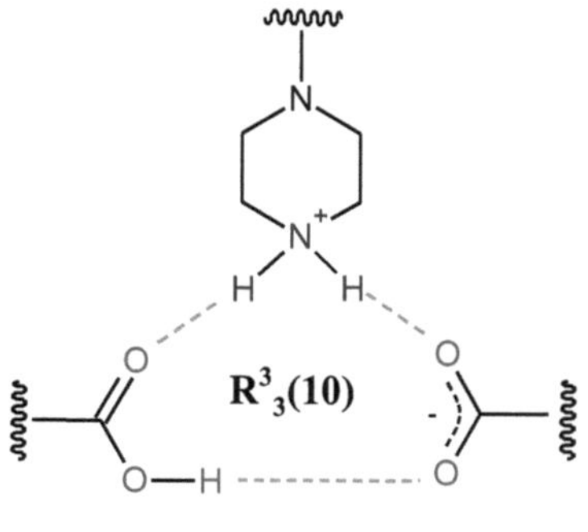	Synthon 14	HIGSUN
$C_2^2(9)$	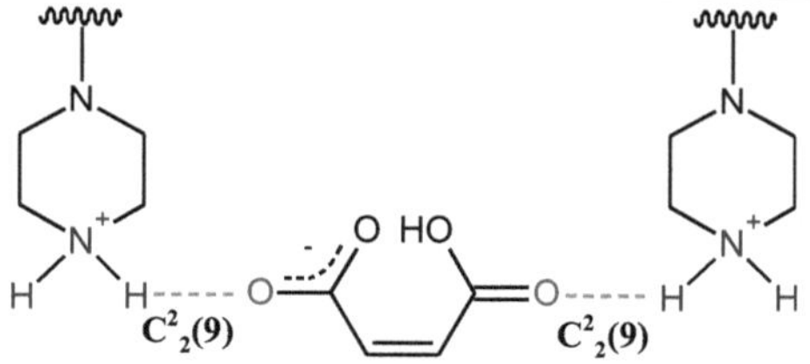	Synthon 15	QUKHIP01 HIGSAT
$C_4^3(8)$	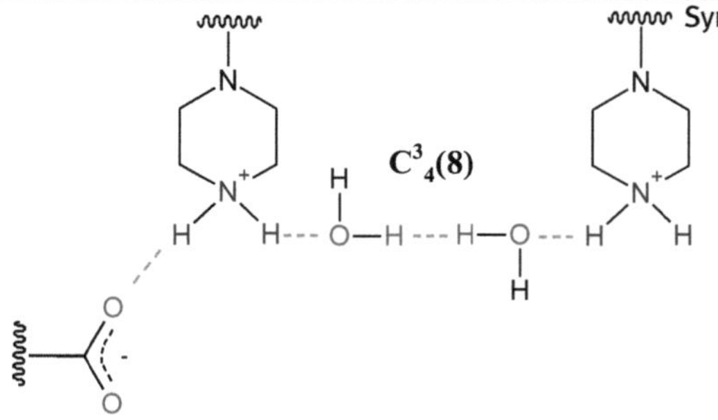	Synthon 16	QABTOF
$C_4^4(12)$	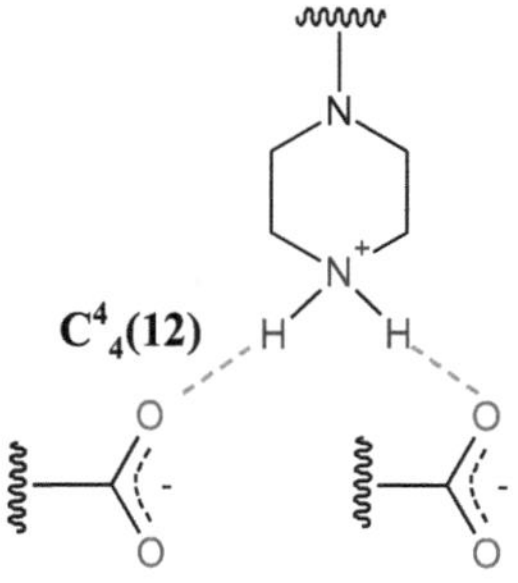	Synthon 17	LOQKOT

Tab. 2.2 (continued).

Graph-set notation	Schematic illustration of synthon	Synthon type	CSD Refcode
$C_4^4(24)$	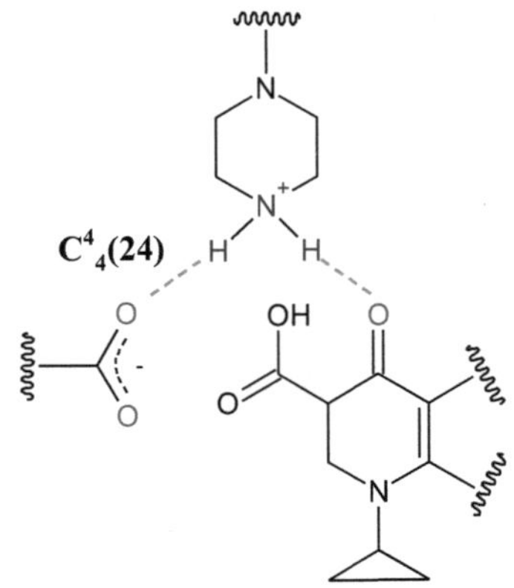	Synthon 18	LOQLAG
$C_4^4(28)$	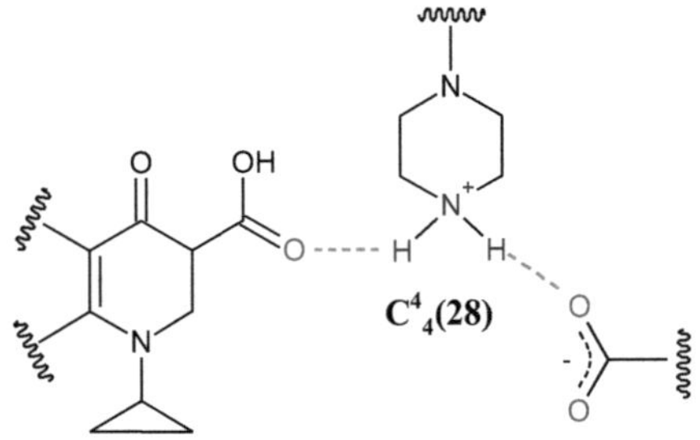	Synthon 19	UPAYOA
$D(2)$	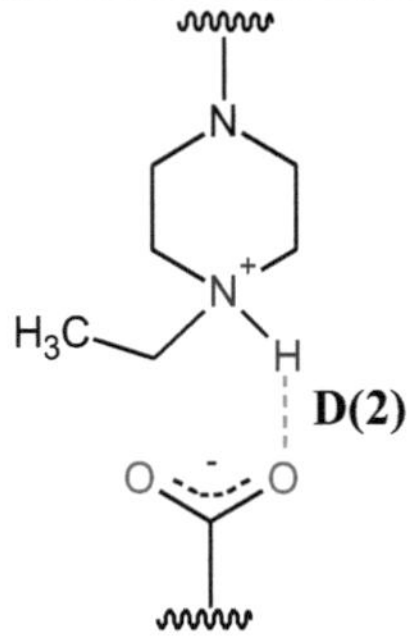	Synthon 20	PILTAH PILSOU PILSUA

form of Graph-Set notation: $R^a_d(2 \cdot a + d)$. It might be reasonable to assume that the relationship between the number of donors, acceptors and size of the ring motif seems most favourable in terms of geometry and the energy of intermolecular interactions. For example, the $R^3_4(10)$ motif found in ciprofloxacin malonate dihydrate (Synthon 8) also obeys the proposed condition. Furthermore, motifs with a $R^a_d(2 \cdot a + d)$ common formula are observed in ciprofloxacin and norfloxacin salts with saccharin (PEFZIL) and acesulfamate (WEFLEA, WEFLIE). All these facts confirm the assumption that the $R^a_d(2 \cdot a + d)$ arrangement corresponds to a stable configuration of hydrogen bonds in the fluoroquinolone salts. The $R^6_6(16)$ ring motif (Synthon 9) is less frequently adopted. This synthon can be envisaged as Synthon 1 extended by two water molecules. Similarly, the $R^4_6(12)$ ring motif (Synthon 10) is a "hydrated" version of Synthon 4. The remaining ring motifs shown in Table 2.2 are unique as they are observed only once in fluoroquinolone salts. It should be noted that chain motifs in the fluoroquinolone salts are found only in a small number of structures.

2.2.3 Polymorphism in fluoroquinolone multi-component crystals

Despite the considerable diversity of the reported fluoroquinolone multi-component crystalline forms, only two of them are polymorphic. According to Cruz-Cabeza et al. [28], 37% of anhydrous salts and 28% of hydrated salts deposited in the CSD have two or more polymorphic modifications. This fact indicates that polymorphic behaviour of fluoroquinolone multi-component crystals has not been thoroughly studied, and further investigations are needed to explore the polymorph propensity of these systems. The first description of polymorphism was done by Romanuk et al. who studied two polymorphs of the ciprofloxacin salt with saccharin [16]. However, only the crystal structure of form II was reported, while form I was characterised only by spectroscopic methods (IR and solid-state NMR spectroscopy). As mentioned above, two polymorphic modifications of the ciprofloxacin salt with maleic acid are reported [26, 27] and these are the only polymorphs for which structural information is available for both forms. The crystal structures of form I [26] and form II [27] of ciprofloxacin maleate are different in terms of both the hydrogen bonding network and packing arrangements. In form I, the ciprofloxacin and maleate ions are linked by hydrogen bonds to form a stable $R^2_4(8)$ ring motif, while in form II the components are assembled into less frequently observed chain structures (Synthon 15 in Table 2.2). The packing arrangement of the ciprofloxacin ions in form II is typical of fluoroquinolone multi-component crystals and can be described as infinite columnar stacks held by $\pi \cdots \pi$ stacking interactions. So, the overall structure comprises alternating layers containing π-stacks of the drug molecules and the hydrogen bonded acid molecules (Figure 2.5b). In the case of form I, no infinite columnar stacks of the ciprofloxacin molecules are found. Instead, two maleate ions occupy the space between the pairs of the π-stacked ciprofloxacin molecules, which prevents conventional columnar packing (Figure 2.5a). The uncommon packing arrangement in form I is accompanied by a greater calculated density, i.e. 1.496 · cf. 1.416 g cm^{-3} for form II, which suggests a greater thermodynamic stability cf. form II.

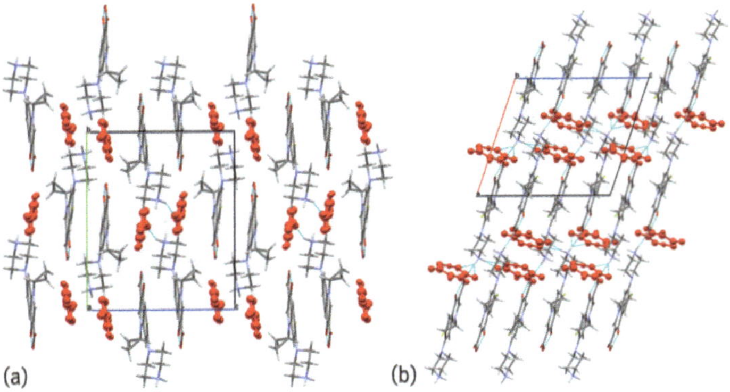

Fig. 2.5: Packing arrangements of (a) form I and (b) form II of ciprofloxacin maleate (1 : 1). The maleate ions are coloured *red*. (Reprinted with permission from (Surov AO, Churakov AV, Perlovich GL. Three Polymorphic Forms of Ciprofloxacin Maleate: Formation Pathways, Crystal Structures, Calculations, and Thermodynamic Stability Aspects Cryst. Growth Des., 2016, 16 (11), 6556–6567) Copyright (2017) American Chemical Society).

2.2.4 Conformational analysis

The fluoroquinolone molecules comprise two conformationally rigid fragments, namely, the quinolone moiety and the piperazinyl ring. These fragments are connected via an N–C bond which has a relatively low-energy barrier to rotation and should, in principle, provide some degree of conformational flexibility. Therefore, the main conformational variations in the fluoroquinolone ions are described by the torsion angle, τ, which is responsible for the mutual orientation of the piperazinyl and quinolone moieties. Figure 2.6 shows the distribution of τ in all known multi-component crystals of ciprofloxacin, norfloxacin and enrofloxacin.

It is evident from Figure 2.6 that τ is clustered into two tight regions with maxima at −169 and 176°. In both regions, the torsion angle deviates by no more than ±20 °C from the value corresponding to the maximum occurrence. This distribution pattern is a consequence of steric hindrance associated with piperazinyl ring rotation. Therefore, the degree of conformational freedom of the piperazinyl moiety is considerably limited, which leads to the conclusion that fluoroquinolone compounds should be generally considered as conformationally rigid molecules.

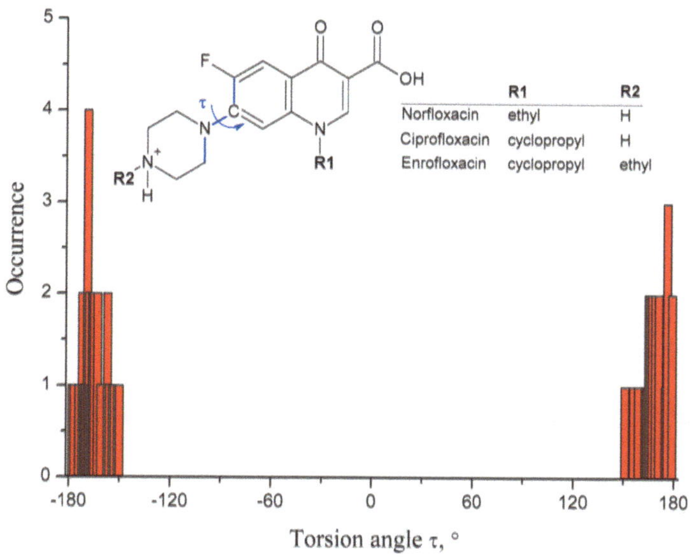

Fig. 2.6: The distribution of torsion angles, τ, in ciprofloxacin, norfloxacin and enrofloxacin ions.

2.2.5 Analysis of free volume in unit cells

It is of interest to analyse the reason(s) behind hydrate formation in fluoroquinolone multi-component crystals. It has been assumed that salt crystallisation results in defects/imperfections in the form of non-enveloping free volume. The accumulation of free volume imparts instability to the crystal lattice. In other words, a two-component crystal is thermodynamically stable until the moment when the volume of defects overcomes a critical value. In order to stabilise the system thermodynamically, water molecules are incorporated into the crystal to fill voids. To verify this hypothesis, the free volume of the unit cell per structural unit of salt/co-crystal ([API + CF] (m : n)) has been calculated without water molecules using the following equation:

$$\beta = [(V_{\text{cell}}/Z-(m\cdot V^{\text{vdw}}(\text{API})+n\cdot V^{\text{vdw}}(\text{CF}))/(m\cdot V^{\text{vdw}}(\text{API})+n\cdot V^{\text{vdw}}(\text{CF}))]\cdot 100\% \quad (2.1)$$

where V_{cell} is the unit cell volume, $V^{\text{vdw}}(\text{API})$ and $V^{\text{vdw}}(\text{CF})$ are the van der Waals volumes of API and CF, respectively, and Z is the number of structural units.

Figure 2.7a–c shows the dependence of the number of water molecules in a unit cell versus the β-parameter for the selected fluoroquinolones. It is evident that as the free volume per structural unit of API and CF increases, the number of water molecules rises as well. It can be supposed that there are transitional zones with "critical" content of the free volume, above which it is thermodynamically advantageous to accumulate water molecules in the crystal. For example, for ciprofloxacin the region $40 < \beta < 45\%$ corresponds to the incorporation of one water molecule in the unit cell, at $45 < \beta < 49\%$ two water molecules are accumulated, at $49 < \beta < 54\%$,

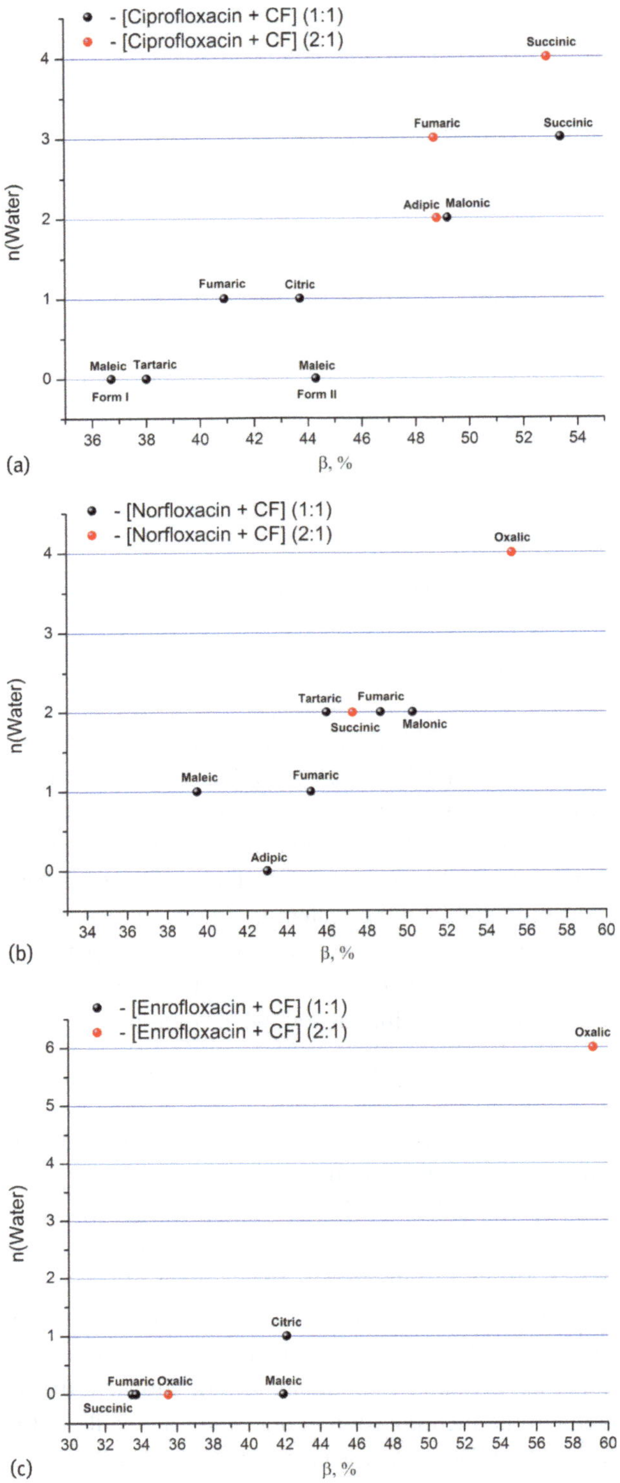

Fig. 2.7: The dependence of the number of water molecules in unit-cells, n(water), on the free volume per structural unit of API and co-former (β): (a) ciprofloxacin, (b) norfloxacin and (c) enrofloxacin.

three water molecules and at 54% < β, four. Naturally, the boundaries of the "transitional" zones are not clearly distinguished/defined and can overlap each other, for example, the zones with accumulation of two and three, as well as three and four water molecules. Special attention should be paid to polymorphic forms I and II of the ciprofloxacin + maleic acid salt. The second form has an excess of free volume and, as a consequence, is thermodynamically unstable. Analogous regularities are observed for the other two compounds, Figure 2.7b, c, regardless of their stoichiometry.

2.3 Analysis of melting points

In order to evaluate the melting points of co-crystals for the selected fluoroquinolones, an approach was applied using correlation equations that had been developed earlier [29]. For this purpose the relevant literature was analysed and melting points of two-component co-crystals were selected based on the presence of glycolic, malonic, maleic, adipic, succinic, fumaric, benzoic and 4-hydroxybenzoic acids/anions. The obtained data led to the following correlation equation:

$$T_{\mathrm{fus}}(\mathrm{CC}) = A + B \cdot T_{\mathrm{fus}}(\mathrm{CF}) \tag{2.2}$$

where $T_{\mathrm{fus}}(\mathrm{CF})$ is the melting point of the individual co-formers. It should be noted that the sample included co-crystals both of different stoichiometry and various polymorphic modifications. In the case of different polymorphic forms, each modification was considered as a self-contained co-crystal. Co-crystals with different stoichiometry were grouped into clusters and analysed independently. Then, the clusters were consolidated. If the correlation parameters of a consolidated cluster were better than those of the individual clusters, the equation for the consolidated cluster was chosen. If – not, the equation of the initial cluster with the best correlation parameters was used. Table 2.3 summarises the analysis. The application of the equations above enables an estimation of the melting point of the selected salt/co-crystal of a fluoroquinolone (Table 2.4). As mentioned above, many salts/co-crystals of the fluoroquinolones crystallise as hydrates. In these cases, the melting points correspond to dehydrated forms. It is evident that the calculated temperatures are in good agreement with the experimentally determined values for salts/co-crystals of ciprofloxacins. A satisfactory prediction is observed for norfloxacins, while the experimental data for enrofloxacins are hardly predictable. In all probability, these discrepancies are connected with the decomposition of the components of the salts/co-crystals at fusion. Thus, the proposed approach is only useful in designing novel ciprofloxacin salts/co-crystals.

With the above results in mind, it was also of interest to analyse the melting points of ciprofloxacin salts in relation to the fusion temperatures of the co-former (Figure 2.8). As has been noted earlier, many salts/co-crystals of the fluoroquinolones crystallise as hydrates and so the melting points used in the analysis correspond to their dehydrated

Tab. 2.3: Coefficients of the correlation equation (2.2) for the clusters including the considered compound as one of the components of the salt/co-crystal.

Co-former (CF)	(API:CF)	A	B	R[a]	SD[b]	n[c]
Glycolic acid	(1 : 1)	192 ± 37	0.454 ± 0.078	0.9588	8.51	5
Malonic acid	(1 : 1)	−21 ± 106	0.911 ± 0.215	0.8658	20.0	8
Maleic acid	(1 : 1) (2 : 1)	208 ± 27	0.478 ± 0.056	0.9547	9.1	9
Adipic acid	(1 : 1) (2 : 1)	203 ± 27	0.529 ± 0.062	0.9112	13.5	17
Succinic acid	(2 : 1)	163 ± 46	0.602 ± 0.101	0.7866	18.3	24
Fumaric acid	(1 : 1)	308 ± 22	0.325 ± 0.049	0.9105	8.30	11
Benzoic acid	(1 : 1)	121 ± 37	0.654 ± 0.084	0.9202	17.4	13
4-Hydroxybenzoic acid	(1 : 1)	133 ± 28	0.690 ± 0.061	0.9306	14.3	22

[a] Pair correlation coefficient
[b] Standard deviation
[c] The number of points in the cluster

forms. Figure 2.8 represents all the ciprofloxacin salts for which the melting temperatures are described in the literature. The blue line corresponds to the bi-sector whereas the red, horizontal-dashed line corresponds to the melting point of ciprofloxacin. It is clear that no correlation is evident. Most of the two-component crystals have melt-

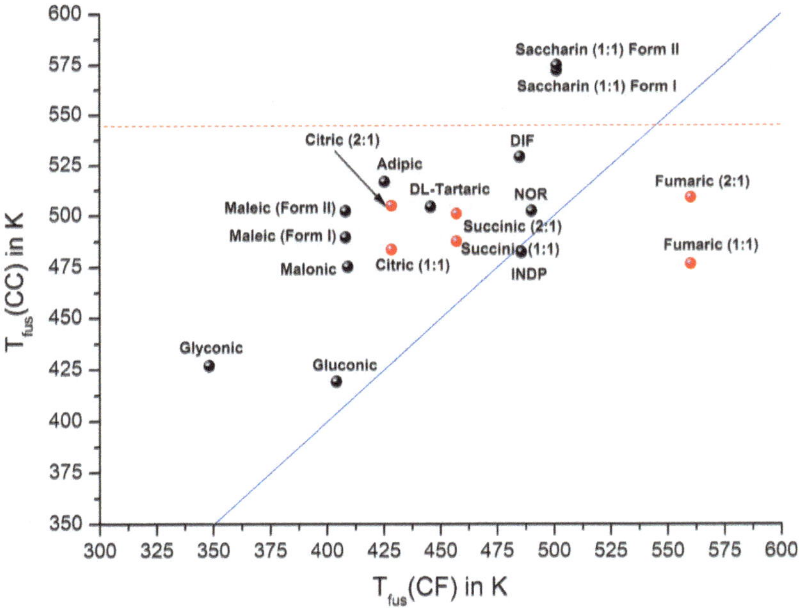

Fig. 2.8: Melting points of ciprofloxacin salts versus the melting points of the co-former. The blue line corresponds to the bi-sector whereas the red, horizontal dashed line corresponds to the ciprofloxacin melting point.

Tab. 2.4: Comparative analysis of the melting points of salts/co-crystals [API + CF] obtained by means of the correlation equation (T_{fus}^{calc}(CC) [K]) and experimentally (T_{fus}^{exp}(CC) [K]).

CF	(API:CF)	Ciprofloxacin			Norfloxacin			Enrofloxacin		
		T_{fus}^{calc}(CC)	T_{fus}^{exp}(CC)	$\|\Delta T_{fus}\|$	T_{fus}^{calc}(CC)	T_{fus}^{exp}(CC)	$\|\Delta T_{fus}\|$	T_{fus}^{calc}(CC)	T_{fus}^{exp}(CC)	$\|\Delta T_{fus}\|$
Glycolic acid	(1 : 1)	438.9	426.9 [22]	12.0	416.2	388.5 [22]	27.7	–	–	–
Malonic acid	(1 : 1)	474.5	475.3[a] [15]	0.8	–	–	–	–	–	–
Maleic acid	(1 : 1) (2 : 1)	468.0	502.2[b] [27]	34.2	–	–	–	446.9	498.7 [24]	51.8
		468.0	489.5[c] [26]	21.5	–	–	–	–	–	–
Adipic acid	(1 : 1) (2 : 1)	490.7	516.7[a] [26]	26.0	464.2	487.0 [23]	22.8	–	–	–
Succinic acid	(2 : 1)	490.4	501.3[a] [19]	10.9	–	–	–	–	–	–
Fumaric acid	(1 : 1)	484.8	477.0[a] [26]	7.8	–	–	–	470.4	540.6 [24]	70.2
Benzoic acid	(1 : 1)	–	–	–	443.9	455.0[a] [15]	11.1	–	–	–
4-Hydroxybenzoic acid	(1 : 1)	–	–	–	473.7	500.0[a] [23]	26.3	–	–	–

[a] Melting point has been obtained from hydrate after dehydration process
[b] Polymorphic form II
[c] Polymorphic form I

ing points below that of the API and above that of the co-former. The only exception is found in the saccharin salts (forms I and II) with fusion temperatures higher than both ciprofloxacin and the co-former. The opposite trend is seen in the salts of ciprofloxacin with fumaric acid (2 : 1) and (1 : 1), which exhibits melting points lower than for each of ciprofloxacin and the co-former. As it has been shown earlier that the melting points of compounds (including co-crystals) are directly proportional to their Gibbs sublimation energies [29]. Thus, salts/co-crystals of ciprofloxacin with minimal fusion temperatures are good candidates for systems with optimal solubility, i.e. if the solvation terms of the co-formers do not differ significantly. Among such salts are [ciprofloxacin + gluconic acid (1 : 1)] and [ciprofloxacin + glyconic acid (1 : 1)]. The [ciprofloxacin + maleic acid (1 : 1)] and [ciprofloxacin + saccharin (1 : 1)] salts each have two polymorphic forms with differences in melting points being 13 and 3 K, respectively. The salts [ciprofloxacin + citric acid], [ciprofloxacin + succinic acid] and [ciprofloxacin + fumaric acid] form crystals with various stoichiometries, i.e. (2 : 1) and (1 : 1). The melting point differences between the (2 : 1) and (1 : 1) salts are 21 K for citric acid, 13 K for succinic acid and 32 K for fumaric acid. Thus, the melting point differences in ciprofloxacin salts of various stoichiometry are generally greater than the analogous differences for polymorphic forms.

Figure 2.9 shows the experimental melting point data of ciprofloxacin salts formed with dicarboxylic acids versus the length of carbon chain length of the dcarboxylic

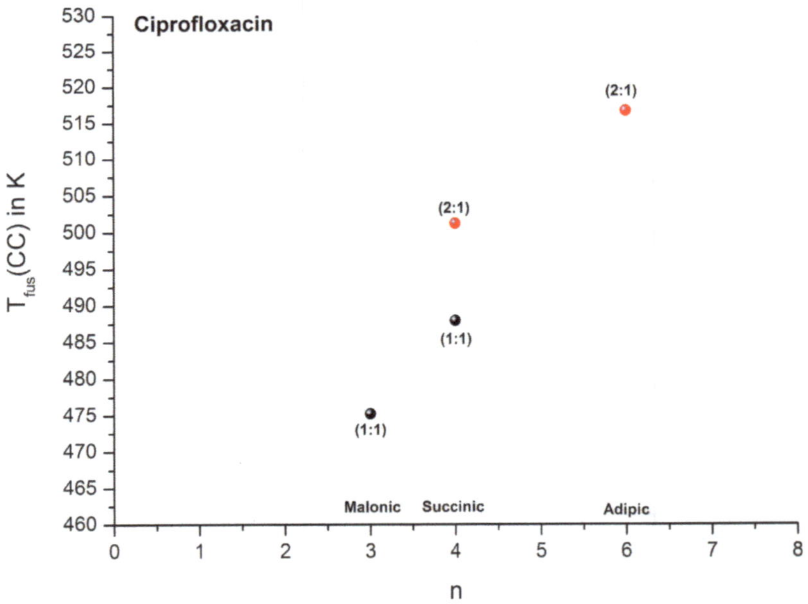

Fig. 2.9: Experimental data of melting points of ciprofloxacin salts with dicarboxylic acids versus carbon chain length of dicarboxylic acids.

acids. At $n = 4$ (succinic acid), there is a step which corresponds to a change in the stoichiometry of the salts, i.e. from (1 : 1) to (2 : 1).

2.4 Dissolution behaviour of fluoroquinolones

Aqueous solubility is a crucial parameter for pharmaceutical salts and co-crystals as it correlates with the oral bioavailability of drugs [30, 31]. A literature analysis shows that the lengthening of the carbon aliphatic chain increases the molecule's hydrophobicity and diminishes its acidity, which affects in a negative fashion the ability of dicarboxylic acids to enhance API solubility. Actually, the investigation of the influence of the dicarboxylic acid carbon chain length in the salts and co-crystals of fluoroquinolones upon their solubility parameters shows that more carbons in the chain reduces the potential API solubility gain by co-crystal/salt formation (Figure 2.10) [32].

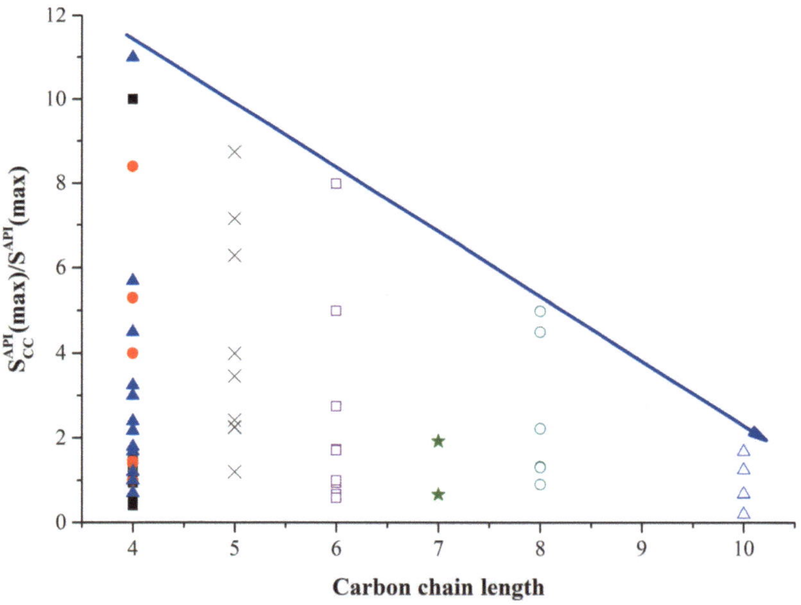

Fig. 2.10: Co-crystal-API solubility ratio as a function of carbon chain length of dicarboxylic acids. The co-crystals and salts with different acids as co-formers are represented by the following symbols: (■) – fumaric; (●) – maleic; (▲) – succinic; (×) – glutaric; (□) – adipic; (∗) – pimelic; (○) – suberic; (△) – sebacic acids.

This tendency explains the fact that dicarboxylic acids with carbon chain lengths less than five, i.e. oxalic, malonic, citric, maleic, succinic, tartaric and fumaric acids, are usually selected as co-formers for synthesising new co-crystals or salts with fluoroquinolones in order to increase API solubility. The only exception is adipic acid with six carbon atoms in the chain.

The aqueous solubilities of fluoroquinolone are highly pH-dependent owing to their amphoteric entity. Therefore, the solubility of ciprofloxacin salts were investigated in buffer solutions with pH = 1.2 and 6.8, which simulate the acidic environment of the stomach and the neutral environment of the intestines, respectively. The dissolution profile for ciprofloxacin at pH = 1.2 (Figure 2.11) shows a solution mediated conversion of the compound which can be interpreted by the formation of a ciprofloxacin hydrate, which is more stable than pure ciprofloxacin under the experimental conditions. The DSC (Differential Scanning Calorimetry) analysis of the residual sample derived from the experiment confirms this hypothesis [26]. The ciprofloxacin salts with dicarboxylic acids are less soluble in acidic conditions in contrast to the hydrochloride salts of ciprofloxacin. The same results were obtained for other salts of ciprofloxacin and norfloxacin by Nangia et al. [15]. The reduction in solubility is usually very near to two fold, but it can decrease more than five fold.

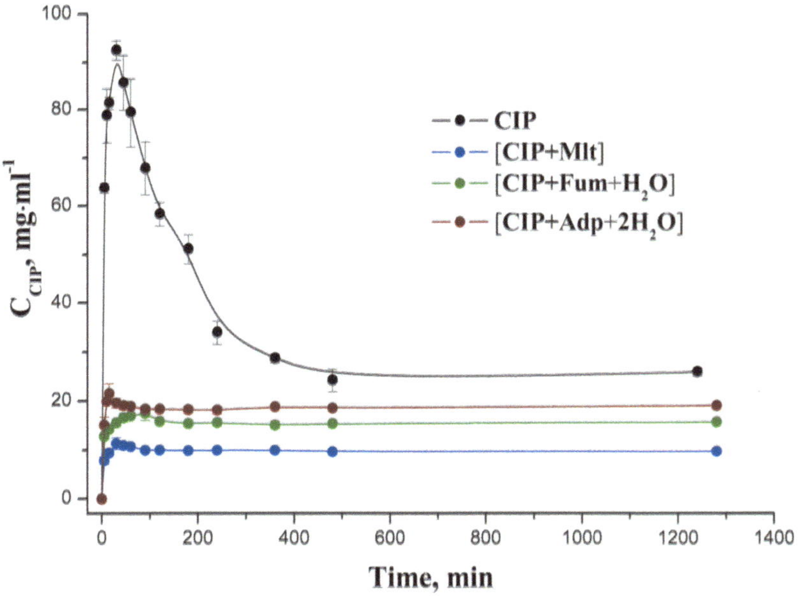

Fig. 2.11: Dissolution profiles at 28 °C for pure ciprofloxacin and salts at pH = 1.2. (Reprinted from The European Journal of Pharmaceutical Sciences, Vol. 77, Surov A.O., Manin A.N., Voronin A.P., Drozd K.V., Simagina A.A., Churakov A.V., Perlovich G.L., Pharmaceutical salts of ciprofloxacin with dicarboxilic acids, Pages No. 112–121, Copyright (2017), with permission from Elsevier).

The results of the dissolution experiments for the salts formed with ciprofloxacin and dicarboxylic acid in buffer solutions at pH = 6.8 are shown in Figure 2.12. Increasing the pH of the solution adversely affects the solubility of fluoroquinolones. The solubility of ciprofloxacin in base buffer decreases by 300 times compared to the ciprofloxacin solubility in the acidic buffer solution, i.e. from 25 to 0.09 mg ml^{-1}. Similar behaviour is observed for the solubility of norfloxacin. Increasing of buffer pH reduces the norfloxacin solubility more than 100 times, from 31.31 to 0.27 mg ml^{-1}.

The ciprofloxacin salts reveal a significant enhancement of solubility compared to pure ciprofloxacin under similar conditions. Figure 2.12 shows that the [CIP + Fum + H_2O] salt has more than 33 times greater solubility than pure ciprofloxacin. The solubility for form I of [ciprofloxacin + maleate] rises by a factor of 20, while [CIP + Adp + 2 H_2O] shows only a moderate, seven fold improvement in solubility.

The relevant literature mainly contains solubility data about salts and co-crystals of fluoroquinolones for buffers with pH values ranging from 6.8 to 7.4. All the experimental data for salts and co-crystals of ciprofloxacin, norfloxacin and enrofloxacin are presented in Table 2.5. The solubilities of fluoroquinolones in salts and co-crystals with dicarboxylic acids increase by up to three fold. The exceptions are [NOR + adipate + adipic acid] (2 : 1 : 1) and [CIP + malonate + H_2O] (1 : 1 : 2), where the solubility of each increases by more than 100 fold.

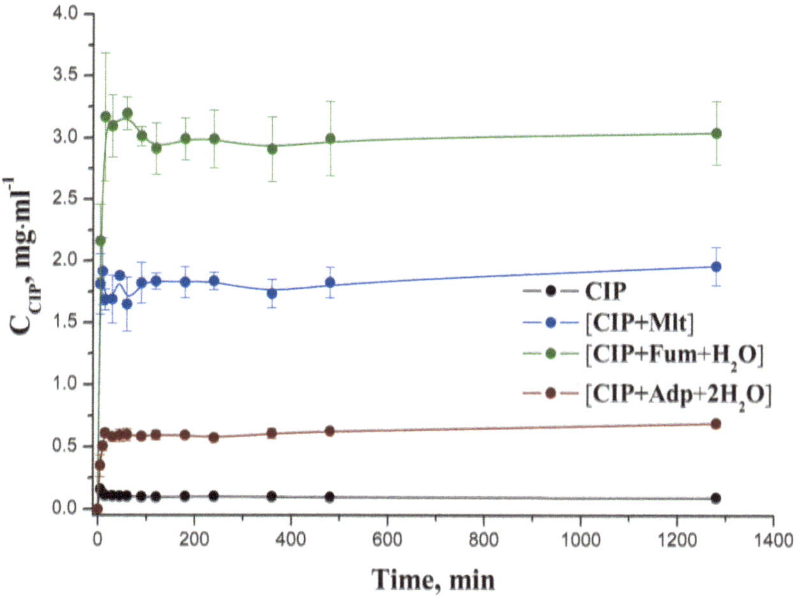

Fig. 2.12: Dissolution profiles at 28 °C for pure ciprofloxacin and salts at pH = 6.8. (Reprinted from The European Journal of Pharmaceutical Sciences, Vol. 77, Surov A.O., Manin A.N., Voronin A.P., Drozd K.V., Simagina A.A., Churakov A.V., Perlovich G.L., Pharmaceutical salts of ciprofloxacin with dicarboxilic acids, Pages No. 112–121, Copyright (2017), with permission from Elsevier).

Tab. 2.5: Experimental values of melting temperatures (T_{fus}), solubilities (S^{API}) and relative changes of the solubility of the API in salt/co-crystals compared to the individual API (S^{API}_{CC}/S^{API}).

Compound	T_{fus}, K	S^{API}, mg ml^{-1}	S^{API}_{CC}/S^{API}	Ref.
CIP	544.9	0.09[a]	-	[15, 26]
[CIP + malonate + H_2O] (1 : 1 : 2)	475.3	8.47[a]	94.1	[15]
[CIP + maleate] (1 : 1)	489.5	1.8[a]	20	[26]
[CIP + tartrate] (1 : 1)	504.2	0.72[a]	8	[15]
[CIP + adipate + H_2O] (2 : 1 : 4)	516.7	0.63[a]	7	[26]
[CIP + fumarate + H_2O] (1 : 1 : 1)	477.0	3.0[a]	33.3	[26]
[CIP + citrate + H_2O] (1 : 1 : 1)	484.0	1.36[a]	15.1	[15]
NOR	493.7	0.27[a]	-	[15]
[NOR + oxalate + H_2O] (1 : 0.5 : 2)	519.5	2.54[a]	9.4	[15]
[NOR + malonate + H_2O] (1 : 1 : 2)	~ 473.2	3.9[b]	14.4	[37]
[NOR + maleate + H_2O] (1 : 1 : 1)	~ 488.2	9.8[b]	36.3	[37]
[NOR + succinate + H_2O] (1 : 0.5 : 1)	~ 493.2	6.6[b]	24.4	[37]
[NOR + tartrate + H_2O] (1 : 1 : 2)	524.4	4.67[a]	17.3	[15]
[NOR + adipate + adipic acid] (2 : 1 : 1)	487.0	38.21[c]	141.5	[23]
ENR	499.8	0.615[d]	-	[24]
[ENR + dioxalate + H_2O] (2 : 1 : 6)	517.8	16.21[d]	26	[24]
[ENR + maleate] (1 : 1)	498.7	4.7[d]	8	[24]
[ENR + disuccinate + succinic acid] (2 : 1 : 1)	485.9	20.65[d]	34	[24]
[ENR + difumarate + fumaric acid] (2 : 1 : 1)	540.6	9.09[d]	15	[24]

[a] solubility was measured in phosphate buffer with pH = 6.8
[b] solubility was measured in water
[c] solubility was measured in deionised water (pH = 6.4)
[d] solubility was determined in water (pH = 6.8)

As is well known, the melting temperatures of molecular crystals are usually used to rate/estimate/imitate the crystal lattice energy, as in the general solubility equation of Yalkowsky and Valvani [33]. Moreover, it has been revealed that there is a linear correlation between the Gibbs energy of sublimation and the melting temperatures of structurally similar compounds [29, 34]. A rise in the melting temperature leads to an increase in Gibbs energy of sublimation, which must decrease the solubility value. So, it can be assumed that for co-crystals and salts, an increase in melting temperature leads to a decrease in solubility. Figure 2.13 summarises the analysis of solubility and melting points change in salts and co-crystals of fluoroquinolones. The results are presented in coordinates of the melting temperature ratio ($T_{fus}(CC)/T_{fus}(API)$) versus their solubility ratio (S^{CC}/S^{API}).

Indeed, the increase in the melting temperature of a fluoroquinolone API when in a salt or in a co-crystal lowers the solubility of the API, with the notable exception of [NOR + adipate + adipic acid] (2 : 1 : 1).

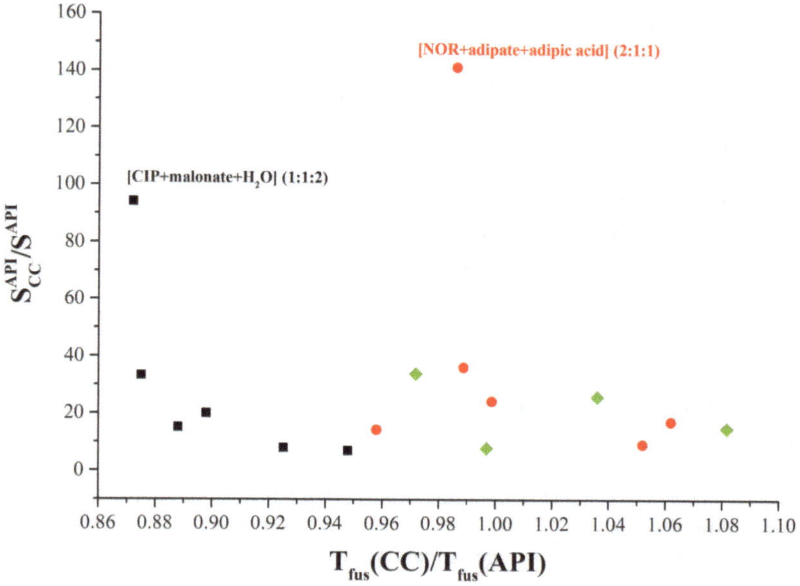

Fig. 2.13: Influence of API melting temperature change in co-crystals and salts of dicarboxylic acids on their solubility. The co-crystals and salts with different APIs are presented by the following symbols: (■) – ciprofloxacin; (●) – norfloxacin; (◆) – enrofloxacin.

2.5 Conclusions

Crystallographic data for co-crystals and salts derived from fluoroquinolones and dicarboxylic acids have been summarised and analysed. The available thermochemical data (melting points of individual components and their salts/co-crystals) have been systematically analysed and correlation equations proposed for the prediction of the melting points of co-crystals. Finally, the dissolution behaviour of different solid forms of fluoroquinolones was summarised and an inverse correlation between melting point and solubility was established for some APIs.

Acknowledgement: This work was supported by the Russian Science Foundation (No. 14-13-00640). We thank "the Upper Volga Region Centre of Physicochemical Research" for technical assistance with TG and XRPD experiments.

Bibliography

[1] King DE, Malone R, Lilley SH. New classification and update on the quinolone antibiotics. Am. Fam. Physician 2000, 61, 2741–2748.

[2] Appelbaum P, Hunter P. The fluoroquinolone antibacterials: past, present and future perspectives. Int. J. Antimicrob. Agents 2000, 16, 5–15.

[3] Pallo-Zimmerman LM, Byron JK, Graves TK. 2010. Fluoroquinolones: then and now. Compend. Contin. Educ. Vet. 2010, 32, E1–E9.
[4] Sharma PC, Jain A, Jain S, Pahwa R, Yar MS. Ciprofloxacin: review on developments in synthetic, analytical, and medicinal aspects. J. Enzyme Inhib. Med. Chem. 2010, 25, 577–589.
[5] Emmerson AM, Jones AM. The quinolones: decades of development and use. J. Antimicrob. Chemo. 2003, 51, 13–20.
[6] Holmes B, Brogden RN, Richards DM. Norfloxacin. A review of its antibacterial activity, pharmacokinetic properties and therapeutic use. Drugs 1985, 30, 482–513.
[7] Oliphant CM, Green GM. Quinolones: A comprehensive review. Am. Fam. Physician 2002, 65, 455–464.
[8] Hu J, Wang W, Zhu Z, Chang H, Pan F, Lin B. Quantitative structure activity relationship model for prediction of genotoxic potential for quinolone antibacterials. Environ. Sci. Technol. 2007, 41, 4806–4812.
[9] Ross DL, Riley CM. Aqueous solubilities of some variously substituted quinolone antibacterials. Int. J. Pharm. 1990, 63, 237–250.
[10] Breda SA, Jimenez Kairuz AF, Manzo RH, Olivera ME. Solubility behavior and biopharmaceutical classification of novel high-solubility ciprofloxacin and norfloxacin pharmaceutical derivatives. Int. J. Pharm. 2009, 371, 106–113.
[11] Takagi T, Ramachandran C, Bermejo M, Yamashita S, Yu LX, Amidon GL. A provisional biopharmaceutical classification of the top 200 oral drug products in the United States, Great Britain, Spain, and Japan. Mol. Pharm. 2006, 3, 631–643.
[12] Kawabata Y, Wadab K, Nakatanib M, Yamada S, Onoue S. Formulation design for poorly water-soluble drugs based on biopharmaceutics classification system: Basic approaches and practical applications. Int. J. Pharm. 2011, 420, 1–10.
[13] Turel I, Golobic A. Crystal structure of ciprofloxacin hydrochloride 1.34- hydrate. Anal. Sci. 2003, 19, 329–330.
[14] Li S, Wong S, Sethia S, Almoazen H, Joshi YM, Serajuddin AT. Investigation of solubility and dissolution of a free base and two different salt forms as a function of pH. Pharm. Res. 2005, 22, 628–635.
[15] Reddy JS, Ganesh SV, Nagalapalli R, Dandela R, Solomon KA, Kumar KA, Goud NR, Nangia A. Fluoroquinolone salts with carboxylic acids. J. Pharm. Sci. 2011, 100, 3160–3176.
[16] Romanuk CB, Garro Linck Y, Chattah AK, Monti GA, Cuffini SL, Garland MT, Baggio R, Manzo RH, Olivera ME. Crystallographic, thermal and spectroscopic characterization of a ciprofloxacin saccharinate polymorph. Int. J. Pharm. 2010, 391, 197–202.
[17] Romanuk CB, Garro Linck Y, Chattah AK, Monti GA, Manzo RH, Olivera ME, Characterization of the solubility and solid-state properties of saccharin salts of fluoroquinolones. J. Pharm. Sci. 2009, 98, 3788–3801.
[18] Garro Linck Y, Chattah AK, Graf R, Romanuk CB, Olivera ME, Manzo RH, Monti GA, Spiess HW. Multinuclear solid state NMR investigation of two polymorphic forms of Ciprofloxacin-saccharinate. Phys. Chem. Chem. Phys. 2011, 13, 6590–6596.
[19] Paluch KJ, McCabe T, Müller-Bunz H, Corrigan OI, Healy AM, Tajber L. Formation and physicochemical properties of crystalline and amorphous salts with different stoichiometries formed between ciprofloxacin and succinic acid. Mol. Pharmaceutics. 2013, 10, 3640–3654.
[20] Bag PP, Ghosh S, Khan H, Devarapalli R, Reddy CM. Drug–drug salt forms of ciprofloxacin with diflunisal and indoprofen. CrystEngComm. 2014, 16, 7393–7396.
[21] Vitorino GP, Sperandeo NR, Caira MR, Mazzieri MR. A supramolecular assembly formed by heteroassociation of ciprofloxacin and norfloxacin in the solid state: co-crystal synthesis and characterization. Cryst. Growth Des. 2013, 13, 1050–1058.

[22] Florindo C, Costa A, Matos C, Nunes SL, Matias AN, Duarte CMM, Rebelo LPN, Branco LC, Marrucho IM. Novel organic salts based on fluoroquinolone drugs: Synthesis, bioavailability and toxicological profiles. Int. J. Pharm. 2014, 469, 179–189.
[23] Xu Y, Jiang L, Mei X. Supramolecular structures and physicochemical properties of norfloxacin salts. Acta Crystallogr. B 2014, 750–760.
[24] Karanam M, Choudhury AR. Structural landscape of pure enrofloxacin and its novel salts: enhanced solubility for better pharmaceutical applicability. Cryst. Growth Des. 2013, 1626–1637.
[25] Grothe E, Meekes H, Vlieg E, ter Horst JH, de Gelder R. Solvates, salts, and cocrystals: a proposal for a feasible classification system. Cryst. Growth Des. 2016, 16, 3237–3243.
[26] Surov AO, Manin AN, Voronin AP, Drozd KV, Simagina AA, Churakov AV, Perlovich GL. Pharmaceutical salts of ciprofloxacin with dicarboxylic acids. Eur. J. Pharm. Sci. 2015, 77, 112–121.
[27] Zhang G, Zhang L, Yang D, Zhang N, He L, Du G, Lu Y. Salt screening and characterization of ciprofloxacin. Acta Crystallogr. B 2016, 72, 20–28.
[28] Cruz-Cabeza AJ, Reutzel-Edens SM, Bernstein J. Facts and fictions about polymorphism. Chem. Soc. Rev. 2015, 44, 8619–8635.
[29] Perlovich GL. Thermodynamic characteristics of cocrystal formation and melting points for rational design of pharmaceutical two-component systems. CrystEngComm 2015, 17, 7019–7028.
[30] Weyna DR, Cheney ML, Shan N, Hanna M, Zaworotko MJ, Sava V, Song S, Sanchez-Ramos JR. Improving solubility and pharmacokinetics of meloxicam via multiple-component crystal formation. Mol. Pharm. 2012, 9, 2094–2102.
[31] Sanphui P, Tothadi S, Ganguly S, Desiraju GR. Salt and cocrystals of sildenafil with dicarboxylic acids: solubility and pharmacokinetic advantage of the glutarate salt. Mol. Pharm. 2013, 10, 4687–4697.
[32] Manin AN, Surov AO, Churakov AV, Perlovich GL. Crystal structures, thermal analysis, and dissolution behavior of new solid forms of the antiviral drug arbidol with dicarboxylic acid. Crystals 2015, 5, 650–669.
[33] Yalkowsky SH, Valvani SC. Solubility and Partitioning I: Solubility Nonelectrolytes in Water. J. Pharm. Sci., 1980, 69, 912–922.
[34] Perlovich GL, Ryzhakov AM, Tkachev VV, Hansen LKR, Raevsky OA. Sulfonamide molecular crystals: structure, sublimation thermodynamic characteristics, molecular packing, hydrogen bonds networks. Cryst. Growth. Des. 2013, 13, 4002–4016.
[35] Etter MC. Encoding and decoding hydrogen-bond patterns of organic compounds. Acc. Chem. Res. 1990, 23, 120–126;
[36] Bernstein J, Davis RE, Shimoni L, Chang N-L. Patterns in hydrogen bonding: functionality and graph set analysis in crystals. Angew. Chem. Int. Ed. Engl. 1995, 34, 1555–1573.
[37] Basavoju S, Bostrom D, Velaga SP. Pharmaceutical Cocrystal and Salts of Norfloxacin. Cryst. Growth Des. 2006, 6, 2699–2708.

Anna Krawczuk and Marlena Gryl

3 Qualitative and quantitative crystal engineering of multi-functional co-crystals

3.1 Introduction

In recent years, an increasing emphasis on the development of multi-functional crystals and co-crystals has been observed. In our understanding, a multi-functional material should be characterised by the following features:
- the presence of at least two significant physical/chemical properties,
- thermal and mechanical stability when no external stimuli (i.e. temperature, electric/magnetic field, stress, etc.) are applied
- the possibility of technological/industrial application

For example, triglycine sulphate (TGS) [1] crystallises in the polar space group $P2_1$ and was found to be an excellent pyroelectric and ferroelectric material [2] widely used as a part of infrared detectors and formerly applied in pyroelectric vidicon tubes operating at room temperature. Furthermore, when doped with other organic (e.g. urea [3], alanine [4]) or inorganic (e.g. sodium bromide [5]) compounds, the physical properties were enhanced significantly. Such materials achieve a higher figure of merit compared with pure TGS.

The entire process of designing, validating and adopting a particular material for practical use is like walking up a flight of stairs (Figure 3.1). In the case of a multi-functional co-crystal, a few major steps need to be overcome to reach the ultimate goal. The first step is the choice of appropriate molecular components to ensure certain structural features of a potential multi-functional co-crystal. For example, to guarantee a

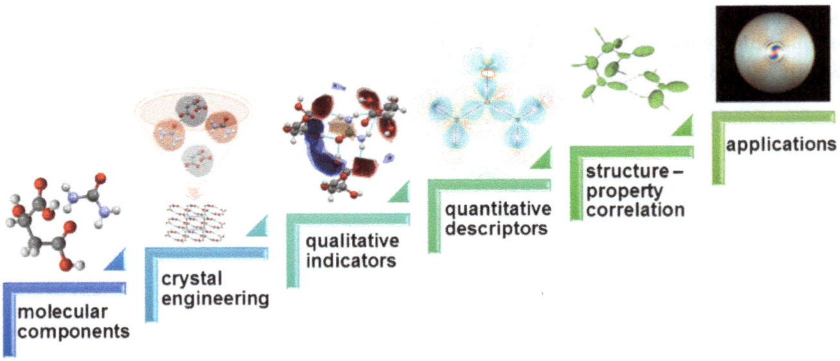

Fig. 3.1: From the design to applications of a multi-functional co-crystal.

https://doi.org/10.1515/9783110464955-003

noncentrosymmetric crystal structure, it is advisable to use chiral molecules, as their presence in the crystal structure excludes a centre of symmetry.

The next step requires an understanding of how the already chosen building blocks interact with each other and which of the interactions will be at the origin of the desired physical properties. For that purpose, various crystal engineering tools can be used, either based on qualitative or quantitative system analysis. Whereas the first method is based mainly on geometrical and statistical criteria, quantitative crystal engineering offers the quantification of observable properties, e.g. electron density (ED). Recent studies showed [6–9] that the use of well-defined physical descriptors significantly increases the amount of available data on the structural features of a material, thus decreasing the level of uncertainty and ensuring better modelling of a material.

The next level of successful multi-functional co-crystal design involves the correlation between structure and physical/chemical property, i.e. quantitative structure-property relationship, QSPR. At this point quantum-mechanical tools can be exploited to predict given properties. This is a relatively fast way to screen whether the obtained material is likely to display useful properties for use in materials science. At this point there are two possible paths to choose: i) both the theoretical prediction and then experimental validation confirms the material is ready for application (final step) and the main goal is achieved or ii) one is forced to return to the first and/or second step of the stairs (Figure 3.1) and embrace a different strategy.

In this chapter the most popular and commonly used tools of qualitative and quantitative crystal engineering will be introduced and it will be demonstrated how these could be used to successfully design multi-functional co-crystals.

3.2 How to design a multi-functional co-crystal

3.2.1 Co-crystals – a problematic case of multi-component solids

The term co-crystal was first introduced by Wöhler in 1844 [10] and popularised by Etter in 1991 [11] to describe multi-component organic solids. Co-crystals have become increasingly important in the pharmaceutical industry to modify the solubility and bioavailability of active pharmaceutical ingredients (APIs) [12]. The scope of possible co-crystal applications has grown exponentially. Nowadays, they are produced to improve material performance by modifying the mechanical [13], energetic [14] or optical [15] properties of materials.

Despite the vast applications of multi-component materials, the overall definition of a co-crystal is still controversial. A variety of possible co-crystallisation products have led to several more or less accurate descriptions of what should constitute a co-crystal. In the scientific literature, co-crystals have variously been described as molecular complexes, molecular compounds, lattice compounds, adducts, composite crystals, etc. In 2003 Desiraju [16] proposed to replace the often problematic term co-crystal with molecular complex. This claim was rebutted by Dunitz [17], who showed that this term molecular complex is deeply rooted in chemistry and has too broad a sense.

This scientific debate led to an article by Bond in 2007 [18] who highlighted problematic cases of co-crystals and solvates, and proposed to retain the original definition of a co-crystal as a "multi-component molecular crystal". This description also includes solvates, molecular complexes, clathrates, inclusion compounds and solid solutions. In 2011, the United States Food and Drug Administration published guidelines for co-crystals containing APIs with a very narrow definition of a co-crystal as being composed of two or more molecules combined in a crystal lattice. This regulation excluded a large family of so called ionic co-crystals [19], which in the crystal structure contains organic molecules alongside inorganic ions. This provoked further discussion in the scientific community to standardise the nomenclature. In 2012, Aitipamula et al. [20] published an article stating that co-crystals are "solids that are crystalline single phase materials composed of two or more different molecular and/or ionic compounds generally in a stoichiometric ratio, which are neither solvates nor simple salts." For the purposes of this chapter, this definition of a co-crystal is adopted.

3.2.2 Crystal engineering strategies

Crystal engineering is a form of non-covalent synthesis where the products are crystalline materials with desired and predictable properties. The most commonly accepted definition is still the one provided by Desiraju [21]:

> "Crystal engineering (...) the understanding of intermolecular interactions in the context of designing new solids with desired physical and chemical properties".

The first step of a successful crystal engineering process should be the choice of the building blocks. The rationale behind selecting components should be based on the required material properties. There is no general recipe for how to design a multi-component solid for a specific application. What then, should be taken into account? If the goal is crystal symmetry, like in the case of many physical properties requiring a polar and/or chiral crystal structure, components (often chiral molecules) need to be selected that will generate a noncentrosymmetric crystal structure [22]. Another option could be co-crystallisation with the use of chiral solvents. A polar crystal structure gives hope for pyroelectric, ferroelectric, piezoelectric, linear and non-linear optical properties [23]. Chiral crystals are worth considering as piezoelectric or optical materials [24]. Of course it needs to be remembered that crystal symmetry is a necessary but not a sufficient condition for a bulk effect to take place.

With the aim to tune the mechanical performance of a material, e.g. plastic-like bending crystals, components should be selected that possess pseudo-spherical functional groups such as alkyl, $-OCH_3$, $-Cl$ and others. Those molecules should form layers facilitated by van der Waals interactions involving those functional groups [13, 25].

When selecting molecular building blocks, the self-assembly processes that usually lead to a single component crystal structure need to be considered. In other words, a co-former needs to be found that would be "a more attractive partner" for binding than the parent molecule [26]. This requires an assessment of the complementarity of the molecules, both energetically and geometrically. To predict *binding propensities*, crystal structures can be dissected into smaller parts considering either supramolecular synthons [27] or tectons [28] as structural motifs. Tectons are molecules whose interactions dominate in the crystal structure and induce molecular self-assembly into an organised network with specific functional features. One of the simplest tectons is the water molecule, which forms strong hydrogen bonds. Supramolecular synthons are defined as structural units within a crystal assembled from functional groups connected by weak interactions, often hydrogen bonds. The spatial arrangements of interactions can be easily classified and thus, to some extent, predicted. The search for common synthons in known structures containing a selected molecular building block can suggest alternate candidates as the co-former. This statistical analysis greatly depends on synthon modularity.

An approach based either on synthons or tectons (Figure 3.2) can benefit from the empirical hydrogen bond formation guidelines formulated by Etter [29]. Three general rules state that: "all good proton donors and acceptors are used in hydrogen bonding, six-membered-ring intramolecular hydrogen bonds form in preference to intermolecular hydrogen bonds and the best proton donors and acceptors remaining after intramolecular hydrogen bond formation form intermolecular hydrogen bonds". Good proton donors are those in carboxylic acids, amides, anilines, urea, imides and phen-

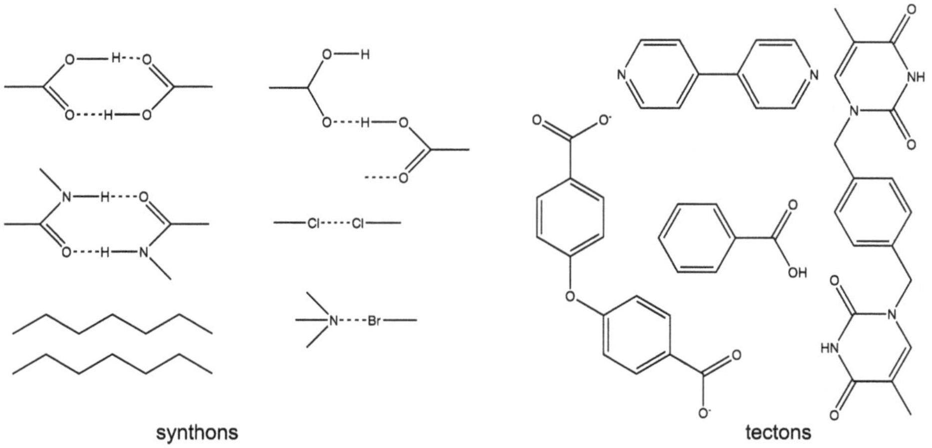

Fig. 3.2: Examples of synthons and tectons often referred to in the literature.

ols. Good proton acceptors are acid and amide carbonyl groups, phosphoryls, nitroxides, sulfoxides and amine nitrogen atoms.

3.3 Qualitative crystal engineering tools

3.3.1 Cambridge Structural Database

The Cambridge Structural Database (CSD) [30] provides a variety of ready made tools for crystal engineering purposes. Solid state modules incorporated in Mercury software [31] and integrated with the CSD are capable of searching hydrogen bond motifs in crystal structures. This can be based on synthon identification using the 3DSEARCH algorithm. The program has various possibilities for the selection of organic functional groups and, based on its analysis, returns the frequency of a particular motif's occurrence. Taking the crystal structure of 4-hydroxybenzoic acid monohydrate as an example (Figure 3.3) [32], the $R_2^2(8)$ ring system of hydrogen bonds can be readily identified. This is the only hydrogen bonding interaction between organic molecules; the water molecules play the role of linkers allowing the formation of a 3D architecture.

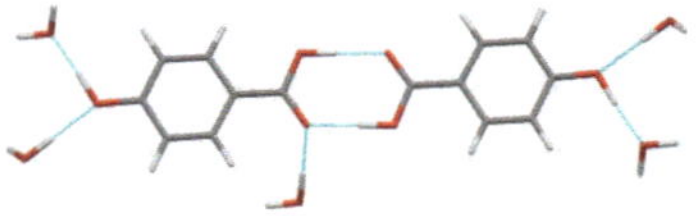

Fig. 3.3: Hydrogen bond motifs in the structure of 4-hydroxybenzoic acid monohydrate.

Molecules of 4-hydroxybenzoic acid can be considered as an example of a push-pull organic NLO-chromophore, which could be used in the design of multi-functional co-crystals. However, a molecule capable of forming preferable interactions with the acid, cf. acid···acid interactions, needs to be found. The search could be simplified by investigating possible functional groups capable of forming hydrogen bonds with the carboxylic group. Candidate functional groups could be carboxamide, oxime, hydroxyl and ester. The frequency of occurrence of heteromeric motifs (between different functional groups) could be then compared with the percentage of homomeric units in the CSD database. The results of this search are presented in Table 3.1.

Based on these statistics, oxime, carboxamide and hydroxyl functional groups are likely to form interactions with carboxylic acids to a greater extent than carboxylic acid···carboxylic acid formation. Indeed, in the CSD co-crystals of 4-hydroxybenzoic acid with carboxylic acid···carboxamide and carboxylic acid···hydroxyl interactions are known (Figure 3.4).

Another approach towards the selection of co-crystal building blocks is through an analysis of hydrogen bond propensities. This method is based on assessing hydrogen bond formation, taking into account factors like steric hindrance. The results are the probabilities of forming a hydrogen bond in a specific environment, encompassing

Tab. 3.1: Frequency of occurrence of carboxylic acid homo- and hetero-interactions in the CSD. L1 – number of structures with motif; L2 – number of structures where the motif could occur (real and hypothetical).

Interaction between functional groups	L1/L2	% frequency of occurrence
COOH–COOH	6899/23546	29.3%
COOH–R_1COOR_2	157/1146	13.7%
COOH–R_3COONH_2	261/495	52.7%
COOH–R_4CNOH	27/41	65.9%
COOH–R_5OH	1423/2807	50.7%

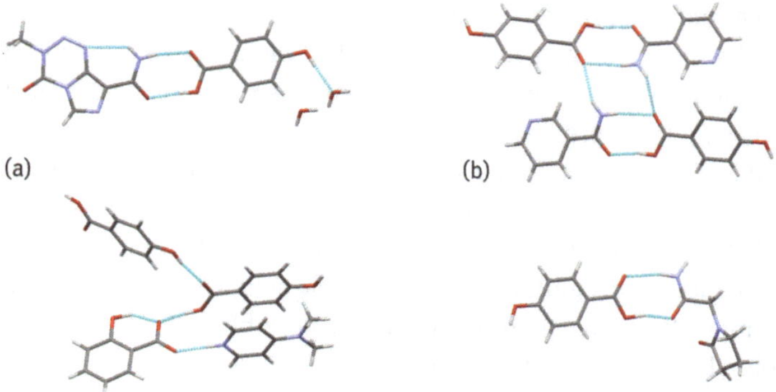

(a) (b) (c) (d)

Fig. 3.4: Hydrogen bond motifs in co-crystals containing 4-hydroxybenzoic acid with (a) temozolomide monohydrate [33], (b) nicotinamide [34], (c) 4-(dimethylamino)pyridinium salicylate [35] and (d) 2-(oxo-1-pyrrolidinyl)acetamide [36].

functional groups found in the vicinity of the formed interaction. This method is not limited by having a large number of structures containing specific functional groups in the CSD [30].

The L-malic acid theophylline co-crystal provides an example [37]. The asymmetric unit comprises one L-malic acid molecule and one theophylline molecule connected via an O–H···O hydrogen bond (Figure 3.5), which in this crystal phase has the greatest propensity to form (0.54). If we consider the target molecule (primary component) as L-malic acid and the co-former as theophylline, this O–H···O interaction is classified as an AB type. On the other hand, the AA interaction between the two L-malic acid molecules has a propensity of 0.45.

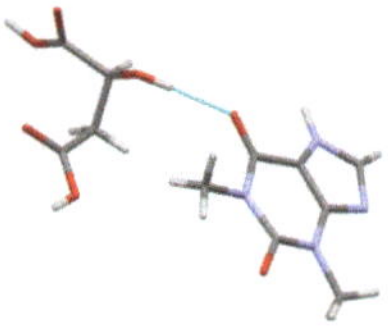

Fig. 3.5: The asymmetric unit in the L-malic acid theophylline co-crystal.

The difference between AB and AA formation propensities (0.09) is called the Multi-Component Score (MC Score). If this parameter is positive, it can be assumed that the formation of a hetero-interaction is more favourable than the formation of a homo-interaction. A positive value indicates the feasibility of co-crystal formation between the components. This method can be used to predict the possibility of hetero-interactions and thus, can be treated as an indicator for crystal engineering purposes.

A similar tool, also incorporated in the Mercury software, is called molecular complementarity. It is based on an algorithm designed to find similar molecules capable of forming intermolecular interactions using indicators like shape and polarity. This method was developed by Fábián [38] who performed a statistical analysis on multiple co-crystal structures in a search for the factors that determine molecular recognition. The choice of theophylline molecule as a target exemplifies the usage of this tool. A quick search for complementary molecules in the small database comprising the most common co-crystal formers, CCF's (provided by Mercury molecular libraries), leads to the results summarised in Table 3.2. Out of 89 selected molecules, 60 possess appropriate properties allowing them to be considered as suitable co-formers for theophylline. Among the 89 molecules, 12 are known to form co-crystals with theophylline. The method fails for only two molecules: glutaric acid and sorbic acid, which each have a hit rate equal to 0%, which eliminates those molecules as possible CCFs. Co-crystals of both acids and theophylline had already been discovered [39, 40].

A final attractive method for qualitative crystal engineering is based on the analysis of full interaction maps. This method combines the information on intermolecular contacts stored in the IsoStar library [41] with an approach developed for identifying interaction sites in proteins (SuperStar) [42, 43]. When a molecule is chosen for analysis, the first step involves the splitting of the molecule into so-called central groups, which are actually the selected functional groups of the target. The next step is the generation of 3D scatterplots using the CSD contact search and then the conversion of the plots into scaled density maps. Full Interaction Maps are generated by combining the density maps formed for the functional groups. In this way, the probabilities of finding specific donor/acceptor sites in close vicinity of the molecule in question can be analysed.

Returning to theophylline, Figure 3.6a displays an interaction map for the molecule. The blue regions show the possible donor sites, whereas acceptor sites are indicated by the red regions. The orange regions indicate weaker interactions C–H…π or π…π. When the co-crystal of theophylline with L-malic acid is analysed, will see how the regions of high interaction likelihoods are filled with suitable donors and acceptors from both the target and co-former molecules (Figure 3.6b).

Tab. 3.2: Co-former selection for theophylline based on the molecular complementarity tool [31].

Co-former	Hit Rate %	Co-former	Hit Rate %
(+)-camphoric acid	100	citric acid	100
(-)-camphorsulfonic acid	0	ethylparaben	100
1-hydroxyethylidene-1,1-diphosphonic acid	0	folic acid	0
2-amino-5-methylbenzoic acid	100	fumaric acid	100
3-methylpyridine	0	gentisic acid	100
4-acetamidobenzoic acid	0	glutaric acid	0
4-aminobenzoic acid	100	glycine	100
4-hydroxybenzoic acid	100	glycolic acid	100
theophylline	100	hesperetin	100
D-alanine	0	hippuric acid	100
D-glucuronic acid	100	hydrocinnamic acid	100
D-pantothenol	100	imidazole	100
EDTA	100	isonicotinamide	100
L-arginine	100	ketoglutaric acid	100
L-aspartic acid	100	lactobionic acid	0
L-aspartic acid z	0	lactose	100
L-glutamic acid	100	maleic acid	100
L-glutamic acid z	0	malic acid	100
L-glutamine	100	malonic acid	100
L-glutathione	100	maltitol	0
L-lactic acid	0	mannitol	0
L-leucine	100	methanesulfonic acid	0
L-mandelic acid	100	methylparaben	100
L-methionine	100	monobutyrin	100
L-phenylalanine	100	nicotinamide	100
L-proline	0	oxalic acid	100
L-serine	100	pamoic acid	0
L-tartaric acid	100	phthalamide	100
L-tryptophan	100	pimelic acid	0
L-tyrosine	100	piperazine	100
N-ethylacetamide	100	propylparaben	0
acesulfame	100	pyrazine	100
acetic acid	100	riboflavin	100
acetophenone oxime	100	saccharin	100
acetylenedicarboxylic acid	100	sorbic acid	0
adipic acid	0	suberic acid	0
alitame	0	succinic acid	100
apigenin	100	t-butylamine	0
azelaic acid	0	t-butylhydroxyanisole	0
benzoic acid	100	theophylline	100
biotin	100	thymidine	0
caprolactam	100	triphenylacetic acid	0
capsaicin	0	urea	100
cholic acid	0	valerolactam	100
		xanthine	100

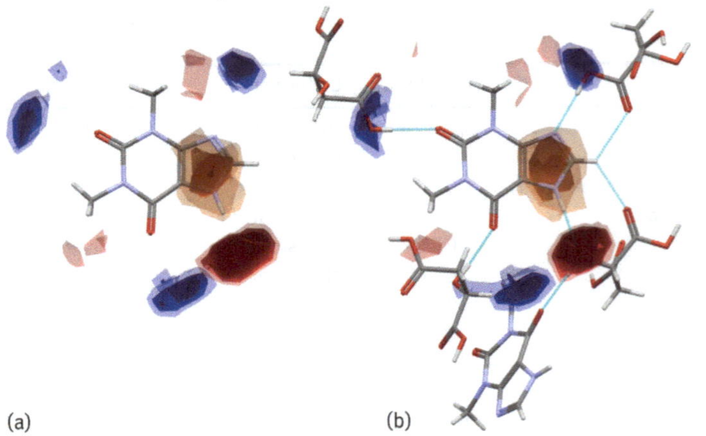

Fig. 3.6: Full Interaction Map for (a) an isolated theophylline molecule and (b) the theophylline L-malic acid co-crystal.

3.3.2 Hirshfeld surfaces and fingerprint plots

Different intermolecular interactions can be made visible through Hirshfeld surfaces [44] (Figure 3.7) and the analysis of 2D fingerprint plots [45] (Figure 3.8). In addition, the properties of a Hirshfeld surface such as volume, surface area and packing index can be evaluated for a chosen molecule in the crystal structure. Here, a description of how these methods can be used in co-crystal design is given.

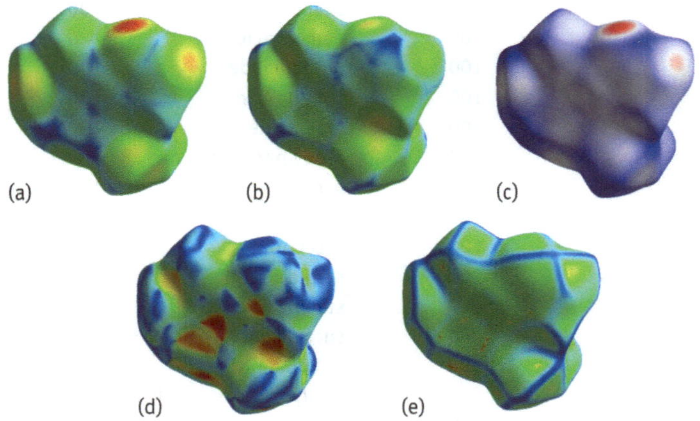

Fig. 3.7: Hirshfeld surfaces of theophylline mapped with a) d_i, b) d_e, c) d_{norm}, d) shape index and e) curvedness.

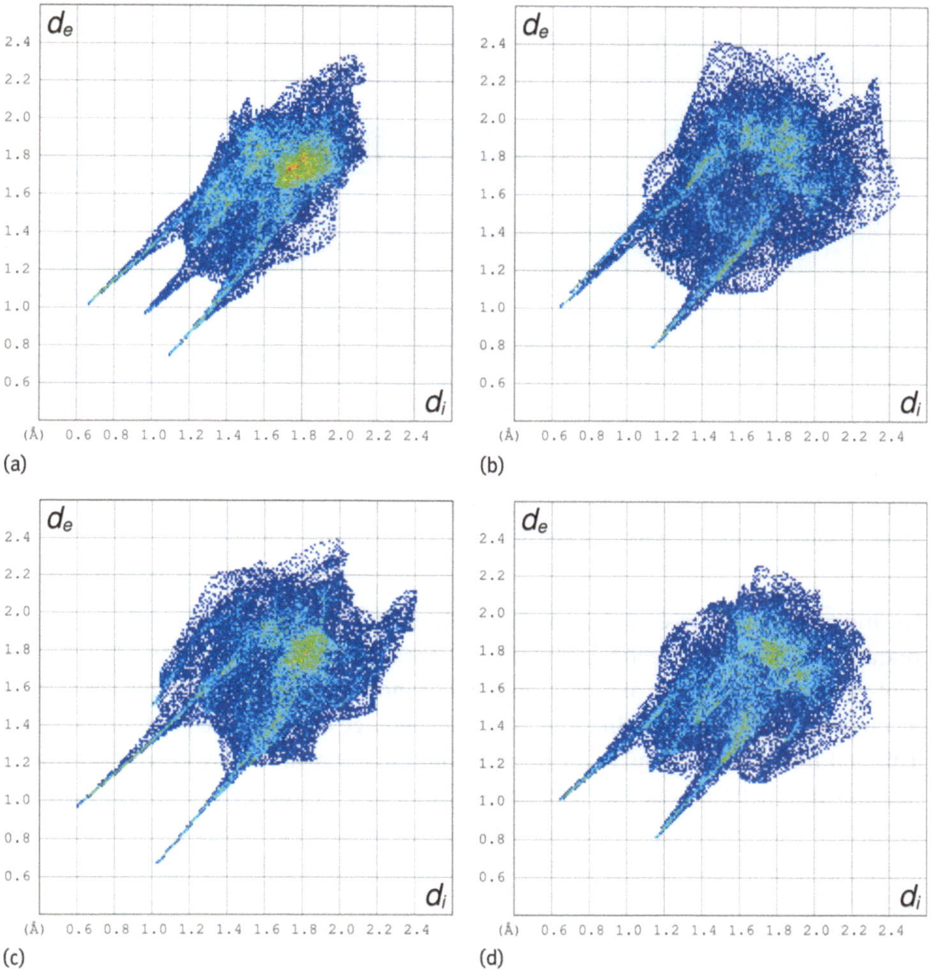

Fig. 3.8: Fingerprint plots for co-crystals of 4-hydroxybenzoic acid with (a) temozolomide monohydrate [33], (b) acid nicotinamide [34], (c) 4-(dimethylamino)pyridinium salicylate [46] and (d) 2-(oxo-1-pyrrolidinyl)acetamide [47].

Distances between the surface and the nearest nucleus inside the surface (d_i) and the one outside the surface (d_e) can be mapped (Figure 3.7a, b) and their combination gives a fingerprint plot, which is unique for a molecule in that particular environment.

Taking into account the different sizes of the atoms, the values of d_i and d_e can be can normalised to van der Waals radii. This gives the d_{norm} values (Figure 3.7c), which are the most useful indicators of the percentage contributions of all intermolecular contacts to the surface. The curvature functions, from which the shape index (Figure 3.7d) and curvedness (Figure 3.7e) can be computed are of particular interest for crystal engineering. The shape index shows the close contact fragments, both posit-

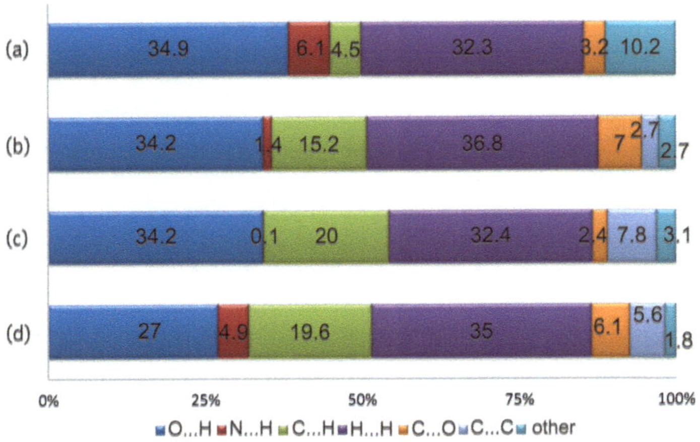

Fig. 3.9: Percentage contribution of each interaction to the total surface in co-crystals of 4-hydroxybenzoic acid with (a) temozolomide monohydrate, (b) nicotinamide, (c) 4-(dimethylamino)pyridinium salicylate and (d) 2-(oxo-1-pyrrolidinyl)acetamide.

ive and negative, between neighbouring molecules. Curvedness helps in deducing the planarity of the molecule in the context of potential stacking interactions. Fingerprint plots give the possibility of comparing two or more co-crystal structures containing a common target molecule in terms of the percentage of intermolecular interactions. This feature could be useful in the design of new material as it allows the prediction of possible interactions based on similarities found in the plots. Figure 3.8 shows fingerprint plots calculated for 4-hydroxybenzoic acid in different chemical environments. The percentage contributions of each interaction to the total surface are shown in Figure 3.9.

3.4 Crystal engineering via electron density (ED) studies

As already pointed out, the classification and recognition of chemical bonding in a crystal and its role in molecular self-assembly is essential knowledge for the design of materials with desired physical properties. However, a question arises, i.e. whether classical X-ray diffraction experiments are sufficient to gain full insight into the nature of the intermolecular interactions responsible for the formation of three-dimensional crystals and thus, the physical properties of the material. It should be emphasised that the basis of conventional X-ray diffraction analysis assumes that the total ED of a molecule (here called promolecule) is well described by a sum of spherically averaged densities of individual neutral atoms at a position R_k:

$$\rho_{\mathrm{IAM}}(r) = \sum_k \rho_k^0(r - R_k)$$

This definition (independent atom model, IAM) entails two important consequences: it prevents the transfer of electrons between atoms and it denies the contraction and expansion of the valence shell; thus any quantitative information on the deformation of ED within a solid is lost. A schematic drawing of the difference between actual aspherical ED distribution and promolecule density is given in Figure 3.10.

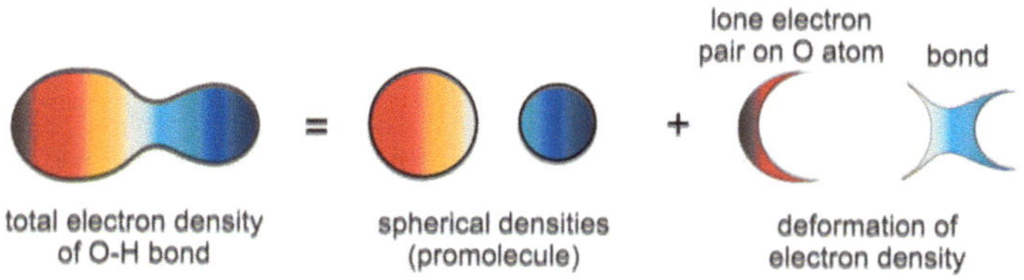

Fig. 3.10: Illustration of the deformation of electron density (ED) due to formation of an O–H bond.

This simple illustration shows that in order to correctly reconstruct the ED of the interacting atoms, more advanced models than IAM are necessary. Nowadays, there are a number of approaches dealing with this issue. However, since the main goal here is to present some quantitative tools offered by ED studies for crystal engineering purposes, the reader is referred to high quality textbooks [5, 48–51] and review articles [53, 54]. A summary of concepts is given here, on how to determine ED either through experimentation or from quantum mechanical calculations, and how this information can be useful for crystal engineers.

In the field of experimental charge density analysis, the most adopted and widely used method for modelling ED is the multipolar model by Hansen and Coppens [55], in which atomic density is partitioned into three components, one defining spherical core density $\rho_c(r)$ and two describing spherical $\rho_v(\kappa r)$ and aspherical $\rho_v(\kappa' r)$ valence density:

$$\rho_{\mathrm{atom}}(r) = P_c\rho_c(r) + P_v\kappa^3\rho_v(\kappa r) + \sum_{l=0}^{l_{\max}} \kappa'^3 R_l\left(\kappa' r\right) \sum_{m=0}^{l_{\max}} P_{lm\pm}y_{lm\pm}(\theta, \phi)$$

where the last summand defines $\rho_v(\kappa' r)$. In the above equation, $y_{lm\pm}$ represents the spherical harmonic angular functions (usually truncated at the hexadecapolar level), R_l is the radial function, κ and κ', are the expansion/contraction parameters of spherical and aspherical valence density and P_c, P_v $P_{lm\pm}$ represent the population parameters of both shells. The atomic density (pseudo-atom density) offers an accurate model for describing molecular ED including bond formation, polarisation of atoms and charge transfer between molecules within the crystalline state. However, it has to be stressed that in order to correctly model ED with the multipolar approach, the following requirements for X-ray diffraction experiments need to be fulfilled:

- high resolution data – for accurate fixing atomic position and thermal parameters ($d < 0.5$ Å)
- high quality data – majority of reflections with $I > 2\sigma(I)$
- low temperature (at least 100 K) – to ensure minimal thermal motion and better deconvolution of the ED
- large redundancy (> 10) – for extensive and complete datasets, as well as a proper account for aspherical distribution of ED
- high quality crystals – accurate corrections of measured diffracted intensities

Although multipolar ED is not a quantum mechanical function, it can be compared to ED computed ab initio by the use of quantum chemical methods. This leads to an important outcome for crystal engineering purposes. In many cases, chemists face the problem of growing good quality single crystals appropriate for measurements of a certain physical property, such as refractive indices, optical activity or first and high-order electric susceptibilities. Thus, before engaging in any experiments, it is desirable to predict whether the magnitude of a given effect is sufficient to be applicable in materials science. The prediction can be achieved by launching periodic boundary calculations that allow the mimicking of ED in a crystal. One of the most common packages used for this purpose is the CRYSTAL14 package [56, 57], which enables the estimation of the crystal orbital periodic Hartree–Fock or DFT wave functions. Many studies have shown that by using the coupled perturbed Kohn–Sham approach [58], it is possible to calculate reasonable values (comparable with those obtained from experiment) of some physical properties of a material, such as refractive indices [59], dielectric constant [60] and piezoelectricity [61]. Those examples prove that theoretical calculations are powerful tools that allow more rapid screening of designed materials and help to identify the most promising candidates for further experimental studies on a chosen physical property. Examples will be given later in the chapter (Section 3.5).

Another advantage for crystal engineering purposes that comes from experimentally or theoretically derived ED, is that the deformation of ED of a given functional group is similar in different molecules and in different crystalline states. This means that by exporting multipolar parameters from one system to another, it is possible to reconstruct the ED of any molecule of choice. Based on this assumption, in the last decade numerous databanks [62–69] have been developed and successfully applied for building up the ED of major biologically active molecules such as amino acids, drugs and proteins. Furthermore, since exportability of multipolar parameters holds for functional groups, Hathwar with co-workers [70, 71] proposed a library scheme (supramolecular synthon based fragments approach, SBFA) only for structural fragments that represent supramolecular synthons commonly used in crystal engineering. This database provides information on how to design and fabricate new synthons in order to form a given hydrogen bonded network within a crystal and thus, at least in principle, enhance certain physical properties exhibited in that assembly.

3.4.1 Quantitative descriptors for crystal engineering

3.4.1.1 Bond path, bond critical point and Laplacian of electron density (ED)

By far the most common and most powerful method to show how electronic charge is distributed throughout real space and thus, how chemical bonding can be "visualised" in the solid state, is Bader's quantum theory of atoms in molecules (QTAIM) [72]. This approach is based on the assumption that every molecule can be uniquely partitioned into atomic basins that are bound by surfaces of local zero flux (S) in the gradient field. Atoms defined in this way can exchange charge and momentum across their interatomic surfaces (IAS). The topology of ED, $\rho(r)$, can be explored by the analysis of its gradient vector field defined in Cartesian coordinates, $\hat{\imath}$, $\hat{\jmath}$, $\hat{k}$ as follows:

$$\nabla\rho(r) = \hat{\imath}\frac{\partial\rho}{\partial x} + \hat{\jmath}\frac{\partial\rho}{\partial y} + \hat{k}\frac{\partial\rho}{\partial z}$$

$\nabla\rho(r)$ is represented by so-called trajectories which indicate an increase of ED. Since ED exhibits a maximum at the position of each nucleus, trajectories terminate at each of them and must originate at the minima of ED. The same situation holds in case of saddle points where, in at least one direction, minimum or maximum of ED is observed. An example of the trajectories of the density gradient partitioning of the urea molecule superimposed on the ED contours is given in Figure 3.11. The zero-flux surfaces forming the boundaries of atomic basins are indicated by thick purple lines and the thin purple lines represent the gradient trajectories that terminate at the nuclei (nuclear attractors). When two atoms are involved in chemical bonding, they are linked by a bond path (BP, black solid lines in Figure 3.11), which is a unique line corresponding to maximum concentration of ED. It crosses the interatomic surface at a bond critical point (BCP, black points) where the gradient of ED is equal to zero.

The existence of both a BCP and a BP between two atoms is a necessary and sufficient condition for the existence of chemical bonding [73]. However, as pointed out by Bader [74], not every BP represents a bond defined in accord with the Lewis model (e.g. covalent bond), and therefore, the presence of BPs and BCPs corresponds to all kinds of interactions which can occur between two atoms in the solid state. In principle, the length of a BP should coincide with the internuclear axis (bond length). Yet, in many cases BPs are found to be curved or chemically strained and the BP length exceeds the bond length. This is especially profound in the case of hydrogen bond formation and within strongly strained cyclic molecules. Examples illustrating both cases are given in Figure 3.12.

The position of the BCP on the BP depends on the electronegativity difference between bonded atoms. More electronegative atom is characterised by a larger atomic basin, thus the boundary surface and hence, BCP, shift towards the less electronegative atom. This is clearly seen in Figure 3.11, which gives a comparison of atomic basins for carbon and oxygen atoms in a carbonyl group. Not only is the atomic basin of carbonyl-oxygen significantly larger, but the shift of BCP towards the less electronegative carbon atom is also more profound.

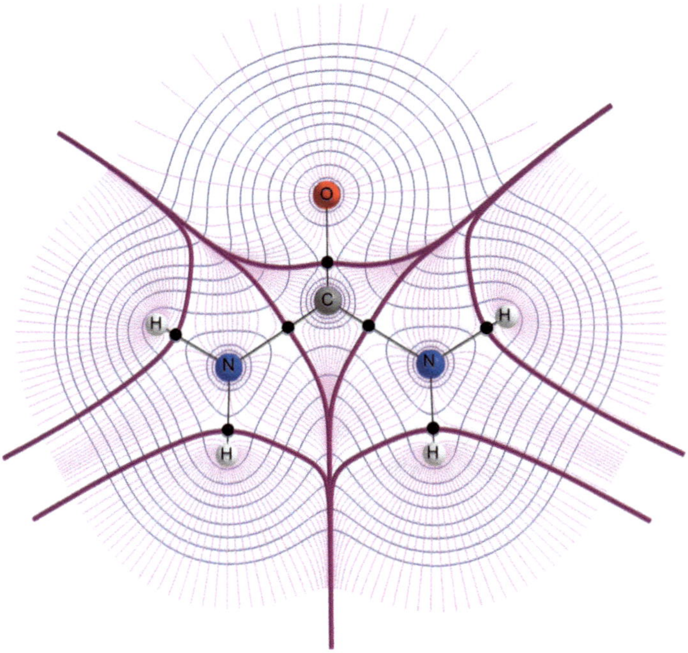

Fig. 3.11: The QTAIM partitioning of molecular urea from the theoretical wave function. The trajectories of the gradient of the electron density (ED; *thin purple lines*) of the urea molecule are superimposed on the ED contours (*blue lines*). The zero-flux surfaces forming the boundaries of the atomic basins are indicated by *thick purple lines* and the bond critical points are marked as *black solid points*.

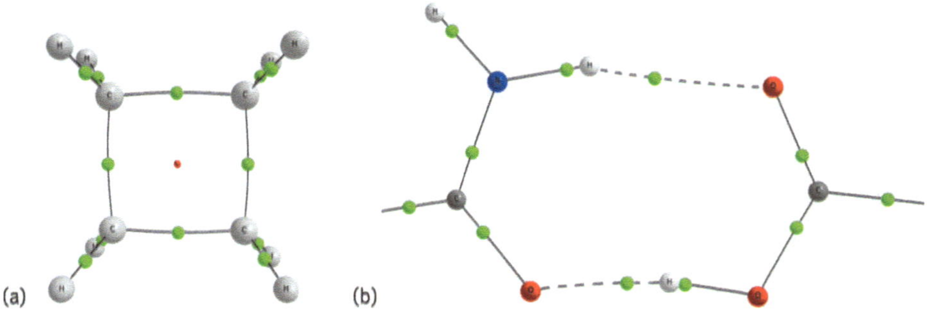

Fig. 3.12: Molecular graphs of (a) cyclobutane and (b) a fragment of the heteromer of urea and oxalic acid in the co-crystal of urea and oxalic acid showing bond paths (*black solid* and *dashed lines*), bond critical points (*small green dots*) and ring critical points (*red dots*). Notice the curved bond paths due to strain in the case of cyclobutane and for the co-crystal of urea and oxalic acid due to hydrogen bond formation (*dashed lines*).

Due to the dominant character of core electrons in the total density, subtle changes in the distribution of paired electrons, as well as lone-pairs, are better assessed with the use of second-order derivatives of ED. One of the most useful and extremely important second-order electronic properties evaluated at the BCP is the Laplacian of ED, $\nabla^2\rho(r)$. Its sign and magnitude provide convenient classification of any type of chemical bonding. The Laplacian is a trace of diagonalised Hessian matrix and represents the sum of three curvatures of the ED ($\lambda_1 < 0$, $\lambda_2 < 0$, $\lambda_3 > 0$) at the BCP. The negative eigenvalues measure the degree of ED contraction perpendicular to the BP towards the BCP, and the positive eigenvalues indicate the contraction of ED from the BCP towards neighbouring nuclei. Depending on the dominating contribution, negative or positive, ED can be either locally concentrated in the region of BCP ($\nabla^2\rho(r) < 0$) or concentrated separately on each of the atomic basins ($\nabla^2\rho(r) > 0$, depletion of ED at BCP). The first case corresponds to shared shell interactions such as covalent bonds (open shell, Figure 3.13a) and the latter corresponds to highly polarised bonds such as ionic (Figure 3.13b) or closed shell interactions (Figure 3.13c).

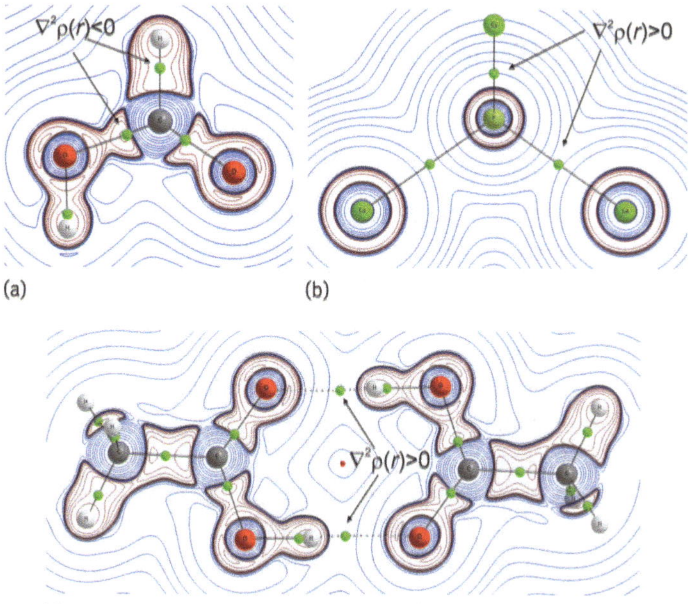

Fig. 3.13: Contour plots of the Laplacian distribution in (a) the formic acid molecule with only covalent bonds and (b) calcium fluoride with ionic bonds and (c) the acetic acid dimer where hydrogen bonds are marked with dashed lines and represent closed shell interactions. Contours are given at logarithmic intervals in $-\nabla^2\rho(r)$eÅ^{-5}.

3.4.1.2 Energy densities

Espinosa et al. [75] developed another convenient way, based on electronic energy densities introduced by Cremer and Kraka [76], to characterise weak interactions present in the crystalline state. The relationship between kinetic $G(r_{\mathrm{BCP}})$ and potential $V(r_{\mathrm{BCP}})$ energy densities at the BCP affords the opportunity to visualise how ED around the BCP is affected by the formation of a weak electrostatic interaction. Krawczuk and Macchi [8] showed that the total energy density, $H(r_{\mathrm{BCP}})$ $(= G(r_{\mathrm{BCP}})+V(r_{\mathrm{BCP}}))$, describes subtle borders of molecules approaching hydrogen bond formation, much better than a Laplacian analysis.

By definition, energy densities require a full, first order density matrix to be computed. Thus, they cannot be retrieved from an X-ray diffraction experiment, since ED is only a trace of the density matrix. However, this difficulty can be bypassed thanks to Abramov's functional [77], which allows an estimation of kinetic energy density based only on the ED and its derivatives, and therefore, makes it available from experimental determination. Kinetic energy density can be thus defined as:

$$G(r_{\mathrm{BCP}}) = \frac{3}{10}\left(3\pi^2\right)^{2/3}\rho^{5/3}(r_{\mathrm{BCP}}) + \frac{1}{6}\nabla^2\rho(r_{\mathrm{BCP}})$$

and the potential energy density can be then obtained from the local virial theorem following the equation:

$$V(r_{\mathrm{BCP}}) = \frac{\hbar^2}{4m}\nabla^2\rho(r_{\mathrm{BCP}}) - 2G(r_{\mathrm{BCP}})$$

Once both energy densities are known, it is possible to differentiate shared and closed shell interactions. If the local potential energy density is in excess of local kinetic energy density (which leads to negative values of $H(r_{\mathrm{BCP}})$), the interaction can be classified as shared shell. Furthermore, the greater the absolute value of $V(r_{\mathrm{BCP}})$, the more covalent is the interaction. On the other hand, $H(r_{\mathrm{BCP}})$ becomes positive for closed shell interactions because the kinetic energy density is more dominant and increases, relatively, per electron. For closed shell interactions, the ratio $G(r_{\mathrm{BCP}})/\rho(r_{\mathrm{BCP}})$ is usually greater than 1 (in au). However, one must be careful with the interpretation of the ratio, especially for complexes of 3d transition metal atoms, as this has been found to be inadequate for the classification [78]. Instead, it is better to use the bond index, $|V(r_{\mathrm{BCP}})|\,/G(r_{\mathrm{BCP}})$, which gives reliable results for a variety of bonded interactions (closed shell, shared shell and those involving transition metals). Thus, a bonded interaction is indicated as closed shell when the ratio $|V(r_{\mathrm{BCP}})|\,/G(r_{\mathrm{BCP}}) < 1$, shared shell when $|V(r_{\mathrm{BCP}})|\,/G(r_{\mathrm{BCP}}) > 2$ and intermediate when the ratio falls in the range between 1 and 2. In Table 3.3 the energy density criteria are summarised for chosen shared shell, intermediate and closed shell interactions within synthons present in co-crystal of urea and oxalic acid [79].

The potential energy density at the BCP has further application in qualitative crystal engineering, in particular in delineation of the characteristics of hydrogen bond formation. It has been shown [75], that the energetic response of the hydrogen bond

Tab. 3.3: Topological and energetic properties of $\rho(r)$ calculated at bond critical points of several interactions in the synthon present in the co-crystal of urea and oxalic acid. The electron density (ED) is given in e/bohr^3, the Laplacian is given in e/bohr^3 and energy densities are given in Hartree.

Examples	$\rho\,(r_{BCP})$	$\nabla^2\rho\,(r_{BCP})$	$V\,(r_{BCP})$	$G\,(r_{BCP})$	$H\,(r_{BCP})$	$\lvert V\,(r_{BCP})\rvert\,/G\,(r_{BCP})$	Type of interaction
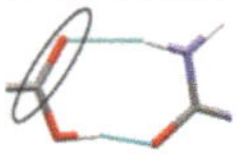	0.41	−0.67	−1.31	0.57	−0.74	2.3	Shared shell
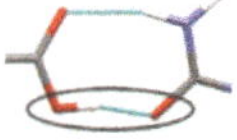	0.07	0.16	−0.07	0.06	−0.01	1.3	Intermediate
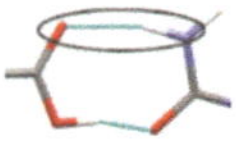	0.02	0.07	−0.02	0.03	0.01	0.9	Closed shell

to the force applied on electrons in the area of BCPs can be expressed by $V(r_{BCP})$ multiplied by a factor defined in volume atomic units:

$$E_{HB} = \frac{1}{2} V\,(r_{BCP})$$

Thus, the knowledge of the potential density energy can be used in designing materials based on hydrogen bond networks, especially when there is a tendency to form different polymorphic forms. By knowing which hydrogen bond is energetically more stable, thus preferable in molecular self-assembly, it would be possible, in principle, to set the experimental crystallisation conditions to induce the formation of a certain polymorph.

3.4.1.3 NCI index

To facilitate the discussion of the interatomic interactions in particular non-covalent interactions, it is very useful to introduce the recently developed NCI (non-covalent interaction) index [80, 81], which enables the visualisation of regions of space involved in either attractive or repulsive non-covalent interactions. The NCI method uses the reduced density gradient (RDG), $s(r)$, defined in the following way:

$$s(r) = \frac{\lvert \nabla\rho(r) \rvert}{2\,(3\pi)^{1/3}\,\rho(r)^{4/3}}$$

If a weak non-covalent interaction is present, characteristic spikes are observed on the $s(r)$ vs $\rho\,(r)$ plots in the low density and low gradient regions. Augmenting this relationship with the sign of a second eigenvalue λ_2 of the Hessian matrix, it is possible not only to determine the strength of an interaction ($\rho(r)$ value) but also to distinguish a stabilising chemical interaction ($\lambda_2 < 0$, bonded interaction) from destabilising ones ($\lambda_2 > 0$, non-bonded interaction). Thus, when ED multiplied by the sign of

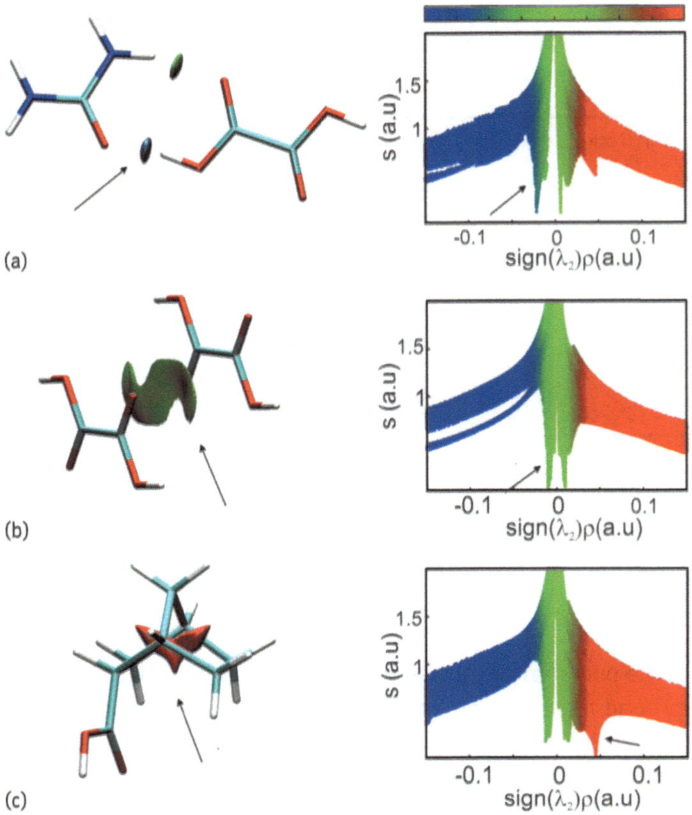

Fig. 3.14: RDG surfaces (*left*) and plots of the reduced density gradient versus the electron density (ED) multiplied by the sign of the second Hessian eigenvalue (*right*) in the (a) heterodimer of urea and oxalic acid connected via strong O–H⋯O (*blue region*) and moderate N–H⋯O hydrogen bonds, (b) oxalic acid dimer showing the carbonyl⋯carbonyl interaction (*broad green region*) and (c) sterically crowded bicyclo[2.2.1]heptane-2-carboxylic acid (*red region*). NCI surfaces correspond to $s = 0.4$ au and a colour scale of $-0.04 < \rho < 0.04$ au.

λ_2 is mapped onto the RDG surface, the nature and the strength of any non-covalent interaction can be revealed (Figure 3.14).

The NCI scheme complements QTAIM. While the analysis of bond paths is usually a localised and discontinuous description (non-covalent interactions are by nature delocalised), the NCI index offers a description of the interactions in terms of extended and flat RDG isosurfaces, with characteristic features present even if the ED at the critical point is zero. The NCI characteristics are tied to critical points of the density gradient field. Calculated isosurfaces are able to reveal i) the ways critical points are connected in a real space and ii) the character of these relations. Non-covalent strong interactions (e.g. strong hydrogen bonds) are represented by small disc shaped RDG domains, whereas broad multi-form domains denote much weaker interactions (e.g.

van der Waals, π. . .π and others). Furthermore, it has been shown by Saleh et al. [82] that RDG surfaces can be accessed either through quantum mechanical calculations or based on data obtained directly from experiment. Thus, the NCI approach, supported by QTAIM, provides quantitative insight into RDG surfaces, gives reliable chemical descriptions of the nature of any intra- and intermolecular interactions and offers new possibilities for self-assembled materials, especially in the field of biologically active molecules.

3.4.1.4 Distributed atomic polarisabilities

The electric polarisability of a molecule has long been used to predict and understand chemical reactivity, intermolecular interactions and some physical properties of materials. It has entered many quantitative expressions describing the physical and chemical behaviour of matter, such as long range intermolecular induction and dispersion forces, reaction kinetics and linear optics. There were some attempts to associate polarisability with some characteristics of electronic structure such as global softness [83], acidity and basicity [84, 85] or some properties such as electronegativity [86]. The prediction of reliable values of molecular polarisabilities by quantum chemical methods has made significant contributions and added new vigour to some fields like collision induced spectroscopy [87, 88], Raman scattering [89] or intensively developing non-linear optics [91–93].

From a crystal engineering point of view, there is particular interest in the deformation of atomic polarisabilities with respect to an isolated molecule and how this deformation influences macroscopic physical properties. Some of the properties originate from the polarisability of either individual atoms (e.g. optical activity) or functional groups (e.g. refractive indices). Some time ago, Krawczuk et al. [94, 95] developed a program PolaBer, which calculates distributed atomic polarisabilities from electric field perturbations of partitioned ED distribution. The software is based on QTAIM space partitioning of the molecular ED in terms of atomic domains (Ω), already introduced earlier in the chapter (Of course, other schemes would be equally suitable for this purpose. However, QTAIM is preferable because it provides a consistent definition of chemical bonding).

Atomic partitioning enables the extraction of atomic dipole moments, which are constituted of two parts: the polarisation $\mu_{\mathrm{p}}(\Omega)$ inside the atomic basin and charge translation $\mu_{\mathrm{c}}(\Omega)$ which originates from distributing atomic charge to all related bond critical points (BCP):

$$\mu(\Omega) = \mu_{\mathrm{p}}(\Omega) + \mu_{\mathrm{c}}(\Omega) = -\int_{\Omega} [r - R_{\mathrm{O}}]\,\rho(r)dr + \sum_{\Omega'}^{N_{BCP(\Omega)}} \left[R_{\Omega} - R_{\mathrm{BCP}}\left(\Omega|\Omega'\right)\right] Q\left(\Omega|\Omega'\right)$$

where $Q\left(\Omega|\Omega'\right)$ is the charge contributed to atom Ω by the bond to atom Ω′, $N_{BCP(\Omega)}$ is the number of BCPs connected to the atom Ω, R_{O} is an arbitrary origin of the coordinate system, R_Ω is the position vector of atom Ω and $R_{BCP}(\Omega|\Omega')$ is the position vector of the BCP between atom Ω and Ω′.

The direction and magnitude of $\mu_c(\Omega)$ depend on the type and number of groups bonded to the selected atom. Whereas the polarisation component is accessible directly from ED partitioning, the charge translation part requires that a set of equations be solved while adopting certain constraints, first introduced by Keith [96] and further developed by Krawczuk et al. [94]. Taking into account the characterisation of chemical bonding within a given system, the most important constraint used within PolaBer is the weighting scheme, which is inversely proportional to the bond strength measured by the value of ED at the BCP between atoms Ω and Ω′:

$$\lambda\left(\Omega|\Omega'\right) = \frac{1}{\rho\left(\Omega|\Omega'\right)}$$

This method avoids drastic changes to the shape and values of atomic polarisabilities caused by the presence of weak electrostatic interactions. With no weight put on any of the bonds, in particular in the presence of a ring, the formation of a weak bond creates a massive discontinuity of the atomic moments, hence creating polarisabilities, which is in fact chemically unreasonable. Weak interactions have only a small impact on the bond charge partitioning, but are not negligible.

The components of atomic polarisability tensor can be further calculated as derivatives of the dipolar ED with respect to the applied field:

$$\alpha_{ij}(\Omega) = \lim_{E\to 0} \frac{\mu_i^{E_j}(\Omega) - \mu_i^0(\Omega)}{E_j}$$

where $\mu_i^{F_j}(\Omega)$ is the atomic dipolar component along the i direction computed within a given electric field (0 or ε) in direction j (ε is of course as small as possible, preferably corresponding to optical frequencies).

Since atomic polarisabilities are described by symmetric second order positive tensors, it is possible to visualise them as ellipsoids, in the same real space as molecules assuming that 1 Å (atomic coordinates unit) responds to 1 Å^3 (polarisability unit); however, usually a scaling factor is applied for visualisation purposes (usually 0.4 Å^{-2}, unless stated otherwise). The main axes of ellipsoids have dimensions of volumes and indicate anisotropy of polarisability, whereas the size of the ellipsoids is proportional to the total atomic polarisability.

The atomic polarisabilities are remarkably sensitive to the local chemical environment. For example, the atomic polarisability tensors can be analysed when a molecule is moved from the isolated state, through a dimer, to supramolecular assembly (Figure 3.15). The analysis of atomic polarisabilities in isolated urea and oxalic acid molecules (Figure 3.15a) reveals a number of characteristic features. In general, polarisability ellipsoids are strongly elongated towards stronger bonds (compare for example C–O and C–N bonds). Furthermore, for more electronegative atoms the ellipsoids are more elongated along the bonds (compare oxygen and nitrogen atoms with

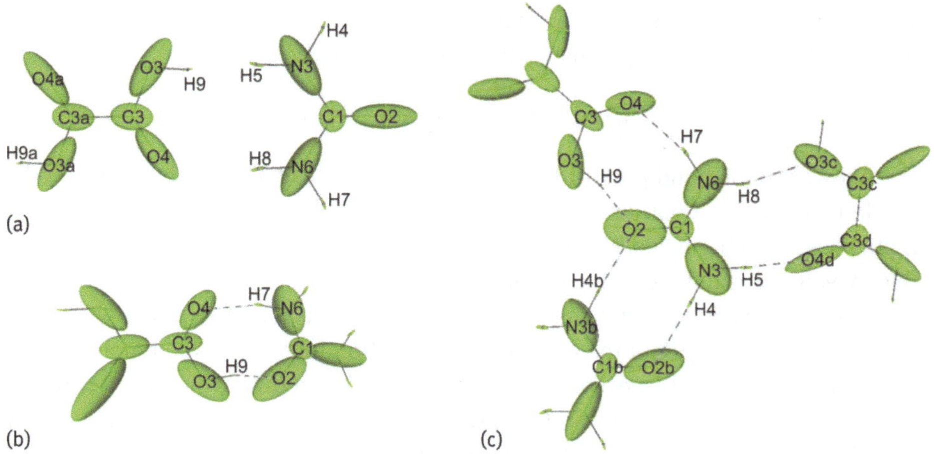

Fig. 3.15: Graphical representation of atomic polarisabilities of oxalic and urea molecules in the (a) isolated state, (b) heterodimer and (c) supramolecular assembly. The scaling factor for atomic polarisabilities is 0.4 $Å^{-2}$. The dashed lines indicate hydrogen bonds between molecules. For clarity only labels of functional groups involved in hydrogen bond formation are given. Symmetry codes: (a) $-x, -y+1, -z$, (b) $-x+2, -y+1, -z+1$, (c) $-x+1, y+1/2, -z+1/2$, (d) $x+1, -y+3/2, z+1/2$.

carbon atoms polarisabilities). The same behaviour holds for hydrogen atoms. However due to their small electronic population, the ellipsoids of the hydrogen atoms are very small. Nonetheless, the largest component of the polarisability tensor is along the H–X bond. When urea and oxalic acid molecules start to interact with each other through O–H···O and N–H···O hydrogen bonds (Figure 3.15b), the polarisabilities of donors and acceptors (O3, N6 and O2, O4 respectively) are modified in orientation and are stretched towards the directions of hydrogen bonds. Also, ellipsoids representing polarisabilities of hydrogen change their orientation, with respect to isolated state, towards acceptors of both hydrogen bonds. When a larger cluster is considered (taken from co-crystal structure of urea and oxalic acid) where one of the molecules, e.g. the central urea of Figure 3.15c, is involved in all possible interactions, the changes of tensors are more pronounced. Thus, changes in the orientation of the atomic polarisability ellipsoids depend not only on the number of possible interactions but also on their geometry. For example, the polarisability ellipsoid of oxygen atom O2, which is an acceptor of two hydrogen bonds (the angle between hydrogen bonds is ca 90°), adopts disc-like shape. The same happens with nitrogen atoms of amino groups. The behaviour of atomic polarisability ellipsoids is different, as in the case of the N3–H5...O4d hydrogen bond. Hydrogen-H5 is oriented in an almost linear fashion with respect to the O4d–C3d bond, which, in contrast to the O2, N3 N6 atoms, causes pronounced elongation of oxygen and hydrogen atom polarisabilities.

The subtle changes in the orientation of atomic polarisabilities due to the presence of weak electrostatic interactions can be quantified using an indicator called bond polarisability [97]. Having precise values of atomic polarisabilities, it can be defined as a projection of α_Ω and $\alpha_{\Omega'}$ tensors along the $\Omega\Omega'$ bond:

$$\alpha_{\Omega-\Omega'} = r^{\mathrm{T}}_{\Omega\Omega'} \cdot (\alpha_\Omega + \alpha_{\Omega'}) \cdot r_{\Omega\Omega'}$$

In other words, the bond polarisability measures how feasible the polarisation of the ED along a bond is, upon any change within a given system, a strategy particularly applicable to electric fields. Moreover, $\alpha_{\Omega-\Omega'}$ could be a quantum mechanical observable; however it is not measured experimentally.

When the oxalic acid molecule is considered as well as its heteromer with urea, the bond polarisabilities of the C3–O3 and C3=O4 bonds in an isolated molecule are 16.04 and 17.20 Bohr^3 respectively, whereas in the heterodimer they are 14.59 and 16.72 Bohr^3. The decrease of bond polarisabilities proves that the presence of any weak interaction cannot be neglected, especially when a material needs to be designed based on highly polarisable atoms, which influence macroscopic physical properties such as electric susceptibility and thus, birefringence or optical activity.

Another important outcome from the distributed atomic polarisability approach for crystal engineering purposes is the transferability of the functional group polarisabilities between different systems, which was first studied by Dos Santos et al. [98]. Based on the research of the series of amino acids and their aggregates, it has been proven that the polarisabilities of the common functional groups (group polarisabilities) are very similar in different chemical environments. Accurate quantum mechanical methods show that the polarisabilities of functional groups in small model molecules are sufficiently similar to those in larger and therefore, computationally more expensive molecules. Furthermore, group transferability supports the identification of functional groups that influence the enhancement of refractivity on the molecular level and thus, refractive indices on the macroscopic level. This enables material design in the selection of appropriate molecules to fabricate effective optical devices in terms of rapid screening of the best building block candidates.

3.4.1.5 Other quantitative indicators for chemical bonding

In the literature there are other sophisticated tools (in many cases available only through the use of quantum mechanical methods) to reveal and characterise the nature of chemical bonding. Here, only a brief summary of a few of these are given, i.e. those that are suitable for quantitative crystal engineering and that could be more extensively used to predict the physical properties of a material in the future.

- *Source function (SF)* – first introduced by Bader and Gatti [99] and extensively discussed in the last decade by many authors [100–102]. According to this approach, ED at any point in space can be determined by contributions of local source functions (LS) to all other points in the same space:

$$\rho(r) = \int \mathrm{LS}(r, r')dr'$$

Further integration of local source functions over all atomic basins (defined with QTAIM) allows a representation of the ED as a sum of atomic source functions $S(r, \Omega)$:

$$\rho(r) = S(r, \Omega) + \sum_{\Omega \neq \Omega'} S(r, \Omega')$$

Such a decomposition gives a measure of the contribution of other atoms or functional groups to the density at any point. For example, when projected along a BP, it gives information of which regions of the atoms involved in chemical bonding are accumulating or removing electronic charge at the BCP. Based on this, a source function could be further used for the understanding of mechanisms of some chemical reactions. The chemical interpretation of the SF was carefully examined by Farrugia and Macchi [103]. In studies on acetamide, thiocoumarin and carbonyl compounds, attention was drawn to the fact that while the source function may be helpful in the interpretation of chemical bonding, it should be used with caution, since the interpretation is not always straight forward and the obtained results might be over-interpreted (especially when only a single reference point is considered such as BCP).

– *Localisation* $\lambda(\mathrm{A})$ *and delocalisation* $\delta(\mathrm{A, B})$ *indices* [104, 105] – the localisation index gives a measure of how many electrons are localised in the atomic basin (defined with QTAIM), whereas the delocalisation index provides a quantitative measure of the electron pairs actually shared between two atomic basins; this requires a second density matrix. Both indices can be defined as follows:

$$\lambda(\mathrm{A}) = \int_{\mathrm{A}}\int_{\mathrm{A}} \rho_{2,\mathrm{xc}}(r_1, r_2)\, dr_1 dr_2$$

$$\delta(\mathrm{A, B}) = 2\int_{\mathrm{A}}\int_{\mathrm{B}} \rho_{2,\mathrm{xc}}(r_1, r_2)\, dr_1 dr_2$$

where $\rho_{2,\mathrm{xc}}$ is the exchange-correlation density. The localisation and delocalisation indices always sum up to the total number of electrons in a given system. Thus, they provide, following Matta & Arabi [7]: "*a bookkeeping of the whereabouts of electrons in a molecule.*". If two atoms A and B are bonded, in the sense that there is a BCP, BP, and interatomic surface identified, then the delocalisation index (two-electron property) measures the bond order and as Matta & Hernandez–Trujillo [106] suggested, it is exponentially related to ED at the BCP. Both indices find applications in predicting NMR J coupling [107], as a local aromaticity measure [108] and as an identifier of some halogen bonds, in particular those with iodine [109].

– *Interacting quantum atoms* (IQA) [110–112] – this approach entails an energy partitioning scheme which follows the QTAIM formalism that a molecule can be partitioned into topological atoms. Every molecular IQA energy can be formulated as:

$$E_{\text{IQA}} = \sum_{\text{A}} E_{\text{IQA}}^{\text{A}} = \sum_{\text{A}} E_{\text{intra}}^{\text{A}} + \frac{1}{2} \sum_{\text{B} \neq \text{A}} V_{\text{inter}}^{\text{AB}}$$

where $E_{\text{IQA}}^{\text{A}}$ represents the atomic energy of atom A, $E_{\text{intra}}^{\text{A}}$ is the intra-atomic or self-energy of atom A, and $V_{\text{inter}}^{\text{AB}}$ describes the interatomic interaction energy between two atoms, A and B. Since the self-energy part requires a non-diagonal first order density matrix and the interatomic part is a two-electron property, they are not experimentally accessible (experimentally determined ED is only a trace of the first order density matrix). The intra-atomic component is further separated into the kinetic energy of electrons T^{A}, electron–electron repulsion potential energy $V_{\text{ee}}^{\text{AA}}$ and electron–nucleus attractive potential energy $V_{\text{en}}^{\text{AA}}$ within atom A. A similar partitioning scheme is applied to the interatomic part, where energy is represented by a sum of the electron–nucleus, $V_{\text{en}}^{\text{AB}}$, and the nucleus–electron, $V_{\text{ne}}^{\text{AB}}$, attractive energies along with nucleus–nucleus, $V_{\text{nn}}^{\text{AB}}$, and electron–electron, $V_{\text{ee}}^{\text{AB}}$, repulsive energies. The $V_{\text{ee}}^{\text{AB}}$ term is finally divided into Coulombic $V_{\text{Coul}}^{\text{AB}}$ and exchange-correlation $V_{\text{XC}}^{\text{AB}}$ energies. All the energy terms have clear physical meanings and can be correlated with the nature and type of any chemical bond present in the structure. Thus, the IQA energy decomposition scheme is a powerful and chemically intuitive (however, computationally very expensive) tool giving a consistent picture of the physical atomistic components in various chemical bonds. It has been already tested and applied in various systems to characterise for example hydrogen bonds [113] or halogen bonds [114] and many other interactions.

– *Electron localisability indicator* (ELI) – introduced by Kohout [115, 116], and based on ω-restricted space partitioning (ωRSP) [117] in which the volume is partitioned into non-overlapping space filling regions (*micro-cells* when ω is sufficiently small) that are assumed to contain the same portion of electron pairs (the probability of finding an electron pair is the same in all regions). ELI reflects the discrete distribution of electron populations. In particular, ELI is a measure of the local pairing of same spin electrons (ELI-D, pair restricted). It can be further associated with the topology of ED, for instance the location of ELI-D critical points and their signatures are related to the topological properties of ED. This interplay is a bridge between one-electron properties and the properties derived from the pair density (i.e. electron localisability and electron pairing), and thus supplies another quantum mechanical tool for distinguishing and characterising various types of chemical bonding. However, it has to be stressed that so far, this indicator has been mainly applied to the characterisation of covalent bonds, rather than to weak electrostatic interactions [118].

3.5 Multi-functional co-crystals – examples

In the previous sections, many of the available indicators and methods used in modern qualitative and quantitative crystal engineering were summarised. These tools support the efficient engineering of materials exhibiting at least one meaningful physical property. In the literature there are only few examples of successfully designed multi-functional co-crystals. The use of the know-how approach for the generation of new functional materials requires a combination of multiple experimental and theoretical techniques. The proposed approach can be considered innovative and at the same time very challenging.

In this section examples of reverse crystal engineering [119] techniques are presented to show that the described co-crystal phases could have been efficiently predicted using the tools discussed in previous sections. An attempt to also correlate the observed structural features with materials properties will be made as well as a summary of their potential application in materials science. The starting components (building blocks) are: urea (U), barbital (BB), melamine (MEL), L-malic acid (LMA), L-tartaric acid (LTA), 4-hydroxybenzoic acid (PHBA) and benzotriazole (B) (Figure 3.16). The first task will be dividing the molecules into functional groups to select best co-formers.

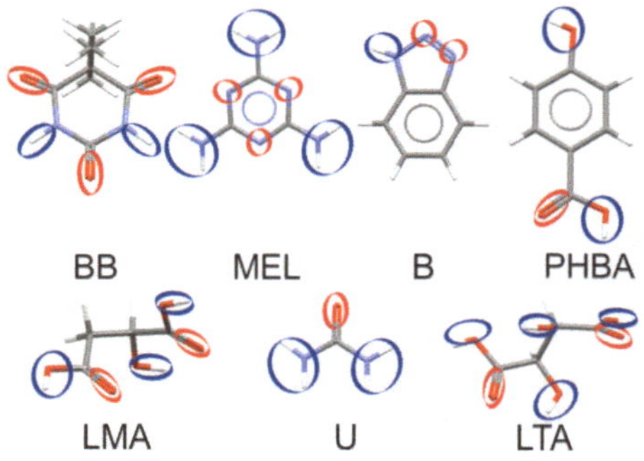

Fig. 3.16: Molecular building blocks for crystal engineering of multi-functional co-crystals: barbital (BB), melamine (MEL), benzotriazole (B), 4-hydroxybenzoic acid (PHBA), L-malic acid (LMA), urea (U and L-tartaric acid (LTA). The possible donor and acceptor sites of molecules are indicated with blue and red ellipses, respectively.

To assess the probability of a particular interaction, a full interaction analysis for the chosen target molecules has been performed: urea, melamine and 4-hydroxybenzoic acid (Figure 3.17). The distribution of donor and acceptor sites allowed the confirmation of the existence of the following co-crystals: urea with both L-malic [80, 120] and L-tartaric acids [121], 4-hydroxybenzoic acid with benzotriazole [122–125] and melamine with barbital [126, 127]. Considering the acid molecules, it is easily seen that the carboxamide (from urea)...carboxylic acid interaction statistically exceeds the likelihood of the homomeric carboxylic acid···carboxylic acid formation (Section 3.1). The results of the molecular complementarity tool and propensity calculations are summarised in Table 3.4. Except for melamine and barbital, all other target co-former pairs are similar in terms of shape or polarity (based on the 100% hit score for mo-

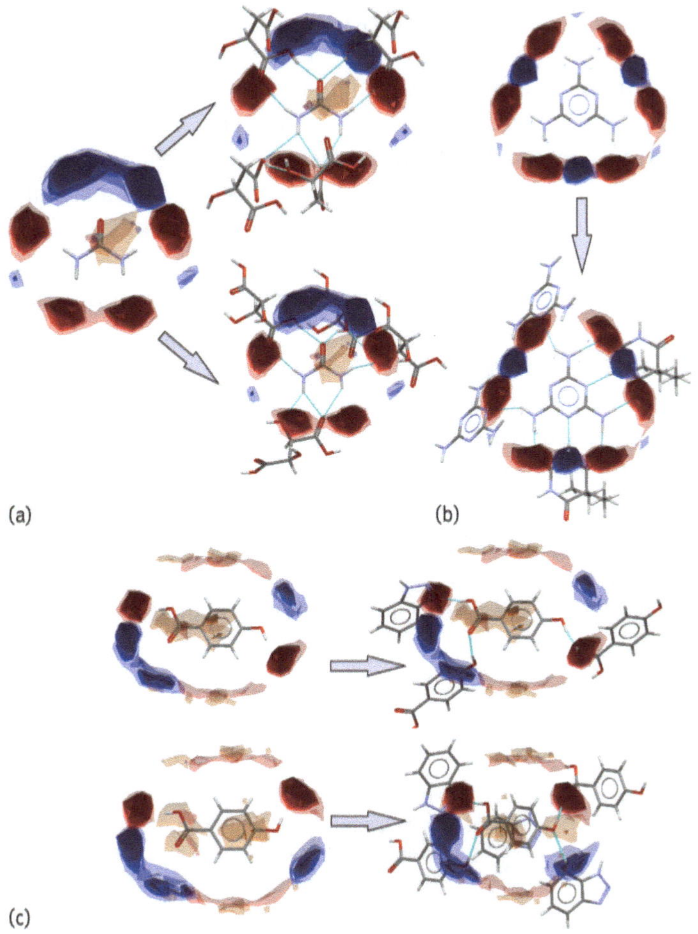

Fig. 3.17: Full interaction maps for the selected target molecules and co-crystals: (a) from urea molecule to crystal structure of ULMA and ULTA; (b) from melamine to the crystal structure of MELBB; (c) from 4-hydroxybenzoic acid to two polymorphs of BPHBA.

Tab. 3.4: Molecular complementarity and multi-component score for the selected target and co-former molecules.

Co-crystal	Molecular complementarity (%)	MC Score
ULMA	100	−0.06
ULTA	100	−0.06
BPHBA	100	0.02
MELBB	0	0.00

lecular complementarity). The MC score explains this effect very well. The probability of AA and AB interactions in MEL-BB are exactly the same, which means that the molecules are indeed complementary partners firmly joined (under specific experimental conditions). ULMA and ULTA each have a slightly negative MC score, which is associated with a high predisposition for the formation of urea···urea interactions. However, there are no such interactions in the examined co-crystals. This shows that statistical tools should be used with caution and preferably combined with different methods to reach consistent final conclusions.

The identified co-crystals are built of chemically and biologically important molecules, which make them attractive for multiple applications in materials science and technology. First of all, they display relevant (non-)linear optical properties and in particular they have potential as second harmonic generation (SHG) materials. The data in Table 3.5 indicates that the powder SHG of these co-crystals is at least comparable to that of KDP. Moreover, since the co-crystals crystallise in noncentrosymmetric space groups, they can be considered as potential pyroelectric, piezoelectric or ferroelectric materials. For example, some studies of the piezoelectric properties of ULTA [128] showed that although the effect is not as outstanding, as for example for Rochelle's salt, it is still comparable to that exhibited by triglycine sulphate (TGS) [23] and thus, is not negligible. Furthermore, some of the discussed co-crystals are interesting from a pharmaceutical point of view. The co-crystal of melamine and barbital contains a pharmaceutically active ingredient (API), whereas one of the polymorphic forms of BPHBA ($Pna2_1$ crystal phase) is known to exhibit anti-microbial properties [125].

In the following, the focus will be upon evaluating the structural features of individual co-crystals and evaluating a QSPR analysis using the crystal engineering tools described earlier.

Tab. 3.5: Selected linear and non-linear properties of selected co-crystals; n_α, n_β n_γ – refractive indices, $\overline{n}$ – averaged refractive index, Δn – linear birefringence.

Co-crystal	n_α	n_β	n_γ	$\overline{n}$	Δn	SHG efficiency (relative to KDP)
ULMA	1.44	1.52	1.53	1.50[a]	0.09	3.10
ULTA	–	–	–	1.44[b]	0.074	3.10
BPHBA[b]	–	–	–	–	–	1.02/3.07[a]
MELBB[d]	1.45	1.52	1.59	1.52	0.14	1.86

[a] measured at 532 nm [131]
[b] calculated at 800 nm using reflectance R; $n = (-R + 1) \pm 2\,[(R)^{1/2}/(R-1)]$ [130]
[c] measurements performed for the $Pna2_1$ crystal phase, two values of relative SHG efficiency reported in [124] and in [125] are for powdered samples. The measurement is thus dependent on many factors including particle size and texture. The obtained results should be averaged from multiple measurements. Unfortunately, the authors did not mention detailed information on sample preparation.
[d] measured at 589 nm [126]

The first material to be considered is the 1 : 1 co-crystal between urea and L-malic acid (ULMA, Figure 3.18), which crystallises in the $P2_1$ polar space group. Molecules of L-malic acid form infinite helices along 2_1 screw axis, due to O12-H12. . . O3 hydrogen bonds. The helices are interconnected along [001] via O3–H3O···O51 and N–H···O bonds with urea molecules. The urea molecules together with helical chains form layers parallel to [001] that are tied into a three-dimensional framework by weak C–H···O interactions. This particular layout of hydrogen bonds can be directly correlated with previously measured refractive indices [129].

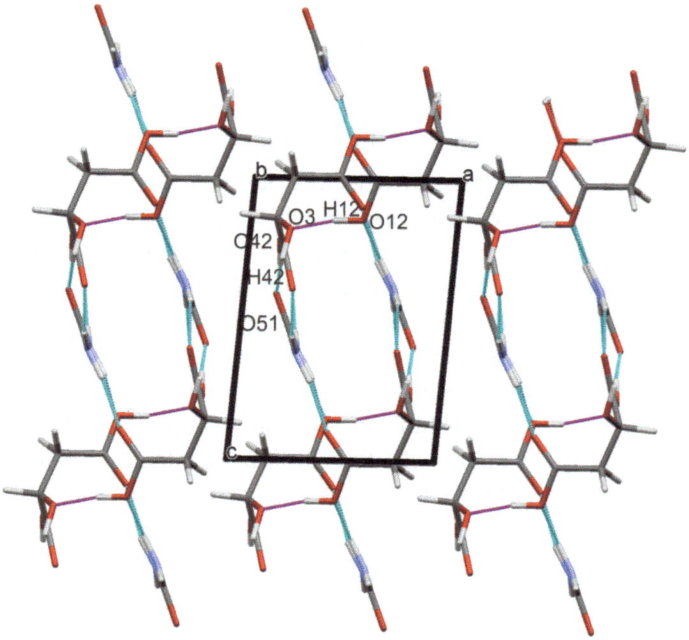

Fig. 3.18: A view of the molecular packing in ULMA viewed along [010]. Hydrogen bonds of the type O–H···O between acid molecules are indicated with dashed magenta lines with the remaining hydrogen bonds indicated with cyan dashed lines. The strongest hydrogen bonds are labelled.

For that purpose, some of the results of ED studies can be utilised along with a full QTAIM topological analysis as reported by Krawczuk et al. [80]. There is special interest in how the strength of an interaction influences the value of the refractive index along a certain direction. The topological properties of the non-covalent interactions together with energetic criteria are collected in Table 3.6, indicating that the strongest hydrogen bonds are those building the helical chains along [010] (row 6 in Table 3.6) and the O–H···O hydrogen bonds connecting the chains along [001] via urea molecules (rows 5 and 8 in Table 3.6). These interactions are identified as intermediate between closed and shared shell characters ($1 < |V(r_{BCP})|/G(r_{BCP}) < 2$). In the directions of

Tab. 3.6: Topological characteristics of intermolecular interactions in ULMA; the experimental values are taken from [82]. $\rho(r_{BCP})$eÅ^{-3}, $\nabla^2\rho(r_{BCP})$/eÅ^{-5}, $G(r_{BCP})$ and $V(r_{BCP})$ are the local kinetic and local potential energy density (eÅ^{-3}), respectively, $H(r_{BCP})$ is electronic local energy density (eÅ^{-3}). Donor atoms are given in brackets.

		$\rho\,(r_{BCP})$	$\nabla^2\rho\,(r_{BCP})$	$G\,(r_{BCP})$	$V\,(r_{BCP})$	$H\,(r_{BCP})$	$\lvert V\,(r_{BCP})\rvert\,/\,G\,(r_{BCP})$
1	(N–)H521...O11	0.09	1.27	0.07	−0.06	0.02	0.86
2	(N–)H511...O11	0.11	0.91	0.06	−0.06	0.00	1.00
3	(N–)H522...O41[a]	0.12	0.63	0.05	−0.06	−0.01	1.20
4	(N–)H521...O12[b]	0.06	0.94	0.05	−0.04	0.01	0.80
5	(O–)H42...O51[c]	0.37	5.35	0.40	−0.43	−0.03	1.08
6	(O–)H12...O3[d]	0.33	2.30	0.24	−0.31	−0.08	1.29
7	(N–)H512...O41[c]	0.07	2.06	0.11	−0.07	0.04	0.64
8	(O–)H30···O51[a]	0.29	2.07	0.20	−0.26	−0.06	1.30
9	(C–)H2B···O42[e]	0.08	0.84	0.05	−0.04	0.01	0.80

Symmetry codes: (a) $-x+1, y+1/2, -z$; (b) $-x+1, y+1/2, -z$; (c) $-x+1, y-1/2, -z+1$; (d) $-x+1, y-1/2, -z$; (e) $-x, y+1/2, -z$; (f) $-x, y-1/2, -z$.

propagation of the above hydrogen bonds, the refractive indices have the highest values (Table 3.5), whereas along [100], the refractive index is the smallest due to the formation of only very weak closed shell C-H···O interactions (row 9 in Table 3.6). Moreover, the strength of the hydrogen bonds also determines the thermal properties of the co-crystal. It has been proven [131] that the value of thermal conductivity is highest along [010] (along 2_1 screw axis) and that the high anisotropy of thermal parameters correlates well with measured mechanical hardness anisotropy.

A related example of a co-crystal where weak non-covalent interactions are key factors responsible for the anisotropy of optical properties is found in the co-crystal of L-tartaric acid and urea (ULTA). Although L-tartaric acid is chemically comparable to L-malic acid, the structure of ULTA is quite different (Figure 3.19).

ULTA crystallises in the orthorhombic and chiral space group $P2_12_12_1$, in which two structurally distinct helices are identified, whose mutual relationship creates a third dimension in the crystal. One helix, along [100], is built of L-tartaric acid molecules connected by strong O–H···O bonds, whereas the second one extends along [001] and is based on alternating urea and L-tartaric acid molecules linked via O–H···O and N–H···O hydrogen bonds. The third-dimension is created by the interplay of both helices. In order to quantify all hydrogen bonds present in the structure and correlate them with the reported in the literature optical properties, a QTAIM analysis was performed supported by the NCI approach (Figure 3.20). Both methods indicated that the strongest hydrogen bond is the one between the carboxylic-O1–H1 group and carbonyl-O2 of the urea molecule. The $\rho(r_{BCP})$ and $\nabla^2\rho(r_{BCP})$ values at the BCP are 0.03 and 0.12 e/Bohr^5, respectively. The interaction is considered to be intermediate in strength (see dark blue spikes in Figure 3.20) as the ratio of $|V(r_{BCP})|/G(r_{BCP})$ is 1.07. Hydrogen bonds between tartaric acid molecules as well as N–H···O interactions are

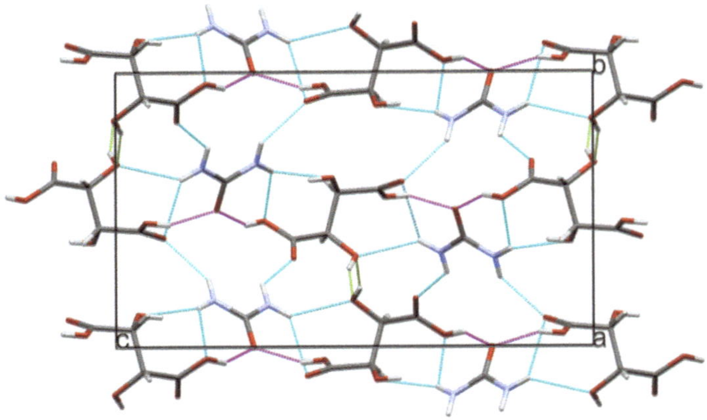

Fig. 3.19: Molecular packing viewed along the a-direction in the co-crystal of L-tartaric acid and urea. Hydrogen bonds of the type O–H···O between tartaric acid molecules are marked with green dashed lines, between tartaric acid and urea with magenta dashed lines and hydrogen bonds of the type N–H···O are marked with cyan dashed lines.

similar in strength and nature. They are characterised as pure closed shell since the ratio between the magnitude of potential and kinetic energy densities is equal to 0.86 and 0.78, respectively (for N–H···O, the averaged value was taken since more than one hydrogen bond was identified). Naturally, in the co-crystal of ULTA other interactions are present (see green spikes in Figure 3.20). However, they are not crucial for the macroscopic properties of that material.

Although there are no experimental values for the refractive indices along the main crystallographic directions in the literature (only the average value is given cal-

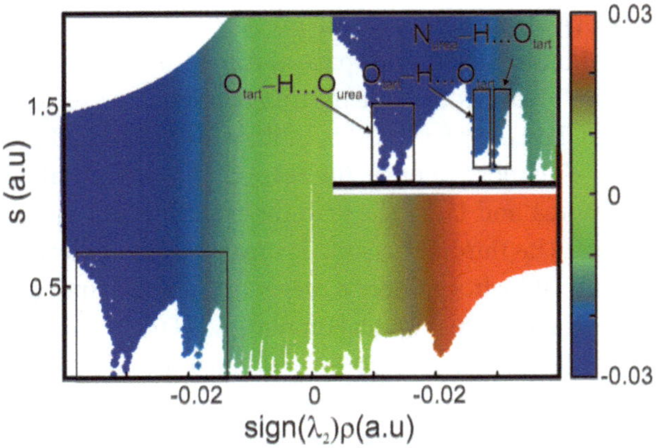

Fig. 3.20: Plots of the reduced density gradient versus the electron density (ED) multiplied by the sign of the second Hessian eigenvalue for a fragment of the ULTA structure where the interactions of tartaric acid molecules are taken into account.

culated via reflectance [130]), it can be concluded with a relatively high probability that the greatest refractive index will be along [001] due to the strongest interactions in that direction. The remaining indices should fulfil the following relationship: $n_3 > n_1 > n_2$, where $n_1 \parallel [100]$, $n_2 \parallel [010]$, $n_3 \parallel [001]$.

The third example of a multi-functional material in this section is that of the benzotriazole 4-hydroxybenzoic acid co-crystal (BPHBA). As discussed in Section 3.1, 4-hydroxybenzoic acid is considered an interesting target molecule for the design of optical materials. PHBA can also be found in natural products, e.g. in the indigenous cinnamon tree, and as such, its role in drug discovery and development has been a subject of many studies [132, 133]. The crystal structure of its hydrate (Figure 3.3) is centrosymmetric and thus not interesting from the viewpoint of non-linear optical properties of even order. Co-crystallisation in this case is a means to modify the properties of the target molecule in a way that will change the symmetry, to non-centrosymmetric, and at the same time orient molecules in a way that will maximise the desired optical effect.

In the literature there are two forms of the benzotriazole 4-hydroxybenzoic acid co-crystal that differ in their stoichiometry (Figure 3.21). Both crystal phases crystallise in polar space groups: (I) – Pn and (II) – $Pna2_1$. The second differentiating factor in the two forms is in the relative orientation of the hydroxyl group with respect to the carboxylic group of the acid. This causes variations in the utilisation of available donor and acceptor sites. In (I), each acid molecule is connected via O–H···O hydrogen bonds with two neighbouring acid molecules and via O–H···N bonds with one co-former. In (II), the molecules are joined by pairs of PHBA:benzotriazole interactions. To understand the formation of (I) and (II), a Hirshfeld surface analysis was performed (Figure 3.22) along with a QTAIM analysis (Figure 3.23). Both Hirshfeld surfaces and fingerprint plots highlight differences in the chemical environments of 4-hydroxybenzoic acid in the co-crystals. Among the studied weak interactions, the contribution of O···H and N···H hydrogen bonds varies the most, 4.5% and 3.5%, respectively. For (II),

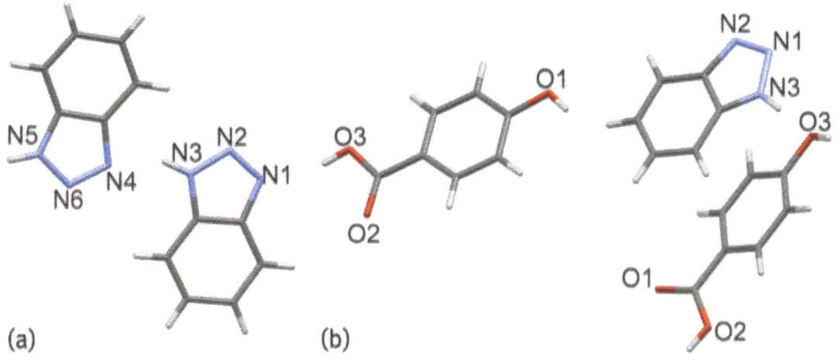

Fig. 3.21: Contents of the asymmetric units of the different co-crystals formed between benzotriazole and *p*-hydroxybenzoic acid: (a) 2B-PHB (I) and (b) B-PHB (II).

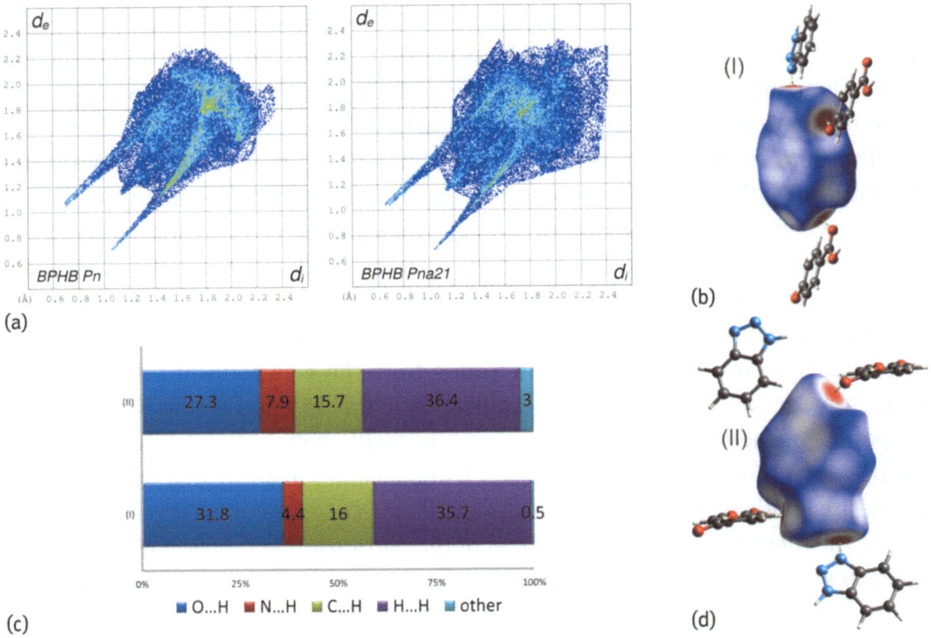

Fig. 3.22: (a) Fingerprint plots, (b) Hirshfeld surface for (I), (c) percentage contributions to the surface and (d) Hirshfeld surface for (II).

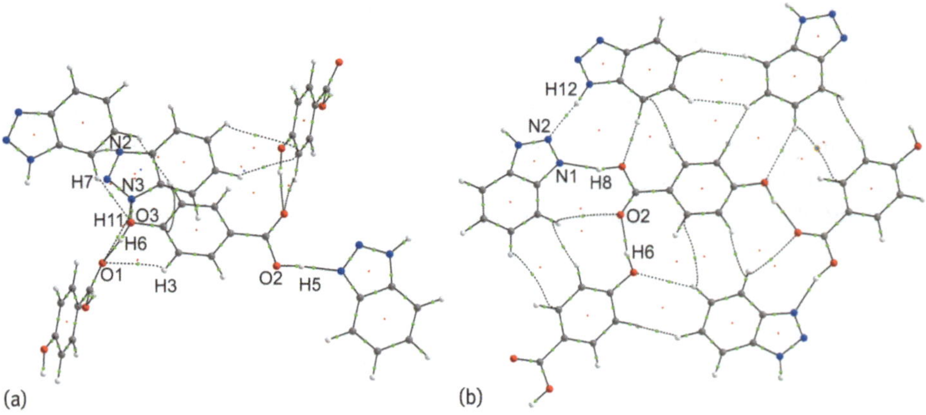

Fig. 3.23: Molecular graph for (a) 2B-PHB (I) and (b) B-PHB (II), with bond (3,-1) and ring (3, +1) critical points as green and red spheres, respectively.

a much wider spread of points, especially at the higher values of d_i and d_e, than for (I) is apparent. This could be attributed to $\pi...\pi$ interactions present in (II).

The QTAIM analysis of BCPs and energetic criteria based on local potential and kinetic energy density allowed a classification of the weak interactions. In Figure 3.22, interactions involving 4-hydroxybenzoic acid in both co-crystals are compared. In both circumstances there are O–H···O and O–H···N hydrogen bonds, showing intermediate interactions between closed shell and shared shell. Both geometric and energetic criteria indicate these interactions are stronger for (II), which could be correlated with its higher stability. Moreover, the structure of (II) is stabilised by $\pi\ldots\pi$ interactions between the constituents from adjacent layers (Table 3.7). The rather small SHG response of (II), despite the presence of NLO-attractive building blocks, can be associated with molecular packing. The c whromophores are not preferentially oriented and thus, the dipole moments are partially cancelled. To enhance the observed properties a different molecular pair may need to be selected.

Tab. 3.7: Properties of the Bond Critical Points for (I) and (II): $\rho(r_{BCP})$/au, $\nabla^2\rho(r_{BCP})$/au, d – internuclear separations (Å), d_1, d_2 – distance between BCPs and atom 1, 2, respectively (Å), $V(r_{BCP})$/a.u., $G(r_{BCP})$/a.u. local kinetic and local potential energy density, respectively, $H(r_{BCP})$/a.u. – total energy density.

	d	d_1	d_2	$\rho(r_{BCP})$	$\nabla^2\rho(r_{BCP})$	$V(r_{BCP})$	$G(r_{BCP})$	$H(r_{BCP})$	$\|V(r_{BCP})\|/G(r_{BCP})$
(I)									
H8...N1	1.73	0.55	1.18	0.05	0.11	−0.04	0.04	−0.01	1.23
H6...O2	1.74	0.59	1.15	0.04	0.13	−0.03	0.03	0.00	1.03
(II)									
H6...O1	1.71	0.56	1.15	0.04	0.13	−0.04	0.04	0.00	1.05
H5...N2	1.68	0.52	1.16	0.05	0.09	−0.05	0.04	−0.01	1.35
H7...O3	2.35	0.92	1.43	0.01	0.04	−0.01	0.01	0.00	0.86
H11...O1	2.57	1.08	1.51	0.01	0.03	−0.01	0.01	0.00	0.85
H3...O1	2.64	1.12	1.54	0.01	0.03	−0.01	0.01	0.00	0.84
N3...O3	3.24	1.63	1.61	0.01	0.02	−0.01	0.01	0.00	0.84
C5...C9	3.43	1.82	1.81	0.01	0.01	−0.01	0.01	0.00	0.78
N2...C10	3.51	1.75	1.78	0.01	0.01	−0.02	0.01	0.00	0.77

The last example of a potentially multi-functional material is that of a melamine barbital co-crystal – MELBB (Figure 3.24). This crystal phase was designed based on the knowledge of multiple polar crystal structures containing each co-former. The electron density and topological properties were analysed to reveal the factors influencing the mutual recognition of these two molecules and promoting crinkled tape formation.

The analysis confirmed that the formation of the co-crystal is guided by two factors: displacement of ED prior to crystallisation towards a mesomeric form of the barbital molecule and an energetic criterion resulting in a lock-key molecular recognition. In order to obtain a co-crystal, it is necessary to overcome the tendency of single component molecules to combine into a homomeric molecular crystal. The displacement of ED differentiates the ability of barbital-O2A and -O4A atoms to form hydrogen bonds. This is evident in the different values of net atomic charges obtained

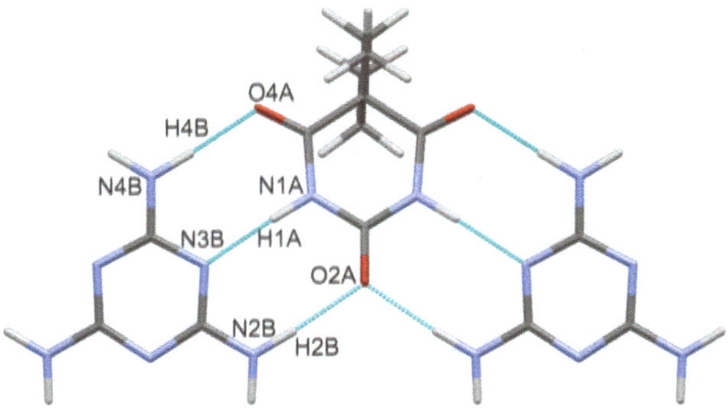

Fig. 3.24: The trimer motif in the melamine and barbital (MEL-BB) co-crystal.

from the topological analysis of ED for the isolated molecules (−1.12 for both O2A and O4A) and for the co-crystal (−1.09 for O2A and −1.19 for O4A).

The barbital and melamine molecules are interlocked by three moderately strong hydrogen bonds, forming two fused $R_2^2(8)$ rings. The N1A-H1A···N3A interaction is the strongest, showing intermediate character between closed and shared shell interactions. This molecular recognition is reflected in the optical properties. The N1A-H1A···N3A interaction is at the basis of the infinite tape motif running along [100], which is also the direction with the highest refractive index.

In order to validate the electron density results and to correlate the structural features of MELBB with its optical properties, the atomic and bond polarisabilities were calculated with the PolaBer software. The atomic polarisabilities of melamine and barbital, in the isolated state, were compared with those taken from the crystal structure

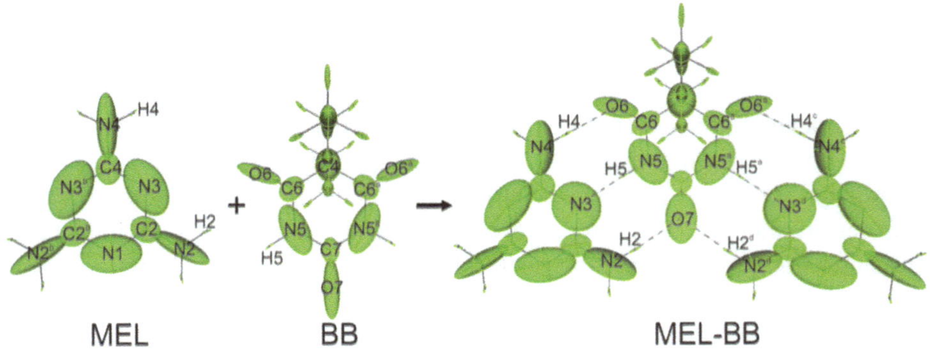

Fig. 3.25: Atomic polarisability ellipsoids for isolated molecules of melamine (MEL) and barbital (BB), and within a fragment taken from the MELBB co-crystal. Dashed lines indicate hydrogen bond formation. The scaling factor for atomic polarisabilities is 0.4 Å^{-2}. Polarisabilities are calculated at the CAM-B3LYP/d-aug-cc-pVDZ level of theory. Symmetry codes: (a) $1-x, y, z$; (b) $-x, y, z$; (c) $1+x, y, z$.

Tab. 3.8: Bond polarisabilities of hydrogen bonds in MELBB. For comparison, the values of the N-H covalent bonds in the isolated state are given. Notice the increase of electron density (ED) polarisation of the covalent bond as the hydrogen bond is being formed.

	Bond polarisability $\alpha_{\Omega-\Omega'}$ /Bohr^3	
	Isolated molecule	MEL-BB
H5...N3	–	16.4
H2...O7	–	14.1
H4...O6	–	12.7
N5–H5	10.0	14.0
N2–H2	10.7	13.2
N4–H4	10.7	12.9

(Figure 3.25). In the crystal, both molecules are involved in the formation of three hydrogen bonds: N5–H5...N3, N2–H2...O7 and N4–H4...O6, which causes significant changes of the atomic polarisability ellipsoids. This is particularly dramatic for the N3 atom, which changes from an egg-like into a disc-like shape (the polarisability tensor becomes less anisotropic in comparison to the isolated state). Such a strong transition arises from the fact that the N3 atom is an acceptor of the strongest hydrogen bond. In fact, the examination of bond polarisabilities in Table 3.8 confirms that the greatest increase in ED polarisation occurs along the vector describing this interaction.

3.6 Concluding remarks

In this chapter, an introduction to the various tools and approaches that support modern crystal engineering of new multi-functional materials has been presented. It was demonstrated that going beyond the standard determination of molecular geometry and molecular packing via ED analysis improves the understanding of the structural origins of macroscopic physical properties. However, it should be stressed that none of the described methods are sufficient to successfully design a multi-functional co-crystal. The statistical methods offered by the CSD may lead to wrong conclusions, such as in the case of melamine and barbital where the complementarity tool does not predict co-crystal formation. On the other hand, the QTAIM approach may be blind to some interactions crucial for three-dimensional packing and therefore, physical properties, such as in the case of carbonyl···carbonyl interactions in the co-crystal of oxalic acid and urea. Moreover, ab initio quantum chemical methods are sensitive to the choice of functionals and basis sets; thus predicted physical properties need to be critically analysed and should be verified experimentally. Nonetheless, reverse crystal engineering of known multi-functional co-crystals demonstrated that by simultaneous use of various qualitative and quantitative descriptors, it may be possible to efficiently predict how molecules will interconnect into three-dimensional structures.

Further, such studies can provide insight into the relationship between the structure and physical properties (QSPR) of a material.

Given recent progress in the development of crystal engineering tools and their wide range of applications, it is anticipated these tools will lead to better modelling of molecular co-crystals and will increase the probability of obtaining materials with at least one relevant physical property.

Acknowledgement: The authors are grateful to Professor Katarzyna Stadnicka for valuable hints and fruitful discussions. This research was partially supported by PL-Grid Infrastructure. This research was also supported in part by the National Science Centre, grant number 2014/15/D/ST5/02608 "Crystal Engineering of Linear and Non-Linear Optical Materials – From Atomic and Molecular Properties to 3D Structural Architectures".

Bibliography

[1] Kay MI, Kleinberg R. The crystal structure of triglycine sulfate. Ferroelectrics 1973, 5, 45–52.
[2] Lal RB, Batra AK. Growth and properties of triglycine sulfate (TGS) crystals: Review. Ferroelectrics 1993, 143, 51–82.
[3] Chang J-M, Batra AK, Lal RB. Growth and properties of urea-doped triglycine sulfate (UrTGS) crystals. J. Cryst. Growth 1996, 158, 284–288.
[4] Lock PJ. Doped triglycine sulfate for pyroelectric applications. Appl. Phys. Lett. 1971, 19, 390–391.
[5] Theresita Shanthi N, Selvarajan P, Mahadevan CK. Studies on triglycine sulfate (TGS) crystals doped with sodium bromide (NaBr) grown by solution method. Indian J. Sci. Technol. 2009, 2, 49–52.
[6] Stalke D. Meaningful structural descriptors from charge density. Chem. Eur. J. 2011, 17, 9264–9278.
[7] Matta CF, Arabi AA. Electron-density descriptors as predictors in quantitative structure–activity/property relationships and drug design. Future Med. Chem. 2011, 3, 969–994.
[8] Krawczuk A, Macchi P. Charge density analysis for crystal engineering. Chem. Cent. J. 2014, 8, article no. 68.
[9] Spackman MA. Charge densities and crystal engineering in Gatti C, Macchi P., ed. Modern Charge-Density Analysis, New York, Springer, 2012, 553–572.
[10] Wöhler F. Untersuchungen über das chinon. Annalen 1844, 51, 145–163.
[11] Etter MCJ. Hydrogen bonds as design elements in organic chemistry. J. Phys. Chem. 1991, 95, 4601–4610.
[12] Steed JW. The role of co-crystals in pharmaceutical design. Trends Pharmacol. Sci. 2013, 34, 185–193.
[13] Reddy CM, Krishna GR, Ghosh S. Mechanical properties of molecular crystals—applications to crystal engineering. CrystEngComm 2010, 12, 2296–2314.
[14] Aakeröy CB, Wijethunga TK, Desper J. Crystal engineering of energetic materials: co-crystals of ethylenedinitramine (EDNA) with modified performance and improved chemical stability. Chem. Eur. J. 2015, 21, 11029–11037.

[15] Gryl M. Charge density and optical properties of multicomponent crystals containing active pharmaceutical ingredients or their analogues. Acta Crystallogr. B 2015, 71, 392–405.

[16] Desiraju G. Crystals and co-crystal. CrystEngComm 2003, 5, 466–467.

[17] Dunitz J. Crystals and co-crystal – a second opinion. CrystEngComm 2003, 5, 506.

[18] Bond A. What is a co-crystal? CrystEngComm 2007, 9, 833–834.

[19] Braga D. Grepioni F, Maini L, Prosperi S, Gobetto R, Chierotti MR. From unexpected reactions to a new family of ionic co-crystals: the case of barbituric acid with alkali bromides and caesium iodide. Chem. Commun. 2010, 46, 7715–7717.

[20] Aitipamula S, Banerjee R, Bansal AK, Biradha K, Cheney ML, Choudhury AR, Desiraju GR, Dikundwar AG, Dubey R, Duggirala M, Ghogale PP, Ghosh S, Goswami PK, Goud NR, Jetti RKR, Karpinski P, Kaushik P, Kumar D, Kumar V, Moulton B, Mukherjee A, Mukherjee G, Myerson AS, Puri V, Ramanan A, Rajamannar T, Reddy CM, Rodriguez Hornedo N, R. D. Rogers, Row TNG, Sanphui P, Shan N, Shete G, Singh A, Sun CC, Swift JA, Thaimattam R, Thakur TS, Thaper RK, Thomas SP, Tothadi S, Vangala VR, Variankaval N, Vishweshwar P, Weyna DR and Zaworotko MJ. Polymorphs, salts, and cocrystals: What's in a name? Cryst. Growth Des. 2012, 12, 2147–2152.

[21] Desiraju GR. Crystal Engineering: The Design of Organic Solids. Elsevier, Amsterdam, 1989.

[22] ITC International Tables for Crystallography, Vol. D Physical Properties of Crystals edited by Authier A. Kluwer Academic Publisher, Dordrecht 2003.

[23] Newnham, RE. Properties of Materials. New York, Oxford University Press, 2005.

[24] Wampler RD, Begue NJ, Simpson GJ. Molecular design strategies for optimizing the nonlinear optical properties of chiral crystals. Cryst. Growth Des. 2008, 8, 2589–2594.

[25] Reddy CM, Gundakaram RC, Basavoju S, Kirchner MT, Padmanabhan KA, Desiraju GR. Chem. Commun. 2005, 3945–3947.

[26] Aakeröy, CB. Is there any point in making co-crystals? Acta Crystallogr. B 2015, 71, 387–391.

[27] Desiraju GR. Supramolecular synthons in crystal engineering – a new organic synthesis. Angew. Chem., Int. Ed. Engl., 1995, 34, 2311–2327.

[28] Wuest J. Use of hydrogen bonds to control molecular aggregation. Self-assembly of three-dimensional networks with large chambers. J. Am. Chem. Soc. 1991, 113, 4696–4698.

[29] Etter M. Encoding and Decoding Hydrogen-Bond Patterns of Organic Compounds. Acc. Chem. Res. 1990, 23, 120–126.

[30] Macrae CF, Edgington PR, McCabe P, Pidcock E, Shields GP, Taylor R, Towler M, van de Streek J. J. Appl. Crystallogr. 2006, 39, 453–457.

[31] Groom CR, Bruno IJ, Lightfoot MP, Ward SC. The Cambridge Structural Database. Acta Crystallogr. B 2016, 72, 171–179.

[32] Hsieh T-J, Su C-C, Chen C-Y, Liou C-H, Lu L-H. Using experimental studies and theoretical calculations to analyze the molecular mechanism of coumarin, *p*-hydroxybenzoic acid, and cinnamic acid. J. Mol. Struct. 2005, 741, 193–199.

[33] Babu NJ, Sanphui P, Nangia A. Crystal engineering of stable temozolomide cocrystals. Chem. Asian J. 2012, 7, 2274–2285.

[34] Lemmerer A, Adsmond DA, Bernstein J. An investigation of the hydrogen-bond preferences and co-crystallization behavior of three didonor compounds. Cryst. Growth. Des. 2011, 11, 2011–2019.

[35] Lou B, Perumalla SR, Sun CC, Significant expansion of the solid state landscape of salicylic acid based on charge-assisted hydrogen bonding interactions. Cryst. Growth Des. 2015, 15, 24–28.

[36] Vishweshwar P, McMahon JA, Peterson ML, Hickey MB, Shattock TR, Zaworotko MJ. Crystal engineering of pharmaceutical co-crystals from polymorphic active pharmaceutical ingredients. Chem. Commun. 2005, 4601–4603.

[37] Friščić, T, Fábián L, Burley JC, Reid DG, Duer MJ, Jones W. Exploring the relationship between cocrystal stability and symmetry: is Wallach's rule applicable to multi-component solids? Chem. Commun. 2008, 1644–1646.

[38] Fábián L. Cambridge Structural Database analysis of molecular complementarity in cocrystals. Cryst. Growth Des. 2009, 9, 1436–1443.

[39] Trask AV, Motherwell WDS, Jones W. Physical stability enhancement of theophylline via cocrystallization. Int. J. Pharm. 2006, 320, 114–123.

[40] Childs SL, Stahly GP, Park A. The salt-cocrystal continuum: The influence of crystal structure on ionization state. Mol. Pharmaceutics, 2007, 4, 323–328.

[41] Bruno IJ, Cole JC, Lommerse JPM, Rowland RS, Taylor R and Verdonk ML. IsoStar: A library of information about nonbonded interactions. J. Comput-Aided Mol. Des. 1997, 11, 525–537.

[42] Verdonk ML, Cole JC, and Taylor R. SuperStar: A knowledge-based approach for identifying interaction sites in proteins. J. Mol. Biol. 1999, 289, 1093–1108.

[43] Verdonk ML, Cole JC, Watson P, Gillet V, Willett P. Superstar: improved knowledge-based interaction fields for protein binding sites. J. Mol. Biol. 2001, 307, 841–859.

[44] Spackman MA, Jayatilaka D. Hirshfeld surface analysis. CrystEngComm 2009, 11, 19–32.

[45] Spackman MA, McKinnon JJ. Fingerprinting intermolecular interactions in molecular crystals. CrystEngComm 2002, 4, 378–392.

[46] Lou B, Perumalla SR, and Sun CC, Significant expansion of the solid state landscape of salicylic acid based on charge-assisted hydrogen bonding interactions. Cryst. Growth Des. 2015, 15, 24–28.

[47] Vishweshwar P, McMahon JA, Peterson ML, Hickey MB, Shattock TR, Zaworotko MJ. Crystal engineering of pharmaceutical co-crystals from polymorphic active pharmaceutical ingredients. Chem. Commun. 2005, pp.4601–4603.

[48] Coppens P. X-Ray charge densities and chemical bonding. Oxford, Oxford University Press, 1997.

[49] Gatti C, Macchi P. Modern Charge-Density Analysis. New York, Springer, 2012.

[50] Stalke D. Electron Density and Chemical Bonding I Experimental Charge Density Studies. Berlin, New York, Springer, 2012.

[51] Stalke D. Electron Density and Chemical Bonding II Theoretical Charge Density Studies. Berlin, New York, Springer, 2012.

[52] Tsirelson VG, Ozerov RP. Electron Densities and Bonding in Crystals. Bristol, Philadelphia, Institute of Physics, 1996.

[53] Macchi P. Modern charge density studies: The entanglement of experiment and theory. Cryst. Rev. 2013, 19, 58–101.

[54] Koritsanszky TS, Coppens P. Chemical applications of X-ray charge-density analysis. Chem. Rev. 2001, 101, 1583–1627.

[55] Hansen NK, Coppens P. Testing aspherical atom refinements on small-molecule data sets. Acta Crystallogr. A 1978, 34, 909–921.

[56] Dovesi R, Orlando R, Erba A, Zicovich-Wilson CM, Civalleri B, Casassa S, Maschio L., Ferrabone, De La Pierre M, D'Arco P, Noel Y, Causa M, Rerat M, Kirtman B. CRYSTAL14: A program for the ab initio investigation of crystalline solids. Int. J. Quantum Chem. 2014, 114, 1287–1317.

[57] Dovesi R, Saunders VR, Roetti C, Orlando R, Zicovich-Wilson CM, Pascale F, Civalleri B, Doll K, Harrison NM, Buch IJ, D'Arco P, Llunell M, Causa M, Noel Y. CRYSTAL14 User's Manual, University of Torino, Torino, 2014.

[58] Kohn W, Sham LJ. Self-consistent equations including exchange and correlation effects. Phys. Rev. A 1965, 140, 1133–1138.

[59] Chimpri AS, Gryl M, Dos Santos LHR, Krawczuk A, Macchi P. Correlation between accurate electron density and linear optical properties in amino acid derivatives: L-Histidinium hydrogen oxalate. Cryst. Growth. Des. 2013, 13, 2995–3010.
[60] Darrigan C, Rérat M, Mallia G, Dovesi R. Implementation of the finite field perturbation method in the CRYSTAL program for calculating the dielectric constant of periodic systems. J. Comput. Chem. 2003, 24, 1305–1312.
[61] Noel Y, Llunell M, Orlando R, D'Arco P, Dovesi R. Performance of various Hamiltonians in the study of the piezoelectric properties of crystalline compounds. The case of BeO and ZnO. Phys. Rev. B 2002, 66, 214107.
[62] Jelsch C, Pichon-Pesme V, Lecomte C, Aubry A. Transferability of multipole charge-density parameters: Application to very high resolution oligopeptide and protein structures Acta Crystallogr D 1998, 54, 1306–1318.
[63] Volkov A, Li X, Koritsanszky T, Coppens P. Ab initio quality electrostatic atomic and molecular properties including intermolecular energies from a transferable theoretical pseudoatom databank. J Phys. Chem. A 2004, 108, 4283–4300.
[64] Dittrich B, Hubschle CB, Luger P, Spackman MA. Introduction and validation of an invariom database for amino-acid, peptide and protein molecules. Acta Crystallogr. D 2006, 62, 1325–1335.
[65] Volkov A, Messerschmidt M, Coppens P. Improving the scattering-factor formalism in protein refinement: application of the University at Buffalo Aspherical-Atom Databank to polypeptide structures. Acta Crystallogr D 2007, 63, 160–170.
[66] Dittrich B, Strumpel M, Spackman MA, Koritsanszky T: Invarioms for improved absolute structure determination of light-atom crystal structures. Acta Crystallogr. A 2006, 62, 217–223.
[67] Dominiak PM, Volkov A, Li X, Messerschmidt M, Coppens PA. Theoretical databank of transferable aspherical atoms and its application to electrostatic interaction energy calculations of macromolecules. J. Chem. Theory Comput. 2007, 3, 232–247.
[68] Hubschle CB, Luger P, Dittrich BJ. Automation of invariom and of experimental charge density modelling of organic molecules with the preprocessor program Invariom Tool. J. Appl. Cryst. 2007, 40, 623–627.
[69] Zarychta B, Pichon-Pesme V, Guillot B, Lecomte C, Jelsch C. On the application of an experimental multipolar pseudo-atom library for accurate refinement of small-molecule and protein crystal structures. Acta Crystallogr. A 2007, 63, 108–125.
[70] Hathwar VR, Thakur TS, Guru Row TN, Desiraju GR. Transferability of multipole charge density parameters for supramolecular synthons: a new tool for quantitative crystal engineering. Cryst. Growth Des. 2011, 11, 616–623.
[71] Hathwar VR, Thakur TS, Dubey R, Pavan MS, Guru Row TN, Desiraju GR. Extending the supramolecular synthon based fragment approach (SBFA) for transferability of multipole charge density parameters to monofluorobenzoic acids and their cocrystals with isonicotinamide: importance of C–H···O, C–H···F, and F···F intermolecular regions. J. Phys. Chem. A 2011, 115, 12852–12863.
[72] Bader RFW. Atoms in Molecules: A Quantum Theory. Oxford, Oxford University Press, 1990.
[73] Bader RFW. A bond path: A universal indicator of bonded interactions. J. Phys. Chem. A 1998, 102, 7314–7323.
[74] Bader RFW. Bond paths are not chemical bonds. J. Phys. Chem. A 2009, 113, 10391–10396.
[75] Espinosa E, Molins E, Lecomte C. Hydrogen bond strengths revealed by topological analyses of experimentally observed electron densities. Chem. Phys. Lett. 1998, 285, 170–173.
[76] Cremer D, Kraka E. Chemical bonds without bonding electron density – does the difference electron-density analysis suffice for a description of the chemical bond? Angew. Chem. 1984, 23, 627–628.

[77] Abramov YA. On the possibility of kinetic energy density evaluation from the experimental electron-density distribution. Acta Crystallogr. A 1997, 53, 264–272.
[78] Gibbs GV, Spackman MA, Jayatilaka D., Rosso KM, Fox DF. Bond length and local energy density property connections for non-transition-metal oxide-bonded interactions. J. Phys. Chem. A 2006, 110, 12259–12266.
[79] Johnson ER, Keinan S, Mori-Sánchez P, Contreras-García J, Cohen AJ, Yan W. Revealing noncovalent interactions. J. Am Chem. Soc. 2010, 132, 6498–6506.
[80] Krawczuk A, Gryl M, Pitak M, Stadnicka K. Electron density distribution of urea in co-crystals with rigid and flexible dicarboxylic acids. Cryst. Growth Des. 2015, 15, 5578–5592.
[81] Contreras-García J, Johnson ER, Keinan S, Chaudret R, Piquemal J-P, Beratan DN, Yang W, NCIPLOT: A program for plotting noncovalent interaction regions. J. Chem. Theory Comput. 2011, 7, 625–632.
[82] Saleh G, Gatti C, Lo Presti L, Contreras-García J. Revealing non-covalent interactions in molecular crystals through their experimental electron densities. Chem. Eur. J. 2012, 18, 15523–15536.
[83] Vela A, Gázquez JL. A relationship between the static dipole polarizability, the global softness, and the fukui function. J. Am. Chem. Soc. 1990, 112, 1490–1492.
[84] Hehre WJ, Pau CF, Headley AD, Taft RW. A scale of directional substituent polarisbility parameters from ab initio calculations of polarizability potentials. J. Am. Chem. Soc. 1986, 108, 1711–1712.
[85] Headley AD. Substituent effects on the basicity of dimethylamines. J. Am. Chem. Soc. 1987, 109, 2347–2348.
[86] Nagle JK. Atomic polarizability and electronegativity. J. Am. Chem. Soc. 1990, 112, 4741–4747.
[87] Elliasmine A, Godet JL, Le Duff Y, Bancewicz T. Collision-induced depolarized light scattering by CF4. Mol. Phys. 1997, 90, 147–157.
[88] Elliasmine A., Godet JL, Le Duff Y, Bancewicz T. Isotropic collision-induced light scattering by gaseous CF4. Phys. Rev. A 1997, 55, 4230–4237.
[89] Gough KM, Yacowar MM, Cleve RH, Dwyer JR. Analysis of molecular polarizabilities and polarizability derivatives in H_2, N_2, F_2, CO, and HF, with the theory of atoms in molecules. Can. J. Chem. 1996, 74, 1139–1144.
[90] Cole JM Organic materials for second-harmonic generation: advances in relating structure to function. Phil. Trans. R. Soc. Lond. A 2003, 361, 2751–2770.
[91] Chantal D, Dupuis M. Nonlinear optical properties of organic solids: ab initio polarizability and hyperpolarizabilities of nitroaniline derivatives. Chem. Phys. Lett. 1990, 171, 209–216.
[92] Seidler T, Stadnicka K, Champagne B. Evaluation of the linear and second-order NLO properties of molecular crystals within the Local Field Theory: Electron correlation effects, choice of XC functional, ZPVA contributions, and impact of the geometry in the case of 2-methyl-4-nitroaniline. J. Chem. Theory Comput. 2014, 10, 2114–2124.
[93] Seidler T, Krawczuk A, Champagne B, Stadnicka K. QTAIM-based scheme for describing the linear and nonlinear optical susceptibilities of molecular crystals composed of molecules with complex shapes. J. Phys. Chem. C 2016, 120, 4481–4494.
[94] Krawczuk-Pantula A, Perez D, Stadnicka K, Macchi P. Distributed atomic polarizabilities from electron density. 1. Motivations and theory. Trans. Amer. Cryst. Ass. 2011, 42, 1–25.
[95] Krawczuk A, Perez D, Macchi P. Polaber: a program to calculate and visualize distributed atomic polarizabilities based on electron density partitioning. J. Appl. Cryst. 2014, 47, 1452–1458.
[96] Keith TA. Atomic Response Properties. In Matta CF. Boyd RJ., ed. The Quantum Theory of Atoms in Molecules: from Solid State to DNA and Drug Design. Weinheim, Wiley-VCH, 2007, 61–94.

[97] Macchi P, Krawczuk A. The polarizability of organometallic bonds. Comput. Theo. Chem. 2015, 1053, 165–172.
[98] Dos Santos LHR, Krawczuk A, Macchi P. Distributed atomic polarizabilities of amino acids and their hydrogen-bonded aggregates. J. Phys. Chem. A 2015, 119, 3285–3298.
[99] Bader RFW, Gatti C. A Green's function for the density. Chem. Phys. Lett. 1998, 287, 233–238.
[100] Gatti C, Cargoni F, Bertini L. Chemical information from the source function. J. Comput. Chem. 2003, 24, 422–436.
[101] Gatti C, Bertini L. The local form of the source function as a fingerprint of strong and weak intra- and intermolecular interactions. Acta Crystallogr. A 2004, 60, 438–449.
[102] Overgaard J. Expanding the usage of the source function to experimental electron densities. Acta Crystallogr. B 2016, 72, 169–170.
[103] Farrugia L, Macchi P. On the interpretation of the source function. J. Phys. Chem. A 2009, 113, 10058–10067.
[104] Bader RFW, Stephens ME. Spatial localization of the electronic pair and number distributions in molecules. J. Am. Chem. Soc. 1975, 97, 7391–7399.
[105] Fradera X, Austen MA, Bader RFW. The Lewis model and beyond. J. Phys. Chem. A 1999, 103, 304–314.
[106] Matta C, Hernández-Trujillo J. Bonding in polycyclic aromatic hydrocarbons in terms of the electron density and of electron delocalization. J. Phys. Chem. A 2009, 107, 7496–7504.
[107] Matta CF, Hernández-Trujillo J, Bader RFW. Proton spin-spin coupling and electron delocalization. J. Phys. Chem. A 2002, 106, 7369–7375.
[108] Poater J, Fradera X, Duran M, Sola M. The delocalization index as an electronic aromaticity criterion: Application to a series of planar polycyclic aromatic hydrocarbons. Chem. Eur. J. 2003, 9, 400–406.
[109] Bartashevich S, Nasibullina E, Tsirelson VG. Electron delocalization indices as criteria for the identification of strong halogen bonds of iodine. J. Struct. Chem. 2015, 56, 1223–1225.
[110] Blanco MA, Martín Pendás A, Francisco E. Interacting quantum atoms: A correlated energy decomposition scheme based on the quantum theory of atoms in molecules. J. Chem. Theory Comput. 2005, 1, 1096–1109.
[111] Francisco E, Martín Pendás A, Blanco MA. A molecular decomposition scheme for atoms in molecules. J. Chem. Theory Comput. 2006, 2, 99–102.
[112] Maxwell P, Martín Pendás A, Popelier P. Extension of the interacting quantum atoms (IQA) approach to B3LYP level density functional theory (DFT). Phys. Chem. Chem. Phys. 2016, 18, 20986–1000.
[113] Guevara-Vela JM, Garcia-Revilla M, Hernández-Trujillo J, Christiansen O, Francisco E, Martín Pendás A, Rocha-Rinza T. Hydrogen-bond cooperative effects in small cyclic water clusters as revealed by the interaction quantum atoms approach. Chem. Eur. J. 2013, 19, 14304–14315.
[114] Syzgantseva OA, Tognetti V, Joubert L. On the physical nature of halogen bonds: A QTAIM study. J. Phys. Chem. A 2013, 117, 8969–8980.
[115] Kohout M. Bonding indicators from electron pair density functionals. Faraday Discuss. 2007, 1, 1096–1109.
[116] Kohout M. A measure of electron localizability. Int. J. Quantum Chem. 2004, 97, 651–658.
[117] Martín Pendás A, Kohout M, Blanco MA, Francisco E. Beyond standard charge density topological analyses in Gatti C, Macchi P., ed. Modern Charge-Density Analysis, New York, Springer, 2012, 303–358.
[118] Wagner FR, Kohout M, Grin Y. Direct space decomposition of ELI-D: Interplay of charge density and pair-volume function for different bonding situations. J. Phys. Chem. A 2008, 112, 9814–9828.

[119] Macchi P. Crystallographic approaches for the investigation of molecular materials: Structure property relationships and reverse crystal engineering. CHIMIA Int. J. Chem. 2014, 68, 31–37.

[120] de Matos Gomes E, Venkataramanan V, Nogueira E, Belsley M, Proenca F, Criado A, Dianez MJ, Estrada MD, Perez-Garrido S. Synthesis, crystal growth and characterization of a new nonlinear optical material – urea L-malic acid. Synt. Mat. 2000, 115, 225–228.

[121] Wentao Y, Mengkai L, Fanqing M. Crystal and molecular structure of urea–(+) tartaric acid. Mater Res. Bull. 1996, 31, 1127–1131

[122] Wang L, Zhao L, Xue RY, Lu XF, Wen YH, Yang Y. Construction of interesting organic supramolecular structures with synthons cooperation in the cocrystals of 1H-benzotriazole and hydroxybenzoic acids. Sci. Chin Chem. 2012, 55, 2515–2522.

[123] Thirunavukkarasu A, Silambarasan A, Chakkaravarthi G, Mohankumar R, Umarani PR. 1H-Benzotriazole-4-hydroxybenzoic acid (1/1). Acta Crystallogr. E 2013, 69, o1605.

[124] Silambarasan A, Krishna Kumar M, Thirunavukkarasu A, Mohan Kumar R, Umarani PR. Synthesis, crystal growth, solubility, structural, optical, dielectric and microhardness studies of Benzotriazole-4-hydroxybenzoic acid single crystals. J. Cryst. Growth 2015, 420, 11–16.

[125] Sathya P, Vidyalaksmi Y, Pugazhendhi S, Gopalakrishnan R. Benzotriazole *p*-hydroxybenzoic acid: physicochemical and biological evaluation of an organic cocrystal. Mater. Res. Innovations, 2016, 1 DOI: 10.1080/14328917.2016.1200843.

[126] Gryl M, Seidler T, Stadnicka K, Matulková I, Němec I, Tesařová N, Němec P. The crystal structure and optical properties of a pharmaceutical co-crystal – the case of the melamine–barbital addition compound. CrystEngComm 2014, 16, 5765–5768

[127] Gryl M, Cenedese S, Stadnicka K. Crystal engineering and charge density study of pharmaceutical nonlinear optical material: Melamine-Barbital co-crystal. J. Phys. Chem. C, 2015, 119, 590–598.

[128] Fanqing M, Mengkai L, Xin Y, Wentano Y, Yuguo W. The preparation and characterization of electric property of a new organic crystal: urea-(+) tartaric acid. Mater. Res. Bull. 1996, 31, 1121–1125.

[129] Dixit VK, Vanishri S, Bhat HI, de Matos Gomes E, Belsley M, Santinha C, Arunmozhi G, Venkataramanan V, Proena F, Criado A. Crystal growth and characterization of a new nonlinear optical material: Urea L-Malic Acid. J. Cryst. Growth. 2003, 253, 460–466.

[130] Vinothkumar P, Rajeswari K, Mohan Kumar R, Bhaskaran A. Structural, optical, thermal and mechanical properties of Urea tartaric acid single crystals. Spectrochim. Acta Part A 2015, 145, 33–39.

[131] Deepthy A, Vanishri S, Ambika D, George SD, Nampoori VPN, Bhat HL, de Matos Gomes E, Belsley. Photoacoustic investigations on thermal anisotropy in urea L-malic acid single crystals. Mater. Res. Bull. 2008, 43, 1641–1648.

[132] Orjala J, Erdelmeier CAJ, Wright AD, Rali T, Sticher O. Five new prenylated *p*-hydroxybenzoic acid derivatives with antimicrobial and molluscicidal activity from piper aduncum leaves. Planta Med. 1993, 59, 546–551.

[133] Soni MG, Carabin IG, Burdock GA. Safety assessment of esters of *p*-hydroxybenzoic acid (parabens). Food Chem. Toxicol. 2005, 43, 985–1015.

Kohei Johmoto and Hidehiro Uekusa

4 Control of photochromism in *N*-salicylideneaniline by crystal engineering

4.1 Introduction

Chromism is a reversible colour change induced by an external stimulus. Chromism in materials may be achieved by many methods, such as light irradiation (photochromism) [1, 2], heating or cooling (thermochromism) [3, 4], changing solvent polarity (solvatochromism) [5, 6], pressure application (piezochromism) [7, 8] and vapour application (vapochromism) [9, 10]. Such transformations are important because they can be utilised as sensors for various stimuli. Photochromic materials are expected to serve as optical storage elements or colour display elements [1], photo induced motors [11] and biological sensors [12], and many studies have been reported with those applications as motivations. From the structural point of view, chromism in the solid-state is a dynamic phenomenon in which the molecular structure/conformation or arrangement of a molecule changes in the solid-state under external stimuli. Thus, with the above in mind, chromism has garnered considerable attention in a variety of fields.

In this chapter, the photochromism of *N*-salicylideneaniline (SA, Figure 4.1) in the crystalline state is described. Specifically, the relationship between the molecular conformation of SA and the chromism it exhibits is described. The control of the conformation and properties using crystal engineering techniques is also emphasised.

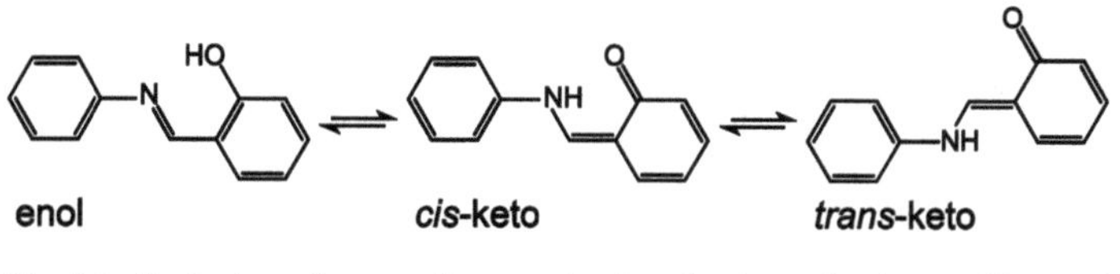

Fig. 4.1: Photochromism reaction mechanism (tautomerisation and isomerisation) in SA.

4.2 Photochromism of SA and its reaction mechanism

N-Salicylideneaniline (SA, Figure 4.1) is one of the oldest known photochromic compounds by UV irradiation, first reported by Alfred Senier in 1909 [13]. The chromism of SA derivatives were reported in 1912 and found to be dependent on the molecular structure and the crystalline form [14]. For example, when the hydroxyl group of SA

https://doi.org/10.1515/9783110464955-004

was substituted with a methoxy group, the resulting molecule did not exhibit photochromism, which clearly indicated that a proton transfer from the hydroxyl group was critical to chromism [15, 16]. Many important studies were carried out by Cohen and Schmidt in the 1960s. They summarised the photochromism of SA as follows [17]: i) SA shows photo- or thermochromism exclusively; ii) SA takes the enol form in the crystal; iii) in thermochromic crystals an intra-molecular proton transfer from the hydroxyl group to the N-atom occurs, inducing the *cis*-keto form, which causes the colouration; and iv) the activation energy of the reverse reaction is higher in photochromism than that in thermochromism, and therefore, the coloured species formed via photochromism should be the *trans*-keto form. In solution, the coloured species is known to be in the *trans*-keto form [18–20]. All SA derivatives have now been shown to exhibit thermochromism [21] so the observed photochromism and thermochromism of SA are no longer considered exclusive properties.

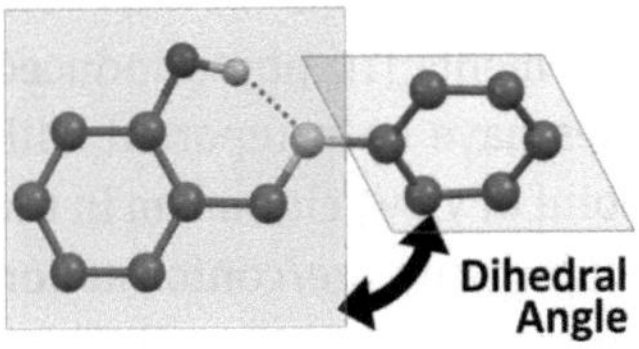

Fig. 4.2: The key dihedral angle in the twisted conformation of SA.

A series of topochemical studies in 1964 revealed important structural details regarding the molecular conformation of SA derivatives in their crystals, which affected their photochromism and thermochromism [16, 17, 22–26]. When the SA molecule exhibits a twisted conformation, where the dihedral angle between the two benzene rings is significantly large (Figure 4.2), photochromism is observed. By contrast, when the molecule exhibits a planar conformation, thermochromism is evident. Thus, both properties depend on the molecular conformation. In the 1960s, studies targeting the practical applications of photochromism were also performed, which revealed SA to have very high durability and to exhibit minimal degradation even after 5000 cycles of colouration and fading [27]. The durability was attributed to a small molecular change in the structure owing to enol-keto tautomerisation (Figure 4.1).

Thus, based on previous studies, the coloured species was predicted to be the *trans*-keto form. However, conclusive evidence to confirm the hypothesis was not obtained. In 1985, Kawato et al. reported a new SA derivative, *N*-3,5-di-tert-butylsalicylidene-3-nitroaniline, SA (1), Figure 4.3 (1) [28], which showed photochromism in the crystalline state with a very long lifetime (months). Harada observed the coloured species directly by crystallographic analysis for the first time in 1999 [29]. Such an observation had been attempted but without success because of the very low concentration of coloured species in the crystal, which was too low for direct observation by crystallography. Notably, under UV irradiation, it was found that only a thin layer on the surface of single crystals was coloured, and that the large inner volume of the

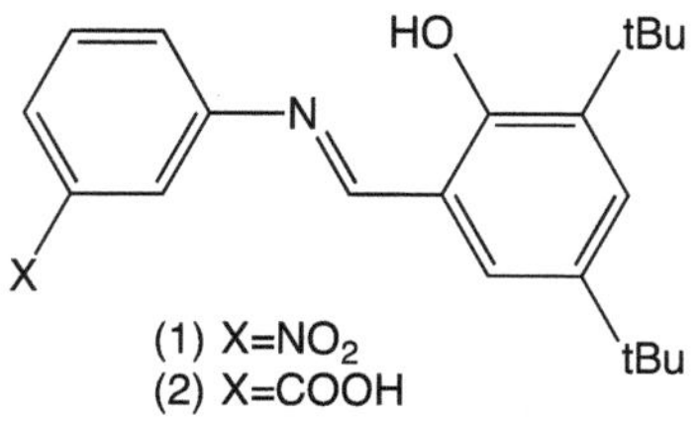

Fig. 4.3: SA derivatives SA (1) and SA (2).

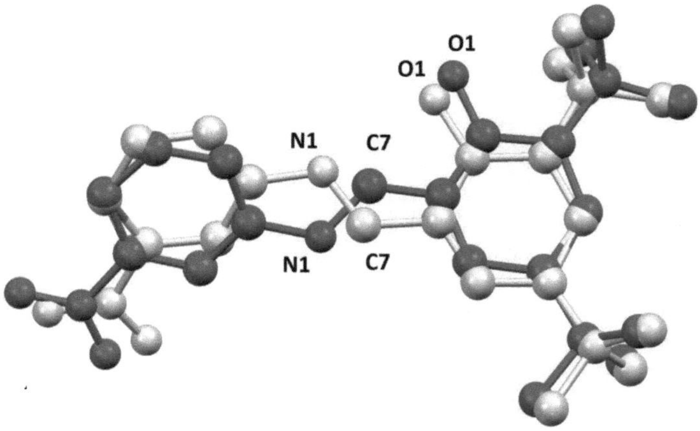

Fig. 4.4: Direct observation of *trans*-keto form (*dark*) with *cis*-enol form (*light*) [29].

crystal was unreacted. In order to overcome this problem, a two-photon excitation technique was applied by short pulse laser irradiation. In this technique, the longer wavelength pulse laser was utilised to penetrate the crystal without absorption, and the photo-reaction proceeded in the centre of the crystal. Eventually, Harada succeeded in observing the *trans*-keto form as a part of disordered structure with the initial enol form (Figure 4.4). It was concluded that photochromism occurs by the formation of the *trans*-keto form via a proton transfer from the enol form.

4.3 Thermal stability of coloured species

The thermal stability of coloured species at different temperatures was reported by Senior [14]. The temperature dependent activation energy barrier of the *trans*-keto and *cis*-keto forms (fading process) was also studied [17]. At –70°C, the *trans*-keto form of *N*-3,5-di-tert-butyl-salicylidene-3-carboxyaniline, SA (2) (Figure 4.3) in the crystal is more stable than that at 20°C, as revealed by the IR observation of the N-H peak in the *trans*-keto form. However, there was no report on the stabilities of coloured species (*trans*-keto form) among different SA derivatives or different crystal forms. SA (2) forms three polymorphic crystals, namely the α, β and γ forms (Figure 4.5) [30]. The α and β forms exhibit photochromism and SA takes on a twisted conformation in which the dihedral

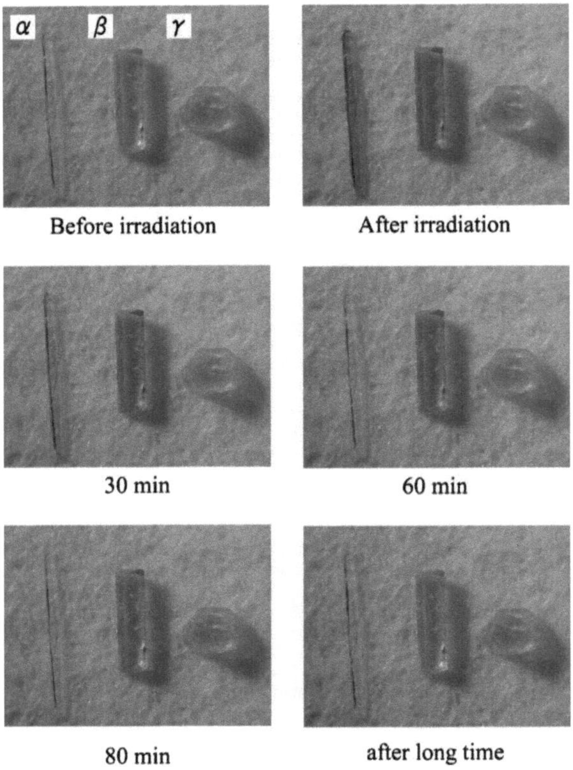

Fig. 4.5: Three polymorphs of *N*-3,5-di-tert-butyl-salicylidene-3-carboxyaniline, SA (2), and the colour lifetime of each. Reprinted with permission from Ref. [30]. Copyright (2009) The Chemical Society of Japan.

angles between the two benzene rings are 61° and 37° for the *α* and *β* forms, respectively. The *γ* form does not exhibit photochromism and the molecular conformation is nearly planar (the dihedral angle is 29°). Notably, even among polymorphic crystals, the photochromism and stability of the coloured species (the colour lifetime during fading) differ depending on the molecular conformation or crystalline environment around the molecule (Figure 4.5).

Because the *β* form has a longer colour lifetime than the *α* form, the *trans*-keto form is likely stabilised by the crystalline environment in the *β* form. When the *trans*-keto form is assumed in the *β* crystal structure, the N–H group of the molecule forms an intermolecular hydrogen bond with the adjacent hydroxyl group, Figure 4.6(b). On the other hand, in the *α* form, there is no strong intermolecular interaction involving the N–H group, Figure 4.6(a), indicating that the *trans*-keto form molecule was not stabilised by its crystalline environment. The stabilisation of the *trans*-keto species by hydrogen bonding was also observed for *N*-3,5-di-tert-butyl-salicylidene-3-nitroaniline, SA (1), Figure 4.6(c), which displays the longest colour lifetime (*trans*-

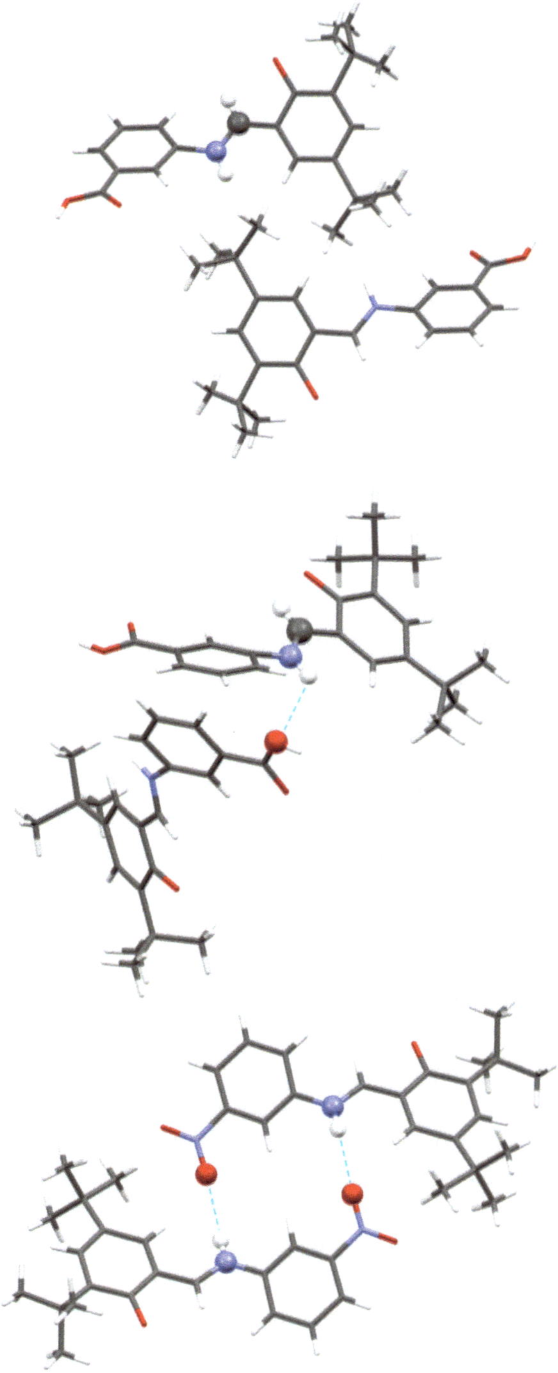

(a)

(b)

(c)

Fig. 4.6: The crystalline environments of *trans*-keto form of SA (1) and SA (2). (a) SA (2)-α, (b) SA (2)-β, and (c) SA (1).

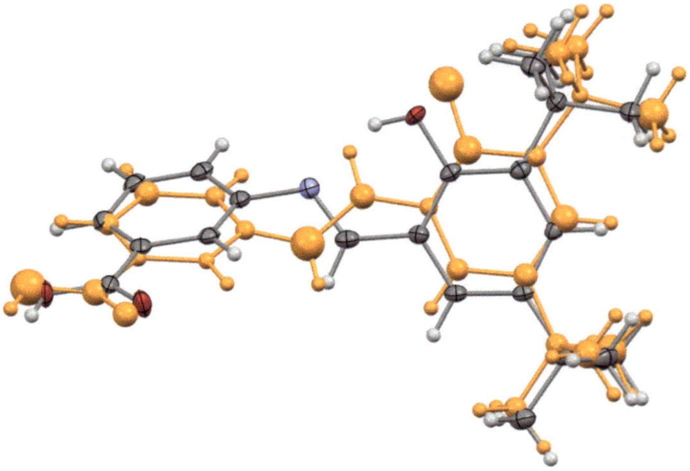

Fig. 4.7: Direct observation of *trans*-keto form of *N*-3,5-di-tert-butylsalicylidene-3-carboxyaniline, SA (2) [30].

keto form) among many SA derivatives [29]. Mikami calculated the stabilisation energy of the *trans*-keto form of SA (1) as −39.4 kJ/mol [31]. In this crystal structure, two symmetry-related *trans*-keto molecules form N–H···O(nitro) hydrogen bonds to generate a stable dimeric aggregate, Figure 4.6 (c). Therefore, the stabilisation of the *trans*-keto form in the coloured crystal is caused by the crystalline environment. It is also notable that in the *α* form, the *trans*-keto species was observed in the coloured crystal as the similar disordered structure to SA (1), Figure 4.7 [30]. However, the UV irradiation technique was employed instead of the two photon technique. This observation confirmed that the *trans*-keto form caused the colouration of the crystal.

4.4 Relationship between molecular structure and photochromism

In SA, the salicylimine group is planar due to the intramolecular hydrogen bond between the OH group and N; thus, the only conformational variable is the free rotation around the single bond between the N atom of the imine and the aniline moiety, Figure 4.8. The most stable molecular structure has a dihedral angle of 44° between the salicyl and aniline rings according to theoretical calculations [32]. This is because of the steric repulsion between H atoms in the ortho-position of the aniline ring and the azomethine group (Figure 4.8) [33]. However, in the crystal, the dihedral angle spans from 0 to 90°, forming the planar or twisted conformations and the molecular conformation is always affected by the intermolecular interactions in the crystal.

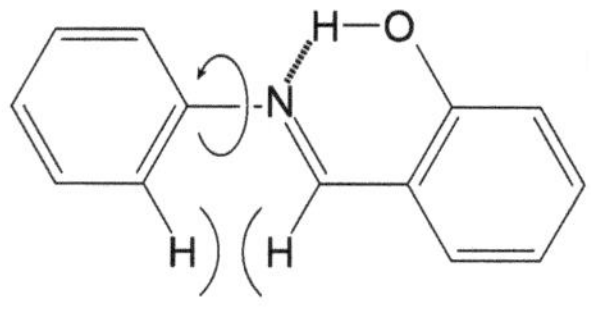

Fig. 4.8: Intramolecular hydrogen bond and steric hindrance in SA.

The molecular conformations found in SA derivatives are critical because only the twisted conformation exhibits photochromism. In general for SA derivatives, photochromism was observed when the dihedral angle was greater than 30°. When the angle was less than 20°, the crystals were non-photochromic and in cases where the dihedral angles were in the intermediate range, i.e. 20–30°, both photo- and non-photochromic crystals existed [34]. A Cambridge Structure Database, (CSD, version 5.33, Feb. 2012) investigation was carried out to reveal that 40 out of 559 SA derivatives exhibited chromism. The photochromic SA derivatives and their dihedral angles are plotted in Figure 4.9.

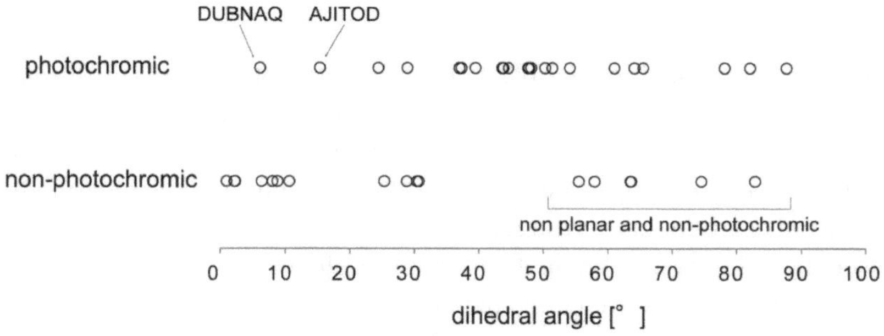

Fig. 4.9: Relationship between photochromism and dihedral angle of SA derivatives.

Most structures shown in Figure 4.9 conform to the proposed dihedral angle thresholds of 30° and 20° for photochromic and non-photochromic crystals, respectively. Two apparently photochromic crystals with dihedral angles less than 20° seemed to be exceptions (refcodes: DUBNAQ [35] & AJITOD [36]). Upon further examination, DUBNAQ was found to be non-photochromic and AJITOD contained molecules in both the twisted (dihedral angle of 49° and photochromic) and planar (15° and non-photochromic) conformations. Also, there were five non-photochromic exceptions with dihedral angles greater than 30°. Some had intermolecular steric hindrance around the hydroxy group, and the others had intramolecular steric hindrance owing to large substituents such as tert-Butyl (tBu) on the ortho-position of the aniline ring. Such steric interactions preclude the molecular motion necessary for the isomerisation from the *cis*-keto to *trans*-keto form (also known as "pedal motion" [37]). Thus, the proposed thresholds of 30° and 20° were confirmed, with only a few exceptions.

The reason for the relationship between planarity and photochromism has not been completely elucidated. Kawato et al. explained it using the idea of *loose packing* [28]. When bulky tBu groups were introduced in SA derivatives, the packing efficiency of the crystal became lower (loose packing), which induced the twisted conformation. However, the idea was not quantitatively confirmed upon examining free volume or intermolecular distances around the molecule. Theoretical investigations were performed by Zgierski and Grabowska for the intermediate structure during the photochromic reaction [32]. From the calculations, the excited-state structure of the *cis*-keto form adopts twisted conformation. So, if the molecular conformation is twisted in both the *trans*-enol and *cis*-keto forms, isomerisation may be facilitated from a twisted *cis*-keto to the *trans*-keto form via the twisted excited-state structure in the course of the photochromic reaction.

4.5 Control of photochromism by manipulation of molecular conformation

As described above, the photochromism of SA derivatives strongly depends on the molecular conformation, which can be affected by the crystalline environment around the molecule. Thus, several attempts to control the photochromism were carried out using different crystal engineering techniques. The first technique was chemical modification. When a bulky group such as tBu is introduced, the molecular arrangement or molecular packing in the crystal and photochromism is altered [28, 38]. However, because this technique deals with different chemical species, other effects (electronic, etc.) should also be considered. The second technique is to form an inclusion compound with, e.g. cyclodextrin [39], deoxycholic acid [40–42] or a polymeric material [43], in order to alter the crystalline environment of the SA derivative. Notably, using this technique, a photochromic inclusion compound was formed even though the original, pure SA crystal was non-photochromic. Unfortunately, the conformation of SA could not be confirmed by crystallographic analysis due to severe disorder of the SA molecule in the crystal. In inclusion compounds, the void space around the molecule becomes rather large, which favours the twisted conformation of the molecule as in solution or in a vacuum. The third technique is to include SA in a MOF (metal–organic framework). When solvent molecules co-existing with SA were exchanged, the molecular conformation and photochromism also changed in order to adapt to the different crystalline environment [44]. This method could be used to control both the molecular conformation and photochromism. However, the change in the conformation was small due to the limited MOF pore space where the SA molecules were included.

4.5.1 Control of photochromism by co-crystal formation

The flexible conformation of SA depends on the surrounding environment, which suggests that the energy difference between the twisted and planar conformations can be governed by intermolecular interactions in the crystal. Therefore, the conformation and by implication photochromism can be controlled by altering the molecular packing in the crystal. One realistic technique to control the conformation is to form a co-crystal. A co-crystal is a multi-component crystal in which a SA derivative and another component (co-former molecule) are present in the same crystalline lattice in stoichiometric amounts. This technique does not require any chemical modifications and can produce various crystal structures by using different components. In this way, numerous molecular conformations can be achieved depending on the crystalline environment.

Previously, an acid-base type co-crystal formation method was reported [45]. Thus, *N*-salicylidene-3-carboxyaniline, SA (3) (3-1 in Figures 4.10 and 4.11), which contains a carboxyl group, was co-crystallised with a co-former containing pyridine and secondary amine functionalities (Figure 4.10) to yield seven co-crystals. The dihedral angle in the original crystal of SA (3) was 27° (Figure 4.11). By forming the co-crystals, crystal structure analysis revealed that the dihedral angles widely changed, spanning a range from 6°(3-1a) to 50°(3-1g). Among the eight structures, four crystals with di-

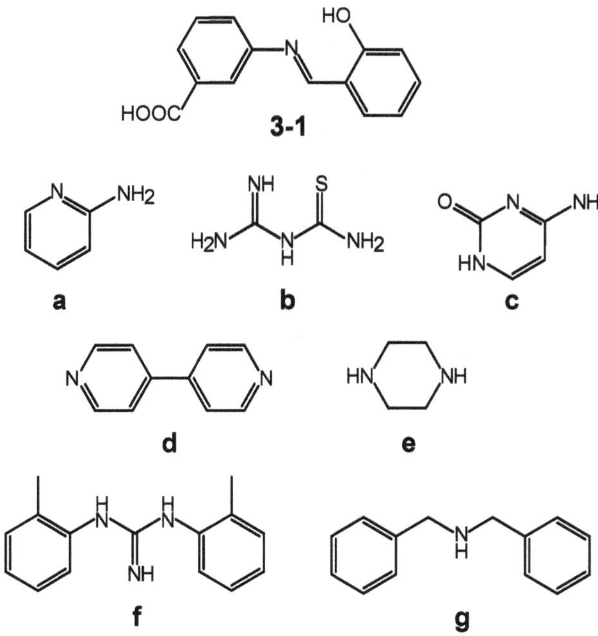

Fig. 4.10: *N*-salycylidene-3-carboxyaniline, SA (3) (3,1), and basic co-formers (a)–(g).

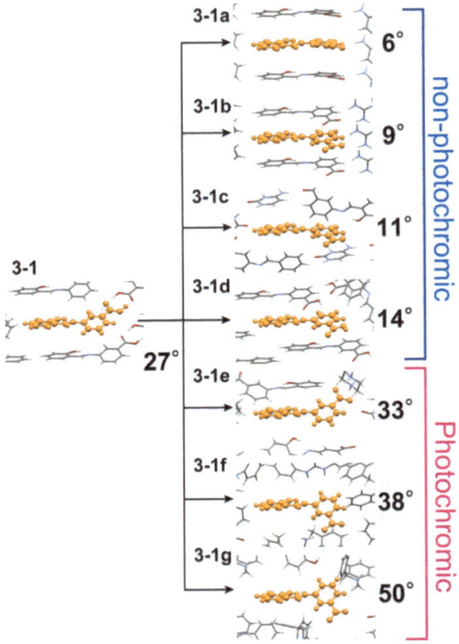

Fig. 4.11: Conformational changes in SA (3) by co-crystallisation showing the dihedral angle. Adapted with permission from Ref. [45]. Copyright (2012) American Chemical Society.

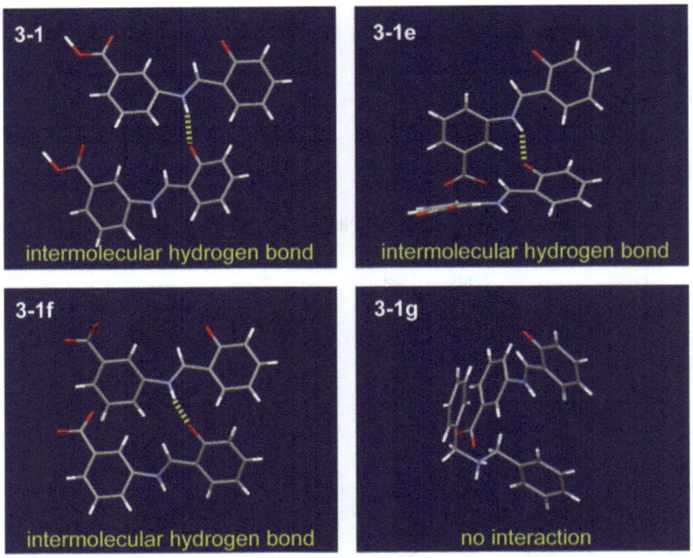

Fig. 4.12: Crystalline environment and hydrogen bonding in simulated *trans*-keto form of SA (3), 3-1, 3-1e, 3-1f, and 3-1g.

hedral angles greater than 27° exhibited photochromism, illustrating the correlation between the twisted conformation and photochromism.

This classification again followed the previously presented threshold of photochromism of 30° to 20°. The colour lifetimes of the crystals were 4.85×10^4, 2.30×10^3, 5.49×10^3, and 4.67×10^2 s for 3-1, 3-1e, 3-1f, and 3-1g, respectively (Figures 4.10 and 4.11). It is noted that the structures with longer lifetimes (3-1, 3-1e, and 3-1f) contained intermolecular hydrogen bonds in their simulated *trans*-keto crystal structures (Figure 4.12). In this case, the molecular conformation (planar to twisted) could be controlled by changing the crystalline environment, and the photochromism could be regulated by altering the molecular conformation. Additionally, the lifetime of the coloured crystal could be also modified by hydrogen bond formation.

4.5.2 Control of photochromism by mixed crystal formation

As described above, SA can adopt various conformations depending on the crystal structure. However, it is not possible to know the conformation prior to crystal structure analysis. In order to overcome this issue and to control the molecular conformation precisely, a mixed crystal technique has been employed. A mixed crystal is a solid solution type crystal in which different compounds are mixed uniformly throughout the crystal. When two compounds form a mixed crystal, the crystal structure is often the same as that of the major component. This approach was utilised to control the photochromic properties of several SA derivatives. Cohen [46] mixed crystals of photochromic *N*-salicylidene-4-bromoaniline, SA (4), and non-photochromic *N*-salicylidene-4-chloroaniline, SA (5), to form mixed crystals. When the mixed crystal contained 65% of SA (4), the crystal exhibited photochromism. Unfortunately, crystallographic analyses were not reported [46].

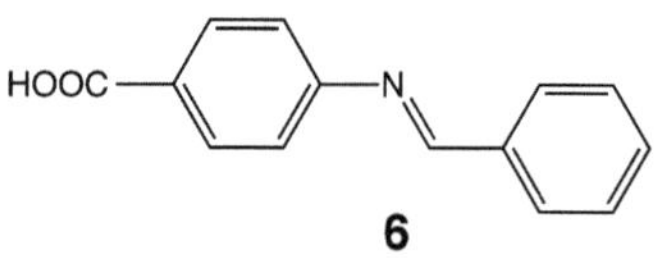

6

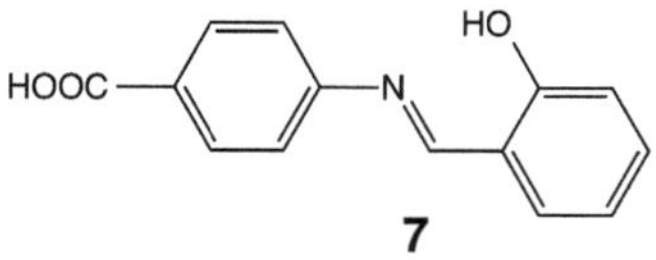

7

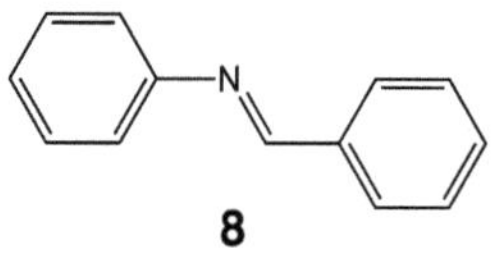

8

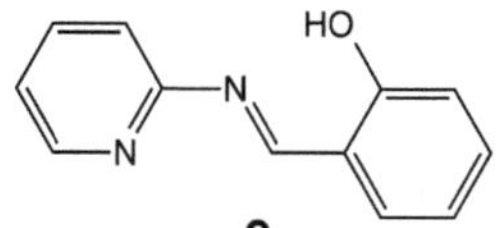

9

Fig. 4.13: Salicylideneaniline derivatives (SA (7), SAP (9)), and benzylideneaniline (BA (6), BA (8)).

In order to obtain more extensive control of the molecular conformation and photochromism, mixed crystals of *N*-benzylidene-4-carboxyaniline, BA (6), and *N*-salycylidene-4-carboxyaniline, SA (7), were examined (Figure 4.13). BA (6) lacks the OH group present in SA and therefore, it does not exhibit photochromism. However, it adopts the twisted conformation in the crystalline state; the dihedral angle is 54.7(1)°. By contrast, SA (7) exhibits the planar conformation in the crystal with the dihedral angle being 5.1(1)°, and does not exhibit photochromism. Thus, on forming mixed crystals of BA (6) and SA (7), BA (6) (major component) can act as a template and the conformation of SA (7) (minor component) may be fitted to that of BA (6). Therefore, the twisted conformation with a dihedral angle of about 55° could be produced. The mixed crystal was obtained from a methanol solution of BA : SA = 95 : 5. Under UV irradiation, the crystal turned red (Figure 4.14). Because BA (6) does not exhibit photochromism, the observed colour change was attributed to the twisted conformation of SA (7). Crystal structure analysis revealed a large residual electron density around H2 on the C2 atom

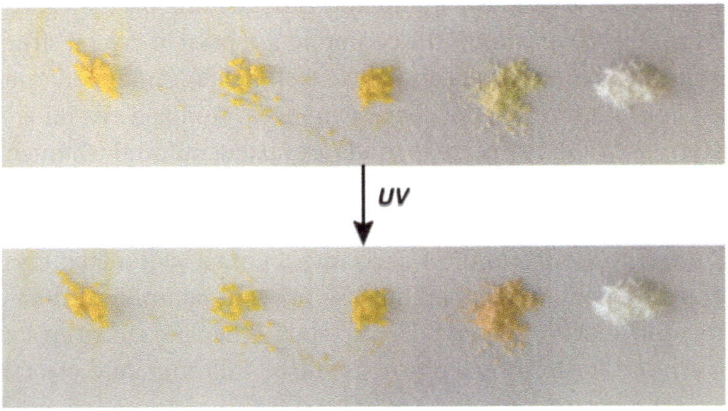

Fig. 4.14: Mixed crystals of BA (6) and SA (7). The ratios are 0 : 100, 5 : 95, 50 : 50, 95 : 5, and 100 : 0 from left to right. Only the fourth sample from the left exhibits photochromism under UV irradiation.

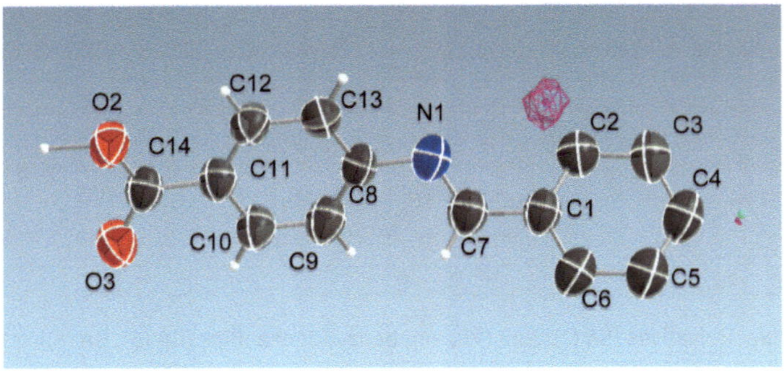

Fig. 4.15: Difference Fourier map of mixed crystal comprised of BA (6) and SA (7). The residual electron density (red nest) is more than 0.3 e/Å^3 around C2 atom.

of BA (6) (Figure 4.15), which was assigned to the OH group of SA (7) with 5.5% occupancy. Therefore, this was modelled as a disordered structure of BA (6) and SA (7) at the same site. In this mixed crystal, the precise control of the molecular conformation of SA (7) was achieved using the BA (6) crystal structure as a template.

A similar mixed crystal strategy was also applied to N-salicylidene-2-aminopyridine, SAP (9) (Figure 4.13). SAP (9) lacks intramolecular H to H steric repulsion because it has a pyridine moiety instead of a benzene ring, and usually exhibits the planar conformation in the crystal [46–48]. In order to force this molecule into the twisted conformation, it was mixed with *N*-benzylideneaniline, BA (8), which exhibits the twisted conformation with a dihedral angle of 64.7(1)°. As expected, the BA (8)/SAP (9) mixed crystal exhibited an isomorphic crystal structure to BA (8). However, SAP (9) could not be observed experimentally owing to its low occupancy. This mixed crystal exhibited photochromism (Figure 4.16), indicating that SAP (9) adopted the twisted conformation like BA (8) to induce photochromism. In these two cases, it is demonstrated that the mixed crystal formation strategy is effective to control the molecular conformation precisely in order to give a specific property such as photochromism to the crystal.

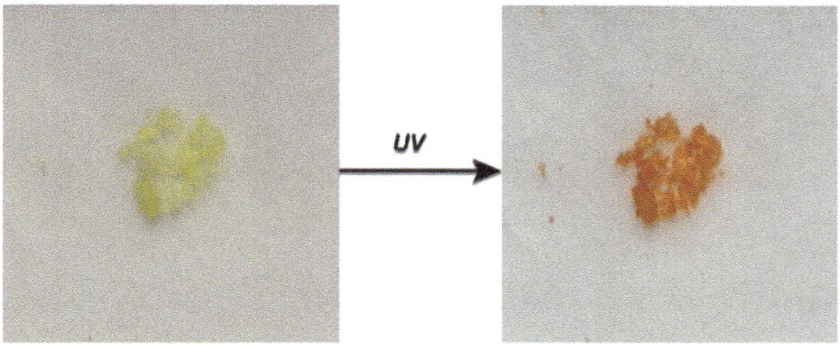

Fig. 4.16: Photographs of the mixed crystal composed of BA (8) and SAP (9) before and after UV irradiation.

4.5.3 Photochromism switching by phase transition

Co-crystal and mixed crystal formation are effective ways to change or precisely control the molecular conformation of a molecule, and have been utilised to control the photochromic properties of several crystals, as described above. A disadvantage of such approaches is that the photochromism cannot be altered after the crystal is formed. Thus, one way to extend the functionality of these photochromic crystals is to change the molecular conformation in the crystal via an external stimulus. Towards this end, an effective approach is to use dehydration/hydration of the crystal to change the crystal structure. Yellow, non-photochromic monohydrate crystals were obtained from the co-crystals of SA (3) and *N*-(4-pyridyl)acetamide from a methyl acetate solution (Figure 4.17). Upon heating to 40 °C, the crystal dehydrated to the yellow, anhydrous phase, and it exhibited photochromism whereby the colour turned orange after UV irradiation. When the anhydrous crystal was rehydrated to form the monohydrate under humidity (97% R.H.), the photochromism vanished. The crystal structure analysis revealed that the crucial dihedral angle in the monohydrate phase crystal was 27.1(1)°, and in the anhydrous phase this increased to 49.1(1)°, indicating significant molecular twisting owing to the phase transition (Figure 4.18). The crystal of the anhydrous phase was obtained by a recrystallisation from THF. The PXRD pattern of the recrystallised sample was exactly the same as that of the anhydrous crystal obtained by dehydration. A similar observation was reported by Kawato et al. for a synthetic SA dimer [49]. In that case, the desolvation and solvation of the crystalline methanol molecule altered the photochromism of the crystal, as confirmed by UV/Vis spectroscopy. However, the crystal structures were not reported and therefore, the mechanistic details are not clear. In this section, it has been demonstrated that the photochromic properties of SA crystals could be altered by heating or by changing the humidity. The reason is clearly explained by crystal structure analyses that showed that the dehydration/hydration phase transition considerably altered the molecular conformation.

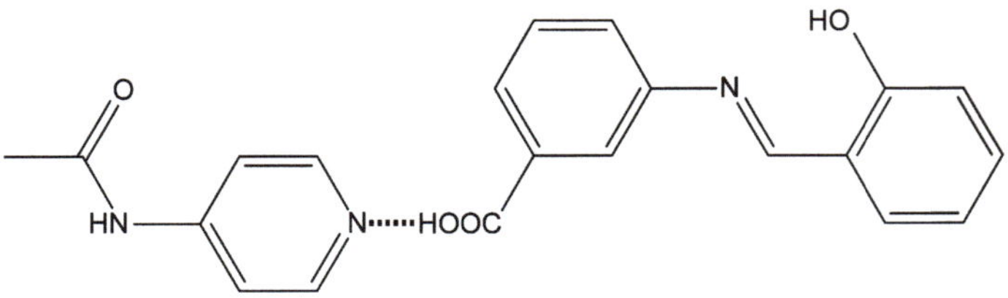

Fig. 4.17: Acid-base co-crystal between SA (3) and *N*-(4-pyridyl)acetamide.

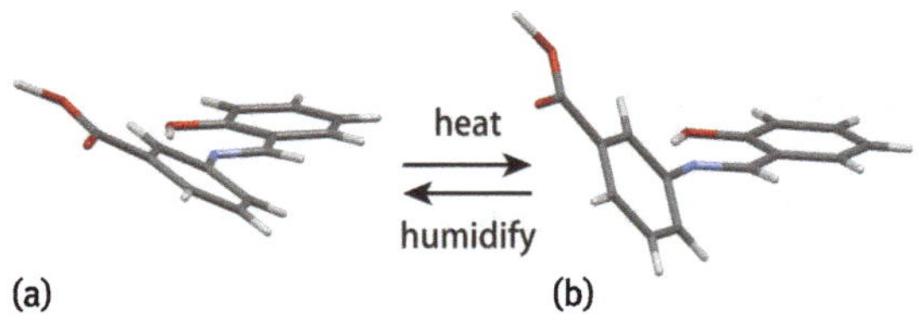

Fig. 4.18: The dynamic conformational transformation of SA in (a) the co-crystal hydrate of SA (3) with *N*-(4-pyridyl)acetamide and (b) the anhydrous form.

4.6 Conclusions

This chapter provides a brief overview of photochromism in SA derivatives. Photochromism is an interesting phenomenon in crystals that can be controlled using various crystal engineering approaches and by changing the molecular conformation, using, for example, co-crystal and mixed crystal techniques. Furthermore, this property can be dynamically switched by changing the crystal structure through dehydration/hydration. In SA derivatives, there is a strong correlation between molecular conformation and photochromism. Because of this feature, the photochromic properties can be controlled by altering the molecular conformation, changing the crystalline environment via crystal engineering techniques. It is also clear that these property modifications of SA crystals can be realised by utilising the flexible nature (softness) of SA molecular conformation. Crystal engineering targeting to such softness of organic crystal and molecule is relatively unexplored. Further crystal engineering research for various chromic organic crystals is expected to be fruitful.

Bibliography

[1] Irie M. Diarylethenes for memories and switches. Chem. Rev. 2000, 100, 1685–1716.

[2] Irie M, Fukaminato T, Matsuda K, Kobatake S. Photochromism of diarylethene molecules and crystals: memories, switches, and actuators. Chem. Rev. 2014, 114, 12174–12277.

[3] Matsumoto S, Higashiyama T, Akutsu H, Nakatsuji S. A functional nitroxide radical displaying unique thermochromism and magnetic phase transition. Angew. Chem. Int. Ed. 2011, 50, 10879–10883.

[4] Naumov P, Ishizawa N, Wang, J, Pejov L, Pumera M, Lee SC. On the origin of the solid-state thermochromism and thermal fatigue of polycyclic overcrowded enes. J. Phys. Chem. A 2011, 115, 8563–8570.

[5] Paley MS, Meehan EJ, Smith CD, Rosenberger FE, Howard SC, Harris JM. Synthesis and characterization of a novel betaine dye: 2,4-dimethyl-6-(2,4,6-triphenyl-N-pyridinio) phenolate. J. Org. Chem. 1989, 54, 3432–3436.

[6] Wender PA, Jeffreys MS, Raub AG. Tetramethyleneethane equivalents: recursive reagents for serialized cycloadditions. J. Am. Chem. Soc. 2015, 137, 9088–9093.

[7] Shan GG, Li HB, Cao HT, Zhu DX, Li P, Su ZM, Liao Y. Reversible piezochromic behavior of two new cationic iridium(III) complexes. Chem. Commun. 2012, 48, 2000–2002.

[8] Pawle RH, Haas TE, Muller P, Thomas SW. Twisting and piezochromism of phenylene-ethynylenes with aromatic interactions between side chains and main chains. Chem. Sci. 2014, 5, 4184–4188.

[9] Kato M. Luminescent platinum complexes having sensing functionalities. Bull. Chem. Soc. Jpn. 2007, 80, 287–294.

[10] Kobayashi A, Hara H, Yonemura T, Chang HC, Kato M. Systematic structural control of multichromic platinum(ii)-diimine complexes ranging from ionic solid to coordination polymer. Dalton Trans. 2012, 41, 1878–1888.

[11] Yamada M, Kondo M, Mamiya J, Yu Y, Kinoshita M, Barrett CJ, Ikeda T. Photomobile polymer materials: towards light-driven plastic motors. Angew. Chem. Int. Ed. 2008, 47, 4986–4988.

[12] Mizuno H, Mal TK, Wälchli M, Kikuchi A, Fukano T, Ando R, Jeyakanthan J, Taka J, Shiro Y, Ikura M, Miyawaki A. Light-dependent regulation of structural flexibility in a photochromic fluorescent protein. Proc. Natl. Acad. Sci. USA 2008, 105, 9227–9232.
[13] Senier A, Shepheard FG. Studies in phototropy and thermotropy. I. Arylidine and naphthylidene-amines. J. Chem. Soc. Trans. 1909, 95, 1943–1955.
[14] Senier A, Shepheard FG, Clarke R. Phototropy and thermotropy. III. Arylideneamines. J. Chem. Soc. Trans. 1912, 101, 1950–1958.
[15] Manchot W. Phenol and quinone isomerism in Schiff's bases of aromatic hydroxyaldehydes. Liebigs Ann. Chem. 1912, 388, 103–135.
[16] Cohen MD, Schmidt GMJ, Flavian S. Topochemistry. VI. Experiments on photochromy and thermochromy of crystalline anils of salicylaldehydes. J. Chem. Soc. 1964, 2041–2051.
[17] Cohen MD, Schmidt GMJ. Photochromy and thermochromy of anils. J. Phys. Chem. 1962, 66, 2442–2446.
[18] Harris L, Kaminsky J, Simard R. The absorption spectrum of malachite green leucocyanide and the mechanism of the dark reaction after photolysis. J. Am. Chem. Soc. 1935, 57, 1151–1154.
[19] Ledbetter JW Jr. Spectroscopic evidence for the enol imine-keto enamine tautomerism of *N*-(*o*- and *p*-hydroxybenzylidene) anils in solution. J. Phys. Chem. 1966, 2245–2249.
[20] Ledbetter JW Jr. Infrared spectra of *N*-aryl imines of *o*-hydroxybenzaldehyde between 2000 and 1500 cm^{-1}. J. Phys. Chem. 1977, 81, 54–59.
[21] Fujiwara T, Harada J, Ogawa K. Solid-state thermochromism studied by variable-temperature diffuse reflectance spectroscopy. A new perspective on the chromism of salicylideneanilines. J. Phys. Chem. B 2004, 108, 4035–4038.
[22] Cohen MD. J, Schmidt GMJ, Flavian S. Topochemistry. VI. Experiments on photochromy and thermochromy of crystalline anils of salicylaldehydes. J. Chem. Soc. 1964, 2041–2051.
[23] Cohen MD, Hirshberg Y, Schmidt GMJ. Topochemistry. VII. The photoactivity of anils of salicylaldehydes in rigid solutions. J. Chem. Soc. 1964, 2051–2059.
[24] Cohen MD, Hirshberg Y, Schmidt GMJ. Topochemistry. VIII. The effect of solvent, temperature, and light on the structure of anils of hydroxynaphthaldehydes. J. Chem. Soc. 1964, 2060–2067.
[25] Bregman J, Leiserowitz L, Schmidt GMJ. Topochemistry. IX. The crystal and molecular structures of *N*-(5-chlorosalicylidene)aniline near 90 and 300 K. J. Chem. Soc. 1964, 2068–2085.
[26] Bregman J, Leiserowitz L, Osaki K. Topochemistry. X. The crystal and molecular structures of 2-chloro-*N*-salicylideneaniline. J. Chem. Soc. 1964, 2086–2100.
[27] Andes RV, Manikowski DM. Photochromism of salicylideneaniline. Appl. Opt. 1968, 7, 1179–1183.
[28] Kawato T, Koyama H, Kanatomi H, Isshiki M. Photoisomerization and thermoisomerization. I: unusual photochromism of *N*-(3,5-di-tert-butylsalicylidene)amines J. Photochem. 1985, 28, 103–110.
[29] Harada J, Uekusa H, Ohashi Y. X-ray analysis of structural changes in photochromic salicylideneaniline crystals. Solid-state reaction induced by two-photon excitation. J. Am. Chem. Soc. 1999, 121, 5809–5810.
[30] Johmoto K, Sekine A, Uekusa H, Ohashi Y. Elongated lifetime of unstable colored species by intermolecular hydrogen bond formation in photochromic crystals. Bull. Chem. Soc. Jpn. 2009, 82, 50–57.
[31] Mikami M, Nakamura S. First-principles study of salicylideneaniline molecular crystals: Tautomerization reaction involving intermolecular hydrogen bonds. Phys. Rev. B 2004, 69, 1–8.
[32] Zgierski MZ, Grabowska A. Photochromism of salicylideneaniline (SA). How the photochromic transient is created: A theoretical approach. J. Chem. Phys. 2000, 112, 6329–6337.

[33] Moustakali-Mavridis I, Hadjoudis E, Mavridis A. Crystal and molecular structure of some thermochromic Schiff bases. Acta Crystallogr. B 1978, 34, 3709–3715.

[34] Johmoto K, Ishida T, Sekine A, Uekusa H, Ohashi Y. Relation between photochromic properties and molecular structures in salicylideneaniline crystals. Acta Crystallogr. B 2012, 68, 297–304.

[35] Özek A, Albayrak Ç, Odabaşoğlu M, Büyükgüngör O. Crystallographic and conformational analysis of (E)-2-[(3-Chlorophenylimino)methyl]-4-methoxyphenol. J. Chem. Crystallogr. 2009, 39, 353–357.

[36] Koşar B, Albayrak Ç, Odabaşoğlu M, Büyükgüngör O. (E)-2-[(4-Chlorophenyl)iminomethyl]-5-methoxyphenol and (E)-2-[(2-chlorophenyl)iminomethyl]-5-methoxyphenol: X-ray and DFT-calculated structures. Acta Crystallogr. C 2009, 65, 517–520.

[37] Harada J, Ogawa K, Tomoda S. Molecular motion and conformational interconversion of azobenzenes in crystals as studied by X-ray diffraction. Acta Crystallogr. B 1997, 53, 662–672.

[38] Amimoto K, Kawato T. Photochromism of organic compounds in the crystal state. J. Photochem. Photobiol. C Photochem. Rev. 2005, 6, 207–226.

[39] Hadjoudis E, Yannakopoulou K, Chatziefthimiou SD, Paulidou A, Mavridis IM. Supramolecular control of photochromism in a β-cyclodextrin/Schiff base system. J. Photochem. Photobiol. A Chem. 2011, 217, 293–298.

[40] Koyama H, Kawato T, Kanatomi H, Matsushita H, Yonetani K. Utilization of crystal lattice cavities of host deoxycholic acid for achieving photochromism of guest salicylideneanilines in the crystal state. J. Chem. Soc. Chem. Commun. 1994, 579–580.

[41] Kawato T, Koyama H, Kanatomi H, Yonetani K, Matsushita H. *N*-Salicylidene-2-aminopyridine as a photochromatic guest for deoxycholic acid crystal lattice cavities. Chem. Lett. 1994, 665–668.

[42] Kawato T, Amimoto K, Maeda H, Koyama H, Kanatomi H. Photochromism of guest N-salicylideneaniline in the rooms of host bile acid derivatives. Mol. Cryst. Liq. Cryst. Sci. Technol. Sect. A. Mol. Cryst. Liq. Cryst. 2000, 345, 57–62.

[43] Hadjoudis E, Vittorakis M, Moustakali-Mavridis I. Photochromism and thermochromism of Schiff bases in the solid state and in rigid glasses. Tetrahedron 1987, 43, 1345–1360.

[44] Haneda T, Kawano M, Kojima T, Fujita M. Thermo-to-photo-switching of the chromic behavior of salicylideneanilines by inclusion in a porous coordination network. Angew. Chemie Int. Ed. 2007, 46, 6643–6645.

[45] Johmoto K, Sekine A, Uekusa H. Photochromism control of salicylideneaniline derivatives by acid-base co-crystallization. Cryst. Growth Des. 2012, 12, 4779–4786.

[46] Cohen MD. Topochemistry. XXVIII. System. 4-chloro-N-salicylideneaniline-4-bromo-N-salicylideneaniline. J. Chem. Soc. (B) 1968, 373–376.

[47] Hadjoudis E. Photochromism and thermochromism of *N*-salicylideneanilines and N-salicylideneaminopyridines. J. Photochem 1981, 17, 355–363.

[48] Hadjoudis E. Photochromic and thermochromic anils. Mol. Eng. 1995, 5, 301–337.

[49] Taneda M, Koyama H, Kawato T. Vapor switching of photochromism of methylenebis{*N*-(3,5-di-tert-butylsalicylidene)aniline} crystals. Chem. Lett. 2007, 36, 354–355.

Jubaraj B. Baruah

5 Quinoline derivatives for multi-component crystals: principles and applications

5.1 Introduction

Co-crystals are crystalline solid materials comprising two or more molecules in the same crystal lattice. Hydrogen bonds between a set of partner molecules (co-formers) are a prime factor for co-crystal formation. Generally, a minimum of two co-formers is required to form a co-crystal. In the absence of a crystal structure of a co-crystal, factors such as differences in spectroscopic or physical properties from the individual parent co-formers are important indicators of co-crystal formation. For example, shifts in infrared absorptions in the O-H stretching region, exchange or shift of labile hydrogen atoms reflected in fluorescence or NMR spectroscopy, colligative properties, phase-related properties, and powder XRD patterns, are some ways to affirm the formation of co-crystals. To aid with the deliberate construction of co-crystals, Etter suggested a few rules relating to the hierarchy of hydrogen bond formation that constitute a useful starting point for building hydrogen bonded assemblies [1]. A key rule suggests that a hydrogen bonded self-assembly utilises all good hydrogen-donors and acceptors. The strengths and directional properties associated with hydrogen bonds are important to provide stability to self-assemblies [2, 3]. Charge-assisted hydrogen bonds may occur when one or both of the hydrogen bond acceptor/donor carries/carry charge. In these circumstances, a linear geometry is conferred upon the hydrogen bond. The rationale of co-crystal design states that multiple hydrogen bonding sites in a molecule increases the complexity. Other crystallisation outcomes have increased demand for pre-designed approaches to constructing supramolecular assemblies, such as polymorphism and having non-equivalent molecules in the asymmetric unit [4–7]. The identification of synthons [8] in crystals is important to extend the range of possibilities of assembling molecules in a predictive manner. Over and above hydrogen bonding, combinations of different weak interactions often participate in the formation of supramolecular architectures. Further, solute-solvent interactions occurring in a solution during crystallisation may influence the conformations of molecules. Such conformational adjustment can yield different polymorphic forms. In addition, solvent molecules may become a part of the solid structure either as inclusion in cavities formed by the self-assembly of parent molecules or self-assemblies formed by the solvent with the host compound. Two component co-crystals are very common but predictable design of the assembly of three or more complementary hydrogen bonding groups during co-crystallisation is difficult. Weak hydrogen bonds

https://doi.org/10.1515/9783110464955-005

compete with conformational changes and such subtle effects influence the selectivity of hydrogen bond formation.

This chapter is dedicated to a discussion of non-covalently linked assemblies of isomeric quinoline in co-crystals possessing two or more components. Quinoline is a constituent of chincona alkaloids, and derivatives of quinoline are useful as drugs and advanced materials. Quinoline was initially obtained from the distillation of coal tar [9] and also extracted by the distillation of cinchonine, quinine and other alkaloids treated with caustic soda [10]. Although quinolines as individual compounds do not have many applications, quinoline derivatives do, having various applications in medicine [11–15], catalysis [16–18] and electronics [19]. The molecular structures of quinolines resemble naphthalene in that one carbon atom of the ring is substituted by nitrogen. There are two isomers of quinoline, having nitrogen in the 1-position of the ring, i.e. quinoline, whereas isoquinoline contains a nitrogen atom at 2-position (Figure 5.1).

1 2

Fig. 5.1: Chemical diagrams for quinoline (1) and isoquinoline (2).

From a supramolecular chemistry point of view, there are three important points about quinolines that require attention: (a) the nitrogen atom in these molecules confers the properties of a base, which is amenable for adduct formation with various hydrogen bond donors, (b) the molecules are planar, hence suitable for π-stacking interactions and (c) the quinoline units are made up of two fused, non-equivalent rings, hence are dipolar. These three aspects play a major role in guiding the supramolecular aspects of quinoline derivatives. From a CSD study [20], it was suggested that about 69% of the 8-functionalised quinoline compounds surveyed possess π–π stacking interactions. Quinoline based structures having π–π stacking interactions also have additional C–H···π interactions significantly contributing to their self-assembled structures. On the other hand, simple quinoline derivatives such as chloroquine provide extensive supramolecular chemistry through self-assembly via N-H···N interactions [21].

5.2 Supramolecular aspects of Quinolines

Quinoline and isoquinoline are liquids, and may be used as solvents. To begin with, the supramolecular chemistry of quinolines starts from solute-solvent interactions. While the present discussion is focused on the development of multi-component crystalline compounds based on quinolines, reference to similar chemistry of the structurally related compound pyridine is essential. Due to close structural similarities between quinoline (pK_a = 4.9) and isoquinoline (pK_a = 5.5) with pyridine

(pK_a = 5.2), they form hydrogen bond synthons in a similar fashion to pyridine [22]. However, the fused dipolar ring-system in quinoline molecules means additional factors need to be taken into account to explain their supramolecular assemblies. For example, the most common co-formers for quinoline are carboxylic acids. Co-crystals formed between carboxylic acids and quinolines are analysed from two perspectives: (a) the matching of pK_a values to form hydrogen bonds between partner molecules and (b) the ability of quinoline to disrupt self-assemblies of the parent carboxylic acids. The expected carboxylic acid-quinoline hydrogen bond motifs are shown in Figure 5.2. Among these three motifs, QA1-QA3, QA1 comprises a discrete N-H···O interaction, QA2 comprises an eight-membered cyclic synthon having two donors and two acceptors, and QA3 is a seven-membered synthon also comprising two donors and two hydrogen bond acceptors. According to graph-set notation, QA2 is representative of the $R_2^2(8)$ motif whereas QA3 is the $R_2^2(7)$ motif [1].

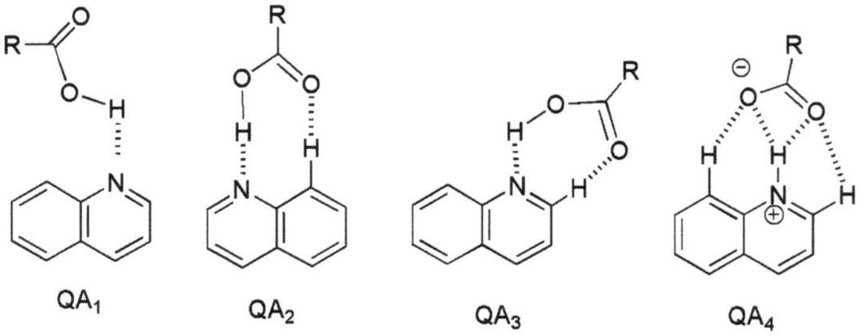

Fig. 5.2: Different hydrogen bonded motifs between quinolone/quinolinium and carboxylic acids.

The lowest-energy orientations of the different motifs of quinoline interacting with formic acid [22] are shown in Figure 5.3. The order of stability between the hydrogen bonded motifs is QA3 > QA2 > QA1. From the energy-optimised structures, the N···H distances are in the order QA1 (1.79 Å) > QA2 (1.76 Å) > QA3 (1.74 Å). Over and above these features, the QA2 and QA3 motifs have additional C-H···O interactions that are absent in QA1. Formic acid projects outwards from the quinoline ring in motif QA1,

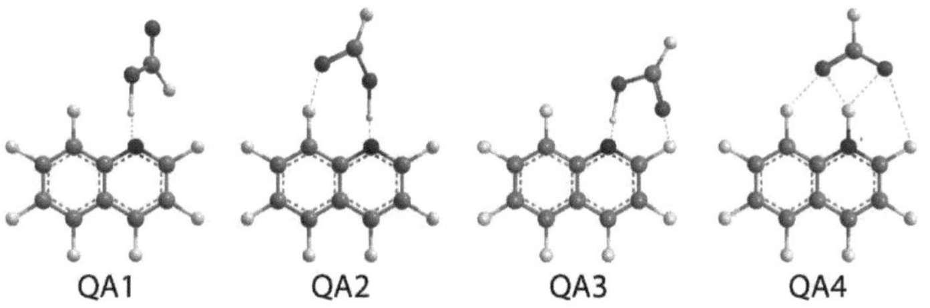

Fig. 5.3: Energy optimised structures of (a)–(c) quinoline: formic acid co-crystals and (d) the quinolinium formate salt.

while it is co-planar in QA2 and QA3. The optimised O-H···N bond angle of QA2 is observed as 166.6° while the analogous bond in QA3 is 179.8°. This is consistent with the stability of motif QA3 over QA2. Although motif QA1 has an O-H···N angle of 179.9°, i.e. the most linear, it is the least stable motif owing to the lack of a complementary C–H···O interaction. B3LYP/6-31+G** level calculations show the difference in energy between the QA3 and QA2 motifs as −0.63 kcal/mol. Motif QA4 is a salt, comprising a quinolinium cation and formate anion. The energy difference between the QA3 and QA4 motifs is −14.87 kcal/mol. Hence, from an energetic point of view, QA4 is the most favourable motif. From the optimised structure of QA4, the two constituents are in the same plane. The calculated N–H bond length was found to be 1.06 Å, which is much longer than a realistic value for an N–H bond; a typical N–H covalent bond length is approximately 0.80 Å. On this basis, the ionic structure suggested for QA4 in Figure 5.3 is considered unrealistic.

Conventionally, the pK_a values of an acid can be used to predict whether co-crystals or salts are formed with appropriate co-formers. A theoretical study on the quinolinium formate salt suggests geometric requirements should be a guiding factor when deciding to adopt a structure, rather than the energy minimum structure. Experiments carried out in solution readily provide evidence for the exchange of a proton between an acid and a base. The question on exchange is not as relevant for a solid state structure which is, to a first approximation is considered static in this context. Under these circumstances, hydrogen bond parameters can be employed to distinguish between a co-crystal and a salt, with the coupling of experimental results with those obtained by theory being highly desirable, especially in borderline cases. As the majority of the crystallisation processes are performed in solution, the pK_a's of quinoline and each co-former become important in determining the formation of a co-crystal or a salt. Empirically, a salt formed between an acid and a base occurs if ΔpK_a (pK_a base − pK_a acid) is greater than 2–3. Although this relationship is conventionally used to predict the formation of a co-crystal or a salt, it is not an absolute rule. This is due to the fact that many crystallisations are from non-aqueous media whereas ΔpK_a is calculated from acid dissociation constants measured in water. Hence, the crystallisation outcomes of the same system from different solvent systems can vary. On the basis of comparing large numbers of structures of co-crystals and salts derived from amines and carboxylic acids, a linear relationship between ΔpK_a and the probability of proton transfer in the pK_a range of −1 to +4 is observed. There exists a crossover point at $\Delta pK_a = 1.3$. There is a greater than 50% probability of salt formation beyond $\Delta pK_a = 1.3$. For general applicability, when $\Delta pK_a > 3$, salts are formed and co-crystals are formed when $\Delta pK_a < 3$ [23]. This background on acid-base reactivity can be analysed by considering some examples of co-crystallisation experiments with quinolines.

5.3 Cocrystals of Quinolines

4-Chlorobenzoic acid forms co-crystal with quinoline [24], **3** in Figure 5.4. The structure of this co-crystal is guided by an O–H···N hydrogen bond. The dihedral angle between the quinoline ring and the aromatic ring of the acid is 44.75 (4)°, as a result of which is there is no hydrogen bonded cyclic motif involving a C–H···O interaction. The same is true in the case of co-crystal **4** formed by fumaric acid and 2-methylquinoline [25] (Figure 5.4).

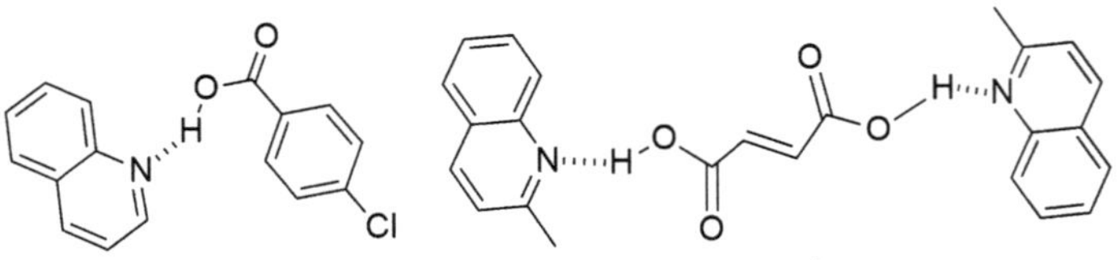

Fig. 5.4: Co-crystals **3** and **4**.

A series of salts of 2-methylquinoline with carboxylic acids such as 3,5-dinitrobenzoic acid, 3-hydroxy-2-naphthoic acid, oxalic acid and citric acid highlighted the interplay of different weak hydrogen bonds in the formation of the respective crystals [25].

The structure of isoquinoline 3-chloro-2-nitrobenzoic acid co-crystal **5** (Figure 5.5) presents interesting features [26], such as the crystallographic disorder in the atoms shared between the donor and acceptor atoms. The disordered hydrogen atoms have unequal probability between two positions with site occupancies at the N and O atoms being 0.30 and 0.70, respectively. On the other hand, isoquinoline forms a salt with 4-chloro-2-nitrobenzoic acid, **6** (Figure 5.5). Here, proton transfer occurs from OH to the nitrogen atom so that the acidic-H atom is located at the N site. Thus, the formation of cyclic or linear motifs comprising hydrogen bonds is dependent on the extent of proton transfer between the donor-acceptor atoms.

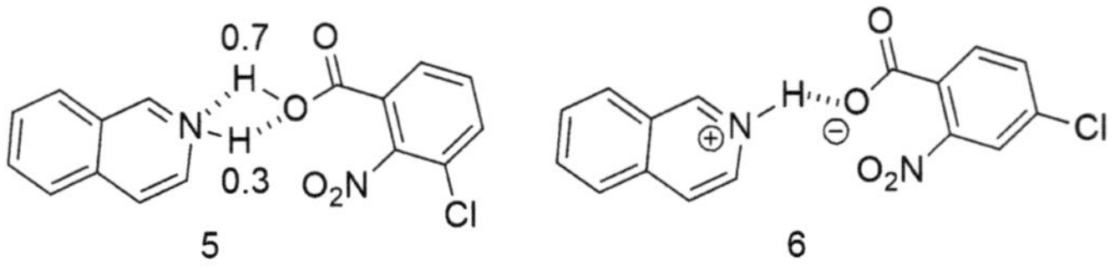

Fig. 5.5: Partial sharing of protons in co-crystal **5** and salt **6**.

These observations of different structural features occurring in a series of structurally closely related molecules open possible avenues for the development of host-guest chemistry of quinoline with functionalised carboxylic acids. Hence, there is room to vary the carboxylic acid to generate new and diverse multi-component systems. Func-

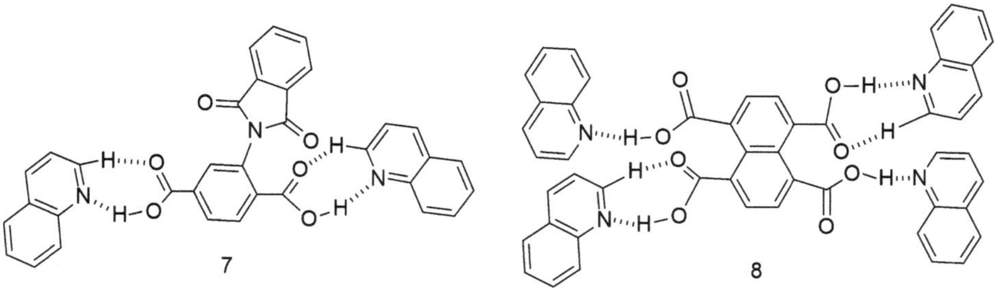

Fig. 5.6: Examples of co-crystals of quinoline with di- and tetra-carboxylic acids, **7** and **8**.

tional groups on quinoline can also play a dominant role in the generation of different assemblies with interesting properties in multi-component quinoline co-crystals, such as in quinoline molecules with multiple functional groups.

Aromatic imides connected to another co-former unit serve as polyfunctional host units that are suitable for co-crystal formation with quinolines. Among the imides, phthalimide or naphthalimides are dipolar having two carbonyl sites that are hydrogen bond acceptors. So, such a unit not only guides the self-assembly through dipole-dipole interactions but also, via the carbonyl bonds, guides the directional properties of such assemblies [27]. Co-crystal **7** (Figure 5.6) has two cyclic motifs mediated by C–H···O and N–H···O hydrogen bonds formed between each carboxylic acid group and a quinoline molecule. On the other hand, the co-crystal of quinoline with naphthalene-1,4,5,8-tetracarboxylic acid, **8**, has both isolated and $R_2^2(7)$ type hydrogen bond motifs [22] (Figure 5.6). There are two different O–H···N interactions involving two quinoline molecules. The two carboxylic acid groups forming the $R_2^2(7)$ rings are diagonally disposed across the fused-ring system. The other carboxylic acids are present at the ends of the other diagonal and each forms an O–H···N interaction. The molecular packing pattern of co-crystal **8** has quinoline molecules interacting with each other by C–H···π interactions, d(C···π) = 3.61 Å. The QA2 pattern was not observed in **8**, possibly due to the presence of strong π–π and O-π interactions among the quinoline molecules.

Naphthalenediimide, **9**, and tethered carboxylic acids, **10** and **11** (Figure 5.7) are effective hosts for a variety of co-crystals with quinolines [28]. For example, **10** forms two types of co-crystals with quinoline. One form has one molecule of **10** connected to

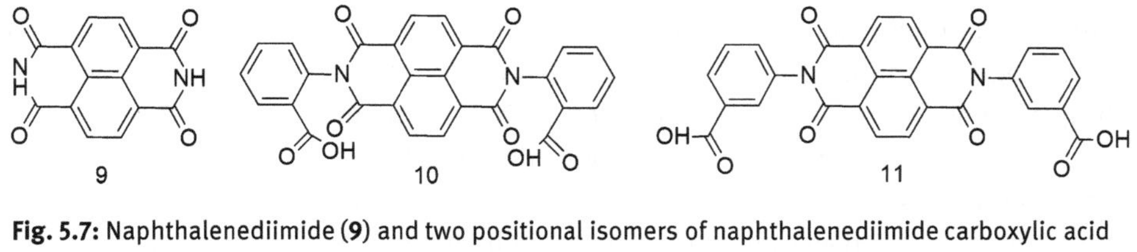

Fig. 5.7: Naphthalenediimide (**9**) and two positional isomers of naphthalenediimide carboxylic acid (**11** and **12**).

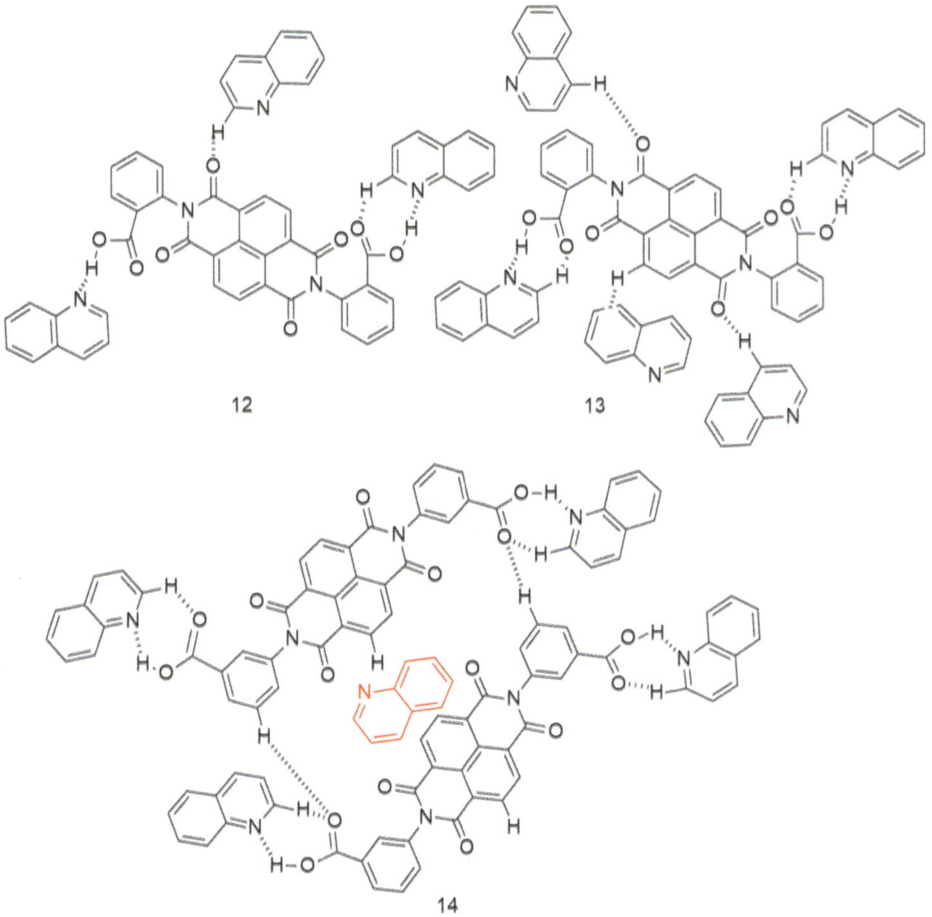

12 13 14

Fig. 5.8: A part of the hydrogen bonded assembly of the two pseudo-polymorphs **12** and **13**, and quinoline encapsulation in co-crystal **14**.

three quinoline molecules and the other has **10** connected to five quinoline molecules. The thermal stabilities of these co-crystals are different, as discussed below. They are also two pseudo-polymorphs of quinoline with **9**, i.e. **12** and **13** (Figure 5.8). In the 1 : 3 co-crystal, the two carboxylic acid groups holds two quinoline molecules by hydrogen bonds and the interstitial voids of the assemblies defined by these motifs are occupied by quinoline molecules, being connected through weak C–H···O interactions with naphthalendiimide or through C–H···π interactions rather than conventional hydrogen bonds (Figure 5.8). For comparison, **13** has two pairs of hydrogen bonds forming $R_2^2(7)$ rings between the quinoline molecules and carboxylic acids whereas in **12**, the quinoline molecules are associated with the carboxylic acid groups of the host molecule via isolated as well as a $R_2^2(7)$ type of hydrogen bonding pattern.

Thermogravimetry of co-crystal **12** shows the loss of three quinoline molecules in a single step at 185 °C, whereas **13** loses five quinoline molecules in two steps, i.e. loses two molecules of quinoline at 160 °C and the remaining three molecules of quinoline are lost at 190 °C. As **10** is a positional isomer of **11**, co-crystal formation by both with quinoline might be expected to follow a similar trend. However, only a 1 : 3 co-crystal, **14**, was observed in the case of **11**. Co-crystal **14** comprises an assembly of two host molecules of **11** connected to four molecules of quinoline through N–H···O interactions. This assembly incorporates a quinoline molecule as illustrated in Figure 5.8. There are two symmetry independent quinoline molecules bound to carboxylic acid groups through O–H···N and C–H···O interactions to make $R_2^2(7)$ arrangements around each carboxylic acid. The co-crystal **14** loses three molecules of quinoline on heating to 148 °C. The differences in the thermal stability of **12** and **14** arise from differences in their packing patterns. The above three examples of quinoline co-crystals have established that the hydrogen bonded patterns predicted from theoretical calculation, namely QA1 [discrete], QA2 [cyclic $R_2^2(8)$] and QA3 [cyclic $R_2^2(7)$] motifs illustrated in Figure 5.3 are observed in the examples **12–14**.

Consistent with the small energy differences between motifs QA1-QA3, two solvates are formed by pyridine with pyromellitic diimide, **15** (Figure 5.9). Compound **15** forms two different types of crystals, one with a 1 : 2 host:guest ratio and the other with a 1 : 3 ratio [29]. The crystallisation of **15** from neat pyridine provided **16a**, whereas crystallisation of **15** from toluene with a limited amount of pyridine provided **16b** (Figure 5.10). Each co-crystal comprises two symmetry independent diimide co-formers in their respective asymmetric units and has two pyridine molecules bound to carboxylic acid groups. The mode of association between co-formers in **16a** is guided by $R_2^2(7)$ hydrogen bonds, whereas **16b** has one symmetry independent diimide with the same pattern but with an additional pyridine linked via a C–H···N interaction. The other molecules of **16b** have two isolated O–H···N interactions along with one pyridine molecule associated via a C–H···N interaction.

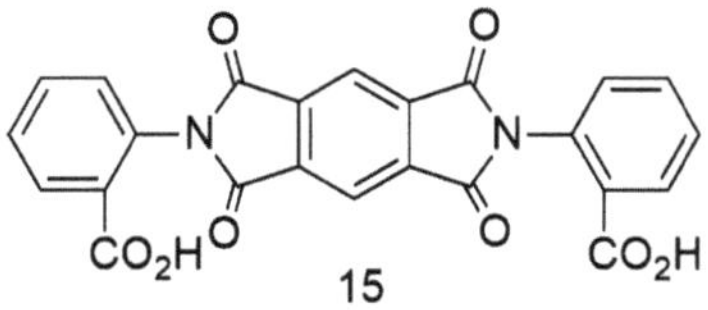

15

Fig. 5.9: A pyromelletic diimide based dicarboxylic acid host, **15**.

The thermal properties of the two pyridine co-crystals are different: co-crystal **16a** loses four pyridine molecules with respect to two host molecules in a single step at 160 °C, whereas co-crystal **16b** loses six pyridine molecules per two host molecules at 110 °C. This difference is a reflection of the different supramolecular motifs present in the co-crystals. The interplay among weak interactions guiding pyridine-carboxylic acid association is crucial in the constructions of a wide range of supramolecular assemblies [30, 31].

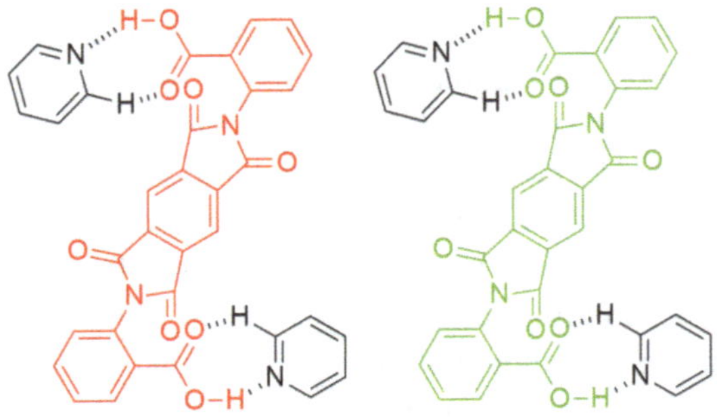

16a

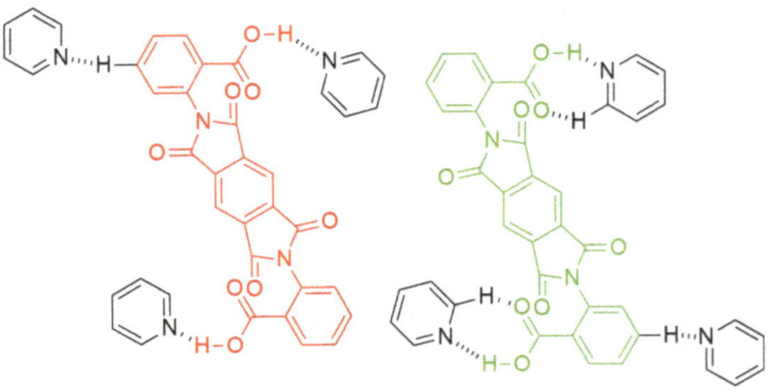

16b

Fig. 5.10: Supramolecular association in two co-crystals, **16a** and **16b**.

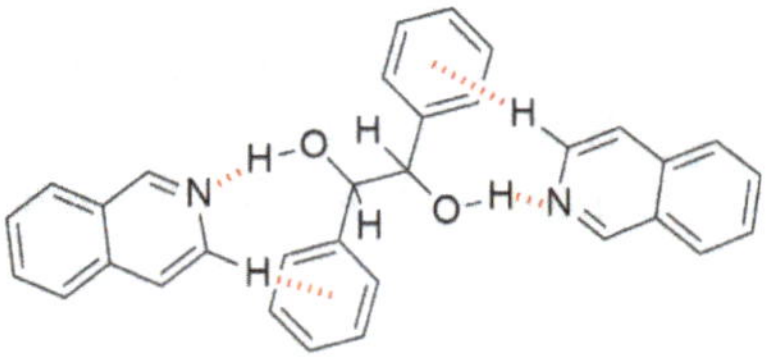

17

Fig. 5.11: Co-crystal of quinoline with *meso*-1,2-diphenyl-1,2-ethanediol, **17**.

Since quinolines form co-crystals with weak acids it is expected that they will also form co-crystals with acidic, OH-containing compounds, and this is borne out by experiment. Since the alcohol-O–H···N hydrogen bond is relatively weak with respect to a carboxylic acid-O–H···N hydrogen bond, additional, supportive interactions are essential to stabilise a co-crystal containing quinoline and an alcohol. For example, in the co-crystal formed between quinoline and meso-1,2-diphenyl-1,2-ethanediol, **17** (Figure 5.11), besides N–H···O hydrogen bonds, the system gains extra stability via C–H···π interactions [32].

Similar features, over and above O–H···N hydrogen bonds, are found in a 2 : 1 co-crystal formed between quinoline and **18** (Figure 5.12). The self-assembly of this co-crystal is primarily guided by O–H···N hydrogen bonds but features several other

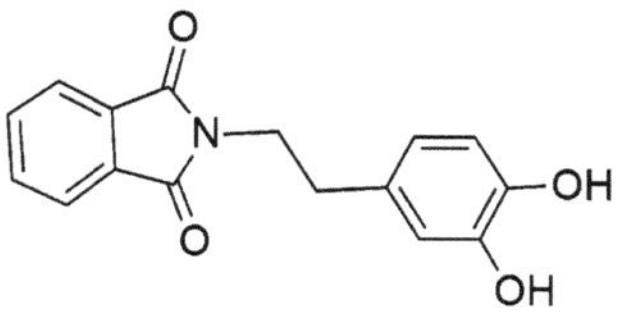

18

Fig. 5.12: Pyromellitic diimide tethered dihydroxybenzene, **18**.

supramolecular interactions. There is marked difference in the host-guest interactions for the independent quinoline molecules [33], with one quinoline molecule being involved in η^2-type C–H···π interactions and other participating in η^1-type C–H···π interactions, as illustrated in the packing pattern of the co-crystal shown in Figure 5.13.

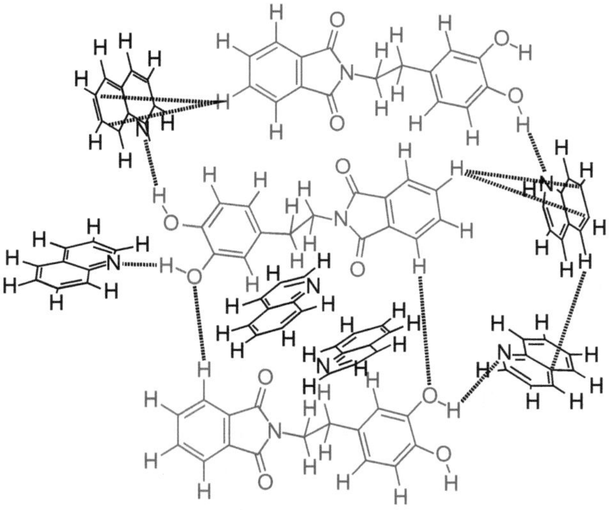

Fig. 5.13: Part of the molecular packing of **19** showing the weak interactions between the host and quinoline molecules.

The combined contributions of hydroxy and carboxylic groups in the self-assembly process of co-crystals of quinoline with hydroxy-carboxylic acid-containing molecules result in the formation of interesting channel-like architectures [34]. For example, dihydroxy-dicarboxylic acid **20** (Figure 5.14) functions as the host in **21**, a 1 : 6 co-crystal with quinoline. The host defines a 3D supramolecular architecture that contains 2D rectangular channels along the *a*-axis of approximate dimensions 11 × 11 Å. The channels are formed by the layered arrangement of host connected to a set of quinoline molecules. With the aid of weak interactions, a set of quinoline molecules is sandwiched inside these channels (Figure 5.15a). Another set of quinoline molecules participates in the enhancement of the supramolecular architecture from 2D to 3D through weak interactions (Figure 5.15b).

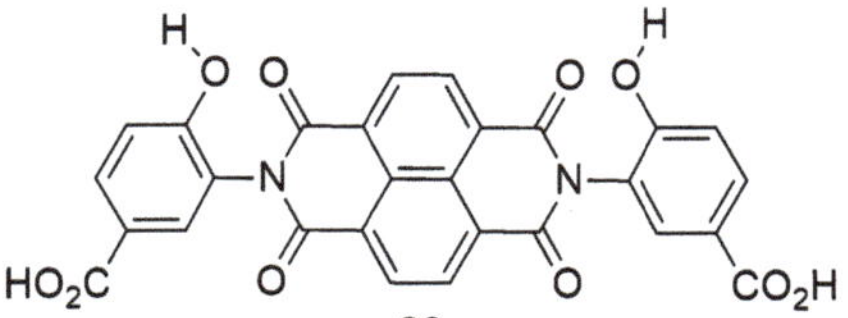

20

Fig. 5.14: A naphthalenediimide tethered to hydroxycarboxylic acid, **20**.

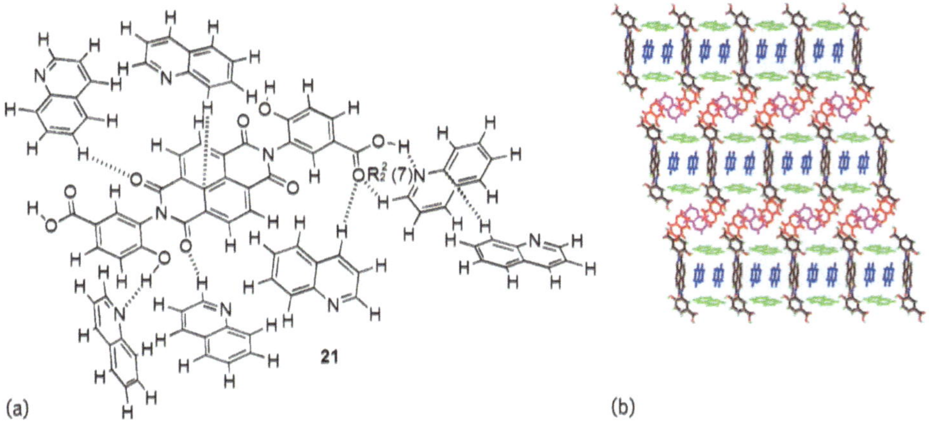

Fig. 5.15: (a) Host-guest interactions in a part of crystal structure of **21** and (b) 3D-assembly of **21** containing guests inside the channels.

The hydrogen-acceptor propensity of quinoline towards multiple functional groups is demonstrated in the co-crystal of 5-nitroquinoline with indole 3-acetic acid [35]. In this case, a 2 : 1 co-crystal, **22**, was formed in which two quinoline nitrogen atoms of two independent molecules hydrogen bond to the carboxylic acid and amine of the indole (Figure 5.16a) to form a ternary aggregate. In this example, the rings of host and guest are π-stacked. 5-Nitroquinoline and indole-2-carboxylic acid also form a 1 : 1 co-crystal, **23**, in which there is a heterosynthon linking the co-formers and a homosynthon to self-assemble a pair of two-molecule aggregates (Figure 5.16b) to generate a quaternary system. This example of different self-assembly patterns for two similar co-crystals depicts a co-crystal system that expands the domain of sub-assemblies to form a robust, larger assembly. Out of **22** and **23**, domain expansion is specific to **23**, and suggests that an adequate choice/ratio of partner molecules can be useful for the domain expansion of sub-assemblies.

Kemp's acid, cyclohexyl 1,3,5-tricarboxylic acid, having a locked conformation, generates tweezer-like structures [36]; the pure form forms a dimeric assembly, **24** (Figure 5.17a). This hydrogen bonded dimer functions as a template [37] to hold 8-aminoquinoline in the form of co-crystal, **25** (Figure 5.17b), or in the case of 8-hydroxyquinoline, in the form of a hydrogen bonded salt, **26** (Figure 5.17c). In the former case, the hydrogen bonds guide assembly formation, while in the latter, electrostatic interactions play a key role.

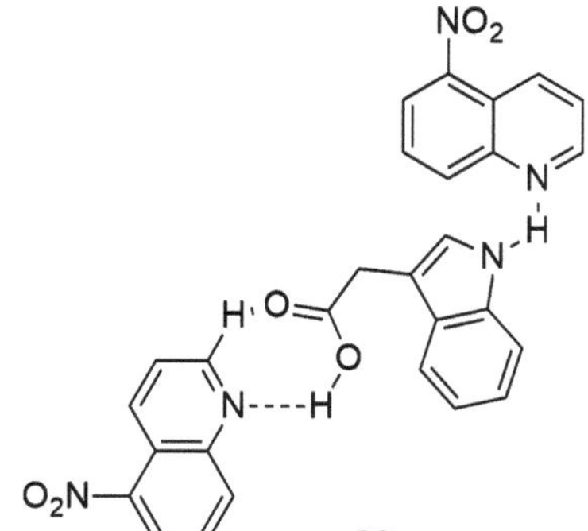

22

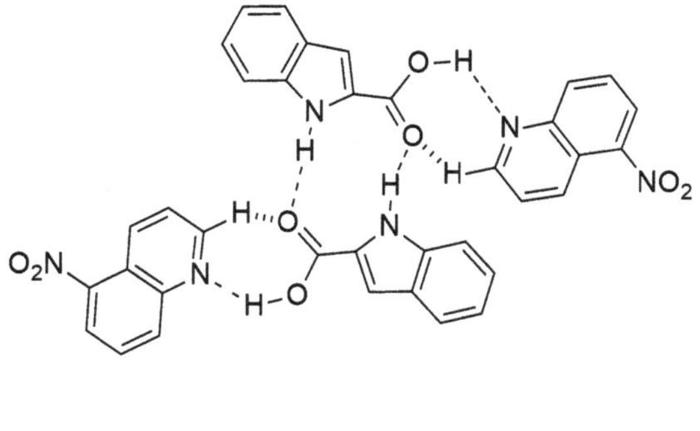

23

Fig. 5.16: Two different co-crystals of 5-nitroquinoline with indole-2-carboxylic acid: **22** (2 : 1) and **23** (1 : 1).

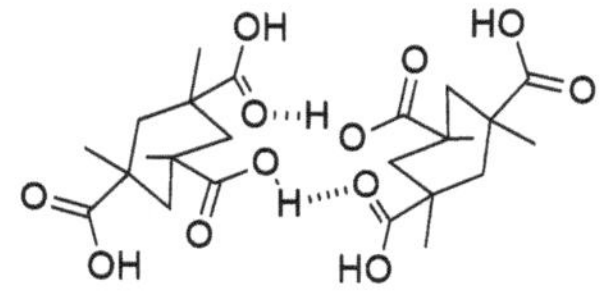

24

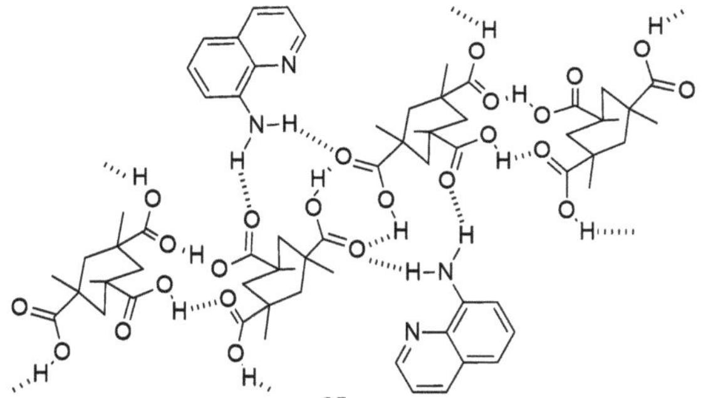

25

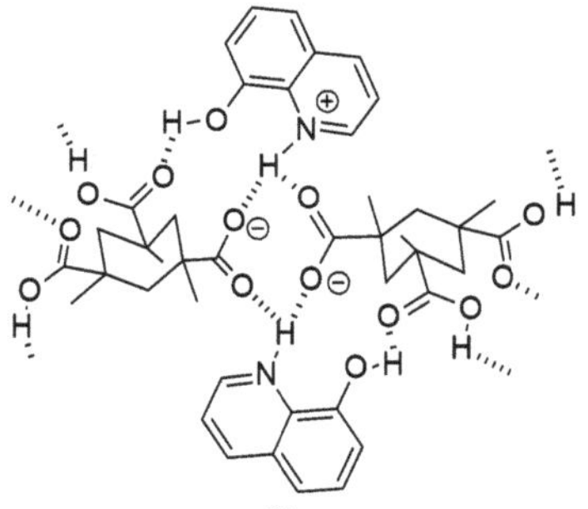

26

Fig. 5.17: (a) Dimeric assembly of Kemp's triacid (**24**), (b) self-assembly in co-crystal of 8-amino-quinoline with Kemp's triacid (**25**) and salt derived 8-hydroxyquinoline and Kemp's triacid (**26**).

5.4 Cocrystals with chiral compounds

Many optical induction processes involve co-crystallisation of chiral quinoline derivatives with racemic mixtures of optically active hydrogen-containing compounds. A vast literature of chiral quinoline derivatives exists [38–43], with and without isolation of co-crystals. Nevertheless, the introduction of a functional group such as amine or hydroxyl in quinolines significantly broadens the scope for generating various supramolecular assemblies. Quinine (**27**) and quinidine (**28**) (Figure 5.18) are two commonly available chiral compounds that are widely used in chiral separation [44]. Co-crystallisation of quinine with saccharin leads to a 2 : 1 salt co-crystal, **29** [45],

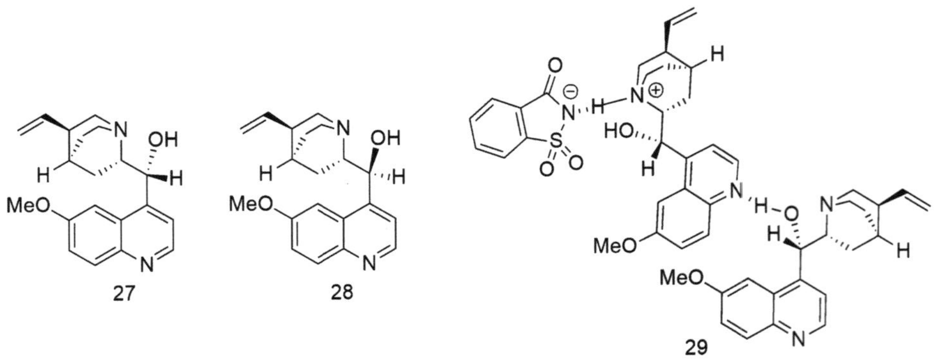

Fig. 5.18: Quinine (**27**), quinidine (**28**) and the salt co-crystal derived from 2 : 1 quinidine saccharine (**29**).

in which the tertiary amine on one of the quinine molecules is protonated and the saccharin-nitrogen is connected to this via a hydrogen bond. The three-component species sees the second, neutral quinine molecule connect via a hydroxyl-O–H···N hydrogen bond, as illustrated in Figure 5.19. The assembly of the just mentioned aggregates leads to zig-zag chain-like structure. The solubility of this salt co-crystal is greater than the marketed drug, the salt quinine sulfate.

Structural aspects of various multi-component systems of quinine with bianthryl [46] and binaphthyl (**30**) derivatives [47], e.g. the 1 : 2 salt of quinidine with binaphthyl (**31**) and the 2 : 1 salt derived from quinidine and bianthryl (**32**) (Figure 5.19), 4-

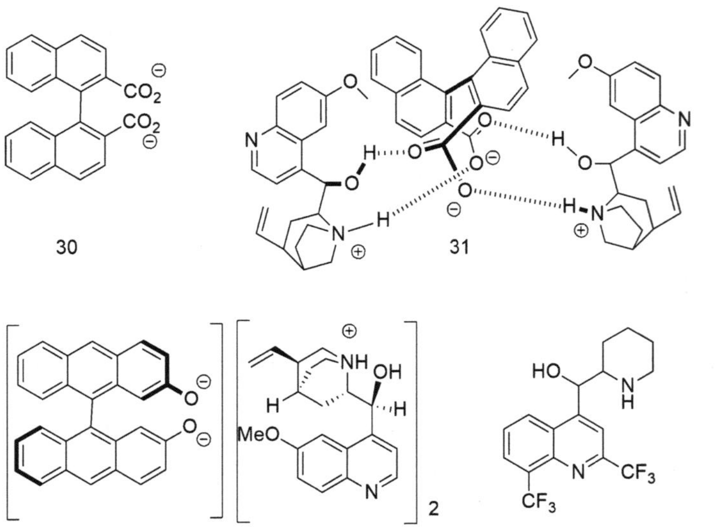

Fig. 5.19: Chemical diagrams for the binaphthyl anion (**30**), 1 : 2 salt of quinidine with binaphthyl (**31**), 2 : 1 salt derived from quinidine and bianthryl (**32**) and Mefloquine (**33**).

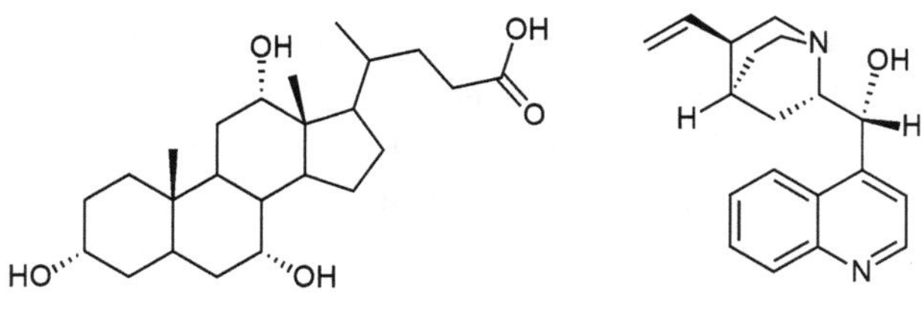

34 35

Fig. 5.20: Chemical diagrams for cholic acid (**34**) and cinchonine (**35**).

hydroxybenzoate [48], salicylic acid [49] and mandelic acid [50] are known. Mefloquine (**33**) (Figure 5.19) is another quinine host that readily forms salts. For example, Mefloquine forms crystalline salts with three isomers namely ortho-, meta- and para-nitrobenzoic acid [51]. In the respective packing patterns of the salt, the orientation of the quinoline ring was found to be guided by the anion.

Different types of spiral, supramolecular architectures are observed in the cocrystals of bile acids with chincona alkaloids [52], e.g. cholic acid (**34**) and chinconine (**35**) (Figure 5.20). Cholic acid has multiple numbers of hydrophobic, aliphatic rings which guide the arrangement during self-assembly. In the crystal, cholic acid mo-

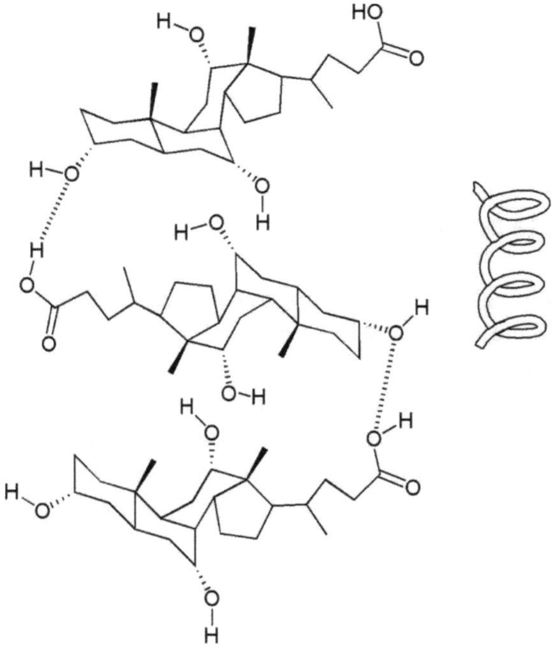

36

Fig. 5.21: A head-to-tail stacking of cholic acid molecules leading to a spiral assembly in **36** (hydrogen atoms are omitted).

lecules are arranged in a head-to-tail manner with connections between the carboxylic acid end (head) and the hydroxyl unit (end) to generate a spiral structure, **36** (Figure 5.21). Such arrangements can assemble molecules of chinconine to form different columnar structures, where the separation between the head and tail of neighbouring molecules can be moderated by changing the functional groups on the bile acid or in the chinconine unit.

Some quinoline derivatives tethered to an amide group have the ability to form gels [53]. The packing arrangements of such amides show different orientations around the guest molecules depending on the substituent. As shown in Figure 5.22, head-to-head or head-to-tail arrangements, encapsulating a water molecule, were observed in two derivatives.

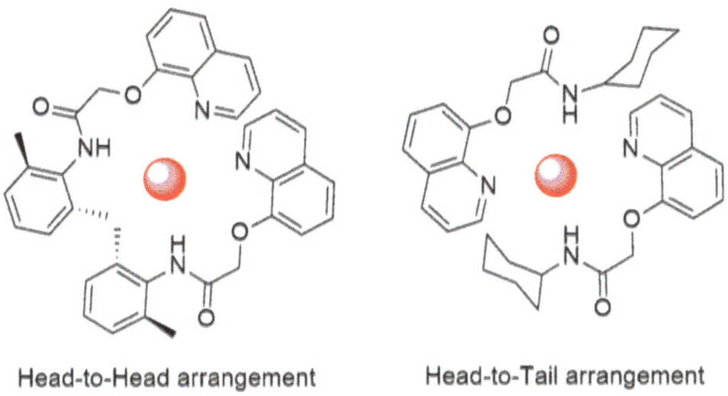

Fig. 5.22: Head-to-head and head-to-tail arrangements of water-assisted self-assemblies (red ball represents a water molecule) for quinoline derivatives tethered with an amide group.

5.5 Self-assemblies of amide bound quinolines

The quinoline derivative **37**, containing an amide group, has a hairpin-like structure. The self-assembly of the perchlorate salt of **37** has a channel-like structure, **38**, having a pore of diameter 7.36 Å (Figure 5.23) [54]. In this particular example, weak interactions, such as C–H···π and C–H···O interactions, contribute significantly to the stability of the channel-like structure. N-[2-(4-Methoxyphenyl)ethyl]-2-(quinoline-8-yloxy)acetamide is a suitable host for aromatic diols such as, 1,4-benzenediol, 1,5-naphthalenediol and 1,5-dihdroxynaphthalenediol [55, 56]; the latter forms a hydrated co-crystal.

The quinoline derivative, **39** (Figure 5.24) on reaction with tetrafluoroboric acid gives rise to a multi-component assembly **40** (Figure 5.24) comprising a salt encapsulating both water and dimethylformamide molecules [54]. The interesting feature of this reaction is that, despite having two sites available for protonation in **39**, only the oxyquinolinate site is protonated by tetrafluoroboric acid. On the other hand, in the diperchlorate salt, **41**, **39** is doubly deprotonated (Figure 5.24).

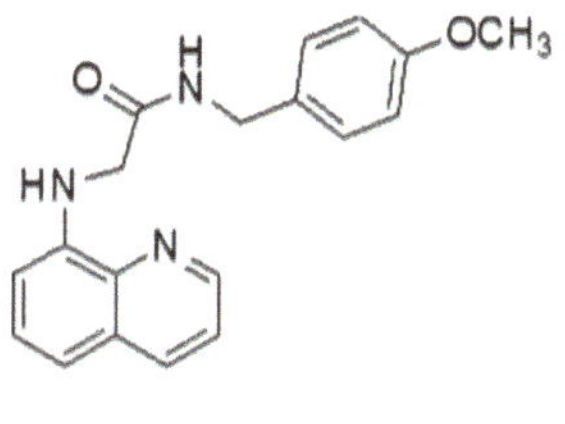

37

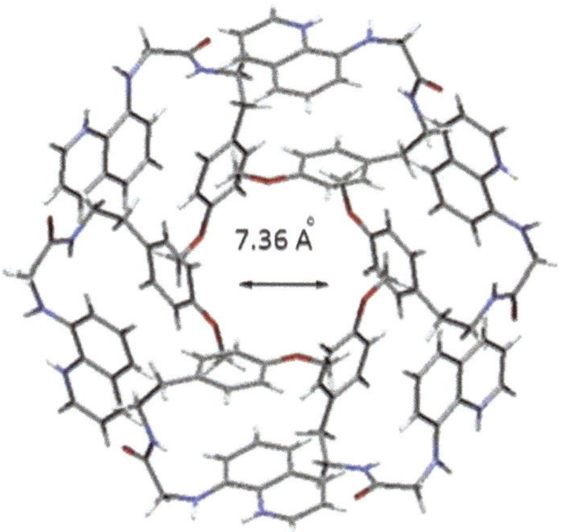

38

Fig. 5.23: Chemical diagram for **37** and the self-assembly in the perchlorate salt, **38** (for clarity, perchlorate anions not shown)

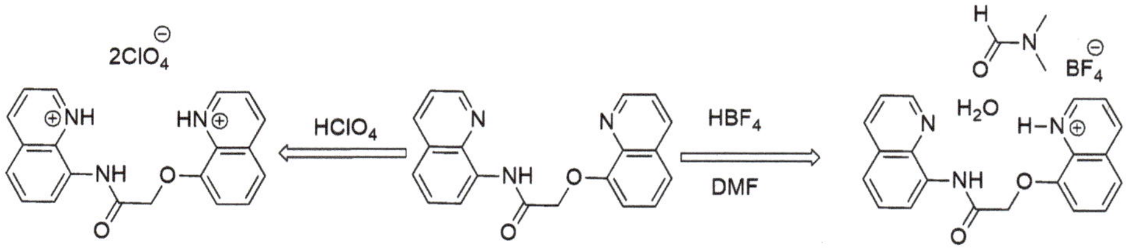

41 39 40

Fig. 5.24: Tetrafluoroborate (**40**) and perchlorate (**41**) salts derived from **39**.

Proton transfer from 1-phenyl-3-(quinolin-5-yl)urea, **42** (Figure 5.25) helps to distinguish isomeric dicarboxylic acids, e.g. maleic acid and fumaric acid, from positional isomers, e.g. terephthalic acid and phthalic acid [57, 58]. One isomer of each pair shows distinct colour change upon interaction with **42.** Co-crystal and salt formation have been established by single crystal X-ray diffraction as well as by solid state ^{13}C-NMR. The solid state NMR spectra of salt **44** show differences in the chemical shifts of the carboxylic acid quaternary carbons.
Perchlorate salts of 1-phenyl-3-(quinolin-5-yl) urea, 1-phenyl-3-(quinolin-8-yl) urea and quinoline-8-yl phenylcarbamate form self-assemblies that contain multiple numbers of symmetry non-equivalent molecules in the asymmetric units [54]. Solid state NMR proved useful for the identification of various heterosynthons and proton transfer processes in quinidine-carboxylic acid systems [59].

Quinoline derivatives **45** and **46** (Figure 5.26), are amide containing compounds that have multiple binding sites and are capable of binding various carboxylic acids. Compound **45** distinguishes hydroxy dicarboxylic acids from their non-hydroxy ana-

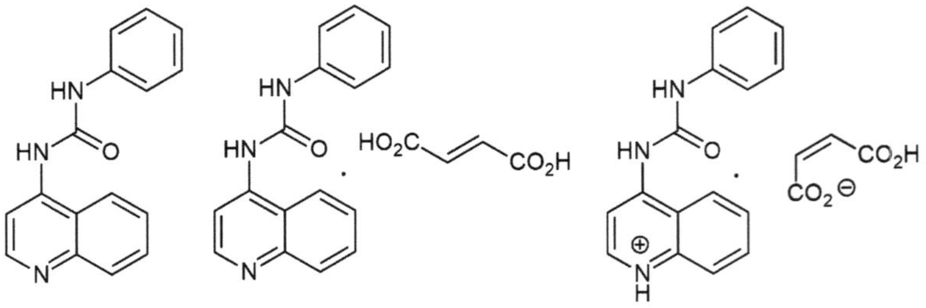

42 43 44

Fig. 5.25: Chemical diagrams for 1-phenyl-3-(quinolin-5-yl)urea (**42**), the co-crystal of **42** with maleic acid, **43**, and the salt derived from the reaction of **42** with fumaric acid, **44**.

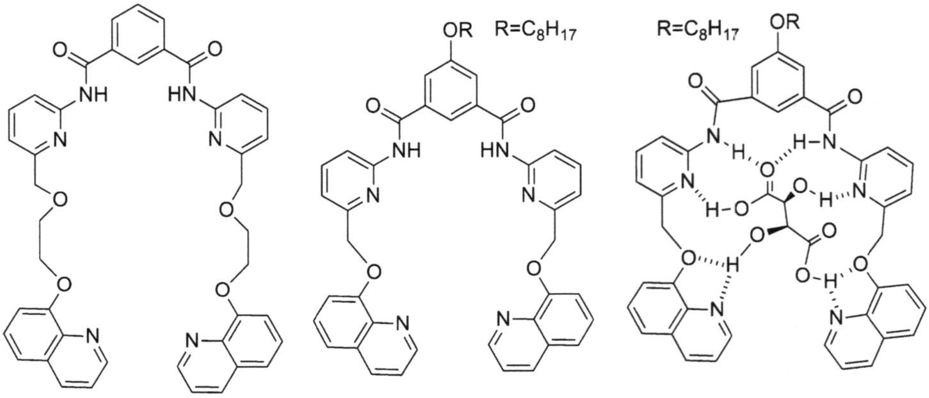

45 46 47

Fig. 5.26: Chemical diagrams for bis-quinoline derivatives, **45** and **46**, designed for dicarboxylic acid binding, and **47** showing the interaction of **46** with tartaric acid.

logues and also aliphatic dicarboxylic acids from aromatic diacids [60]. Compound **45** shows a high affinity towards tartaric acid and while interacting with tartaric acid, the fluorescence emission of compound **46** drastically changes, therefore enabling the detection of tartaric acid in the presence of various other acids. This recognition process uses complementary hydrogen bonds that accommodate tartaric acid within its arms in a scissor-like geometry [61], see **47** in Figure 5.26.

Inter-linking between 2,3-dihydroxynaphthalene appended quinoline derivatives with boron atoms opens an avenue for the construction of host systems, such as **48** (Figure 5.27). Host **48** is a good receptor for the biphosphate anion (**49** in Figure 5.27) and results in colour changes after selective assembly formation with biphosphate [62].

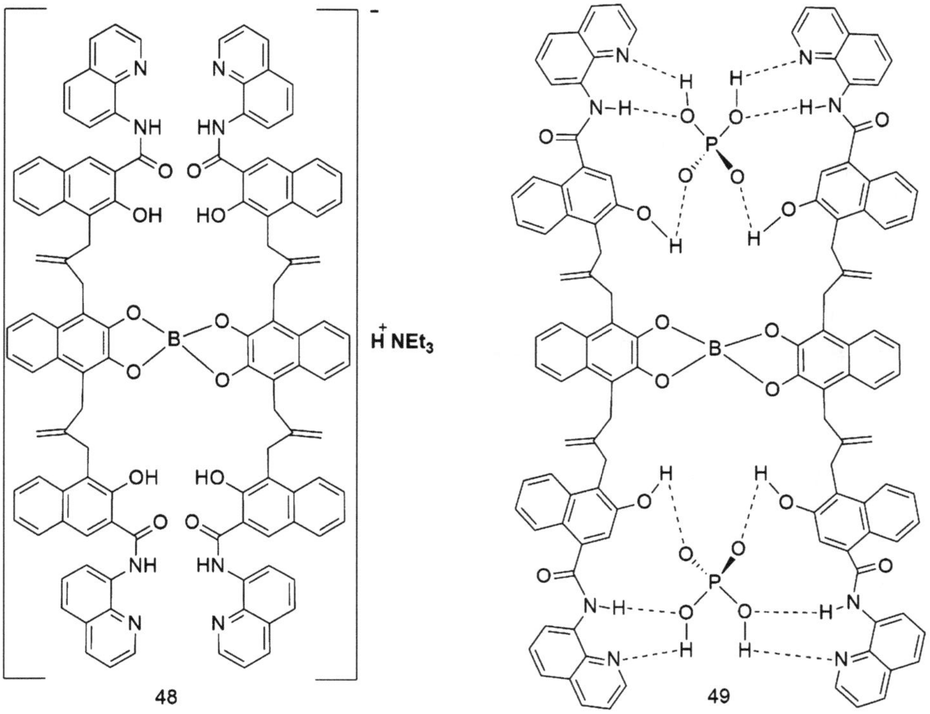

48 49

Fig. 5.27: Boron complex **48** and its host-guest complex with biphosphate anion **49**.

5.6 Interactions of quinolines with biomolecules

The discovery of host-guest complexes of DNA and quinolines was early in the 19th century. Monitoring the changes in the absorption spectrum of a solution of chloroquine and double-stranded DNA helped to ascertain the existence of such complexes [63, 64]. Such host-guest bindings were thought to play a role in anti-malarial action. However, a comparison of the dose dependence and binding ability of different quinoline derivatives with DNA suggests that DNA binding is independent of anti-malarial action. On the other hand, chloroquine inhibits the polymerization of toxic heme released during proteolysis of haemoglobin in the Plasmodium digestive vacuole [65] (Figure 5.28).

Based on such studies, a model for chloroquine blocking the hemozoin polymerisation in the absence of a protein was suggested [66, 67]. A schematic representation showing the end-capping by inclusion of a quinoline derivative to inhibit such polymerisation is shown in Figure 5.29. The proof obtained from such models demonstrated the complexation ability of quinolines with porphyrin complexes such as **50**, as shown in Figure 5.29. In this example, the quinoline compound form end-capped dimers of gallium porphyrin complexes preventing oligomer formation [68].

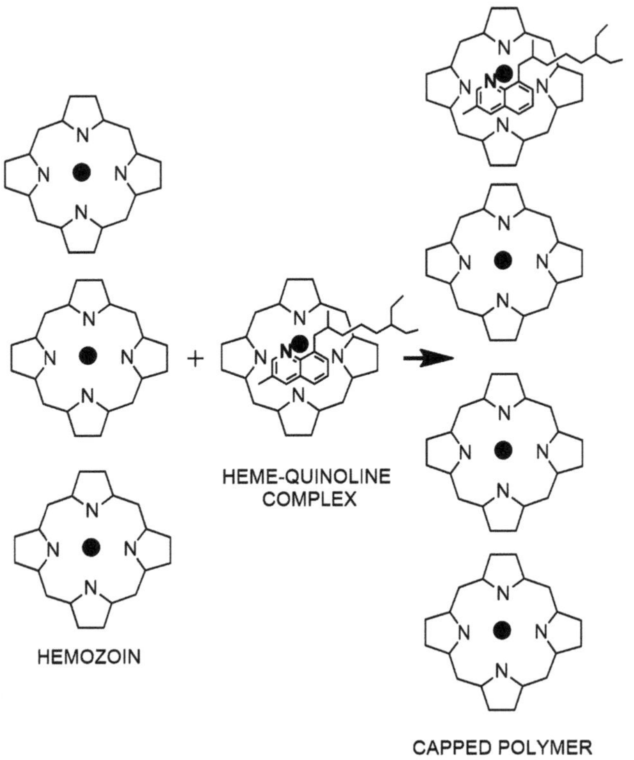

Fig. 5.28: A schematic representation of heme-quinoline interactions (five membered rings are pyrrole units and dark balls are iron ions).

5.7 Self-assemblies of quinoline containing metal complexes

Over and above these examples, quinoline based complexes serve as a template for aromatic guest inclusion through stacking interactions. For example, the thiocyanate isoquinoline nickel complex, **51** (Figure 5.30) forms inclusion complexes [69] with three positional isomers of xylene. In these examples, the xylene molecules occupy the spaces formed between the host molecules, which have extensive regions capable of forming π-interactions.

8-Aminoquinolines forms mixed ligand bi- or trinuclear complexes [70] with manganese. Depending on whether the reaction is performed in solution or in the solid state, mixed ligand binuclear (**52**) or trinuclear manganese(II) carboxylate (**53**) 2-nitrobenzoate complexes with 8-aminoquinolines are observed (Figure 5.31). In these two complexes no guest inclusion takes place, when 3-methylbenzoate and 8-aminoquinoline are reacted with manganese acetate in the solid state, inclusion of 8-aminoquinoline in the interstices of the resulting trinuclear mixed ligand complex is observed.

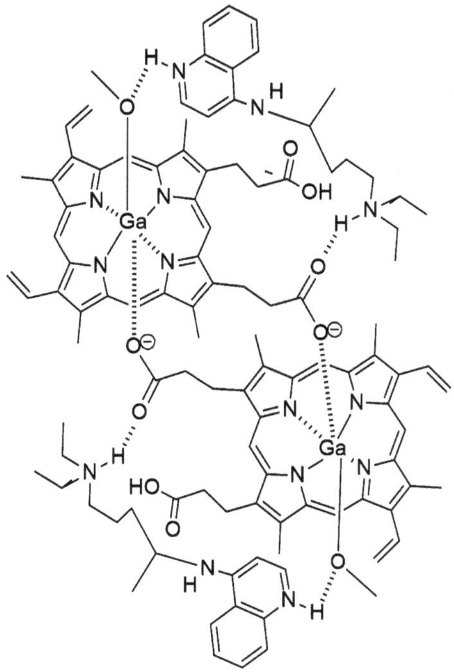

50

Fig. 5.29: End-capped dimeric porphyrin based complex, **50**.

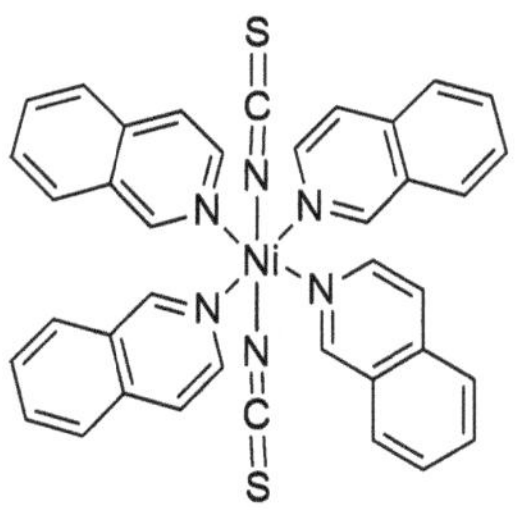

51

Fig. 5.30: Tetra(isolquinoline)bis(thiocayanato)nickel(II), **51**.

The quinolinium nickel(II) 2,6-dipicolinate complexes shown in Figure 5.32 show interesting reactivities and structural features in the solid state [71]. The self-assembled quinolinium bis(2,6-dipicolinate)nickel(II) species exhibit π–π stacking among the quinolinium cations, d(π–π) = 3.72 – 3.74 Å, but no π–π interactions with the dipicolinate residues of the cation are evident. The 8-aminoquinoline replaces the cations of the quinolinium bis(2,6-dipicolinate)nickel(II) by being protonated during the exchange reaction.

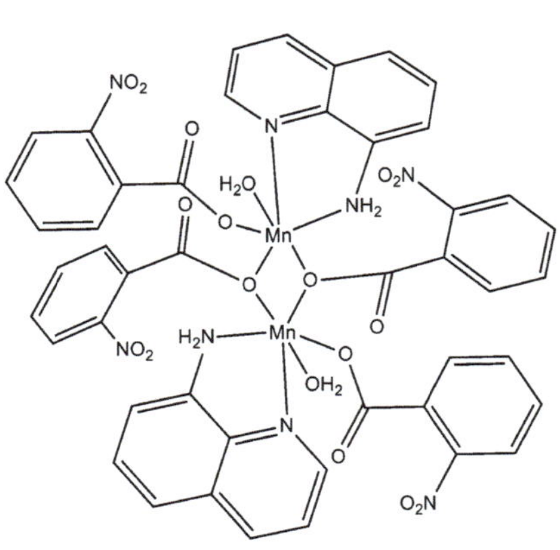

52

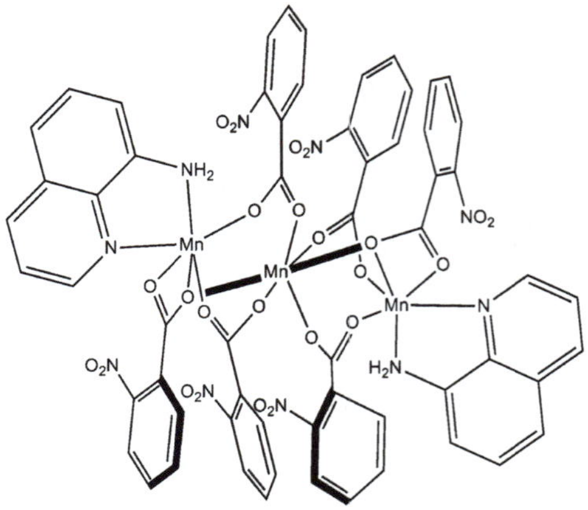

53

Fig. 5.31: (a) Binuclear (**52**) and trinuclear (**53**) mixed ligand complexes (Hydrogen atoms are omitted for clarity of presentation).

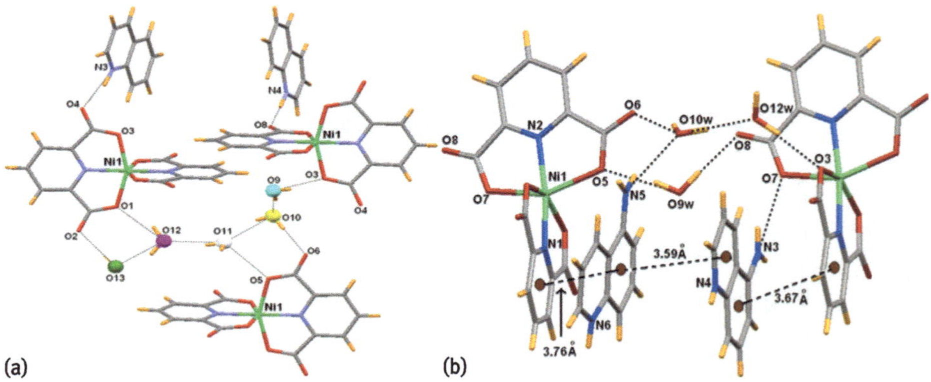

Fig. 5.32: The self-assembled structures of (a) $(HQ)_2[NiL_2].5\ H_2O$ and (b) $(H5AQ)_2[NiL_2].4\ H_2O$ (L = 2,6-dipicolinate, HQ= quinolinium cation and H5AQ = 5-aminoquinolinium cation).

Conclusions

Various self-assemblies of quinoline based compounds have clearly shown the importance of factors such as the (a) strength and directional properties associated with hydrogen bonds, (b) interplay of weak interactions, (c) packing requirements and (d) competitive energy of synthons. The resurgence of interest in multi-component systems suggests that quinoline derivatives will play a role in providing different interactions to connect multiple components. To assemble three chemical components by weak interactions gives rise to multiple possibilities. Therefore, the above factors can be employed to guide their formation as well as selection of partners. Especially in multi-component synthesis, there will be multiple possibilities to form homodimers and heterodimers, which may prevent a third component from participating in a hydrogen bonding scheme. To make a multi-component crystal, the choice of solvent also becomes crucial, as all the components need to be soluble in the same solvent. As the number of components increases, the solubility will decrease compared with binary or single component systems; the reverse may be true in electrostatically guided assemblies. For a sequential growth, the first two components must be more tightly bound than the third component; then a sequential addition can occur. Further, the strength of such synthons should not compete with the interaction of the solvent. Quinoline-carboxylic acid synthons present in multi-component systems are very sensitive to the substituent on the quinoline molecule, mode of preparation and supramolecular environment. These points make quinolines suitable as probes for the study of binding with different substrates for molecular recognition. Quinoline derivatives have widespread properties spanning gel formation to optical materials in which multi-component systems make a definite impact. The understanding of how quinoline based drugs interact with biological systems require further study as such drugs continue to evolve. Due to the analogous structural relationship of quinoline rings with the naphthalene unit, the aggregation-induced optical properties of quinolines have attracted the attention of material scientists. Finally, a focus upon nano-dimensional assemblies through quinoline based compounds will also continue to generate attention in molecular imaging, imprinting, and nano-dimensional assemblies such as optical and magnetic materials.

Bibliography

[1] Etter MC. Encoding and decoding hydrogen bond patterns of organic compounds. Acc. Chem. Res. 1990, 23, 120–126.

[2] Nangia A. in: Organic crystal engineering: Frontiers in crystal engineering (Eds: Tiekink, ERT, Vittal JJ, Zaworotko M, John Wiley & Sons, Ltd, Chichester, UK, 2010.

[3] Desiraju GR, Steiner T. The weak hydrogen bond: In structural chemistry and biology, Oxford University Press, New York, USA, 2001.

[4] Desiraju GR. On the presence of multiple molecules in the crystal asymmetric unit ($Z' > 1$). CrystEngComm 2007, 9, 91–92.

[5] Steed JW. Should solid-state molecular packing have to obey the rules of crystallographic symmetry? CrystEngComm 2003, 5, 169–179.

[6] Steiner T. Frequency of Z' values in organic and organo-metallic crystal structures. Acta Crystallogr. B 2000, 56, 673–676.

[7] Gavezzotti A. Structure and energy in organic crystals with two molecules in the asymmetric unit: causality or chance? CrystEngComm 2008, 10, 389–398.

[8] Desiraju GR. Supramolecular synthons in crystal engineering-a new organic synthesis. Angew. Chem. Int. Ed. Eng. 1995, 34, 2311–2327.

[9] Runge FF. Ueber einige produkte der steinkohlendestillation (On some products of coal distillation). Ann. Physik 1834, 31, 65–78.

[10] Gerhardt C. Untersuchungen über die organischen Basen. Annal. Chem. Pharm. 1842, 42, 310–313.

[11] Klingenstein R, Melnyk P, Leliveld SR, Ryckebusch A, Korth C. Similar structure-activity relationships of quinoline derivatives for antiprion and antimalarial effects. J. Med. Chem. 2006, 49, 5300–5309.
[12] Fang K-C, Chen Y-L, Sheu J-Y, Wang T-C, Tzeng C-C. Synthesis, antibacterial, and cytotoxic evaluation of certain 7-substituted norfloxacin derivatives. J. Med. Chem. 2000, 43, 3809–3812.
[13] Benkovic SJ, Baker SJ, Alley MRK, Woo Y-H, Zhang Y-K, Akama T, Mao W, Baboval J, Rajagopalan PTR, Wall M, Kahng LS, Tavassoli A, Shapiro L. Identification of borinic esters as inhibitors of bacterial cell growth and bacterial methyltransferases, CcrM and MenH. J. Med. Chem. 2005, 48, 7468–7476.
[14] Ablordeppey SY, Fan P, Li S,Clark AM, Hufford CD. Substituted indoloquinolines as new antifungal agents. Bioorg. Med. Chem. 2002, 10, 1337–1346.
[15] Nakamoto K, Tsukada I, Tanaka K, Matsukura M, haneda T, Murai N, Abe S, Ueda N, Miyazaki N, Asada M, Yoshimatsu K, Hata K. Synthesis and evaluation of novel antifungal agents-quinoline and pyridine amide derivatives. Bioorg. Med. Chem. Lett. 2010, 20, 4624–4626.
[16] Blaz E, Pielichowski J. Polymer-supported cobalt (II) catalysts for the oxidation of alkenes. Molecules 2006, 11, 115–120.
[17] Ma WA, Wang ZX, Zinc and aluminum complexes supported by quinoline-based *N*, *N*, *N*-chelate Ligands: synthesis, characterization, and catalysis in the ring-opening polymerization of ε-caprolactone and rac-lactide. Organometallics 2011, 30, 4364–4373.
[18] Zhang S, Sun WH, Xiao T, Hao X. Ferrous and cobaltous chlorides bearing 2,8-bis(imino)quinolines: highly active catalysts for ethylene polymerization at high temperature. Organometallics 2010, 29, 1168–1173.
[19] Curry RJ, Gillin WP. 1.54μm electroluminescence from erbium (III) tris(8-hydroxyquinoline) (ErQ)-based organic light-emitting diodes. Appl. Phys. Lett. 1999, 75, 1380–1382.
[20] Kalita D, Sarma R, Baruah JB. Formation of symmetry non-equivalent molecules in urea and carbamate derivatives: the role of anion. CrystEngComm 2009, 11, 803–810.
[21] Semeniuc RF, Reamera TJ, Smith MD. 8-Quinoline based ligands and their metallic derivatives: A structural and statistical investigation of quinoline π–π stacking interactions. New J. Chem., 2010, 34, 439–452.
[22] Singh D, Bhattacharyya P, Baruah JB. Structural studies on solvates of cyclic imide tethered carboxylic acids with pyridine and quinoline. Cryst. Growth Des. 2010, 10, 348–356.
[23] Ramon G, Davies K, Nassimbeni LR. Structures of benzoic acids with substituted pyridines and quinolines: salt versus co-crystal formation. CrystEngComm, 2014, 16, 5802–5810.
[24] Gotoh K, Katagiri K, Ishida H. 4-Chlorobenzoic acid-quinoline (1/1). Acta Crystallogr. 2010, E66, o3190.
[25] Jin S, Wang D, Huang Y, Fang H, Wang T, Fu P, Ding L. Hydrogen bonded supramolecular framework in organic acid–base adducts: Crystal Structures of five co-crystals/salts assembled from 2-methylquinoline with monocarboxylic acid, dicarboxylic acid, and tricarboxylic acid. J. Mol. Struct. 2012, 1017, 51–59.
[26] Gotoh K, Ishida H. Crystal structures of isoquinoline–3-chloro-2-nitrobenzoic acid (1/1) and isoquinolinium 4-chloro-2-nitrobenzoate. Acta Crystallogr. E 2015, 71, 31–34.
[27] Chen Z, Fimmel B., Wurthner F. Solvent and substituent effects on aggregation constants of perylene bisimide π-stacks: a linear free energy relationship analysis. Org. Biomol. Chem. 2012, 10, 5845–5855.
[28] Singh D, Baruah JB. Different solvates of two isomeric dicarboxylic acids with pyridine and quinoline. CrystEngComm 2009, 11, 2688–2694.
[29] Singh D, Baruah JB. Varieties in symmetry non-equivalent structural arrangements via solvation in 2-(3-methylene-1,3,7-trioxo-6-(2-carboxy-phenyl)-3,5,6,7-tetrahydro-1H-pyrrolo[3,4 f]isoindol-2-yl)benzoic acid. J. Mol. Struct. 2009, 937, 75–80.

[30] Santra R, Ghosh N, Biradha K. Crystal engineering with acid and pyridine heteromeric synthon: neutral and ionic co-crystals. New J. Chem. 2008, 1673–1676.

[31] Santra R, Biradha K. Two dimensional organic brick-wall layers as hosts for the inclusion and study of aromatics ensembles: Acid-pyridine and acid-carbonyl synthons for multi-component materials. Cryst. Growth Des. 2009, 9, 4969–4978.

[32] Olenik B, Boese R, Sustmann R. Cooperative and anticooperative effects in the co-crystals of mono- and diazanaphthalenes with meso-1,2-diphenyl-1,2-ethanediol. Cryst. Growth Des. 2003, 3, 175–181.

[33] Singh D, Baruah JB. Structural study on solvates of dopamine based cyclic imide derivatives. Cryst. Growth Des. 2011, 11, 768–777.

[34] Singh D, Baruah JB. Solid state assemblies of cyclic imides tethered hydroxy benzoic acids with pyridine and quinoline: toward the formation of channels and cavities. Cryst. Growth Des. 2012, 12, 3169–3180.

[35] Lynch DE, Misty N, Smith G, Byriel KA, Kennard CHL. Molecular co-crystals of carboxylic acids: XXXIII the crystal structure of the 1 : 1 adduct of indole-2-carboxylic acid with 5-nitroquinoline. Aust. J. Chem. 1998, 51, 813–818.

[36] Han J, Zang S-Q, Mak TCW. Controlled aggregation of multiple guanidinium ions through a hydrogen-bonded network assembly with deprotonated forms of Kemp's triacid. Chem. Eur. J. 2010, 16, 5078–5088.

[37] Smith G, Wermuth UD, White JM. Molecular recognition involving Kemp's triacid: selectivity towards the 8-substituted quinoline system as seen in the co-crystalline adducts with 8-aminoquinoline and 8-hydroxyquinoline. Chem. Commun. 2000, 2349–2350.

[38] Optical resolution via diastereomeric salt formation, ed. Kozma D, CRC Press, New York, 2002.

[39] Wilen SH, Eliel EL. Tables of resolving agents and optical resolution, Univ. of Notre Dame Press, Notre Dame London, 1971.

[40] Jacques J, Collet A, Wilen SH. Enantiomers, racemates and resolutions, Krieger Publishing Company, Malabar, Florida, 1994.

[41] Woodward RB, Cava PM, Ollis WD, Hunger A, Daeniker HU, Schenker K. The total synthesis of strychnine. Tetrahedron 1963, 19, 247–250.

[42] Larsen S, Lopez de Diego H, Kozma. Optical resolution through diastereomeric salt formation: the crystal structures of cinchoninium (*R*)-mandelate and cinchoninium (*S*)-mandelate at low temperature. Acta Crystallogr. B 1993, 49, 310–316.

[43] Bathori NB, Nassimbeni LR, Oliver CL. Quininium mandelates: a systematic study of chiral discrimination in crystals of diastereomeric salts. Chem. Commun. 2011, 47, 2670–2672.

[44] Báthori NB, Jacobs A, Nassimbeni LR, Sebogisi BK. Quininium malates: Partial chiral discrimination via diastereomeric salt formation. S. Afr. J. Chem. 2014, 67, 160–166.

[45] Bhatt PM, Ravindra NV, Banerjee R, Desiraju GR. Saccharin as a salt former. Enhanced solubilities of saccharinates of active pharmaceutical ingredients. Chem. Commun. 2005, 1073–1075.

[46] Toyota S, Shimasaki T, Tanifuji N, Wakamatsu K. Experimental and theoretical investigations of absolute stereochemistry and chiroptical properties of enantiopure 2,2′-substituted 9,9′-bianthryls. Tetrahedron: Asymmetry 2003, 14, 1623–1629.

[47] Jacobs A, Nassimbeni LR, Sayed A, Weber E. Resolution of 1,10-binaphthyl-2,2′-dicarboxylic acid with quinine: Structure of the intermediate (S)-1,10-binaphthyl-2,2′-dicarboxylate dihydrate diastereomeric salt. J. Chem. Crystallogr. 2011, 41, 854–857.

[48] Khan M, Enkelmann V, Brunklaus G. Heterosynthon mediated tailored synthesis of pharmaceutical complexes: a solid-state NMR approach. CrystEngComm 2011, 13, 3213–3223.

[49] Oleksyn BJ, Serda P. Salt-bridge formation by cinchona alkaloids: quininium salicylate monohydrate. Acta Crystallogr. B 1993, 49, 530–534.

[50] Gjerlov AB, Larsen S. Quininium (S)-mandelate. Acta Crystallogr. C 1997, 53, 1505–1508.

[51] Wardell SMSV, Wardell JL, Skakle JMS, Tiekink ERT. Structural study on melfoquinium salts with three isomeric *n*-nitrobenzoates (n = 2, 3 and 4). Z. Kristallogr. 2011, 226, 68–77.

[52] Sasaki T, Shizuki N, Hiraishi E, Hisaki I, Tohnai N, Miyata M. Construction of multi-component supramolecular architectures of bile acids and cinchona alkaloids through helical-pitch-synchronized crystallization. Org. Biomol. Chem. 2012, 10, 5985–5992.

[53] Karmakar A, Sarma RJ, Baruah JB. Structural aspects and properties of salt and inclusion compounds of 8-hydroxyquinoline based amides, CrystEngComm 2007, 9, 378–388.

[54] Kalita D, Baruah JB. Different spatial orientations of amide derivatives on anion coordination. CrystEngComm 2010, 12, 1562–1567.

[55] Karmakar A, Kalita D, Baruah JB. Structural study on few co-crystals and a salt of quinoline derivatives having amide bond. J. Mol. Struct. 2009, 935, 47–52.

[56] Karmakar A, Baruah JB. N-[2-(4-methoxy-phenyl)-ethyl]-2-(quinolin-8-yloxy)acetamide: a receptor for acid binding, Supramol. Chem. 2008, 20, 667–674.

[57] Kalita D, Baruah JB. Visual distinction of dicarboxylic acids and their salts by 1-phenyl-3-(quinolin-5-yl)urea. J. Mol. Struct. 969, 2010, 75–82.

[58] Kalita D, Baruah JB. 1-Phenyl-3-(quinolin-5-yl)urea as a host for distinction of phthalic acid and terephthalic acid. J. Chem. Sci. 2013, 125, 267–273.

[59] Khan M, Enkelmann V, Brunklaus G. Heterosynthon mediated tailored synthesis of pharmaceutical complexes: a solid-state NMR approach. CrystEngComm 2011, 13, 3213–3223.

[60] Ghosh K, Adhikari S, Chattopadhyay AP, Chowdhury PR. Quinoline based receptor in fluorometric discrimination of carboxylic acids. Beilstein J. Org. Chem. 2008, 4, article no. 52.

[61] K. Ghosh K, Adhikari, S. Fluorescence sensing of tartaric acid: a case of excimer emission caused by hydrogen bond mediated complexation. Tetrahedron Lett. 2006, 47, 3577–3581.

[62] Kameta N, Hiratani K. Phosphate anion-selective recognition by boron complex having plural hydrogen bonding sites. Chem. Commun. 2005, 725–727.

[63] O'Brien RL, Olenick JG, Hahn FE. Reactions of quinine, chloroquine, and quinacrine with DNA and their effects on the DNA and RNA polymerase reactions. Proc. Natl Acad. Sci. U.S.A. 1966, 55, 1511–1517.

[64] Dyke KV, Lantz C, Szustkiewicz C. Interactions of water soluble carcinogen 4-nitroquinoline N-oxide with DNA. Science 1966, 154, 1665–1667.

[65] Slater A, Swiggard W, Orton B, Flitter WD, Goldberg DE, Cerami A. Henderson GB. An iron-carboxylate bond links the heme units of malaria pigment. Proc. Nat. Acad. Sci. USA 1991, 88, 325–329.

[66] Egan TJ. Haemozoin (malaria pigment): a unique crystalline drug target. TARGETS 2003, 2,115–124.

[67] Egan TJ. Physico-chemical aspects of hemozoin (malaria pigment) structure and Formation. J. Inorg. Biochem. 2002, 91 19–26.

[68] Dodd EL, Bohle DS. Orienting the heterocyclic periphery: a structural model for chloroquine's antimalarial activity. Chem. Commun. 2014, 50, 13765–13768.

[69] Wicht MM, Bathori NB, Nassimbeni LR. Isoquinoline-based Werner clathrates with xylene isomers: aromatic interactions vs. molecular flexibility. Dalton Trans. 2015, 44, 6863–6870.

[70] Baruah AM, Karmakar A, Baruah JB. Manganese and cadmium benzoate complexes having 8-aminoquinoline ancillary ligand. Open Inorg. Chem. J. 2008, 2, 62–68.

[71] Das B, Crans DC, Baruah JB. Cation exchange, solvent free synthesis and packing patterns of quinolinium nickel(II) dipicolinates. Inorg. Chim. Acta 2013, 408, 204–208.

Jubaraj B. Baruah

6 N-oxides in multi-component crystals and in bottom-up synthesis and applications

6.1 Introduction

Organic N-oxides are an interesting class of bipolar molecules that have formal N^{+}–O^{-} bond/s. They are biologically important compounds [1], drugs [2] and find applications as ligands in coordination complexes [3]. N-oxides are also extensively used as oxidants and in catalysis [4–7]. Due to their dipolar nature, N-oxides are soluble in polar solvents, including water [8]. The dipolar nature of the N-oxide has been revealed by the crystal structure analysis of pyridine N-oxide, which is a widely used molecule in various spheres of the chemical sciences [9]. Due to the dipolar nature of N-oxides, they can form hydrogen bonds with simple alcohols; these interactions have been studied in solution. For example, the interactions between pyridine N-oxide with methanol is reflected in the blue shift of the first π–π^* transition in spectra recorded in methanol solution with respect to a spectra of a hydrocarbon solution of N-oxide [10]. The N–O bond can participate in linear or bifurcated hydrogen bonds, the latter by the oxygen atom accepting hydrogen bonds from two hydrogen bond donors simultaneously, as illustrated in Figure 6.1 [11]. The hydrogen bond energies of N-oxide fall in the range of 8 to 10 kcal/mol for each bifurcated hydrogen bond, and for a single, linear O–H···O(N-oxide) hydrogen bond, it is about 20 kcal/mol. On the other hand, the energy of heteronuclear N–H···O(N-oxide) interactions is about 4–5 kcal/mol. N-oxides can participate in inter- and intra-molecular hydrogen bonds. The intramolecular hydrogen bond found in picolinic acid N-oxide, shown in Figure 6.1c, shows a very short O···O distance of 2.44 Å, suggesting it to be a very strong hydrogen bond [12]. The co-crystal of pyridine N-oxide with trichloroacetic acid is hydrogen bonded with a short O···O distance of 2.43 Å, also consistent with a very strong hydrogen bond. These observations generate interest in theoretical chemistry [13]. The correlations between vibrational frequencies in different adducts of pyridine N-oxide with water and acids are generally studied by FTIR and NMR spectroscopy, X-ray diffraction, and quantum chemical DFT calculations. Pyridine N-oxide N → O stretching is observed at 1252 cm^{-1}, which changes upon hydrogen bond formation. It is red-shifted by ~18 cm^{-1} upon the interaction of the water with its co-crystal with hydrochloric acid. This band appears at 1198 cm^{-1} in acetonitrile solution and appears at 1196 cm^{-1} in the crystalline phase. The hydrogen bond associated with the N-oxide group of pyridine N-oxide in its co-crystals influences the π-electronic structures as revealed by FTIR and ^{13}C-NMR studies. This also explains the acidifying effect of N-oxide groups on neighbouring protons [14].

https://doi.org/10.1515/9783110464955-006

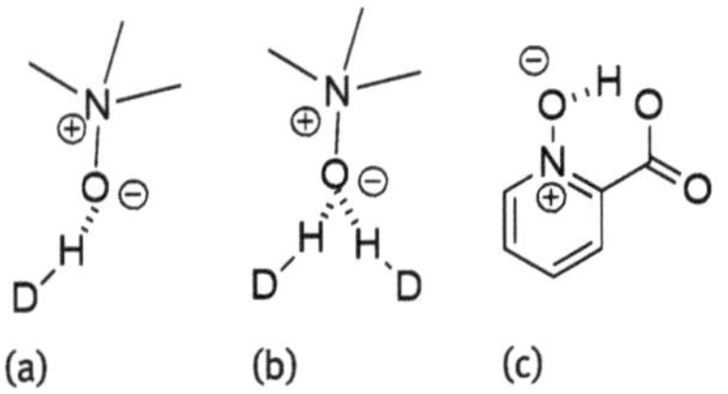

Fig. 6.1: Representations of (a) linear (b) bifurcated (c) intramolecular hydrogen bonding in N-Oxides (D is an electronegative atom such as N and O.).

It is also interesting to note that depending on the functional groups on quinoline N-oxides, such as 8-hydroxyquinoline-N-oxide and 2-carboxyquinoline-N-oxide, different intramolecular hydrogen bonded cyclic motifs can be formed, as shown in Figure 6.2 [15]. The O···O distance in the structure of 2-carboxyquinoline-N-oxide is 2.44 Å and for 8-hydroxyquinoline-N-oxide is 2.54 Å, which are suggestive of having very strong intramolecular hydrogen bonds.

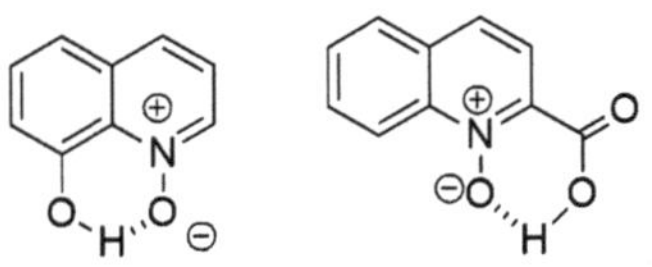

Fig. 6.2: Intramolecular hydrogen bonding in quinoline based N-oxides.

The hydrogen bond forming ability of trimethyl-N-oxide with water affects the folding of proteins and the micellisation of surfactants in aqueous medium. Near-infrared spectroscopic studies on such interactions have revealed that trimethyl-N-oxide modifies the free energy associated with the structural reorganisation of water molecules in these systems [16].

Herein, an overview of the supramolecular chemistry of N-oxides is given, broadly divided into organic and inorganic systems.

6.2 N-oxides in the organic solid state

The N-oxide group acts as a proton acceptor to many hydrogen bond donor molecules. Thus, the competition between different functional groups present in the co-former when forming a co-crystal determines the overall architecture of its self-assembly. It is generally found that hydrogen bonds significantly influence the N–O bond length and its properties, including π-delocalisation. The study of competitive hydrogen bonding [17] between functional groups such as hydroxyl and carboxyl groups present in a co-former, suggests that carboxylic acid groups prefer the formation of acid dimers, while hydroxyl groups have a tendency to form hydrogen bonds with the oxygen atom of N-oxide. However, if a carboxylic acid compound is devoid of a hydroxyl group, it interacts with N-oxide. Among carboxylic acid and amine groups, the former prefers to be a hydrogen donor ahead of the amine donor. Thus, the co-crystal of

4-nitropyridine-N-oxide with 2-hydroxybenzoic acid has O-H(acid)···O(N-oxide) interactions, whereas the self-assembly of the co-crystal between 2,3,4-trihydroxybenzoic acid and 4-nitropyridine-N-oxide has O(N-oxide)-OH(phenolic) contacts. In co-crystal of 4-nitropyridine-N-oxide with 4-aminobenzenosulfonamide, the N-oxide interacts with the NH_2 group as illustrated in Figure 6.3.

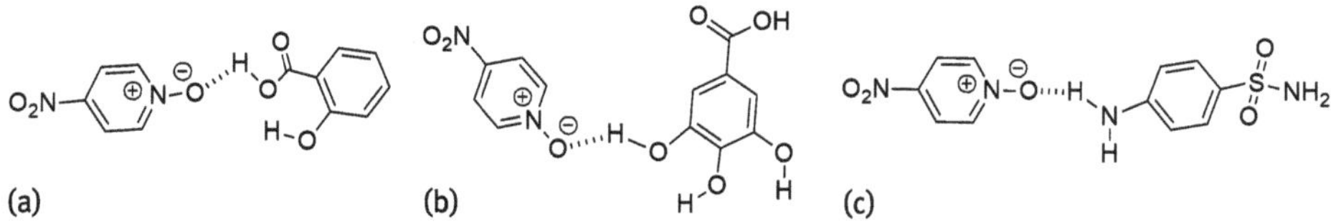

Fig. 6.3: Supramolecular association in the co-crystals of 4-nitropyridine-N-oxide with (a) 2-hydroxybenzoic acid, (b) 3,4,5-benzenecarboxylic acid and (c) 4-aminobenzenesulfonamide.

Conventionally, in biological or synthetic oxidation of aldehydes to carboxylic acids, hydrated aldehydes are proposed as intermediate species. Bifurcated hydrogen bonds are able to stabilise aldehyde hydrates, which are otherwise highly unstable. For example the hydrated form of trichloroacetaldehyde [18] forms cyclic as well as polymeric assemblies through bifurcated hydrogen bonds as illustrated in Figure 6.4.

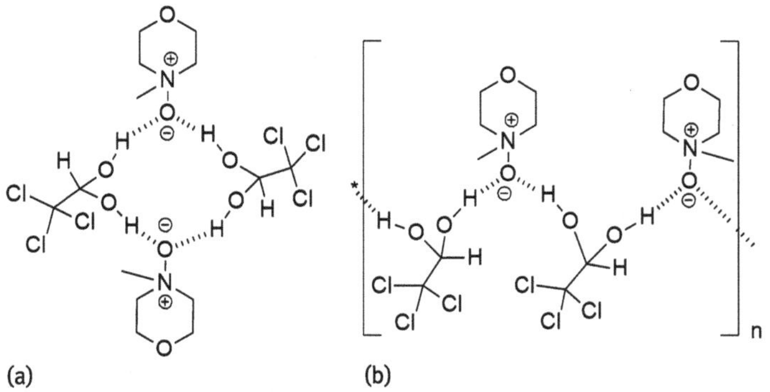

Fig. 6.4: (a) Cyclic and (b) polymeric assembly of trichloroacetaldehyde with N-methylmorpholine N-oxide.

Similarly, pyridine-N-oxide acts as a bridge to link phenylsulphonamide molecules to form a cyclic arrangement [19] as shown in Figure 6.5. In this case, the bifurcated hydrogen bonds involve two N-H donors of sulphonamides. Such cyclic assemblies have extra stability owing to the ionic contribution to the hydrogen bonds. It may be also noted that assemblies formed by bifurcated hydrogen bonds have unequal numbers of donors and acceptors, hence a cyclic motif mediated by the bifurcated hydrogen bond

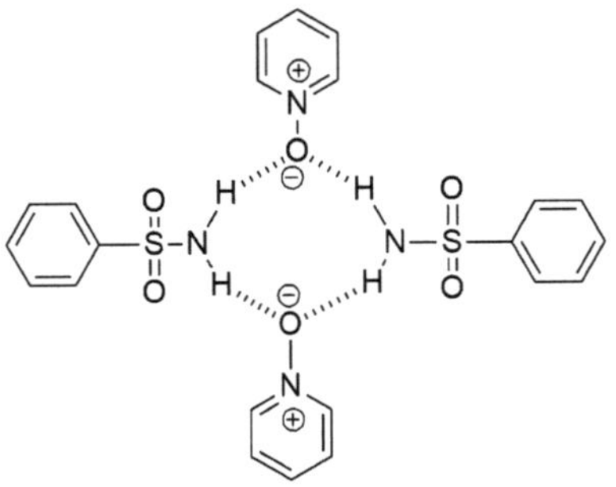

Fig. 6.5: A hydrogen bonded cyclic assembly of pyridine-N-oxide with phenyl sulphonamide.

of N-oxide is usually formed, at least in instances where the molecule has no other hydrogen bonding functionality.

In co-crystals formed between 2,2′-dipyridyl-N,N′-dioxide and different dicarboxylic acids, C-H···O interactions form that provide orientation to bimolecular association between the co-formers. When these interactions are systematic, isosteric synthons are observed in these co-crystals [20]. In a similar vein, amides interact with N-oxide and such interactions facilitate the formation of co-crystals of 4,4′-bipyridine-N,N′-dioxide with barbiturate drugs [21].

Co-crystals of organic N-oxides have deep roots in the basic sciences. For example, trimethylamine-N-oxide belongs to a class of compounds called osmolytes which are accumulated by many prokaryotes, fungi, plants and animals that live under osmotic stress. These organisms accumulate N-oxides to avoid experiencing environmental water stress [22]. Trimethylamine-N-oxide is also used as electron acceptors by some bacteria under anaerobic conditions for the oxidation of organic substrates [23]. Besides these examples, triethylamine-N-oxide is a risk factor for cardiovascular disease in humans [25]. Hence, there are extensive theoretical and experimental studies on this molecule to understand its behaviour in the supramolecular context. Spectroscopic studies have revealed that trimethylamine-N-oxide together with urea decreases the number of defects in the hydrogen bonded structure of water. It has also been suggested that water mediated interactions between trihydrated trimethylamine-N-oxide and urea, as depicted in Figure 6.6, are responsible for the osmoprotectant property of trimethylamine-N-oxide. The three hydrogen bond acceptor sites on the oxygen atom of trimethylamine-N-oxide are provided by water molecules, which, in essence, define a hydration shell. There is no evidence for direct urea and N-oxide

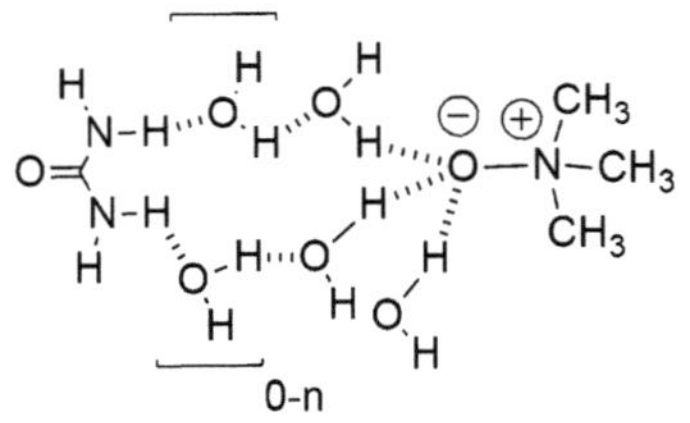

Fig. 6.6: A model showing water mediated assembly between trimethyl-N-oxide with urea in solution.

adduct formation in aqueous medium. Thus, supramolecular association of two compounds balances the defects in the hydrogen bonded structure of water introduced by the individual solutes. This factor is responsible for osmolytic behaviour in biological systems [26].

A halogen bond is a highly directional, non-covalent interaction formed between a polarised A–X bond (where X is halogen and A is more electropositive atom than X) and a nucleophile. Such bonds have contributions from charge transfer and electrostatic interactions [27, 28]. These interactions are commonly observed between N-oxides and polyfluoroalkyl or polyfluorobenzene molecules [29]. Highly volatile liquids such as 1,2-diiodoperfluoroethane and 1,4-diiodoperfluorobutane are stabilised by pyrazine 1-oxide and tetramethylpyrazine-bis-N-oxide through X···O bonds [30]. Such X···O interactions are also useful for the construction of supramolecular self-assemblies of nitroxide free radicals. For example, 1,4-diiodotetrafluorobenzene can assemble with the 4-phenyl-2,2,5,5-tetramethyl-3-imidazolin-1-yloxyl and 4-amino-2,2,6,6-tetramethyl(piperidin-1-yloxyl) radicals [31]. The interactions between the N-oxy radical with halogen in these molecules are functional group dependent. For example, the 2 : 1 self-assembly between the 4-phenyl-2,2,5,5-tetramethyl-3-imidazolin-1-yloxyl radical and 1,4-diiodotetrafluorobenzene occurs via major contributions provided by N···I interactions (Figure 6.7a). In the case of the 4-amino-2,2,6,6-tetramethyl(piperidin-1-yloxyl) radical, the 1 : 1 assembly with 1,4-diiodotetrafluorobenzene results in a supramolecular polymer that features both N–H···I and O···I interactions, as shown in Figure 6.7b [31].

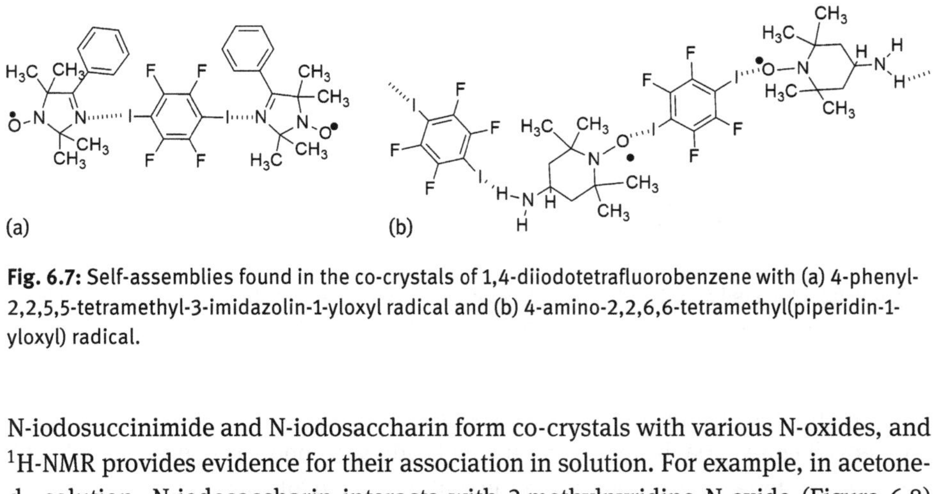

Fig. 6.7: Self-assemblies found in the co-crystals of 1,4-diiodotetrafluorobenzene with (a) 4-phenyl-2,2,5,5-tetramethyl-3-imidazolin-1-yloxyl radical and (b) 4-amino-2,2,6,6-tetramethyl(piperidin-1-yloxyl) radical.

N-iodosuccinimide and N-iodosaccharin form co-crystals with various N-oxides, and ^{1}H-NMR provides evidence for their association in solution. For example, in acetone-d_6 solution, N-iodosaccharin interacts with 2-methylpyridine N-oxide (Figure 6.8) whereby there are definitive shifts in the resonances assigned to the pyridyl protons [32]; the ring protons of the N-iodosaccharin molecule are also affected. The binding constant between 2-methylpyridine-N-oxide and N-iodosaccharin was determined to be 2774 M^{-1}.

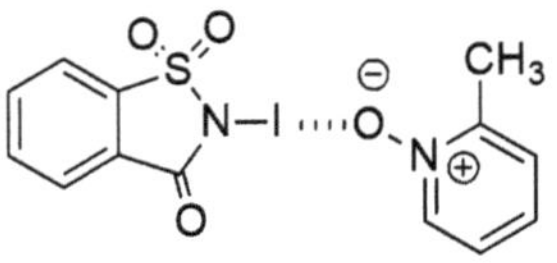

Fig. 6.8: Supramolecular association in the co-crystal of N-iodosaccharin and 2-methylpyridine-N-oxide.

Bifurcated X···O interactions of N-oxide are electrostatic in nature, which has its origin in the electropositive σ-hole in an electron pair of the halogen atom donor. Such interactions are useful in building well-defined polygons in the solid state via bifurcated X···O interactions (Figure 6.9a). For example, I···O interactions are utilised to prepare various supramolecular polygons as shown in Figure 6.9b and c.

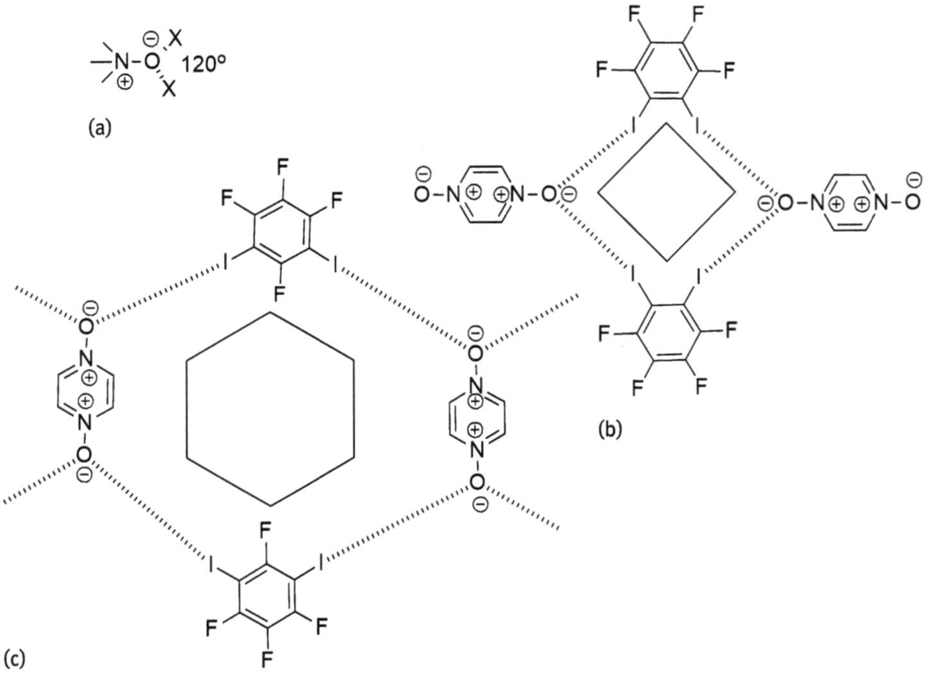

Fig. 6.9: (a) Bifurcated X···O bond. Assemblies through I···O interactions formed between N-oxides and diiodo-tetrafluorobenzene derivatives to form (b) tetragonal and (c) hexagonal arrangements.

4,4′-Bipyridine-N-oxide forms hydrogen bonded networks with 2,2′,6,6′-tetracarboxybiphenyl (Figure 6.10a) by forming bifurcated hydrogen bonds. The self-assembly utilises 2,2′,6,6′-tetracarboxybiphenyl as the central pivot to form architectures which resemble metal-organic frameworks (Figure 6.10b) [35].

Co-crystals of bent N-oxides provide space to form different architectures that are guided by O(N-oxide)···H-O interactions supported by weak C-H···O interactions. For example, two bent N-oxides, i.e. 4-(5-(pyridin-4-yl)furan-2-yl)pyridine 1-oxide and

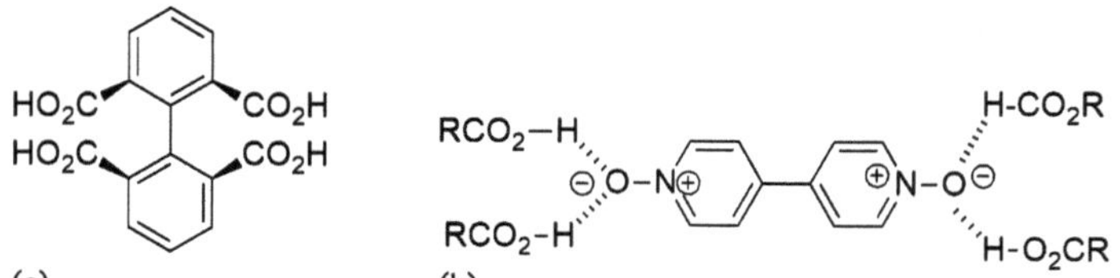

(a) (b)

Fig. 6.10: (a) 2,2′,6,6′-Tetracarboxybiphenyl and (b) interactions of bipryridine-N,N′-dioxide with the carboxylic acid.

4,4′-(furan-2,5-diyl)bis(pyridine 1-oxide) (Figure 6.11a and b) form co-crystals with various dihydroxyaromatic, hydroxycarboxylic and aminocarboxylic acids [36]. The positions of the donor atoms of the molecule play pivotal roles in the packing patterns. The supramolecular assemblies formed by these N-oxides with dihydroxyaromatics, such as 1,3-dihydroxybenzene and 1,3,5-trihydroxybenzene, are two-dimensional. Examples of the variety in packing patterns due to the interplay of directional weak interactions within hydrogen bonded motifs are depicted in Figure 6.11c and d.

Amides and aromatic-N-oxides can form complementary hydrogen bonds. For example, carboxamide and pyridine N-oxide self-assemble to form a helical architecture (Figure 6.12a). In this case, complementary hydrogen bonds participate in the con-

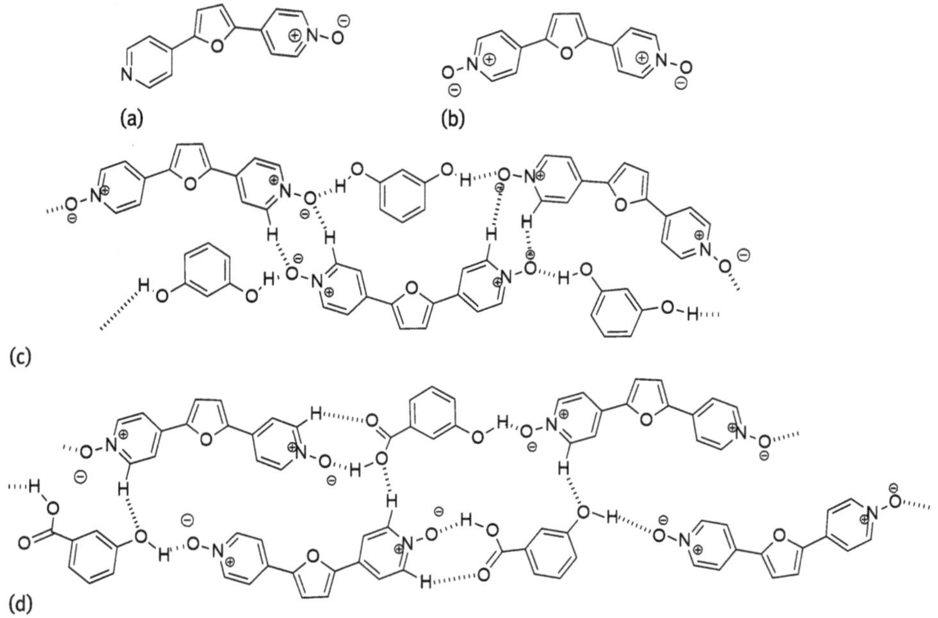

Fig. 6.11: Examples of bent N-oxides (a) 4-(5-(pyridin-4-yl)furan-2-yl)pyridine 1-oxide and (b) 4,4′-(furan-2,5-diyl)bis(pyridine 1-oxide). Self-assembly in co-crystals of the bent N-oxide, 4,4′-(furan-2,5-diyl)bis(pyridine 1-oxide), with (c) 1,3-dihydroxybenzene and (f) 3-hydroxybenzoic acid.

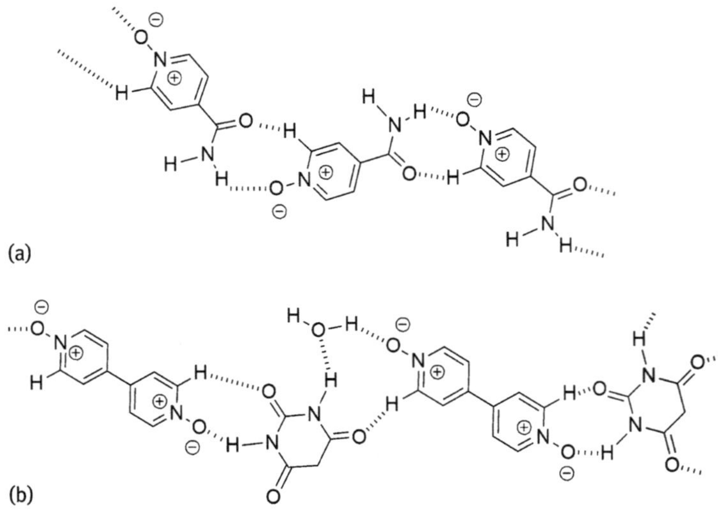

Fig. 6.12: Self-assembly of (a) carboxamide pyridine N-oxide and (b) 4,4′-dipyridyl-N,N′-dioxide with pyrimidine-2,4,6(1*H*,3*H*,5*H*)-trione.

struction of cyclic motifs. Based on similar principles of N-oxide and amide binding it is found that bipyridine-N-oxide is a suitable co-former for self-assembly; one such example is shown in Figure 6.12b.

C_2-2-Methylresorcin-4-arene (Figure 6.13) forms host-guest complexes with pyridine N-oxide and quinoline N-oxide [37]. In solution, 1 : 1 host-guest complexes are formed, while in the solid state, dimeric, capsule-like assemblies with 2 : 3 and 2 : 2 host-guest stoichiometry are observed. 2-Methylpyridine-N-oxide or 4-methoxypyridine-N-oxide are included by C_{ethyl}-2-methylresorcinarene when it adopts a crown conformation and so function as guests. 2-Iodopyridine-N-oxide is trapped by the host through intermolecular N–O···I–C halogen bonds. The halogen bond interactions in these co-crystals are found to be as strong as hydrogen bonds [38]. These examples show that N-oxides, as guests, can guide the conformation of a host to facilitate molecular recognition.

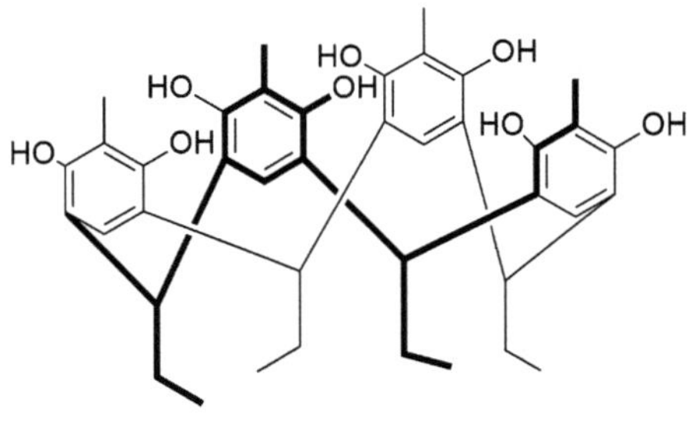

Fig. 6.13: C_2-2-Methylresorcin-4-arene.

Among boronic acids, 1,4-phenyldiboronic acid forms co-crystals with a variety of aromatic N-oxides [39, 40]. The structures feature quite diverse hydrogen bonding motifs. Thus, examples of linear, bifurcated and trifurcated hydrogen bonding have been encountered. For instance, the diboronic acid co-crystal with pyridine is guided by linear O(N-oxide)···H-O interactions (Figure 6.14a). The contribution of C-H···O interactions along with the O(N-oxide)···H-O(boronic acid) hydrogen bonding can generate cyclic motifs. Hence, quinoline N-oxide forms trifurcated hydrogen bonds as shown in Figure 6.14b, whereas isoquinoline self-associates into a dimer to which one O-H group of diboronic acid is hydrogen bonded in an oblique manner (Figure 6.14c). A close inspection of these structures reveals that there are two different geometries observed in the packing patterns, i.e. type 1 and type 2, as shown in Figure 6.14d. In theory, type 2 has greater stability over type 1.

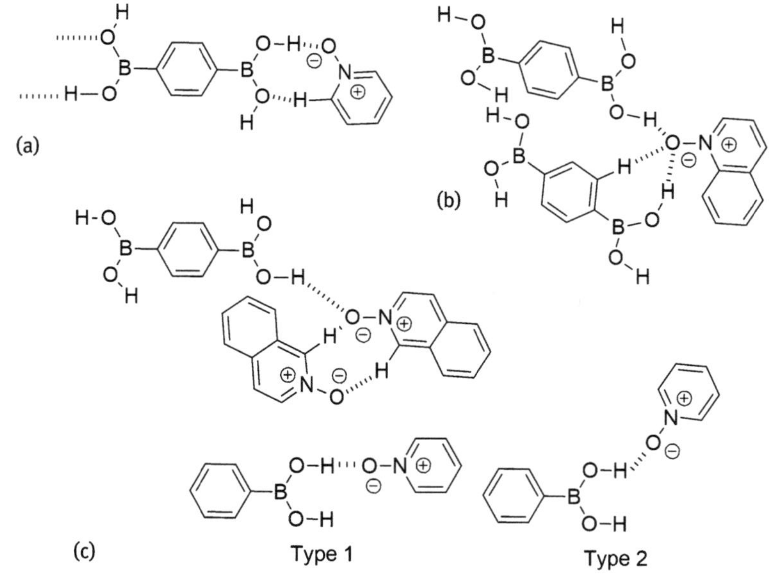

Fig. 6.14: Principal hydrogen bonded motifs in co-crystals of 1,4-phenyldiboronic acid with (a) pyridine-N-oxide, (b) quinoline-N-oxide and (c) isoquinoline-N-oxide. Two types of O(N-oxide)···H-O(boronic acid) interactions: (a) type 1 and (b) type 2 are shown.

A design principle to construct a rotaxane [41] through an N-oxide···amide interaction is represented in Figure 6.15a. An example of a rotaxane suitable for anion binding is shown in Figure 6.15b. The rotaxane has two, diagonally opposite amide functionalities. The thread through the rotaxane, i.e. the rod, is functionalised with N-oxide and amide residues capable of complexing with an anion. As shown in Figure 6.15b, at a given time, the N-oxide forms bifurcated O···HO(amide) hydrogen bonds to stabilise the rotaxane arrangement and to enable the formation of anion···O(carbonyl) interac-

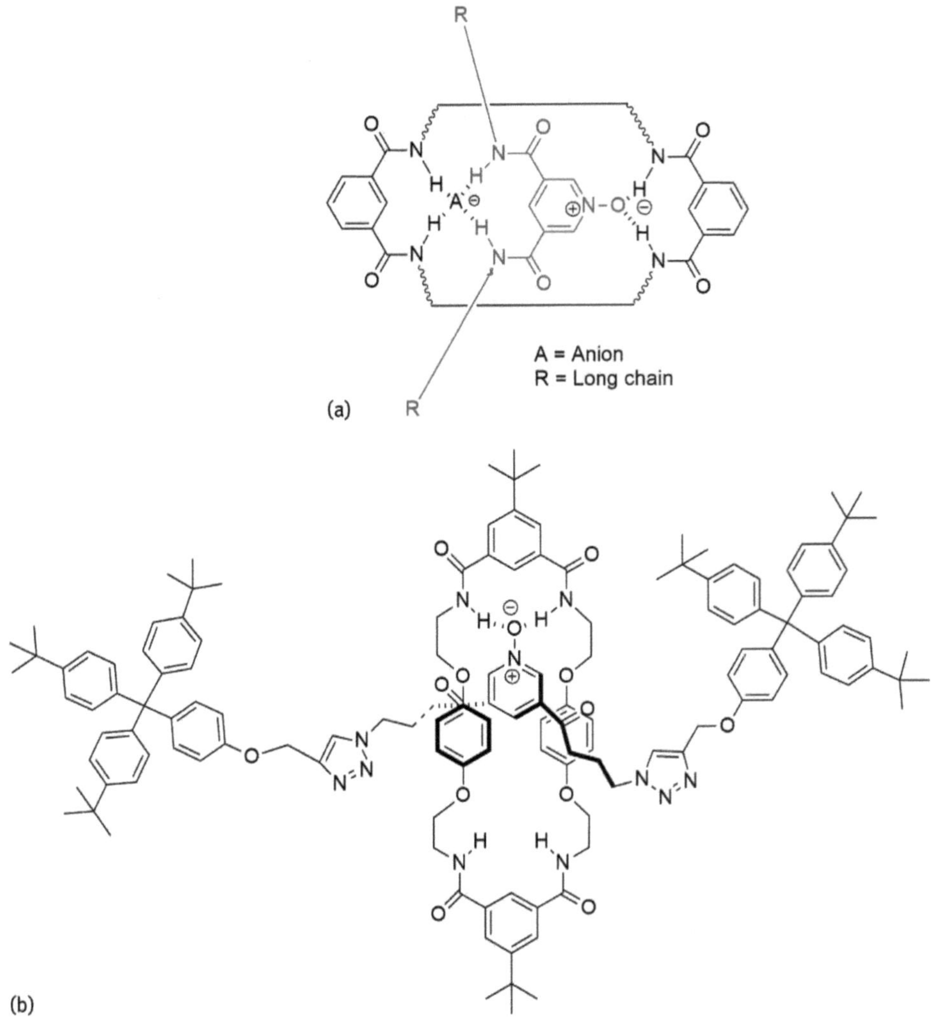

Fig. 6.15: (a) Design principle to construct an anion sensor mediated by N-oxide···amide interactions and (b) a rotaxane where N-oxide can exchange between two positions.

tions. Switching between the hydrogen bonding and complexing sites can be envisaged. To function as an efficient anion sensor, the anion must have complementary geometry and size. Thus, a design principle allows high selectivity in anion binding leading to the possibility of anion sensing applications.

6.3 N-oxides in coordination and self-assembly of metal complexes

The foregoing discussions have illustrated the utilities of N-oxides as partner molecules in the assemblies of multi-component organic molecules. From an inorganic chemistry point of view, N-oxides have also been of great importance. Thus, N-oxides form metal complexes giving rise to different supramolecular architectures by either acting as a formal ligand for a metal or purely as a co-crystal partner, associated via hydrogen bonds. The common binding modes of N-oxides are illustrated in Figure 6.16.

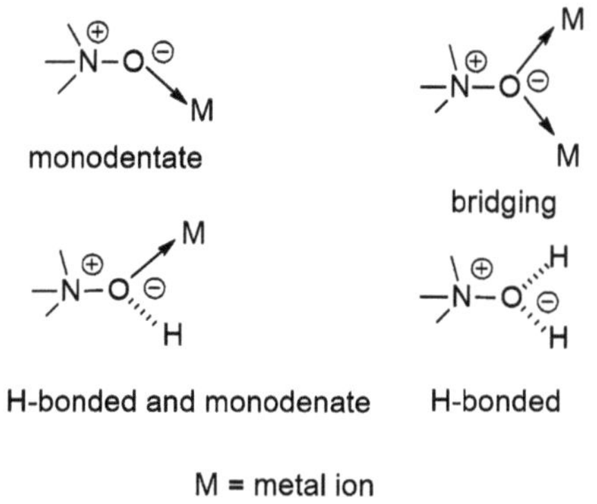

Fig. 6.16: Some binding modes for N-oxides found in metal complexes.

The electron donating ability of amine oxides was initially realised from the reaction of trimethylamine oxide with Lewis acids [42]. Biologically related molecules such as adenine 1N-oxide and adenosine 1N-oxide form complexes with transition metal ions and have generated interest in supramolecular coordination chemistry [43]. Adenine-1N-oxide and adenosine 1N-oxide may chelate first-row transition metal ions as shown in Figure 6.17. In the first example, (Figure 6.17a), the N-oxide group remains uncoordinated, whereas in the latter (Figure 6.17b), the chelate ring includes the N–O bond.

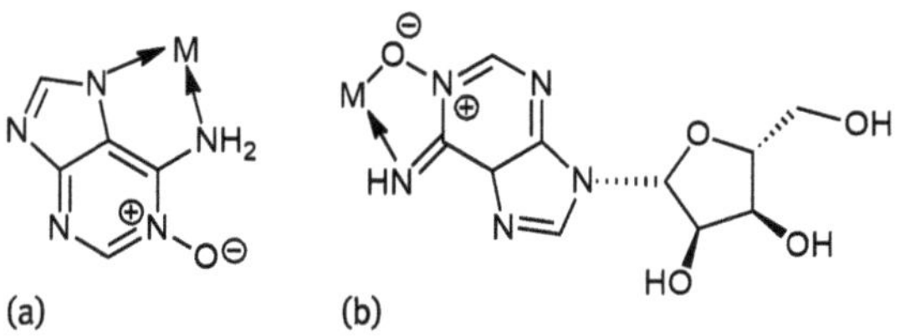

Fig. 6.17: Metal complex formation by (a) adenine 1N-oxide and (b) adenosine 1N-oxide.

Pyridine N-oxide (PNO) binds metal ions either in a monodentate or a μ_2-bridging mode. The coordination ability of pyridine N-oxide can be compared with the coordination ability of oxides such as triphenylphosphine oxide and dimethylsulphoxide [44]. Hence, PNO functions in a μ_2-bridging mode in $[(PNO)CuCl_2]_2$ (Figure 6.18a), whereas both monodentate and μ_2-bridging modes of PNO binding are observed in $[Cu(PNO)_2Cl_2]_2$ (Figure 6.18b) [45–47]. Thus, a change in stoichiometry in closely related complexes allows multiple binding modes of PNO. As an aside, these complexes show interesting antiferromagnetic coupling between the copper(II) ions.

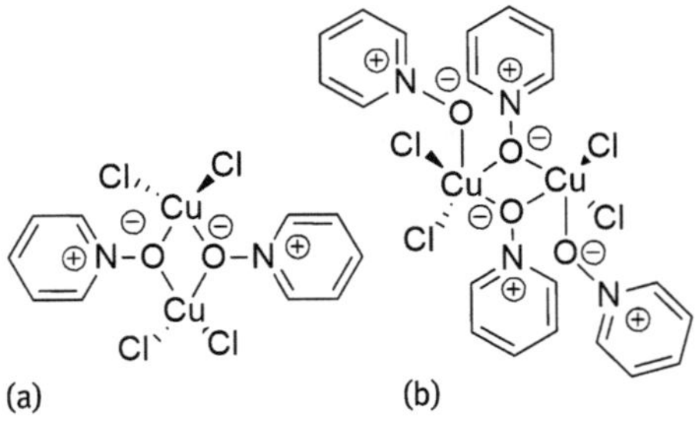

Fig. 6.18: Structures of (a) $[(PNO)CuCl_2]_2$ and (b) $[Cu(PNO)_2C1_2]_2$.

Hexa-aquo cobalt(II) chloride or bromide forms co-crystals with 4,4′-bipyridyl-N,N′-dioxide (4,4′-BPNO). In these co-crystals, two $[Co(H_2O)_6]^{2+}$ cations are bridged by two 4,4′-BPNO molecules through O(N-oxide)···H–O–Co hydrogen bonds. A similar hexaaquo cobalt(II) complex of 4,4′-BPNO but with 1,2,4,5-benzenetetracarboxylate is also known [48]. Such co-crystals are often obtained in hydrated forms. Co-crystals of hexaaqua manganese and zinc complexes with different carboxylates with N-oxides are well known [49–52]; each of them are formed by hydrogen bonding interactions of N-oxide with coordinated water molecules. The competition between coordination of N-oxide or carboxylate ligands can be successfully utilised to prepare co-crystals or coordination polymers. A one-dimensional coordination polymer with PNO bridges of composition $[Mn(RC_6H_4COO)_2(PNO)]_n$ (R = H, 2-NO_2, 4-NO_2, 4-OH, 4-Cl, etc.) can be prepared under ambient conditions; an example, $[Mn(C_6H_5COO)_2(PNO)]_n$, is shown in Figure 6.19. These polymers are antiferromagnetic [52].

2,2′-Bipyridyl-N,N′-dioxide (2,2′-BPNO) can act either as a chelating or a bridging ligand. This is possible because of free rotation around C–C single bond linking the aromatic rings of 2,2′-BPNO (Figure 6.20). Thus, 2,2′-BPNO can coordinate to metal centres either in a cis- or trans-geometry, as shown in Figure 6.21. However, simultaneous μ_2-bridging by both N-oxo groups is not common for this ligand, an observation which may be due to steric factors. The $[La(2,2'\text{-BPNO})_4](ClO_4)_3$ complex has a cubic coordination geometry [53]; a copper(II) benzoate complex with 2,2′-bipyridyl-N,N′-dioxide is a coordination polymer having spiral arrangements [49]. The 2,2′-BPNO ligand forms a one-dimensional coordination polymer with composition $[Cu_2(C_6H_5COO)_4(2,2'-BPNO)]_n$ (Figure 6.21). Each copper(II) centre adopts a square pyramidal geometry, with 2,2′-BPNO coordinated in the apical position. The Cu–Cu distance within the paddle wheel unit is 2.64 Å, while the shortest Cu···Cu distance between paddle wheel units is 8.04 Å.

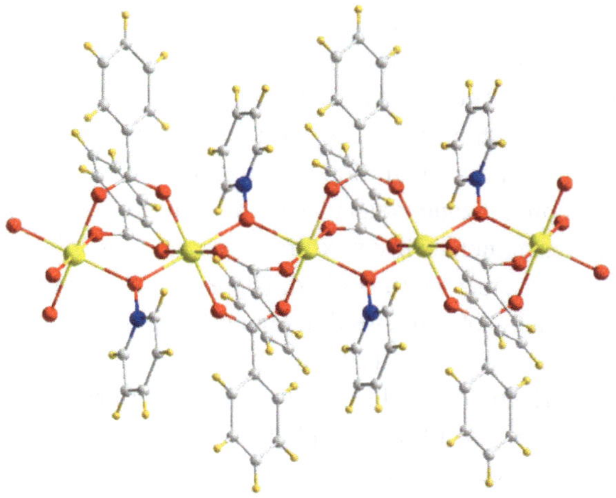

Fig. 6.19: A coordination polymer, $[Mn(C_6H_5COO)_2(PNO)]_n$, of pyridine-N-oxide.

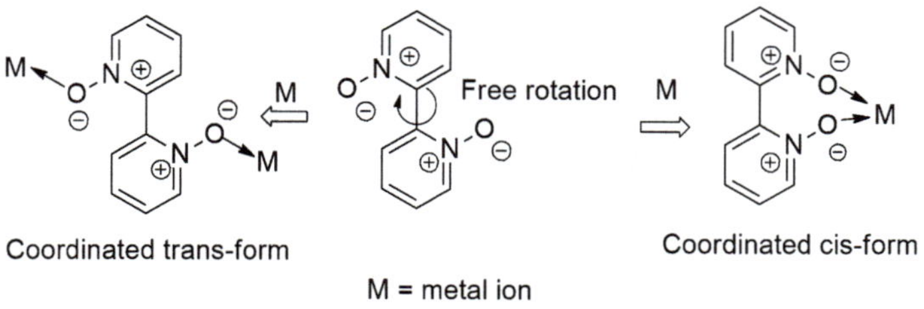

Fig. 6.20: Two coordination geometries observed for 2,2′-bipyridyl-N,N′-dioxide.

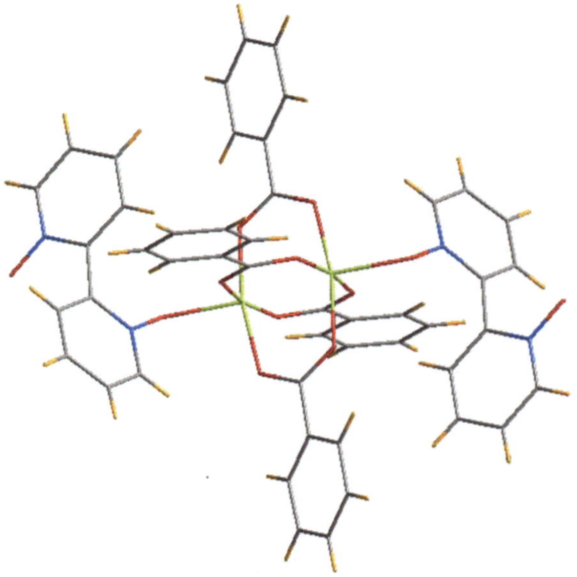

Fig. 6.21: Part of the coordination polymer of $[Cu_2(C_6H_5COO)_4(2,2' - BPNO)]_n$.

A three-dimensional coordination polymer, $[Mn_2(BDC)_2(4,4'\text{–}BPNO)_4 \cdot (DMF)_2]_n$ [54] features coordinating para-benzenedicarboxylate (BDC) and 2,2′-BPNO ligands, the latter through a $\eta^2\text{–}\mu_2:\mu_2$-binding mode. The structure has channels in two different directions, with sizes 13.24 and 17.71 Å, which are filled with DMF molecules (Figure 6.22a). These complexes are useful in the separation of benzene, toluene, ethylbenzene, etc. The intercalation of chlorobenzene molecules in the channels is shown in Figure 6.22b. When the complex was immersed in a mixture of benzene and toluene in a 1 : 1 volume ratio, only benzene was selectively adsorbed. There are also examples of one-dimensional coordination polymers of N-oxides that are held in place through hydrogen bonding to form a grid-like structure. Coordination polymers of 4,4′-BPNO containing different metal ions, $\{[M(H_2O)_4(4,4'\text{-BPNO})][ClO_4]_2{\cdot}2(4,4'\text{-BPNO})\}_n$ (M = Co, Ni, Cu or Zn), has two molecules of 4,4′-BPNO included as solvates, connected by hydrogen bonding with the water-hydrogen atoms (Figure 6.23).

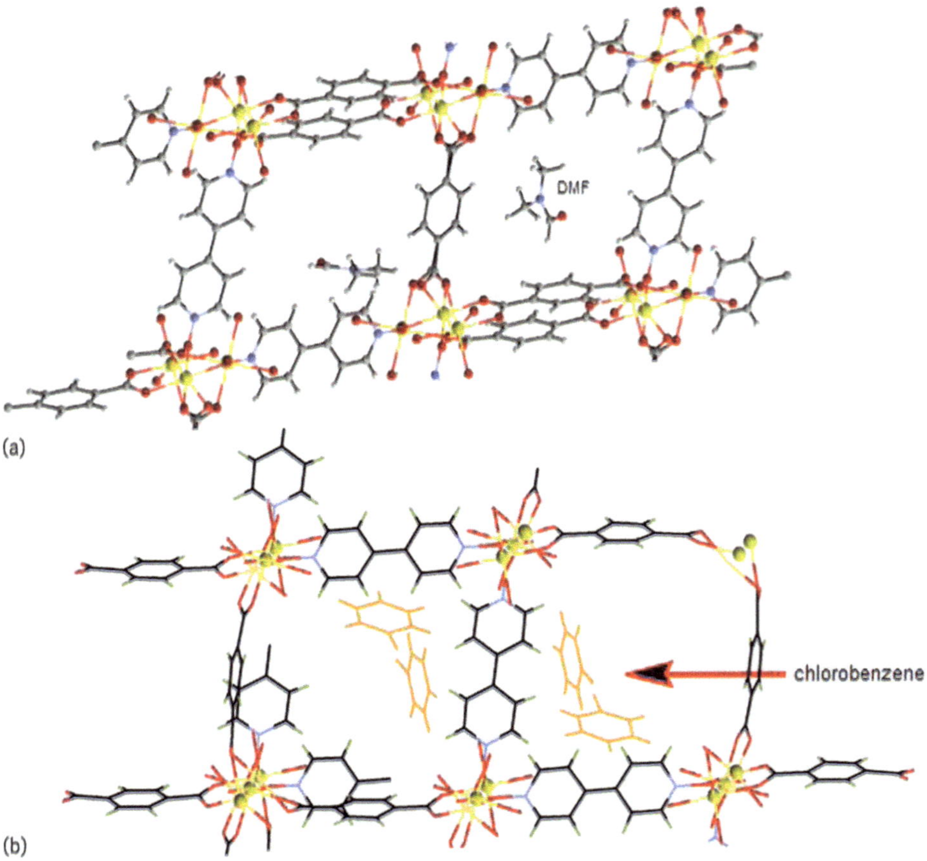

Fig. 6.22: (a) Part of the coordination polymer $[Mn_2(BDC)_2(4,4'\text{–}BPNO)_4 \cdot (DMF)_2]_n$ and (b) intercalation of chlorobenzene molecules in the channels.

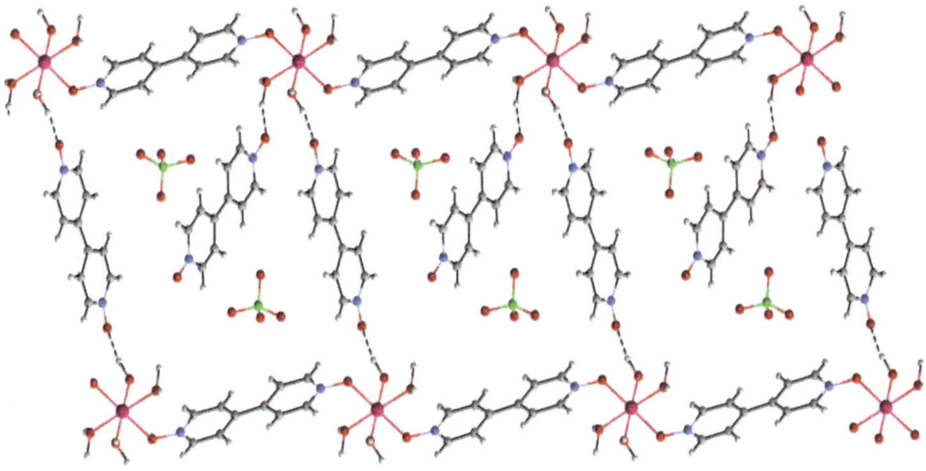

Fig. 6.23: Channel-like structure of $\{[Mn(H_2O)_4(4,4'{-}BPNO)][ClO_4]_2{\cdot}2(4,4'{-}BPNO)\}_n$, showing the inclusion of bipyridine-N-oxide.

A poly-oxometalate based framework, $\{[Ho_4(4{,}4'\text{-}BPNO)_8(H_2O)_{16}BW_{12}O_{40}]{\cdot}2\,H_2O\}$ $(BW_{12}O_{40})_2{\cdot}(Hpz)_2{\cdot}(H_2O)_{11}$, was obtained from multiple ligands such as 4,4′-BPNO and hexahydropyrazine (Hpz) [55], and form nanocages. Four Ho^{3+} ions are connected by four bridging 4,4′-BPNO ligands in an alternate manner to form the main skeleton of the nanocage. Adjacent cages are linked by bridging 4,4′-BPNO ligands to generate a one-dimensional ribbon comprising nanocages (Figure 6.24). Adjacent ribbons are linked through hydrogen bonds and π⋯π stacking interactions. The $(BW_{12}O_{40})^{5-}$ anions and solvent water molecules fill the pores of the crystals.

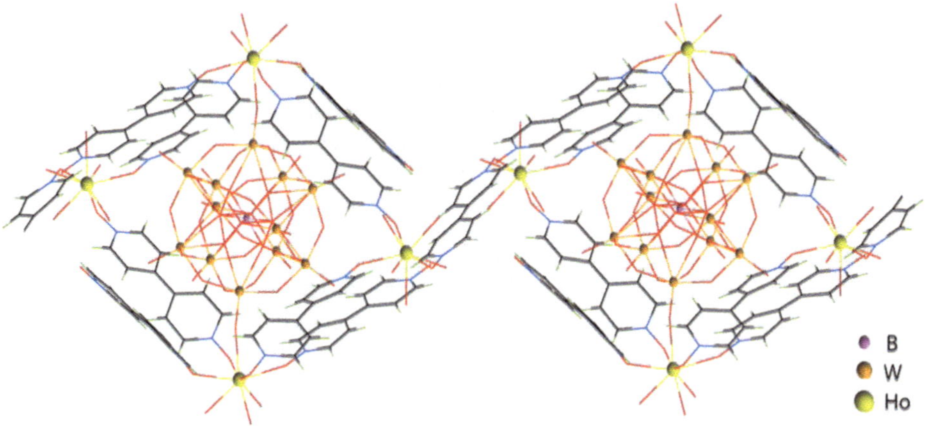

Fig. 6.24: Poly-oxometalate based lanthanide-organic framework.

6.4 Conclusions

The non-covalent assembly of organic and inorganic molecules with N-oxides provides a range of architectures. Selective guest binding is possible in these assemblies. The formation of bifurcated hydrogen bonds in the non-covalent synthesis of N-oxides as hosts or guests play a major contribution in well-defined self-assemblies. On the other hand, the ability of N-oxides to form bridges between metal ions provides room to assemble multi-component systems to form coordination polymers or metal-organic frameworks. Halogen bonds between N-oxides with halogenated organic compounds open up additional avenues for the construction of new pre-designed assemblies. Being polar, N-Oxide molecules are capable of forming strong hydrogen bonds to stabilise self-assemblies through electrostatic contributions and provide useful materials for crystal engineering endeavours.

Bibliography

[1] Monge A, Palop JA, De Cerain AL, Sebador V, Martinez FJ, Sainz Y, Narro S, Garcia E, de Miguel C. Hypoxia-selective agents derived from quinoxaline 1,4-di-N-oxides. J. Med. Chem. 1995, 38, 1786–1792.

[2] Chopra S, Koolpe GA, Tamboong AA, Matsuyamat KN, Ryan KJ, Tran TB, Doppalapudit RS, Riccio ES, Iyer LV, Green CE, Wan B, Franzblau SG, Madrid PB. Discovery and optimization of benzotriazine di-N-oxides targeting replicating and nonreplicating mycobacterium tuberculosis. J. Med. Chem. 2012, 55, 6047–6060.

[3] Roesky HW, Andruh M. The interplay of coordinative, hydrogen bonding and π–π stacking interactions in sustaining supramolecular solid-state architectures. A study case of bis(4-pyridyl)- and bis(4-pyridyl-N-oxide) tectons. Coord. Chem. Rev. 2003, 236, 91–119.

[4] Albini A, Alpegiani M. The photochemistry of the N-oxide function. Chem. Rev. 1984, 84, 43–71.

[5] Gnanamani E, Someshwar N, Sanjeevi J, Ramanathan CR. Conformationally rigid chiral bicyclic skeleton-tethered bipyridine N,N′-dioxide as organocatalyst: asymmetric ring opening of meso-Epoxides. Adv. Synth. Catal. 2014, 356, 2219–2223.

[6] N. Takenaka, R. S. Sarangthem, B. Captain. Helical chiral pyridine N-oxides: a new family of asymmetric catalysts. Angew. Chem. Int. Ed. 2008, 47, 9708–9710.

[7] Tao B, Lo MMC, Fu GC. Planar-chiral pyridine N-oxides, a new family of asymmetric catalysts: exploiting an η^5-C_5Ar_5 ligand to achieve high enantioselectivity. J. Am. Chem. Soc. 2001, 123, 353–354.

[8] Verdejo B, Gil-Ramirez G, Ballester P. Molecular recognition of pyridine N-oxides in water using calix[4]pyrrole receptors. J. Am. Chem. Soc. 2009, 131, 3178–3179.

[9] Ulku D, Huddle BP, Morrow JC. The crystal structure of pyridine 1-oxide. Acta Crystallogr. B 1971, 27, 432–436.

[10] Trombetti A. Hydrogen bonding in pyridine N-oxides. J. Chem. Soc. B 1968, 1578–1580.

[11] Taylor R, Kennard O, Versichel W. Geometry of the nitrogen-hydrogen···oxygen-carbon(N–H···O : C) hydrogen bond. Three-center (bifurcated) and four-center (trifurcated) bonds. J. Am. Chem. Soc. 1984, 106, 244–248.

[12] Steiner T, Schreurs AMM, Kroon J. Strong intramolecular O–H···O hydrogen bonds in quinaldic acid N-oxide and picolinic acid N-oxide. Acta Crystallogr. C 2000, 56, 577–579.
[13] Stare J, Hart M, Daemen L, Eckert J. The very short hydrogen bond in the pyridine N-oxide-trichloroacetic acid complex: an inelastic neutron scattering and computational study. Acta Chim. Slov. 2011, 58, 521–527.
[14] Balevicius V, Aidas K, Svoboda I, Fuess H. Hydrogen bonding in pyridine N-oxide/acid systems: proton transfer and fine details revealed by FTIR, NMR, and X-ray diffraction. J. Phys. Chem. A 2012, 116, 8753–8761.
[15] Desiderato R, Terry JC, Freeman GR, Levy HA. The molecular and crystal structure of 8-hydroxyqunoline N-oxide. Acta Crystallogr. 1971, 27, 2443–2447.
[16] Michele AD, Freda M, Onori G, Santucci A. Hydrogen bonding of water in aqueous solutions of trimethylamine-N-oxide and tert-butyl alcohol: a near-infrared spectroscopy study. J. Phys. Chem. A 2004, 108, 6145–6150.
[17] Rybarczyk-Pirek AJ, Lukomska-Rogala M, Wojtulewski S, Palusiak M. N-Oxide as a proton accepting group in multicomponent crystals: X-ray and theoretical studies on new p-nitropyridine-N-oxide cocrystals. Cryst. Growth Des. 2015, 15, 5802–5815.
[18] Roth AJK, Tretbar M, Stark CBW. Mimicking the active site of aldehyde dehydrogenases: stabilization of carbonylhydrates through hydrogen bonds. Chem. Commun. 2015, 51, 14175–14178.
[19] Goud NR, Babu NJ, Nangia A. Sulfonamide-pyridine-N-oxide cocrystals. Cryst. Growth Des. 2011, 11, 1930–1939.
[20] Baures PW, Wiznycia A, Beatty AM. Hydrogen bonding isosteres: bimolecular carboxylic acid and amine-N-oxide interactions mediated via C–H···O hydrogen bonds. Bioorg. Med. Chem. 2000, 8, 1599–1605.
[21] Babu NJ, Nangia A. Carboxamide–pyridine N-oxide heterosynthon for crystal engineering and pharmaceutical cocrystals. Chem. Commun. 2006, 1369–1371.
[22] Lin T-Y, Timasheff SN. Why do some organisms use a urea-methylamine mixture as osmolyte? Thermodynamic compensation of urea and trimethylamine N-oxide interactions with protein. Biochemistry 1994, 33, 12695–12701.
[23] Hille R. The mononuclear molybdenum enzymes. Chem. Rev. 1996, 96, 2757–2816.
[24] Richardson DJ, Bacterial respiration: a flexible process for a changing environment. Microbiology 2000, 146, 551–571.
[25] Morello JP, Petäjä-Repo UE, Bichet DG, Bouvier M. Pharmacological chaperones: a new twist on receptor folding. Trends Pharmacol. Sci. 2000, 21, 466–469.
[26] Hunger J, Ottosson N, Mazur K, Bonn M, Baker HJ. Water-mediated interactions between trimethylamine-N-oxide and urea. Phys. Chem. Chem. Phys. 2015, 17, 298–306.
[27] Desiraju GR, Ho PS, Kloo L, Legon, AC, Marquardt R, Metrangolo P, Politzer P, Resnati G, Rissanen K. Definition of the halogen bond (IUPAC Recommendations 2013). Pure Appl. Chem. 2013, 85, 1711–1713.
[28] Metrangolo P, Resnati G. (Eds) Halogen Bonding: Fundamentals and Applications, Springer, 2008.
[29] Messina MT, Metrangolo P, Panzeri W, Pilati T, Resnati G. Intermolecular recognition between hydrocarbon oxygen-donors and perfluorocarbon iodine-acceptors: the shortest O···I noncovalent bond. Tetrahedron 2001, 57, 8543–8550.
[30] Aakeroy CB, Wijethunga TK, Benton J, Desper J. Stabilizing volatile liquid chemicals using co-crystallization. Chem. Commun. 2015, 51, 2425–2428.
[31] Boubekeur K, Syssa-Magale J-L, Palvadeauc P, Schöllhorn B. Self-assembly of nitroxide radicals via halogen bonding-directional NO···I interactions. Tetrahedron Lett. 2006, 47, 1249–1252.

[32] Puttreddy R, Jurček O, Bhowmik S, Mäkelä T, Rissanen K. Very strong $^{-}N{-}X^{+}\cdots{}^{-}O{-}N^{+}$ halogen bonds. Chem. Commun. 2016, 52, 2338–2341.
[33] Aakeroy CB, Wijethunga TK, Desper J. Constructing molecular polygons using halogen bonding and bifurcated N-oxides. CrystEngComm 2014, 16, 28–31.
[34] Raffo PA, Cukiernika FD, Baggio RF. The three-component cocrystal 1,3,5- trifluoro-2,4,6-triiodobenzene–pyridine N-oxide–water (1/2/1) built up by halogen bonds, hydrogen bonds and π–π interactions. Acta Crystallogr. C 2015, 71, 84–88.
[35] Roy S, Biradha K. Exploration of salts and cocrystals of with acetic acid, monobasic and dibasic N-heterocycles, and N-oxides. Cryst. Growth Des. 2013, 13, 3232–3241.
[36] Hou G-G, Ma J-P, Wang L, Wang P, Dong Y-B, Huang R-Q. Co-crystallization of oxadiazole-bridged pyridyl-N-oxide building modules with R-aromatics (R = –OH, $-NH_2$ and –COOH). CrystEngComm 2010, 12, 4287–4303.
[37] Beyeh NK, Puttreddy R, Rissanen K. Aromatic N-oxide templates open inclusion and dimeric capsular assemblies with methylresorcinarene. RSC Adv. 2015, 5, 30222–30226.
[38] Puttreddy R, Beyeh NK, Rissanen K. Inclusion complexes of C_{ethyl}-2-methylresorcinarene and pyridine N-oxides: breaking the $C{-}I\cdots{}^{-}O{-}N^{+}$ halogen bond by host–guest complexation. CrystEngComm 2016, 18, 793–799.
[39] Sarma R, Baruah JB. B-π-aromatic and C–H···B interactions in cocrystals of aromatic amine N-oxides with p-phenylenediboronic acid. J. Mol. Struct. 2009, 920, 350–354.
[40] Sarma R, Bhattacharyya PK, Baruah JB. Short range interactions in molecular complexes of 1,4-benzenediboronic acid with aromatic N-oxides Comput. Theo. Chem. 2011, 963, 141–147.
[41] Mercurio JM, Tyrrell F, Cookson J, Beer DP. Neutral [2]rotaxane host systems that recognise halide anions in aqueous solvent mixtures. Chem. Commun. 2013, 49, 10793–10795.
[42] Burg AB, Bickerton JH. Trimethylamine oxide boron fluoride. J. Am. Chem Soc. 1945, 67, 2261.
[43] Perrin DD. Metal complexes with adenine 1-N-oxide and adenosine 1-N-oxide. J. Am. Chem Soc. 1960, 83, 5642–5645.
[44] Carlin RL. Transition metal complexes of pyridine N-oxide, J. Am. Chem Soc. 1961, 83, 3773–3775.
[45] Quagliano JV, Fujita J, Franz G, Phillips DJ, Walmsley JA, Tyree SS. The donor properties of pyridine N-oxide. J. Am. Chem Soc. 1961, 83, 3770–3773.
[46] Sager RS, Williams RJ, Watson WH. Structure of the oxyhyponitrite ion. Inorg. Chem. 1969, 8, 693–694.
[47] Estes ED, Hodgson DJ. Structural characterization of bis[dichloroaquo(pyridine N-oxide)copper(II)]. Inorg.Chem. 1976, 15, 348–352.
[48] Fabelo O, Pasan J, Lioret F, Julve M, Ruiz-Perez C. Structural versatility in cobalt(II) complexes with 1,2,4,5-benzenetetracarboxylic acid (H4bta) and 4,4′-bipyridine-N,N′-dioxide (dpo). CrystEngComm 2007, 9, 815–827.
[49] Sarma R, Boudalis AK, Baruah JB. Aromatic N-oxide bridged copper(II) coordination polymers: synthesis, characterization and magnetic properties, Inorg. Chim. Acta 2010, 363, 2279–2286.
[50] Sarma R, Karmakar A, Baruah JB. Synthesis and characterization of pyridine N-oxide complexes of manganese, copper and zinc. Inorg. Chim. Acta 2008, 361, 2081–2086.
[51] Sarma R, Karmakar A, Baruah JB. N-oxides in metal containing multi components molecular complexes, Inorg. Chem. 2008, 47, 763–765.
[52] Sarma R, Perumal A, Baruah JB. Some aspects of N-oxide bridged manganese (II) coordination polymers, J. Coord. Chem. 2009, 62, 1513–1524.
[53] Karaghouli ARAl, Day R, Wood JS. Crystal structure of tetrakis(2,2′-bipyridine dioxide) lanthanum perchlorate: an example of cubic eight-coordination, Inorg. Chem. 1978, 17, 3702–3706.

[54] Xu G, Zhang X, Guo P, Pan C, Zhang H, Wang C. MnII-based MIL-53 analogues: Synthesis using neutral bridging μ^2-ligands and application in liquid-phase adsorption and separation of C6-C8 aromatics. J. Am. Chem Soc. 2010, 132, 3656–3657.

[55] Dang D, Bai Y, He C, Wang J, Duan C, Niu J. Structural and catalytic performance of a polyoxometalate-based metal-organic framework having a lanthanide nanocage as a secondary building block, Inorg. Chem. 2010, 49, 1280–1282.

Elena V. Boldyreva

7 Multi-component crystals and non-ambient conditions

7.1 Introduction

In the last decade ever more research in the field of organic solid-state chemistry is dedicated to crystals, which include several chemical species, in particular, salts, co-crystals and their solvates [1, 2]. Multi-component crystals are of great interest for basic understanding of the factors that influence the formation of crystal structures from fluid phases under different experimental conditions and for their response to variations in temperature and pressure [1–3]. This is the next level in crystal engineering, since one has different fragments to self-assemble and one must control not only the formation of a certain polymorph, but also of a multi-component crystal vs several separate mono-component phases. Many of these systems are of practical importance, for example in the pharmaceutical industry [4–7], as ferroelectrics [8, 9], non-linear optical materials [10–12] and as energetic materials [13–17].

Non-ambient conditions are widely used when studying multi-component crystals in several contexts: for their synthesis as well as for probing different types of intermolecular interactions by varying temperature and/or pressure. Aspects of multi-component crystals under non-ambient conditions, specifically variable pressure conditions, form the focus of this chapter.

7.2 Synthesis of multi-component crystals at non-ambient conditions

Sometimes, it is thermodynamically preferable to form a multi-component crystal instead of the separate phases of individual components. Alternatively, a multi-component crystal can be formed as a metastable phase only. Non-ambient conditions can be required in both cases, but probably more often in the latter. Fast precipitation can result in multi-component crystals, salts or co-crystals, whereas longer crystallisation times under equilibrium conditions would produce phases of individual components separately. Spray-drying, freeze-drying, anti-solvent crystallisation, crystallisation from super-critical fluids as well as dry or liquid-assisted co-grinding can give the same or different multi-component crystalline phases, which are often metastable, but can be preserved indefinitely [7, 18–38]. When predicting experimental conditions under which a solid phase or multiple solid phases will be formed, it is equally important to consider the thermodynamic phase diagrams and the kin-

https://doi.org/10.1515/9783110464955-007

etics of the transformations [7, 18, 35, 39–41]. In many cases, desired phases form only when nuclei form very fast, i.e. on quick precipitation using anti-solvent techniques, high-temperature spray drying, cryogenic freeze drying and co-grinding [7]. This may be a manifestation of Ostwald's rule according to which the most stable equilibrium form does not precipitate on fast crystallisation, but rather the form that nucleates faster [42, 43]. A comparison of the phases formed when using different techniques, all of which are related to fast precipitation, can give a better insight into the mechanisms of crystallisation. Sometimes, different phases are formed, depending on whether the solution has been mixed on precipitation or not. For example, a bis-glycinium oxalate methanol di-solvate precipitates from a 2 : 1 glycine:oxalic acid solution on anti-solvent precipitation by methanol left to crystallise without stirring, whereas a bis-glycinium oxalate crystallises if methanol is added to the same solution and then mixed actively [44].

Co-grinding of powder mixtures is increasingly often used to produce multi-component crystals [7, 19, 22–26, 28, 35]. Grinding processes can now be monitored both ex situ at different stages of the transformation, as well as in situ using either X-ray diffraction, Raman spectroscopy, or both techniques simultaneously [7, 24, 26, 45–59]. Despite these developments, little is known about the mechanisms of these transformations or the temperatures and pressures generated at the contact between particles at the moment of collision. Further uncertainty surrounds the role of fluid phases, mixing and temperature, or how the type and exact protocol of mechanical action may affect the process. In some situations, a solid–solid reaction may not require mechanical treatment. Instead, one simply needs to ensure particle–particle contacts are formed between reactive species [7, 24]. In fact, it is sometimes possible to separate components by a membrane, through which a volatile component can penetrate in the gas phase, reacting on the surface of a second particle. In such cases, co-grinding simply ensures efficient reactant particle contacts through mixing or removing product layers, thereby refreshing the accessible, reactive surface of the starting material. In other cases, continuous grinding is not required at all. Here, it may suffice to mix a multi-component powder and leave it, as the transformation will start (possibly after some induction period) and proceed to completion [7, 24]. The presence of liquid (for example, even traces of water in air) can facilitate the process enormously [7, 24–26, 28, 54]. By contrast, if the mixture is kept in a dry box, for example under dry N_2, no transformation occurs. This suggests that the transformation does not, in fact, occur in the solid state, but in a fluid, be it a concentrated solution at the interface between solid particles, a mono-layer of liquid on the surface or even a vapour formed under hydrothermal conditions during particle-particle (or particle-milling body) collisions generating pulses of temperature and pressure [24, 60]. A water molecule can form a bridge between the reactive centres and act as a catalyst for the completion of the reaction. Polymeric excipients added to the mixture can also promote the reaction between its components, and one of the reasons for this may be that these polymers influence the water content in the mixture via its release and absorption [61–63]. An alternative is that the polymer acts as a template, a "binding site", at which the two components can be more easily brought into contact and react. Polymers can also increase the lifetime of higher-energy metastable states of the components and facilitate the reaction between the components in this way. Finally, polymeric additives can modify the rheology of the mixture and therefore assist in achieving the multiple inter-particle contacts required for an efficient reaction.

The origin of the fluid may vary. In some cases, samples may absorb moisture from the air. This is a particularly important phenomenon when mixtures of components are hygroscopic, even though their individual components are not. This has been termed contact-assisted hygroscopicity [7]. For systems containing crystal hydrates (or other solvated species), fluid may be formed on decomposition of these fluid-containing phases (also often assisted by the presence of another solid component). The molar content of water in a crystal hydrate is comparable with that of the compound itself. Thus, the presence of water as a potential reactant cannot be neglected. Local concentrations in solution at the interface will be high, a factor that may account for the fact that a single drop of liquid added to the solid mixture can be sufficient for a transformation (liquid-assisted grinding, LAG). In the presence of different liquids, different phases can form [64]. The amount of liquid and the form in which it has been introduced into the system can determine the stoichiometry and the crystal structure of the product [35]. Quite often, the same phases are formed on grinding as on fast anti-solvent crystallisation or spray-drying [24, 35]. Visually, the involvement of liquid in the mechanochemical synthesis can manifest itself as caking: a hard, concrete-like solid is formed from loose, free-flowing starting powder after mechanical treatment. Caking can be prevented by adding a water-absorbing polymer, such as PVP, to the mixture. In this way, the addition of such an "unreactive additive" can make a synthesis possible. A fluid phase can also be formed at the interface of the colliding solid particles, due to eutectic melting [7, 24, 66]. This phenomenon also manifests itself as caking, or a powder may become sticky. Sometimes, one can also detect that a mixture becomes warm to the touch upon grinding. Using thermo-couples, it is possible to follow the temperature at different sites of the milling jar, and this can give a rough estimate of the lower temperature limit that can arise on treatment. It is important to keep in mind that this value is likely significantly lower than the actual temperature at the contact interface between the particles. Estimates given in the literature range from 50 to 1000 °C at these interfaces. Obviously, this value depends on the materials, reactions and the treatment itself [24]. It is even more difficult to estimate the local pressures that can arise at the inter-particle contacts at the moment of collision. The presence of high pressure phases among the reaction products may indirectly indicate that such pressure has been reached during treatment. One must also consider coupling between pressure and heating. However, it is quite often found that the products of mechanical treatment in a mill differ qualitatively from those formed under hydrostatic pressures. Upon milling or grinding, high temperature phases tend to form more frequently than their high pressure counterparts [7, 24, 67].

The presence of fluid phases can also influence a mechanochemical synthesis by affecting mixing. Increasing humidity favours inter-particle bonding and the formation of ordered mixtures with less segregation of the ordered units. When mixed in fluidised beds (liquid–solid, gas–liquid–solids), cohesive powders cannot be fluidised homogeneously because of the tendency to channel. In cases where a reaction does not proceed on grinding of a dry mixture, but becomes possible when a fluid (one that does not dissolve the components) is added, one should consider the possibility that the fluid affects the rheology of the powder. In doing so, the fluid therefore affects mixing.

Mechanical action in an apparatus is not continuous, but comprises multiple short pulses [24, 68]. Each pulse impacts different particles and so even at high frequencies, it is difficult to approximate the mechanochemical reaction (even from the view of a single particle) as being continuous. A consequence is that for mechanochemical reactions the total time that a sample is being ground in a mill is not a kinetic parameter. Significantly different physical processes occur (i) during a mechanical pulse and (ii) between pulses. In the first case, pulses of temperature and pressure are generated. In the second case a relaxation process takes place. One should therefore consider two unique regimes in a mechanochemical process: the time during a mechanical pulse itself (t_1) and the time between pulses (t_2). By changing the pulse frequency, one can affect the ratio t_1 to t_2, and also in this way, the reaction. Quite often, infrequent pulses are more efficient for synthesis, since it is the time between pulses during which a chemical reaction takes place [65, 68, 69]. Another important issue to be considered is that mechanical action can be of different types, ranging from free or restricted impact to pure shear, with intermediate combinations also possible. The efficiency of impact or shear can vary for different reactions. For example, the treatment of a mixture of piroxicam with succinic acid by a series of restricted impacts produces a piroxicam-succinic acid co-crystal. Instead, shear action on the same mixture does not generate any new phases. Moreover, if a piroxicam-succinic acid co-crystal is prepared by co-crystallisation from solution and subsequently subjected to shear, it decomposes into two phases, piroxicam and succinic acid [70]. In a mill, where both shear and impact are possible, a transformation can be incomplete. Since the shear to impact ratios at different sites of a jar are different, different reaction products can form at different sites of the vessel during the ball-milling process [26, 65], and also at different moments in time – first the easily flowing powder sample experiences significant shear, whereas after a compact tablet is formed, the effect of impact becomes dominating [65]. It is also possible to change the impact-shear ratio by changing the type of instrument (using different types of mills, extruders), or by varying the frequency of vibrations using the same mill [65, 68–71].

7.3 Low temperature and high pressure studies of multi-component crystals

Variable-temperature and variable-pressure studies can be very helpful in order to study intermolecular interactions and structure-property relationships in crystals. An analysis of the anisotropy of lattice strain on variations of temperature has been used as a tool of probing the relative strength of different types of hydrogen bonds as shown by the work of Ubbelohde et al. [72–74] in the beginning of this century. During the last decade, studies of the anisotropy of strain on increasing pressure became a very important method for probing various types of intermolecular interactions [75–78].

Multi-component crystals can be studied under variable conditions either in the stable range of the same phase or over polymorphic transitions or solvation/desolvation processes [3]. In the first case, the main idea is to compare the bulk compressibilities and the anisotropy of strain of a multi-component crystal with those of the phases of individual components. This is especially interesting if some of the structure-forming motifs are the same (or similar) in a salt/co-crystal and the individual component structure [12, 79–83]: one can compare the properties of similar motifs in different crystalline environments.

Another research direction is related to modifying the stability of crystal structures with respect to structural rearrangements. Adding second and third components, one can modify or even radically change the crystal structure and, in this way, also the physical properties of a solid in a controlled way. In particular, a structure can become more robust, more elastic, more stable with respect to variations of temperature, pressure, humidity, more soluble or more compressible. The cases when the crystal structures of individual components have the same or similar structural motifs as those of salts/co-crystals, but the presence of another component has a significant effect on the structural stability are of special interest.

7.3.1 Structure stabilisation by a second component

Examples where pressure induced phase transitions are observed for crystals of pure amino acids, but not for their salts, have been provided by cysteinium oxalates. In the crystals of L- and DL-cysteine, the side chains $-CH_2SH$ can form weak hydrogen bonds either with the side chains of the neighbouring molecules (SH···S) or with the terminal carboxy- groups (SH···O) [84–87]. The crystals of the orthorhombic polymorph of L-cysteine show several phase transitions at pressures between 1 GPa and at 3.9 GPa. These transitions are related to the rotation of the side chains and the formation of new types of hydrogen bonds [88, 89]. The monoclinic polymorph of L-cysteine also undergoes a series of phase transitions with increasing pressure, which are related to the switching over of the hydrogen bonds within the –SH···SH···O– structural motif [89]. The crystal structure of DL-cysteine is even less stable with respect to increasing pressure and transforms to another polymorph at about 0.1 GPa (an unusually low pressure for a crystalline amino acid) [90].

In a subsequent study, it was hypothesised that crystal structures of cysteine derivatives could be stabilised with respect to phase transitions by reducing the mobility of the side chain. This was achieved by enhancing the role of the –SH group of the side chain by encouraging the formation of intermolecular hydrogen bonds. The methods employed were (i) by N-acetylation (to remove the strong $-NH_3^+$ donor and negating the influence of strong charge-assisted interactions) and (ii) by co-crystallisation with an acid, e.g. oxalic acid, (to convert the amino acid to a cation to make it a strong anion hydrogen bond acceptor to enhance the importance of potential –SH donors). The crystal structures of three compounds N-acetyl-L-cysteine, DL-cysteinium semi-oxalate, and bis-(DL-cysteinium) oxalate were investigated under conditions of varying of temperature and pressure. Increasing pressure to at least 9.5 GPa did not result in phase transitions for either N-acetyl-L-cysteine or bis-(DL-cysteinium) oxalate. However, in the case of DL-cysteinium semi-oxalate, at 6 GPa, a phase transition occurred cf. to pressure induced phase transitions observed earlier for both monoclinic and orthorhombic L-cysteine at 2.5–3.9 GPa and 1.1–2.5 GPa, respectively, and DL-cysteine at 0.1–5 GPa. The phase transition for DL-cysteinium semi-oxalate was accompanied by a large hysteresis so that the reverse transformation induced decompression was observed at 3.7 GPa only. These changes were correlated with changes in molecular conformations and reorganisation in the N–H···O hydrogen bonds in the crystal [91].

7.3.2 Structure destabilisation by a second component

An opposite effect to that described in Section 7.3.1 is also possible. A multi-component crystal can undergo a structural rearrangement under milder conditions than the crystals of the pure components. Examples include the salts and co-crystals of glycine. Glycine is the simplest amino acid with no side chain. The glycine zwitterions are held together in the crystalline state by N–H···O hydrogen bonds between the terminal amino- and carboxy- groups. Three polymorphs are known under ambient conditions (α-glycine [92, 93], β-glycine [94], γ-glycine [95]), depending on the way the zwitterions are connected. Upon increasing pressure, the crystal structure can undergo a reversible phase transition without destroying the single crystal at low pressure (β-polymorph, 0.76 GPa [96–98]) or a reconstructive phase transition from a single crystal to powder at a higher pressure (γ-polymorph, 3.5 GPa [97, 99–101]), or can remain incredibly robust (α-polymorph, at least up to 23 GPa [102]). It was assumed that the remarkable stability of the α-polymorph is provided by the centrosymmetric double layers linked by N–H···O hydrogen bonds. So, if this motif is destroyed by introducing a second component, one can expect that the new crystal structure will be less stable with respect to pressure than α-glycine.

In fact, phase transitions have been reported for several multi-component crystals containing glycine. Bis-glycinium oxalate undergoes two phase transitions, one at 1.7 and the other at 5 GPa. These are accompanied by the significant rearrangement of hydrogen bonds in the crystal and violent crystal fragmentation, as evidenced by IR-spectroscopy and X-ray powder diffraction [103–105]. Based on the IR-spectroscopic data, it was supposed that the whole hydrogen bond network is radically rearranged so that some N–H. . . O hydrogen bonds are broken while other, weaker N–H. . . O hydrogen bonds are formed, resulting in two non-identical glycine molecules in the asymmetric unit. At 5 GPa, another reorganisation of N–H···O hydrogen bonds takes place [103–105]. Another salt, 1 : 1 glycinium oxalate, does not undergo structural phase transitions upon increasing pressure, though symmetrisation of the O–H···O hydrogen bonds occurs at 8 GPa [106]. No phase transitions have been observed in either of the salts during cooling. A co-crystal of glycine with glutaric acid, GGa, undergoes a single-crystal to single-crystal polymorphic transition at pressures below 0.1 GPa, which is a much lower transition pressure than has been observed for any of the individual amino acids [107]. The same polymorph is formed also upon cooling [107, 108]. The polymorphic transformations upon cooling and with increasing pressure are related to the conformational changes in the flexible glutaric acid: every second glutaric acid molecule in the structure changes its shape. The topology of the hydrogen bonded motifs remains almost the same and hydrogen bonds do not switch to other atoms, although the hydrogen bond lengths do change and some of the bonds become non-equivalent [107, 108]. Co-crystals of glycine with other co-formers, -DL-tartaric and phthalic acids (GTa and GPh, respectively), behave differently than those with glutaric acid. Upon cooling, no phase transitions were observed in GTa or GPh, contrasting the situation with GGa. On hydrostatic compression, both GTa and GPh underwent reversible phase transformations accompanied by fracture. In the high pressure phases, the main structural framework is preserved but the number of crystallographically independent molecules increases. In GTa, dimers are squeezed together so that some hydrogen bonds acquire three-centred character and the interactions of one of the two glycine molecules changes dramatically [109].

High-pressure behaviour was studied also for another multi-component crystal based on glycine, namely, for glycine lithium sulphate [110]. Raman studies and DFT calculations suggested the formation of an intramolecular hydrogen bond at higher pressures. Subsequent to a structural transformation to a new high pressure phase at 9 GPa, the observed spectral changes in the Raman spectra above 14 GPa indicated possible conformational change of glycine from a zwitterionic to a neutral form. At pressures above 18 GPa, the characteristic features in the Raman spectra and the X-ray diffraction patterns suggested transformation to a hydrogen bond assisted polymeric phase with intermediate range order [110].

Another example of making a structure more prone to a high-pressure phase transformation is by adding a second co-former, and in this way an oxalic salt of DL-alaninium, namely DL-alaninium semi-oxalate monohydrate allows for the breaking of homosynthons and radical change to the whole crystal structure [111]. DL-alanine itself does not show any phase transitions with increasing pressure [112]. The molecules in the salt are linked via N–H···O hydrogen bonds into a three-dimensional framework [113, 114]. Similar to the structure of L-alanine [115–118], the CH_3– side chains do not form any hydrogen bonds, and their rotation does not result in any phase transitions either in L- [119–121] or in DL-alanine [112]. In contrast, DL-alaninium semi-oxalate monohydrate, in which the framework found in the pure amino acid is broken, undergoes a single-crystal to single-crystal phase transition at a pressure between 1.5 and 2.4 GPa [111]. Selected hydrogen bonds involving alaninium cations and crystal water molecules switch over and become bifurcated, whereas others are compressed continuously. The transition is accompanied by pronounced discontinuities in the cell parameters and volume versus pressure, although no radical changes in the molecular packing are induced. Remarkably, the rigid homosynthon chain motifs of alaninum cations and of oxalate anions are preserved with increasing pressure, and no proton transfer is observed [111].

7.3.3 A change in the transformation mechanism in the presence of a second component in a similar pressure range

It is also possible that a structural phase transition in a multi-component crystal is observed in the same pressure range as for the crystals of the primary basic component, but the structural mechanism changes. An example of this is provided by a comparison of the effect of pressure on L- and DL-serine [109–115], L-serine monohydrate [116, 117], bis-DL-serinium oxalate dihydrate [118], L-serinium maleate and the L-ascorbic acid–L-serine co-crystal [119].

L-serine has a framework crystal structure [120] similar to those of L-alanine [121], DL-alanine [122] and the orthorhombic polymorph of L-cysteine [84, 85], and its side chain, $-CH_2OH$, is involved in the formation of intermolecular O–H···O hydrogen bonds which are stronger than S–H···O or S–H···S hydrogen bonds. At ambient pressure, O–H···O hydrogen bonds are formed between the $-CH_2OH$ groups of neighbouring chains of molecules. As pressure increases, a cooperative rotation of the side chains results in a structural phase transition at about 5 GPa, so that new O–H···O hydrogen bonds between the side chains and the terminal carboxy- groups are formed. As pressure increases further, the side chains rotate at about 8 GPa, to once again form a new phase with additional bifurcated hydrogen bonds [109, 110, 112–115]. The two reversible phase transitions preserve the integrity of the single crystals, and the interface propagates rapidly through the crystal along one of the crystallographic directions. This is possible since the main structure is held by highly elastic N–H···O hydrogen bonds between the terminal amino- and carboxy- groups, which are preserved, while the side chains rotate. The layered structure of DL-serine in which the $-CH_2OH$ side chains form O–H···O bonds to the carboxy- groups at ambient pressure [111] does not undergo any structural phase transition with increasing pressure at least up to 8.6 GPa [111], though non-monotonic distortion of selected hydrogen bonds in the pressure range below 1–2 GPa, a change in the compression mechanism at 2–3 GPa and evidence of the formation of bifurcated N–H···O hydrogen bonds at 3–4 GPa were reported [115].

The crystal structure of L-serine monohydrate (L-serine monohydrate-I) at ambient pressure contains hydrogen bonded layers of zwitterionic serine molecules linked by hydrogen bonds to water molecules [116, 117]. The water molecules act as donors to oxygen atoms on carboxylate- and alcohol-groups in separate layers. This phase remains stable from ambient pressure to 4.5 GPa; the most prominent structural change in this range is a reduction in the interlayer distance. On increasing the pressure to 5.2 GPa, the structure transforms to a high pressure polymorph, termed L-serine monohydrate-II. During the transition, the serine interlayer distances reduce further and the water molecules rotate so that both donor interactions are made to the same serine layer. The serine ammonium group adopts an eclipsed conformation, reconfiguring the hydrogen bonding within the serine layers. The disruption of hydrogen bonding as water is pushed into the serine layers suggests that a similar process may occur as a first step in the pressure denaturation of proteins. Though the water molecules become coordinatively saturated with respect to hydrogen bonding, and the interlayer serine-serine Coulombic interactions are strengthened, model calculations suggest that overall, the intermolecular interactions are weaker in phase-II than phase-I. According to the authors of the work, the lattice enthalpy becomes more negative through the transition as a result of the smaller PV term applying to the more efficiently packed phase-II structure [117].

Another multi-component crystal containing serine, namely bis-(DL-serinium) oxalate dihydrate, with a three-dimensional network of hydrogen bonds linking oxalate anions, serinium cations and water molecules, shows a reversible phase transition also at about 4–5 GPa, resulting in the formation of multiple disordered domains because of the reorientation of molecules and molecular fragments, which are no longer hold together as robustly as the homosynthon hydrogen bonds hold molecules in L- or DL-serine [118].

7.3.4 Charge (proton) transfer between the components

Pressure is a well-known influence on proton transfer between molecules in mono-component molecular crystals. Both the symmetrisation and, the reverse, an increase of ionicity of the hydrogen bonds (O–H···O or N–H···N) has been reported in the literature [3, 106, 123–128]. An increase in the ionicity of the intermolecular N–H···O hydrogen bonds between the terminal amino- and carboxy- groups with increasing pressure has been reported for L-alanine [129, 130]. A prominent example of a complete intermolecular proton transfer, given to a salt starting as a molecular hydrate was reported for oxalic acid dihydrate [131–133]. Interestingly, this transfer takes place only if the α-polymorph is compressed, but it was not observed for the β-polymorph [132]. Other examples of systems for which the interplay between structural transformations and solid state acid-base (proton transfer) reactions at high pressure that have been studied in relation to paraelectric and ferroelectric properties, are the co-crystals of phenazine with chloranilic, bromanilic and fluoranilic acids [8] and co-crystals of anilic acids and 2,3-di(2-pyridinyl)pyrazine [9], where hydrogen bonds are polarised and neutral molecules are converted into ionised species upon cooling or with increasing pressure.

7.4 Conclusions

Non-ambient conditions are widely used to prepare salts, solvates and co-crystals and to study their properties. As one can see from the examples discussed above, the variety of phenomena that have been observed for a very limited number of systems is enormous. In order to not get lost in the ocean of facts that will only increase as more systems are studied, research could be conducted in the following main-stream directions [3]:

- Use series of co-crystals with the same primary component and variable co-formers to follow the main trends of the response of these structures to increasing pressure. In particular, this approach can be useful for polymorph screening of drugs and for predicting the mechanical robustness of non-linear optical molecular materials as well as for finding ways to stabilise a crystal structure by a carefully chosen co-former;
- Explore the phase diagrams in multi-component systems in order to understand under which conditions a multi-phase crystal is formed at high pressure, at low temperature, and when it decomposes into the separate phases of the pure components. This research would be important for crystal engineering: in particular, for crystal engineering of co-crystals and for understanding the mechanisms of the non-thermal synthesis of multi-component systems, including pharmaceutical co-crystals with and without fluid phases added to the system;
- One can especially consider "multiple Z′ crystals" with chemically identical but symmetrically non-equivalent molecules in a unit cell as a particular case of multi-component crystals and follow the evolution of the Z′ with increasing and decreasing pressure.

Bibliography

[1] Herbstein FH. Crystalline Molecular Complexes and Compounds. Structures and Principles. Oxford University Press 2005, 1273p.

[2] Resnati G, Bombicz P, Kawano M, Boldyreva E. Supramolecular interactions in the solid state. IUCRJ 2015, 2, 675–690.

[3] Boldyreva EV. Multicomponent molecular organic crystals at high pressures. Z. Kristallogr. 2014, 229, 236–245.

[4] Trask AV. An overview of pharmaceutical cocrystals as intellectual property. Mol. Pharm. 2007, 4, 301–309.

[5] Good DJ, Naír R.-H. Solubility advantage of pharmaceutical cocrystals. Cryst. Growth Des. 2009, 9, 2252–2264.

[6] Schultheiss N, Newman A. Pharmaceutical co-crystals and their physicochemical properties. Cryst. Growth Des. 2009, 9, 2950–2967.

[7] Boldyreva EV. Non-ambient conditions in the investigation and manufacturing of drug forms. Current Pharm Des 2016, 22, 4981–5000.

[8] Kumai R, Horiuchi S, Fujioka J, Tokura Y. Ferroelectricity and pressure-induced phenomena driven by neutral ionic valence instability of acid-base supramolecules. J. Am. Chem. Soc. 2012, 134, 1036–1046.

[9] Horiuchi S, Kumai R, Tokura Y. High-temperature and pressure-induced ferroelectricity in hydrogen-bonded supramolecular crystals of anilic acids and 2,3-Di(2- pyridinyl)pyrazine. J. Am. Chem. Soc. 2013, 135, 4492–4500.

[10] Fleck M, Petrosyan AM. Salts of Amino Acids. Crystallization, Structure and Properties. Springer International Publishing. Switzerland 2014.

[11] Ghazaryan VV, Zakharov BA, Petrosyan AM, Boldyreva EV. L-argininium phosphite – a new candidate for NLO-materials. Acta Crystallogr. C 2015, 71, 415–421.

[12] Arkhipov SG, Rychkov DA, Pugachev AM, Boldyreva EV. New hydrophobic L-amino acid salts: maleates of L-leucine, L-isoleucine and L-norvaline. Acta Crystallogr. C 2015, 71, 584–592.

[13] Oswald IDH, Pulham CR. Co-crystallisation at high pressure – an additional tool for the preparation and study of co-crystals. CrystEngComm 2008, 10, 1114–1116.

[14] Bolton O, Matzger AJ. Improved stability and smart-material functionality realized in an energetic cocrystal. Angew. Chem. Int. Ed. 2011, 50, 8960–8963.

[15] Lin H, Zhu S-G, Zhang L, Peng X-H, Chen P-Y, Li H-Z. Intermolecular interactions, thermodynamic properties, crystal structure, and detonation performance of HMX/NTO cocrystal explosive. Int. J. Quantum Chem. 2013, 113, 1591–1599.

[16] Yang Z, Li H, Huang H, Zhou X, Li J, Nie F. Preparation and performance of a HNIW/TNT cocrystal explosive. Propellants, Explosives, Pyrotechnics 2013, 38, 495.

[17] Zhang H, Guo C, Wang X, Xu J, He X, Liu Y, Liu X, Huang H, Sun J. Five energetic cocrystals of BTF by intermolecular hydrogen bond and π-stacking interactions. Cryst. Growth Des. 2013, 13, 679–687.

[18] Padrela L, Rodrigues M A, Velaga S P, Matos H A, Azevedo EG. Formation of indomethacin–saccharin cocrystals using supercritical fluid technology. Eur. J. Pharm. Sci. 2009, 38, 1, 9–17.

[19] Fucke K, Myz SA, Shakhtshneider TP, Boldyreva EV, Griesser UJ. How good are the crystallisation methods for co-crystals – a comparative study of Piroxicam. New J. Chem. 2012, 36, 1969.

[20] Neurohr C, Revelli AL, Billot P, Marchivie M, Lecomte S, Laugier S, Massip S, Subra-Paternault P. Naproxen–nicotinamide cocrystals produced by CO_2 antisolvent. J. Supercrit. Fluids 2013, 83, 78–85.

[21] Lee MJ, Wang IC, Kim MJ, Kim P, Song KH, Chun NH, Park HG, Choi GJ. Controlling the polymorphism of carbamazepine-saccharin cocrystals formed during antisolvent cocrystallization using kinetic parameters. Korean J. Chem. Eng. 2015, 32, 1910–1917.

[22] Trask AV, Jones W. Crystal engineering of organic co-crystals by the solid-state grinding approach. Top. Curr. Chem. 2005, 254, 41–70.

[23] Myz SA, Shakhtshneider TP, Fucke K, Fedotov AP, Boldyreva EV, Boldyrev VV, Kuleshova NI. Synthesis of co-crystals of meloxicam with carboxylic acids by grinding. Mendeleev Comm. 2009, 19, 272–274.

[24] Boldyreva E. Mechanochemistry of inorganic and organic systems: What is similar, what is different? Chem. Soc. Rev. 2013, 42, 7719–7738.

[25] Braga D, Maini L, Grepioni F. Mechanochemical preparation of co-crystals. Chem. Soc. Rev. 2013, 42, 7638–7648.

[26] Michalchuk AAL, Tumanov IA, Boldyreva EV. Complexities of mechanochemistry: elucidation of processes occurring in mechanical activators via implementation of a simple organic system. CrystEngComm, 2013, 15, 6403–6412.

[27] Alhalaweh A, Velaga SP. Formation of cocrystals from stoichiometric solutions of incongruently saturating systems by spray drying. Cryst. Growth Des. 2010, 10, 3302–3305.

[28] Myz SA, Shakhtshneider TP, Tumanov NA, Boldyreva EV. Preparation and studies of the co-crystals of meloxicam with carboxylic acids. Russian Chem. Bull. 2012, 61, 1798–1809.

[29] Alhalaweh A, Kaialy W, Buckton G, Gill H, Nokhodchi A, Velaga SP. Theophylline cocrystals prepared by spray drying: physicochemical properties and aerosolization performance. AAPS Pharm. Sci. Tech. 2013, 14, 265–276.

[30] Lee MJ, Chun NH, Wang IC, Liu JJ, Jeong MY, Choi GJ. Understanding the formation of indomethacin–saccharin cocrystals by anti-solvent crystallization. Cryst. Growth Des. 2013, 13, 2067–2074.

[31] Paudel A, Worku ZA, Meeus J, Guns S, Van den Mooter G. Manufacturing of solid dispersions of poorly water soluble drugs by spray drying: formulation and process considerations. Inter. J. Pharm. 2013, 453, 253–284.

[32] Ober CA, Montgomery SE, Gupta RB. Formation of itraconazole/L-malic acid co-crystals by gas antisolvent cocrystallization. Powder Technol. 2013, 236, 122–131.

[33] Fernández-Ronco MP, Kluge J, Mazzotti M. High pressure homogenization as a novel approach for the preparation of co-crystals. Cryst. Growth Des. 2013, 13, 2013–2024.

[34] Eddleston MD, Patel B, Day GM, Jones W. Cocrystallization by freeze-drying: preparation of novel multicomponent crystal forms. Cryst. Growth Des. 2013, 13, 4599–4606.

[35] Losev EA, Mikhailenko MA, Achkasov AF, Boldyreva EV. The effect of carboxylic acids on glycine polymorphism, salt and co-crystal formation. A comparison of different crystallisation techniques. New J. Chem. 2013, 37, 1973–1981.

[36] Patil SP, Modi SR, Bansal AK. Generation of 1 : 1 carbamazepine : nicotinamide cocrystals by spray drying. Eur. J. Pharm. Sci. 2014, 62, 251–257.

[37] Aitipamula S, Chow PS, Tan RB. Polymorphism in cocrystals: a review and assessment of its significance. CrystEngComm 2014, 16, 3451–3465.

[38] Grossjohann C, Serrano DR, Paluch KJ, O'Connell P, Vella-Zarb L, Manesiotis P, Healy AM. Polymorphism in sulfadimidine/4-aminosalicylic acid cocrystals: solid-state characterization and physicochemical properties. J. Pharm. Sci. 2015, 104, 1385–1398.

[39] Rager T, Hilfiker R. Co-crystal formation from solvent mixtures. Cryst. Growth Des. 2010, 10, 3237–3241.

[40] Rager T, Hilfiker R. Stability domains of multi-component crystals in ternary phase diagrams. Z. Phys. Chem. 2009, 223, 793–813.

[41] Zakharov BA, Ogienko AG, Yunoshev AS, Ancharov AI, Boldyreva EV. Bis(paracetamol) pyridine – a new elusive paracetamol solvate. From modeling the phase diagram to successful single-crystal growth and structure-property relations. CrystEngComm 2015, 17, 7543–7550.

[42] Ostwald W. Studien über die Bildung und Umwandlung fester Körper. 1. Abhandlung: Übersättigung und Überkaltung. Z Phys. Chem. 1897, 22, 289–330.

[43] Threlfall T. Structural and thermodynamic explanations of Ostwald's Rule. Org. Process Res. Devel. 2003, 7, 1017–1027.

[44] Tumanov NA, Boldyreva EV, Shikina NE. Two new structures in the glycine–oxalic acid system. Acta Crystallogr. C 2010, 66, o279–283.

[45] Tumanov IA, Achkasov AF, Boldyreva EV, Boldyrev VV. About the possibilities to detect intermediate stages in mechanochemical synthesis of molecular complexes. Russ. J. Phys. Chem. A 2012, 86, 1014–1017.

[46] Tumanov IA, Achkasov AF, Boldyreva EV, Boldyrev VV. Following the products of mechanochemical synthesis step by step. CrystEngComm 2011, 13, 2213–2216.

[47] Lien Nguyen K, Friščić T, Day GM, Gladden LF, Jones W. Terahertz time-domain spectroscopy and the quantitative monitoring of mechanochemical cocrystal formation. Nature Mater 2007, 6, 206–209.

[48] Halasz I, Puškari A, Kimber SAJ, Beldon PJ, Belenguer AM, Adam F, Honkimäki V, Dinnebier RE, Patel B, Jones W, Štrukil, V, Friščić, T. Real-time in situ powder X-ray diffraction monitoring of mechanochemical synthesis of pharmaceutical cocrystals. Angew. Chem. Int. Ed. 2013, 52, 11538–11541.

[49] Friščić T, Halasz I, Beldon PJ, Belenguer AM, Adams F, Kimber SAJ, Honkimäki V, Dinnebier RE. Real-time and in situ monitoring of mechanochemical milling reactions. Nature Chem. 2013, 5, 66–73.

[50] Halasz I, Kimber SAJ, Beldon PJ, Belenguer AM, Adams F, Honkimäki V, Nightingale RC, Dinnebier RE, Friščić T. In situ and real-time monitoring of mechanochemical milling reactions using synchrotron X-ray diffraction. Nature Protocols 2013, 8, 1718–1729.

[51] Lin H-L, Zhang G-C, Hsu P-C, Lin S-Y. A portable fiber-optic Raman analyzer for fast real-time screening and identifying cocrystal formation of drug-coformer via grinding process. Microchem. J. 2013, 110, 15–20.

[52] Boldyreva EV. Dynamics of mechanochemical processes. In: Howard JAK, Sparkes HA, Raithby PR, Churakov AV (eds). The Future of Dynamic Structural Science. NATO Science for Peace and Security Series A: Chemistry and Biology 2014, 77–89.

[53] Halasz I, Friščić T, Kimber SAJ, Užarević K, Puškarić A, Mottillo C, Julien P, Štrukil V, Honkimäki V, Dinnebier RE. Quantitative in situ and real-time monitoring of mechanochemical reactions. Faraday Discuss 2014, 170, 203–221.

[54] Gnutzmann T, Nguyen Thi Y, Rademann K, Emmerling F. Solvent-triggered crystallization of polymorphs studied in situ. Cryst. Growth Des. 2014, 14, 6445–6450.

[55] Gracin D, Štrukil V, Friščic T, Halasz I, Užarevic K. Laboratory real-time and in situ monitoring of mechanochemical milling reactions by Raman spectroscopy. Angew. Chem. Int. Ed. 2014, 53, 6193–6197.

[56] Mandala VS, Loewus SJ, Mehta MA. Monitoring cocrystal formation via in situ solid-state NMR. J. Phys. Chem. Lett. 2014, 5, 3340–3344.

[57] Batzdorf L, Fischer F, Wilke M, Wenzel K-J, Emmerling F. Direct in situ investigation of milling reactions using combined X-ray diffraction and Raman spectroscopy. Angew. Chem. Int. Ed. 2015, 54, 1799–1802.

[58] Katsenis AD, Puškarić A, Štrukil V, Mottillo C, Julien PA, Užarević K, Pham M-H, Do T-O, Kimber SAJ, Lazić P, Magdysyuk O, Dinnebier RE, Halasz I, Friščić T. In situ X-ray diffraction monitoring of a mechanochemical reaction reveals a unique topology metal-organic framework. Nature Commun. 2015, 6, art. no. 6662.

[59] Julien PA, Užarević K, Katsenis AD, Kimber SA, Wang T, Farha OK, Zhang Y, Casaban J, Germann LS, Etter M, Dinnebier RE, James SL, Halasz I, Friščić T. In situ monitoring and mechanism of the mechanochemical formation of a microporous MOF-74 framework. J. Am. Chem. Soc. 2016, 138, 2929–2932.

[60] Boldyrev VV. Hydrothermal reactions under mechanochemical action. Powder Techn. 2002, 122, 247–254.

[61] Hasa D, Schneider G, Voinovich D, Jones W. Cocrystal formation through mechanochemistry: from neat and liquid-assisted grinding to polymer-assisted grinding. Angew. Chem. Int. Ed. 2015, 54, 7371–7375.

[62] Good D, Miranda C, Rodríguez-Hornedo N. Dependence of cocrystal formation and thermodynamic stability on moisture sorption by amorphous polymer. CrystEngComm 2011, 13, 1181–1189.

[63] Hasa D, Carlino E, Jones W. Polymer-assisted grinding, a versatile method for polymorph control of cocrystallization. Cryst. Growth. Des. 2016, 16, 1772–1779.

[64] Braga D, Giaffreda SL, Grepioni F, Chierotti MR, Gobetto R, Palladino G, Polito M. Solvent effect in a "solvent free" reaction. CrystEngComm 2007, 9, 879–881.

[65] Michalchuk AAL, Tumanov IA, Drebushchak VA, Boldyreva EV. Advances in elucidating mechanochemical complexities via implementation of a simple organic system. Faraday Discuss. 2014, 170, 311–335.

[66] Rastogi RP, Bassi PS. Mechanism of eutectic crystallization. J. Phys. Chem. 1964, 68, 2398–2406.

[67] Boldyreva EV, Shakhtshneider TP, Ahsbahs H, Sowa H, Uchtmann H. A comparative study of the effect of hydrostatic pressure up to 4 GPa on the monoclinic and the orthorhombic polymorphs of paracetamol. J. Therm. Analys. Calorim. 2001, 68, 437–452.

[68] Boldyrev VV. Kinetic factors determining the specifics of mechanochemical processes in inorganic systems. Kinet. Katal. 1972, 13, 1411–1421.

[69] Politov AA, Golyazimova OV. Increasing the energy yield of mechanochemical transformations: Selected case studies. Faraday Discuss. 2014, 170, 345–356.

[70] Tumanov IA, Achkasov AF, Myz SA, Boldyreva EV, Boldyrev VV. Different effect of impact and shear mechanical treatment on mechanochemical cocrystallization of piroxicam and succinic acid. Doklady Chem. 2014, 457, 154–159.

[71] Zyryanov VV. Mechanochemical synthesis of complex oxides. Russ. Chem. Rev. 2008, 77, 105–135.

[72] Robertson JM, Ubbelohde AR. Structure and thermal properties associated with some hydrogen bonds in crystals. I. The isotope effect. Proceedings of the Royal Society of London. Series A. Math. Phys. Sci. 1939, 222–240.

[73] Robertson JM, Ubbelohde AR. Structure and thermal properties associated with some hydrogen bonds in crystals. II. Thermal expansion. Proc of the Royal Society of London. Series A. Math. Phys. Sci. 1939, 241–251.

[74] Ubbelohde AR, Woodward I. Structure and thermal properties of crystals. VI. The role of hydrogen bonds in Rochelle salt. Proc. of the Royal Society of London A. Math. Phys. Eng. Sci. The Royal Society 1946, 185, 448–465.

[75] Boldyreva EV. Anisotropic compression. What can it teach us about intermolecular interactions? In: Boldyreva E, Dera P (eds). High-Pressure Crystallography. From Novel Experimental Approaches to Applications in Cutting-Edge Technologies. Springer: Dordrecht 2010, 147–159.

[76] Cai W, Katrusiak A. Pressure effects on H-ordering in hydrogen bonds and interactions in benzoic acid. CrystEngComm 2012, 14, 4420–4424.

[77] Zieliński W, Katrusiak A. Hydrogen bonds NH···N in compressed benzimidazole polymorphs. Cryst. Growth Des. 2012,13, 696–700.

[78] Sikora M, Katrusiak A. Pressure-controlled neutral–ionic transition and disordering of NH···N hydrogen bonds in pyrazole. J. Phys. Chem. C 2013, 117, 10661–10668.
[79] Arkhipov SG, Zakharov BA, Boldyreva EV. Semi-maleate salts of l- and dl-serinium: The first example of chiral and racemic serinium salts with the same composition and stoichiometry. Acta Crystallogr. C 2013, 69, 5, 517–521.
[80] Rychkov D, Arkhipov S, Boldyreva E. Structure-forming units of amino acid maleates. Case study of l-valinium hydrogen maleate. Acta Crystallogr. B 2016, 72, 160–163.
[81] Arkhipov SG, Losev EA, Boldyreva EV. New 1 : 1 and 2 : 1 salts in the "DL-norvaline–maleic acid" system as an example of assembling various crystal structures from similar supramolecular building blocks, Acta Crystallogr. C 2017, 73, 13–19.
[82] Zakharov BA, Boldyreva EV. DL-alaninium semi-oxalate monohydrate. Acta Crystallogr. C 2011, 67, o47–51.
[83] Zakharov BA, Kolesov BA, Boldyreva EV. Monitoring selected hydrogen bonds in crystal hydrates of amino acid salts: combining variable-temperature single-crystal X-ray diffraction and polarized Raman spectroscopy. Phys. Chem. Chem. Phys. 2011, 13, 13106–13116.
[84] Kerr KA, Ashmore JP. Structure and conformation of orthorhombic L-cysteine. Acta Crystallogr. B 1973, 29, 2124–2127.
[85] Kerr KA, Ashmore JP, Koetzle F. A neutron diffraction study of L-cysteine. Acta Crystallogr. B 1975, 31, 2022–2026.
[86] Görbitz CH, Dalhus B. L-cysteine, monoclinic form, redetermination at 120 K. Acta Crystallogr. C 1996, 52, 1756–1759.
[87] Luger P, Weber M. DL-Cysteine at 298 K. Acta Crystallogr. C 1999, 55, 1882–1885.
[88] Moggach SA, Allan DR, Clark SJ, Gutmann MJ, Parsons S, Pulham CR, Sawyer L. High-pressure polymorphism in L-cysteine: The crystal structures of L-cysteine-III and L-cysteine-IV. Acta Crystallogr. B 2006, 62, 296–309.
[89] Minkov VS, Goryainov SV, Boldyreva EV, Görbitz C-H. Raman study of pressure-induced phase transitions in the crystals of orthorhombic and monoclinic polymorphs of L-cysteine: dynamics of the side-chain. J. Raman Spectrosc. 2010, 41, 1458–1468.
[90] Minkov VS, Tumanov NA, Cabrera RQ, Boldyreva EV. Low temperature/high pressure polymorphism in DL-cysteine. CrystEngComm 2010, 12, 2551- 60.
[91] Minkov VS, Boldyreva EV, Drebushchak TN, Görbitz C.-H. Stabilizing structures of cysteine-containing crystals with respect to variations of temperature and pressure by immobilizing amino acid side chains. CrystEngComm 2012, 14, 5943–5954.
[92] Marsh RE. A refinement of the crystal structure of glycine. Acta Crystallogr. 1958, 11, 654–663.
[93] Jönsson P-G, Kvick Å. Precision neutron diffraction structure determination of protein and nucleic acid components. III. The crystal and molecular structure of the amino acid alpha-glycine. Acta Crystallogr. B 1972, 28, 1827–1833.
[94] Iitaka Y. A new form of glycine. Proc. Japan Soc. 1954, 30, 109–112.
[95] Iitaka Y. The crystal structure of γ-glycine. Acta Crystallogr. 1958, 11, 225–226.
[96] Goryainov SV, Kolesnik EN, Boldyreva EV. A pressure-induced phase transition in β-glycine at 0.76 GPa. Physica B 2005, 357, 340–347.
[97] Dawson A, Allan DR, Belmonte SA, Clark SJ, David WIF, McGregor PA, Parsons S, Pulham CR, Sawyer L. Effect of high pressure on the crystal structures of polymorphs of glycine. Cryst. Growth Des. 2005, 5, 1415–1427.
[98] Tumanov NA, Boldyreva EV, Ahsbahs H. Structure solution and refinement from powder or single-crystal diffraction data? Pros and cons: An example of the high-pressure beta′-polymorph of glycine. Powder Diffr. 2008, 23, 307–316.

[99] Boldyreva EV, Ivashevskaya SN, Sowa H, Ahsbahs H, Weber H.-P. Effect of high pressure on the crystalline glycine: a new high-pressure polymorph formation. Doklady Phys. Chem. 2004, 396, 111–114.
[100] Ivashevskaya SN, Boldyreva EV, Sowa H, Ahsbahs H, Weber H.-P. Effect of high pressure on the crystalline glycine: a new high-pressure polymorph formation. Acta Crystallogr. A 2004, 60, s250.
[101] Boldyreva EV, Ivashevskaya SN, Sowa H, Ahsbahs H, Weber H-P. Effect of hydrostatic pressure on the γ-polymorph of glycine. 1. A polymorphic transition into a new δ-form. Z. Kristallogr. 2005, 220, 50–57.
[102] Murli C, Sharma SM, Karmakar S, Sikka SK. α-Glycine under high pressures: A Raman scattering study. Physica B 2003, 339, 23–30.
[103] Mishra AK, Murli C, Garg N, Chitra R, Sharma SM. Pressure-induced structural transformations in bis (glycinium) oxalate. J. Phys. Chem. B 2010, 114, 17084–17091.
[104] Bhatt H, Murli C, Garg N, Deo MN, Chitra R, Choudhury RR, Sharma SM. High pressure phase transformations in bis (glycinium) oxalate – An infrared absorption study. Chem. Phys. Lett. 2012, 532, 57–62.
[105] Bhatt H, Deo MN, Murli C, Vishwakarma SR, Chitra R, Sharma S M. A temperature dependent infrared absorption study of strong hydrogen bonds in bis (glycinium) oxalate. In the solid state physics symposium 2015, 1731, 040018.
[106] Bhatt H, Murli C, Mishra AK, Verma AK, Garg N, Deo MN, Chitra R, Sharma SM. Hydrogen bond symmetrization in glycinium oxalate under pressure. J. Phys. Chem. B 2016, 120, 851–859.
[107] Zakharov BA, Losev EA, Boldyreva EV. A new high-pressure polymorph of a glycine-glutaric acid co-crystal: the same as the low-temperature form. CrystEngComm 2013, 15, 1693–1697.
[108] Zakharov BA, Losev EA, Kolesov BA, Drebushchak VA, Boldyreva EV. Low-temperature phase transition in glycine–glutaric acid co-crystals studied by single-crystal X-ray diffraction, Raman spectroscopy and differential scanning calorimetry. Acta Crystallogr. B 2012, 68, 287–296.
[109] Kolesnik EN, Goryainov SV, Boldyreva EV. Different behavior of L- and DL-serine crystals at high pressures: Phase transitions in L-serine and stability of the DL-serine structure. Doklady Phys. Chem. 2005, 404, 169–172.
[110] Moggach SA, Allan DR, Morrison CA, Parsons S, Sawyer L. Effect of pressure on the crystal structure of L-serine-I and the crystal structure of L-serine-II at 5.4 GPa. Acta Crystallogr. B 2005, 61, 58–68.
[111] Boldyreva EV, Kolesnik EN, Drebushchak TN, Sowa H, Ahsbahs H, Seryotkin YuV. A comparative study of the anisotropy of lattice strain induced in the crystals of DL-serine by cooling down to 100 K, or by increasing pressure up to 8.6 GPa. Z. Kristallogr. 2006, 221, 150–161.
[112] Drebushchak TN, Sowa H, Seryotkin YV, Boldyreva EV, Ahsbahs H. L-serine III at 8.0 GPa. Acta Crystallogr. E 2006, 62, o4052–4054.
[113] Boldyreva EV, Sowa H, Seryotkin YuV, Drebushchak TN, Ahsbhas H, Chernyshev VV, Dmitriev VP. Pressure-induced phase transitions in crystalline L-serine studied by single-crystal and high-resolution powder X-ray diffraction. Chem. Phys. Lett. 2006, 429, 474–478.
[114] Moggach SA, Marshall WG, Parsons S. High-pressure neutron diffraction study of L-serine-I and L-serine-II, and the structure of L-serine-III at 8.1 GPa. Acta Crystallogr. B 2006, 62, 815–825.
[115] Zakharov BA, Kolesov BA, Boldyreva EV. Effect of pressure on crystalline L- and DL-serine: revisited by a combined single-crystal X-ray diffraction at a laboratory source and polarized Raman spectroscopy study. Acta Crystallogr. B 2012, 68, 275–286.
[116] Johnstone RDL, Francis D, Lennie AR, Marshall WG, Moggach SA, Parsons S, Pidcock E, Warren JE. High-pressure polymorphism in L-serine monohydrate: Identification of driving forces

in high pressure phase transitions and possible implications for pressure-induced protein denaturation. CrystEngComm 2008, 10, 1758–1769.

[117] Johnstone RDL, Lennie AR, Parsons S, Pidcock E, Warren JE. Comparison of the effects of pressure on three layered hydrates: A partially successful attempt to predict a high-pressure phase transition. Acta Crystallogr. B 2009, 65, 731–738.

[118] Zakharov BA, Boldyreva EV. Reversible pressure-induced disordering in bis(DL-serinium) oxalate dihydrate. J. Mol. Struct. 2014, 1078, 151–157.

[119] Arkhipov S, Zakharov B, Boldyreva E. High pressure studies of L-Serine-L-ascorbic acid co-crystal. 23rd Congress and General Assembly of the International Union of Crystallography (IUCr2014), – Montreal, Canada, 5–12 August 2014, 998.

[120] Frey MN, Lehmann MS, Koetzle TF, Hamilton WC. Precision neutron diffraction structure determination of protein and nucleic acid components. XI. Molecular configuration and hydrogen bonding of serine in the crystalline amino acids L-serine monohydrate and DL-serine. Acta Crystallogr. B 1973, 29, 876–884.

[121] Destro R, Marsh RE, Bianchi R. A low-temperature (23 K) study of L-alanine. J. Phys. Chem. 1988, 92, 4, 966–973.

[122] Destro R, Soave R, Barzaghi M. Physicochemical properties of zwitterionic L- and DL-alanine crystals from their experimental and theoretical charge densities. J. Phys. Chem. B 2008, 112, 5163–5174.

[123] Katrusiak A. Stereochemistry and transformation of –OH···O= hydrogen bonds Part I. Polymorphism and phase transition of 1,3-cyclohexanedione crystals. J. Mol. Struct. 1992, 269, 329–354.

[124] Katrusiak A. Macroscopic and structural effects of hydrogen-bond transformations. Crystallogr. Rev 1996, 5, 133–180.

[125] Katrusiak A. Macroscopic and structural effects of hydrogen-bond transformations. Crystallogr. Rev 2003, 9, 91–133.

[126] Szafrański M, Katrusiak A. Thermodynamic behaviour of bistable NH–N hydrogen bonds in monosalts of 1,4-diazabicyclo[2.2.2]octane. Chem. Phys. Lett.2000, 318, 427–432.

[127] Budzianowski A, Katrusiak A. Pressure tuning between NH···N hydrogen-bonded ice analogue and NH···Br polar dab-coHBr complexes. J. Phys. Chem. B 2006,110, 9755–9758.

[128] Sikka SK, Sharma SM. The hydrogen bond under pressure. Phase Transitions 2008, 81, 907–934.

[129] Tumanov NA, Boldyreva EV, Kolesov BA, Kurnosov AV, Cabrera RQ. Pressure-induced phase transitions in L-alanine, revisited. Acta Crystallogr. B 2010, 66, 458–471.

[130] Funnell NP, Dawson A, Francis D, Lennie AR, Marshall WG, Moggach SA, Warren JE, Parsons S. The effect of pressure on the crystal structure of l-alanine. CrystEngComm 2010, 12, 2573–2583.

[131] Casati N, Macchi P, Sironi A. Hydrogen migration in oxalic acid di-hydrate at high pressure? ChemComm 2009, 19, 2679–2681.

[132] Macchi PN, Casati WG, Marshall AS. The α and β forms of oxalic acid di-hydrate at high pressure: A theoretical simulation and a neutron diffraction study. CrystEngComm 2010, 12, 2596–2603.

[133] Bhatt H, Mishra AK, Murli C, Verma AK, Garg N, Deo MN, Sharma SM. Proton transfer aiding phase transitions in oxalic acid dihydrate underpressure. Phys. Chem. Chem. Phys. 2016, 18, 8065–8074.

Kristin M. Hutchins, Jelena Stojaković, Ryan H. Groeneman, and Leonard R. MacGillivray

8 Co-crystals for solid-state reactivity and thermal expansion

8.1 Introduction

Supramolecular chemistry, as defined by Nobel laureate Jean-Marie Lehn [1], is "the association of two or more chemical species held together by intermolecular forces". The field of supramolecular chemistry focuses not only on how to make specific target molecules, but also encompasses how molecules interact with each other, including interactions between identical and different chemical species. The term "supermolecule" [1, 2] describes assemblies of molecules held together by non-covalent interactions, e.g. hydrogen bonds, and relies on molecular recognition, molecular complementarity or self-assembly between molecules [3, 4]. Although the field found its roots in ion recognition [1], host–guest chemistry [5, 6] and coordination chemistry [7, 8], it quickly expanded to include other reversible, non-covalent interactions such as hydrogen bonds, halogen bonds, van der Waals interactions and π–π interactions [9–13]. A multitude of non-covalent interactions have been utilised in supramolecular chemistry, but control over crystal packing remains a significant challenge.

Although initially believed to be a highly rigid and unreactive environment [14], it has become increasingly clear, particularly in the last decade, that the crystalline state can be appreciably flexible so that it can support significant atomic and molecular movements. Crystalline environments have been demonstrated to accommodate reactions including proton transfer, cis-trans isomerisation, pedal motion and bimolecular reactions that result in the formation of covalent bonds. For chemical reactions to occur in the crystalline state, reactant molecules often need to conform to specific geometries, being separated at certain distances to react [15]. If molecules in a crystal do not meet geometric and/or distance requirements, a common approach is to screen for other crystal forms, e.g. polymorphs, solvates. A second approach involves modifying the reactants with auxiliary groups that guide crystal packing into favourable arrangements.

In this regard, co-crystallisation has emerged as a promising approach to modify crystal packing such that targeted properties can be achieved, e.g. reactivity and motion (Figure 8.1) [16]. The process involves combining two or more neutral molecular species to form a crystalline lattice [17]. In the case of reactivity, one component, namely the co-crystal former (CCF), directs a second molecule, i.e. the reactant, into a geometry suitable to react, e.g. promoted by UV radiation. The molecules are usually designed to interact via strong and directional noncovalent forces, e.g. hydrogen and

https://doi.org/10.1515/9783110464955-008

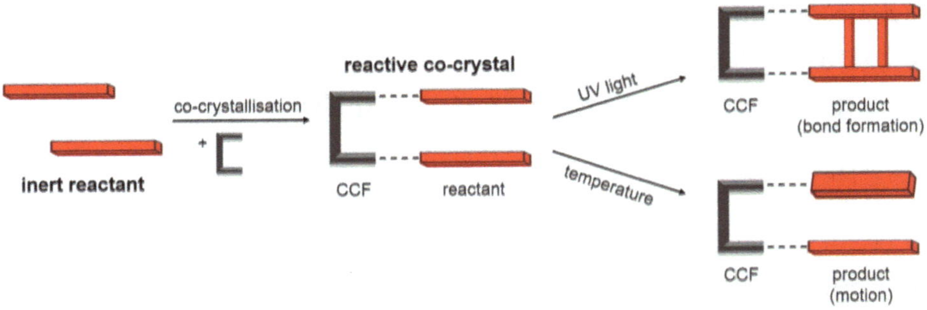

Fig. 8.1: Co-crystallisation approach to solid-state reactivity and motion.

halogen bonds. The CCF may, for example, serve as a template to direct the formation of covalent bonds. The CCF may also furnish crystalline environments that support dynamic movements of the components. A molecule with a functional group known to exhibit motion may be combined with the CCF to form a co-crystal that supports the programmed motion. The ultimate aim is to form solids that exhibit controllable and tuneable thermal expansion (TE) properties. Depending on the magnitude of the TE, the temperature range over which the expansion occurs and/or the direction of expansion, such materials can be of interest for volumetric or linear TE behaviour [18].

In this chapter, we present our efforts to develop co-crystals to control chemical reactivity and TE in molecular solids. Specifically, we provide an account to employ co-crystals based on small-molecule templates to direct intermolecular [2+2] photodimerisations. Crystal packing effects involving olefins have largely prevented chemists from designing solids with controllable reactivity [19, 20]. Here, it is shown how the co-crystal approach can be utilised to synthesise molecules in the solid state by design. We also demonstrate how a similar co-crystal method based on resorcinol can be used to alter crystal packing environments such that molecular motion and TE properties are enabled.

8.2 Co-crystals to control [2+2] photodimerisations

The best studied solid-state reaction is the [2+2] photodimerisation of olefins that affords a cyclobutane ring [21]. Classic examples are based on intermolecular [2+2] photodimerisations of α-trans-cinnamic acid to produce α-truxillic acid and β-trans-cinnamic acid to produce β-truxinic acid [22, 23]. The observation wherein two polymorphs (α and β) produce different stereoisomers upon exposure to light served as an early model of crystal structure–chemical reactivity relationships. Ultimately, crystal structure elucidations of trans-cinnamic acid (Figure 8.2) [24, 25], and derivatives led Schmidt [15, 19, 20] to develop the topochemical postulate, which dictates general geometry requirements for a [2+2] photocycloaddition to occur in a solid. The postu-

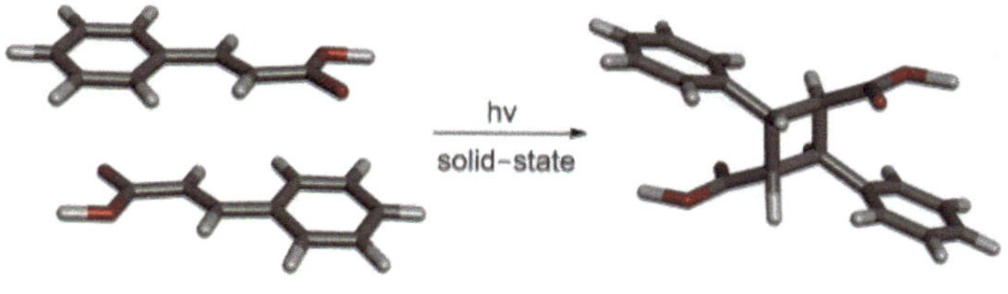

Fig. 8.2: X-ray crystal structures of α-trans-cinnamic acid and α-truxillic acid, highlighting the use of the solid state to facilitate a [2+2] photodimerisation.

late states that two olefinic (C=C) bonds should be arranged parallel and separated on the order of ≤4.2 Å to undergo the cycloaddition reaction.

Since the pioneering work of Schmidt, significant advances have been made to overcome challenges associated with crystal packing to direct the [2+2] photodimerisation in the solid state, with reasonable degrees of success. Intermolecular interactions in supramolecular chemistry and crystal engineering have been exploited to guide olefins into preferred geometries for the photoreaction [26]. The method has involved covalently attaching functional groups that participate in forces such as π–π stacking, charge transfer and hydrogen bonding to control the packing. For example, an olefin can be covalently decorated with Cl– atoms that support Cl···Cl and C–H···Cl contacts. A related approach has involved the formation of face-to-face π–π stacking interactions by introducing favourable electronic attraction between electron-poor and electron-rich rings. Examples include the substitution of aromatic rings with groups that have complementary electronic effects, such as perfluorophenyl–phenyl forces [27, 28] or electron withdrawing, e.g. $-NO_2$, and donating, e.g. –OMe, abilities [29].

8.2.1 Template approach based on resorcinol for [2+2] photodimerisation

In 2000, our group introduced a method using a template molecule based on resorcinol (res) to assemble and pre-organise olefins into a suitable geometry to undergo photodimerisation in the solid state [16]. The template was equipped with two functional groups capable of assembling two olefins into close proximity via hydrogen bonds (Figure 8.3). The method was inspired by earlier works by Aoyama, Ito and Feldman. In the early 1990s, Aoyama et al. [30] employed a res-decorated anthracene for direct stacking of quinones with rings separated on the order of 3.8 Å. The components of the co-crystal were sustained by strong and directional O–H···N hydrogen bonds. In the 1990s, Ito et al. [31] achieved stacking of cinnamates utilising diammonium salts, whereas Feldman and Campbell described a J-shaped hydrogen-bonded dimer of naphthalene dicarboxylic acid that stacked two C=C bonds [32]. Based on the reports, we described a template approach that serves to decouple solid-state reactivity from the effects of long-range molecular packing.

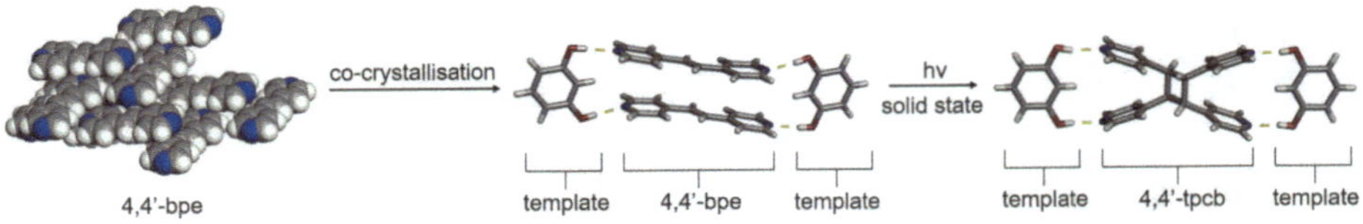

Fig. 8.3: Co-crystallisation applied to 4,4′-bpe to yield photoreactive co-crystal 2(4,4′-bpe) · 2(res) and photoproduct (4,4′-tpcb) · 2(res). *Yellow dashed lines* indicate hydrogen bonds.

Specifically, a small-molecule template based on res was developed wherein the two –OH groups act as hydrogen bond donors. Co-crystallisation of res with trans-1,2-bis-(4-pyridyl)ethylene (4,4′-bpe) formed a discrete four-component supramolecular assembly in the co-crystal 2(4,4′-bpe) · 2(res) sustained by O–H···N hydrogen bonds [16]. The olefins were arranged in a face-to-face manner in close proximity within the assembly, and were separated on the order of 3.70 Å. The olefins were thus positioned to conform to the topochemical postulate. Upon UV irradiation, the bipyridine underwent intermolecular [2+2] photodimerisation stereo-selectively and with a quantitative yield to generate rctt-tetrakis-1,2,3,4-(4-pyridyl)cyclobutane (4,4′-tpcb). The reactivity was therefore borne out of the localised geometry of rationally designed supramolecular assemblies. A covalent modification of the olefinic substrate was not necessary, and the isolation of the photoproduct was facile.

The modularity of the template approach was subsequently addressed. Over the last 15 years, efforts have been made to apply the template method to solid-state photodimerisations of additional symmetrical and unsymmetrical bipyridines n, m'-bpe, where: $n, m = 2, 3$ or 4 [33]. For each reactant, a suitable res-based template may be identified and the corresponding photoproducts, n, m'-tpcb, can be generated (Figure 8.4). The goal is to develop the template approach as a tool for reliable and predictable organic synthesis in solids.

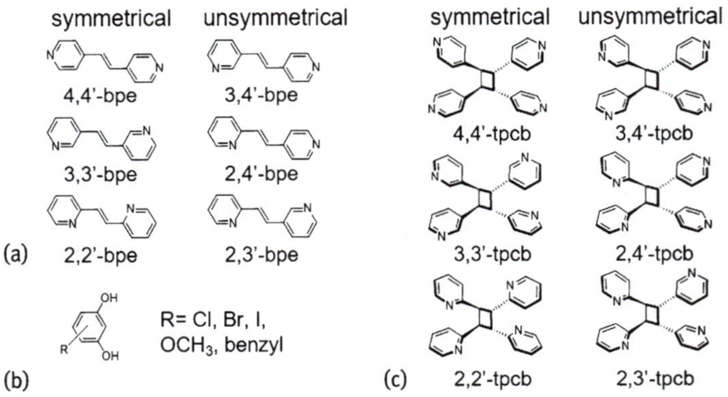

Fig. 8.4: Modularity in solid-state synthesis: (a) olefinic bipyridines n, m'-bpe, (b) res templates and (c) photoproducts n, m'-tpcb.

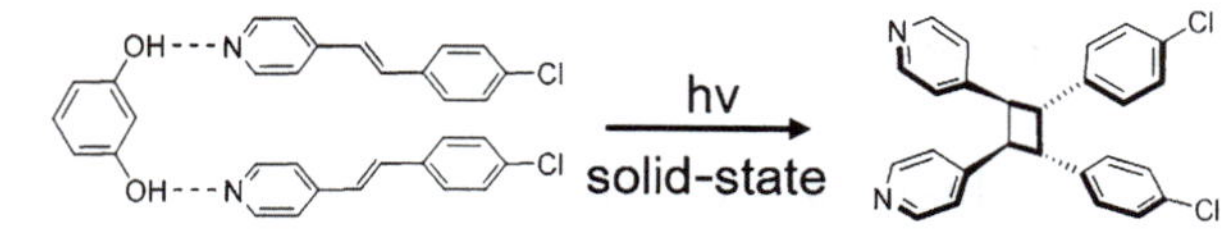

(a) 2(4-Cl-sbz)·(res) 4-Cl-dpcb

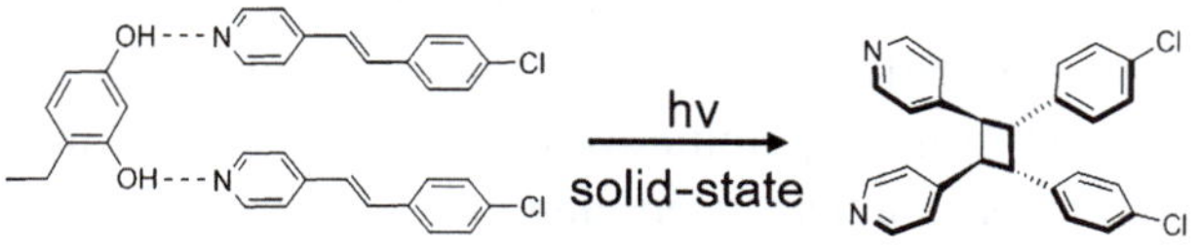

(b) 2(4-Cl-sbz)·(4-Et-res) 4-Cl-dpcb

Fig. 8.5: Three-component hydrogen-bonded assemblies in photoactive 2(4-Cl-sbz) · (res) and 2(4-Cl-sbz) · (4-Et-res) that generate 4-Cl-dpcb.

In addition to bipyridines, monopyridines were studied as reactants in co-crystals with res-based templates. Co-crystallisation of 4-chlorostilbazole (4-Cl-sbz) with either res or 4-ethylresorcinol (4-Et-res), for example, afforded discrete three-component assemblies sustained by O–H···N hydrogen bonds, namely 2(4-Cl-sbz) · (res) and 2(4-Cl-sbz) · (4-Et-res) (Figure 8.5) [34]. The unsymmetrical olefin adopted head-to-head geometry and participated in Cl···Cl forces. The C=C bonds, crossed in 2(4-Cl-sbz) · (res) and parallel in 2(4-Cl-sbz) · (4-Et-res), reacted in both solids to give the head-to-head cyclobutane product, namely rctt-1,2-bis-(4-pyridyl)-3,4-bis(p-chlorophenyl) cyclobutane (4-Cl-dpcb), in near-quantitative yield.

A res-based template was also used to achieve photodimerisation of a monopyridine based on a styrylthiophene (Figure 8.6) [35]. Co-crystallisation of trans-1-(4-pyridyl)-2-(3-thienyl)ethylene (β-pte) with 4,6-diiodoresorcinol (4,6-diI res) afforded discrete three-component assemblies sustained by two O–H···N hydrogen bonds in 2(β-pte) · (4,6-diI res). The olefins in the assemblies were separated at approximately 4.2 Å. Upon UV irradiation, intermolecular [2+2] photodimerisation occurred in up to

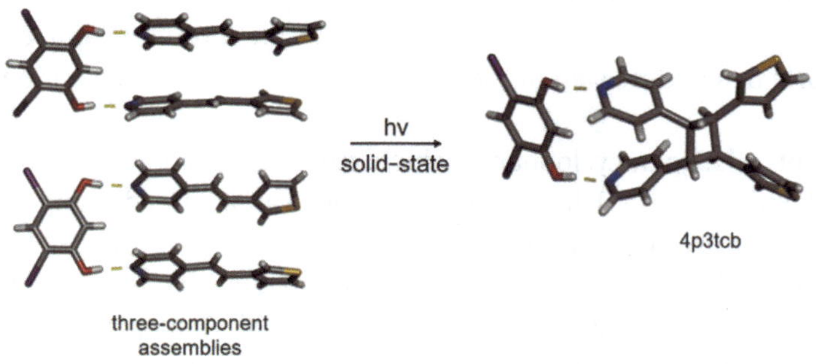

Fig. 8.6: Three-component hydrogen-bonded assemblies in photoactive 2(β-pte) · (4,6-diI res) that generate 4p3tcb. *Yellow dashed lines* indicate hydrogen bonds.

90% yield to afford the head-to-head product rctt-1,2-bis(4-pyridyl)-3,4-bis(3-thienyl) cyclobutane (4p3tcb).

The template approach has also been used to achieve rare cross-photoreactions. In particular, a novel co-crystal solid solution was prepared using res and two structurally similar, yet chemically different, monopyridines, namely 4-Cl-sbz and 4-methylstilbazole (4-Me-sbz) [36]. A co-crystal solid solution was then formed based on (res) $\cdot$ (4–Cl–sbz)$_x$ $\cdot$ (4–Me–sbz)$_{2-x}$ where $x \approx 1$. Upon UV irradiation, the [2+2] photocycloaddition proceeded in near quantitative conversion to give a mixture of cross-products and homodimers, namely heterodimer rctt-1-(4-chlorophenyl)-2-(4-methylphenyl)-3,4-bis(4-pyridyl)cyclobutane (4-Cl,Me-dpcb) and homodimers 4-Cl-dpcb and rctt-1,2-bis-(4-pyridyl)-3,4-bis(4-methylphenyl)cyclobutane (4-Me-dpcb) (Figure 8.7).

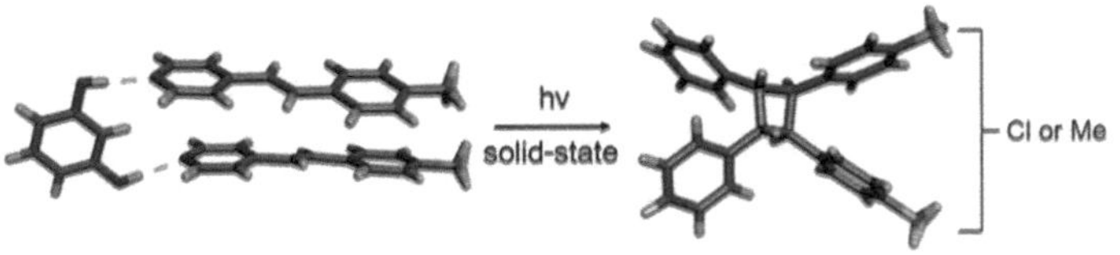

Fig. 8.7: Cross photodimerisation in co-crystal solid solution (res) $\cdot$ (4–Cl–sbz)$_x$ $\cdot$ (4–Me–sbz)$_{2-x}$ to yield cyclobutane products. *Yellow dashed lines* indicate hydrogen bonds.

Further examples that demonstrate general applicability of the template approach involved the syntheses of synthetically-demanding molecules such as [n]-ladderanes and [2.2]paracyclophanes. Both families of molecules require multiple C=C bonds to be assembled and react. In particular, 5-methoxyresorcinol (5-OMe-res) was shown to organise both trans-1,4-(4-pyridyl)-1,3-butadiene (4-py-but) and trans-1,6-(4-pyridyl)-1,3,5-hexatriene (4-py-hex) into co-crystals based on discrete four-component assemblies wherein the C=C bonds lie parallel and separated by <4.2 Å [37]. Upon UV irradiation, each co-crystal produced the corresponding [3]- and [5]-ladderanes stereo-selectively and in quantitative yield (Figure 8.8a). A multi-step preparation of a naturally-occurring ladderane in solution had been reported in low overall yield [38]. Moreover, a particularly demanding step in the total synthesis was the preparation of ladderane backbone.

The preparation of a [2.2]paracyclophane, which required photodimerisation of two sets of C=C bonds separated by a phenyl spacer, was achieved using 4-benzylresorcinol (4-Bn-res) [39, 40]. Co-crystallisation of 4-Bn-res with 1,4-bis[2-(4-pyridyl)ethenyl]benzene (1,4-bpeb) afforded a co-crystal wherein two sets of C=C bonds were organised for double photodimerisation (Figure 8.8b). Analogous to the case of the ladderanes, the solid-state synthesis afforded the [2.2]paracyclophane stereo-selectively and in quantitative yield.

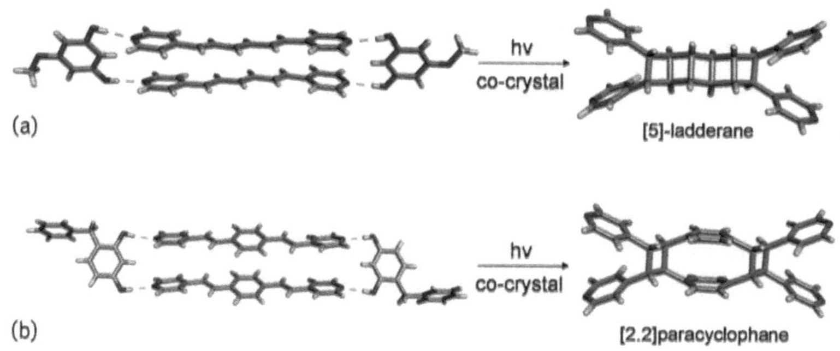

Fig. 8.8: Co-crystallisation and solid-state syntheses: (a) [5]-ladderane in 2(4-py-hex) · 2(5-OMe-res) and (b) [2.2]paracyclophane in 2(1,4-bpeb) · 2(4-Bn-res). *Yellow dashed lines* indicate hydrogen bonds.

In a related work, Santra and Biradha [41] used a tritopic CCF based on phloroglucinol as a template to assemble the diene 1,5-bis(4-pyridyl)-1,4-pentadiene-3-one (1,5-bppo) in 2(1,5-bppo) · 2(phloroglucinol) · 2(CH_3CN). The co-crystallisation yielded a 1D ladder-like network sustained by O–H···N hydrogen bonds, with nearest C=C bonds parallel and separated by 3.60 and 3.85 Å. UV irradiation afforded tricyclo[6.2.0.0]-decane in a stepwise double dimerisation (Figure 8.9).

In more recent work, 5-OMe-res was utilised to assemble the diene (1E,3E)-1,4-di(1H-imidazol-1-yl)buta-1,3-diene (dibd) in 2(dibd) · 2(5-OMe-res). UV irradiation produced a dimer with a single cyclobutane ring. A photoproduct that possesses a double unsaturated eight-membered ring was also obtained (Figure 8.10) [42].

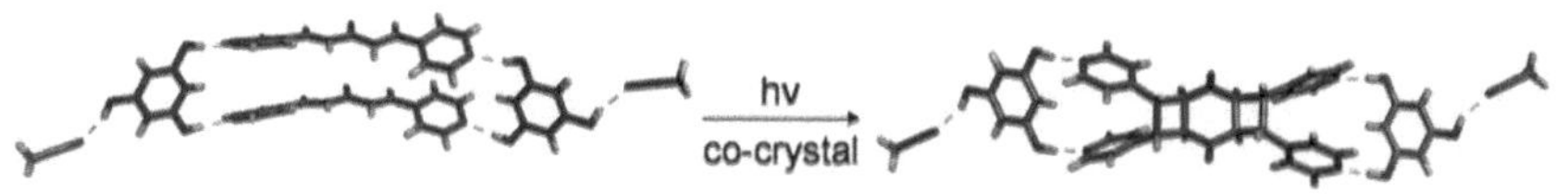

Fig. 8.9: Directed synthesis of tricyclo[6.2.0.0]-decane in 2(1,5-bppo) · 2(phloroglucinol) · 2(CH_3CN) co-crystal. *Yellow dashed lines* indicate hydrogen bonds.

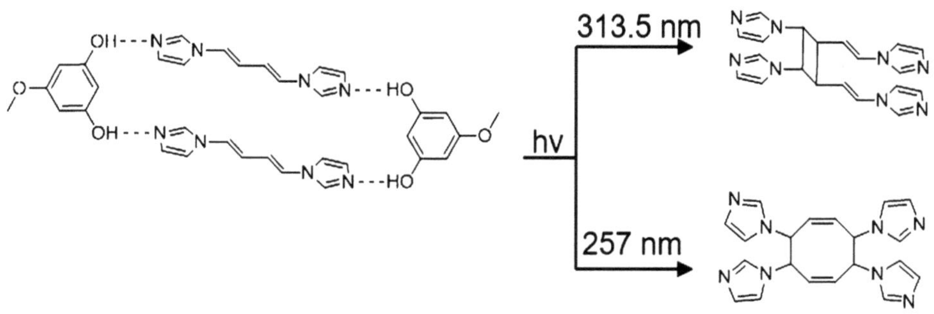

Fig. 8.10: Wavelength-dependent photoreactivity of dibd within 2(dibd) · 2(5-OMe-res).

8.2.2 Expansion of the template approach

Following initial successes involving res templates, MacGillivray's group demonstrated that dicarboxylic acids can serve as templates [43]. Co-crystallisation of 1,8-naphthalenedicarboxylic acid (1,8-nda) with either 4,4′-bpe or 2,2′-bpe yielded co-crystals based on discrete four-component hydrogen-bonded assemblies (Figure 8.11a). Parallel orientations and close proximities of the C=C bonds rendered both co-crystals photoactive. Upon UV irradiation, each solid generated 4,4′- or 2,2′-tpcb in quantitative yield. Succinic acid was later utilised in a similar way to achieve photodimerisation of a thiophene [44].

In related work, the Jones group demonstrated the ability of a triacid to act as a template. Co-crystallisation of tricarballylic acid (tca) with 4,4′-bpe afforded an infinite 1D hydrogen-bonded tape, which contrasts with the discrete assemblies realised with 1,8-nda (Figure 8.11b) [45]. The C=C bonds of 4,4′-bpe were aligned parallel and separated by 3.8 Å. Upon UV irradiation, the cyclobutane 4,4′-tpcb was obtained in 90% yield.

A calix[4]arene tetrabenzoic acid was also shown to direct photodimerisation of 4,4′-bpe within a co-crystal based on a tubular 1D assembly sustained by hydrogen bonds [46]. The tetrabenzoic acid assembled two molecules of 4,4′-bpe into a parallel geometry, separated by 3.8 Å. Upon UV irradiation, 4,4′-tpcb formed in approximately 80% yield [46].

Templates based on res and the di- and polycarboxylic acids relied on identical functional groups to assemble olefins into reactive geometries. To develop unsymmetrical templates, MacGillivray's group demonstrated that heteroditopic Rebek's imide can form both O–H···N and N–H···N hydrogen bonds with 4,4′-bpe (Figure 8.12) [47]. Co-crystallisation of Rebek's imide with 4,4′-bpe yielded a discrete four-component assembly with olefins aligned in a parallel manner and separated by 3.78 Å. Upon UV irradiation, the olefins reacted to give 4,4′-tpcb stereo-selectively and in quantitative

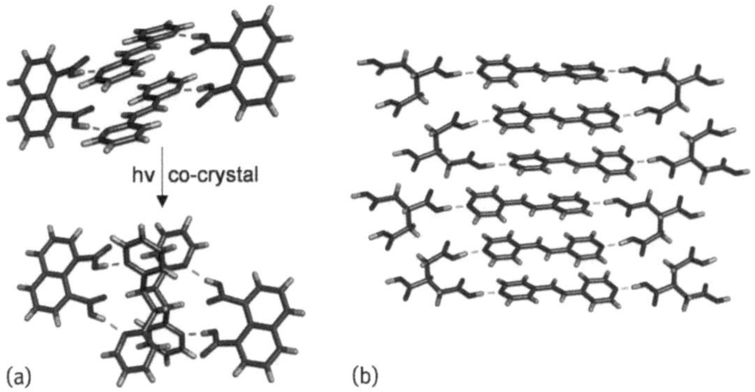

Fig. 8.11: Carboxylic acids as templates in: (a) 2(2,2′-bpe) · 2(1,8-nda), which generates (2,2′-tpcb) · 2(1,8-nda) and (b) 3(4,4′-bpe) · 2(tca). *Yellow dashed lines* indicate hydrogen bonds.

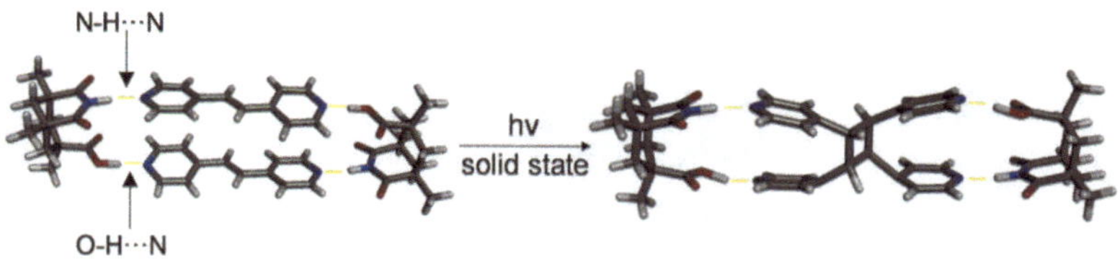

Fig. 8.12: Rebek's imide directs photodimerisation of 4,4′-bpe to 4,4′-tpcb in a partial single-crystal-to-single-crystal transformation. *Yellow dashed lines* indicate hydrogen bonds.

yield. By tuning the UV light, a partial single-crystal-to-single-crystal (SCSC) transformation was also realised. Crystals typically crack during the course of a photoreaction owing to build-ups of stress that result from atomic and molecular movements. In a SCSC transformation, crystallinity is retained, which allows for single-crystal X-ray diffraction data to be collected on a reacted single crystal. Therefore, SCSC reactions offer a direct observation of atomic rearrangements during a cycloaddition.

Our group also demonstrated that templates and reactants serve as hydrogen bond acceptors and donors, respectively. Therefore, 2,3-bis(4-methylenethiopyridyl)naphthalene (2,3-nap) was a template used to direct photodimerisation of fumaric acid (fum). The co-crystal 2(2,3-nap) · 2(fum) was sustained by O–H···N hydrogen bonds, with the olefins aligned in an approximately parallel manner, with a C=C separation of 3.84 Å (Figure 8.13) [48]. The photodimerisation proceeded in SCSC fashion to give rctt-1,2,3,4-cyclobutanetetracarboxylic acid (cbta) stereo-selectively and in up to 70% yield. A similar approach was described by Mei et al. to direct a cycloaddition of fum using 1,8-bis(4-pyridyl)naphthalene (1,8-dpn) to generate cbta in quantitative yield [49].

The ability of "excess" copies of ditopic catechol (cat) – a regioisomer of res – to direct an intermolecular [2+2] photodimerisation was subsequently reported (Figure 8.14) [50]. The idea that cat could stack two olefins to react seemed unlikely

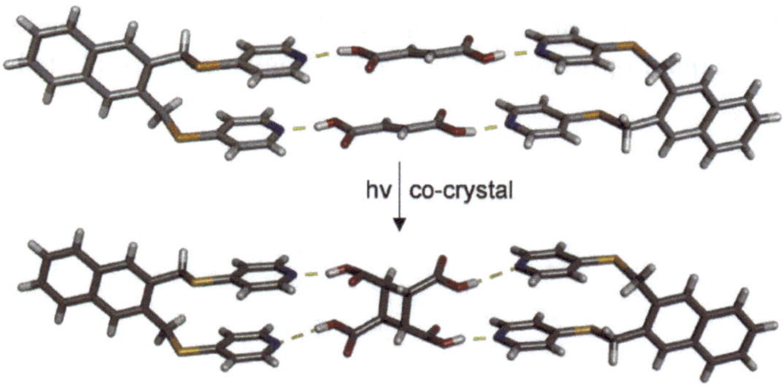

Fig. 8.13: Photodimerisation of 2(2,3-nap) · 2(fum) to yield cbta. *Yellow dashed lines* indicate hydrogen bonds.

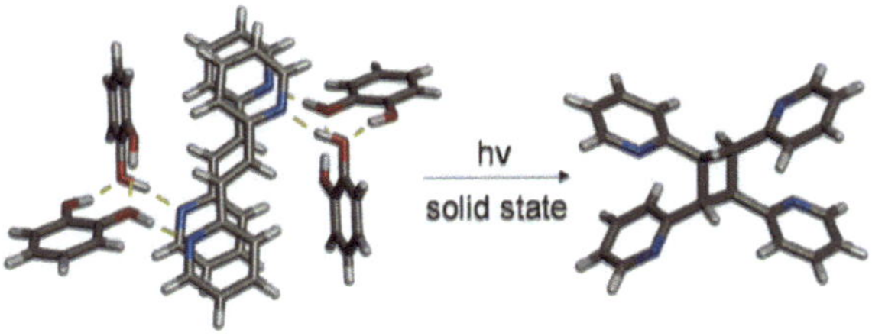

Fig. 8.14: Photodimerisation of 2,2′-bpe in 2(2,2′-bpe) · 4(cat) to afford 2,2′-tpcb. *Yellow dashed lines* indicate hydrogen bonds.

given the close proximity of the two hydroxyl groups of cat. However, cat and 2,2′-bpe yielded a co-crystal, namely 2(2,2′-bpe) · 4(cat), based on a novel, discrete six-component assembly sustained by a total of eight O–H···N and O–H···O hydrogen bonds. The cat molecules self-assembled into dimers that interact with 2,2′-bpe to support face-to-face π stacking. Each cat adopted a different conformation within a dimer, with one cat adopting a syn-anti conformation and second adopting a syn-gauche conformation. The cat dimers assembled 2,2′-bpe such that the C=C bonds lie parallel and separated at 3.71 Å. A partial SCSC photodimerisation reaction confirmed that the cyclobutane ring forms within the discrete hydrogen-bonded structure.

Although ditopic templates assembled olefins into stacked pairs, indolo[2,3-*a*] carbazole (ic) was shown by us to assemble up to four olefins into quadruple stacks [51]. The ic framework exhibits a rigid backbone with two –NH groups separated on the order of 3.10 Å. Initially, a co-crystal based on a cyclic array of four NH···N bonds was expected to direct a photodimerisation of 4,4′-bpe. An included acetonitrile solvent molecule (CH_3CN) was present in the co-crystal 2(ic) · 2(4,4′-bpe) · 2(CH_3CN) (Figure 8.15a). However, modifications to the co-crystallisation conditions yielded unexpected triple and quadruple stacks within 2(ic) · 3(4,4′-bpe) and 2(ic) · 4(4,4′-bpe) respectively (Figure 8.15b, c). The triple and quadruple stacks were supported by weaker C–H···N forces. In all three solids, the face-to-face stacking of 4,4′-bpe and proximities of the C=C bonds conformed to the topochemical postulate. Prolonged UV irradiation showed the quadruple stack in 2(ic) · 4(4,4′-bpe) to undergo photodi-

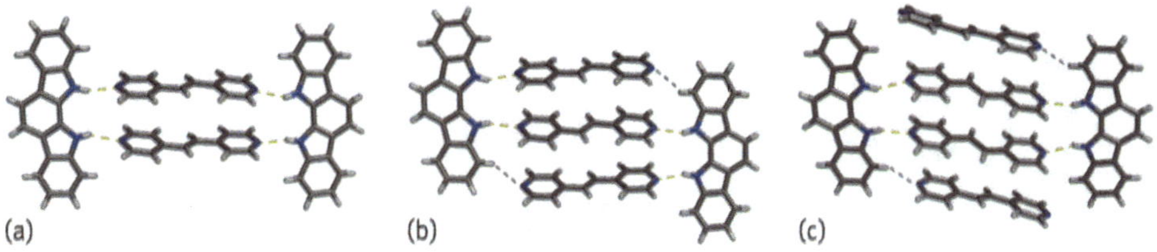

Fig. 8.15: Discrete aromatic stacks: (a) double-stacked 2(ic) · 2(4,4′-bpe) · 2(CH_3CN), (b) triple-stacked 2(ic) · 3(4,4′-bpe) and (c) quadruple-stacked 2(ic) · 4(4,4′-bpe). CH_3CN molecules omitted for clarity. *Yellow dashed lines* indicate hydrogen bonds. *Black dashed lines* indicate C–H···N forces.

merisation to generate 4,4′-tpcb. In the case of double-stacked 2(ic) · 2(4,4′-bpe), the reactivity was likely suppressed by the included solvent, whereas triple-stacked 2(ic) · 3(4,4′-bpe) may have lacked sufficient free space for a dimerisation to occur.

A template that operates via halogen bonds had been reported to direct a [2+2] photodimerisation in the solid state. Halogen bonds are defined by D–X···A, involving donor (D) and acceptor (A) atoms, with the donor and acceptor sharing a halogen (X) atom [12, 52, 53]. The halogen atom acts as an electron density acceptor (Lewis acid) owing to the presence of a region of positive electrostatic potential (σ-hole) [54]. Specifically, Caronna et al. [55] reported a template based on N···I halogen bonds. A tetratopic halogen bond donor, namely pentaerythritol-tetrakis(2,3,5,6-tetrafluoro-4-iodophenyl)ether (tfipe), pre-organised 4,4′-bpe into a face-to-face stacked geometry within 1D ribbons sustained by N···I halogen bonds in 2(4,4′-bpe) · (tfipe) (Figure 8.16). The stacked C=C bonds were disordered and separated by less than 4.5 Å. Upon UV irradiation, 4,4′-tpcb was generated stereo-selectively and in quantitative yield.

Halogen bonding was also applied to sustain [2+2] photodimerisation of vitamin K_3 (menadione). UV irradiation of pure crystals of vitamin K_3 produced a mixture of syn- and anti-dimers. However, co-crystallisation with ditopic halogen bond donors, namely 1,4-dibromotetrafluorobenzene (btfb) and 1,4-diiodotetrafluorobenzene (itfb), afforded co-crystals wherein the C=C bonds were parallel and separated less than 4.2 Å. UV irradiation of the solids primarily produced the syn-photodimer of vitamin K_3 in near quantitative yields (Figure 8.17) [56].

In later work, MacGillivray's group reported a ditopic template that assembled a halogenated olefin into reactive geometry via halogen bonds. Co-crystallisation of 1,8-dpn with trans-1,2-bis(4-iodotetra-fluorophenyl)ethene (ifpe) generated a discrete

Fig. 8.16: Tetratopic halogen bond template tfipe in 2(4,4′-bpe) · (tfipe). *Yellow dashed lines* indicate halogen bonds.

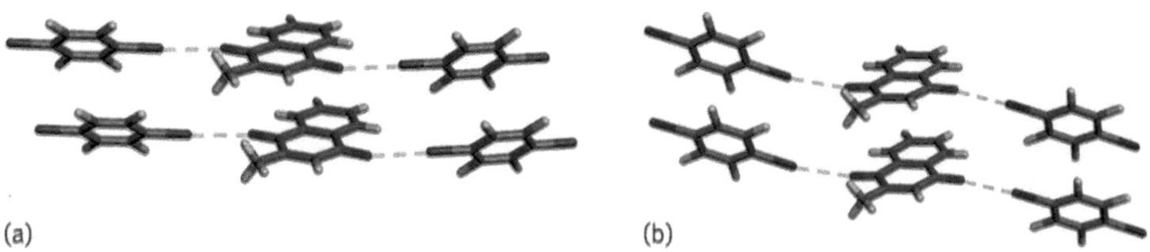

Fig. 8.17: Pre-organisation of vitamin K_3 achieved using co-crystallisation with halogen bond-based templates: (a) btfb and (b) itfb. *Yellow dashed lines* indicate halogen bonds.

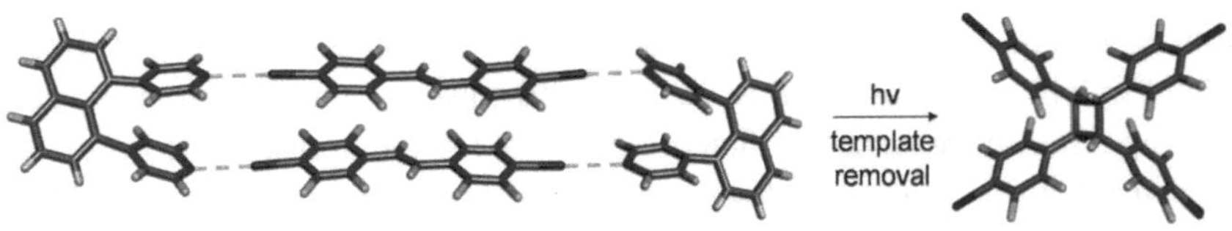

Fig. 8.18: Templated photodimerisation of ifpe in 2(ifpe) · 2(1,8-dpn) to generate ifpcb. *Yellow dashed lines* indicate halogen bonds.

four-component assembly sustained by four N···I halogen bonds in 2(ifpe) · 2(1,8-dpn) (Figure 8.18) [57]. Upon UV irradiation, the halogenated photoproduct rctt-1,2,3,4-tetrakis(4-iodo-2,3,5,6-tetrafluorophenyl)cyclobutane (ifpcb) formed stereoselectively and in quantitative yield. The photoproduct was therefore functionalised with four I atoms and sixteen F atoms. The photoproduct was also shown to assemble to exhibit a rare form of self-inclusion based on 1D channels that were occupied by either solvent molecules or the host itself.

The transfer of molecular information through sequential reactions is crucial in both chemical synthesis and biological processes. Nucleic acids, for example, are products of templated reactions that subsequently function as templates in sequential reactions. The molecules are regarded as reciprocal templates and provide a high-order means of controlling reactivity in complex biological processes, e.g. replication [58].

The feasibility of reciprocal templates to direct reactivity in the solid state was originally demonstrated via successive templated [2+2] photodimerisations. In particular, Saha and MacGillivray [59] showed that the product of a templated photodimerisation can be used as a template in a second photoreaction. Fum had been reported to undergo templated [2+2] photodimerisation to give cbta using either 2,3-nap or

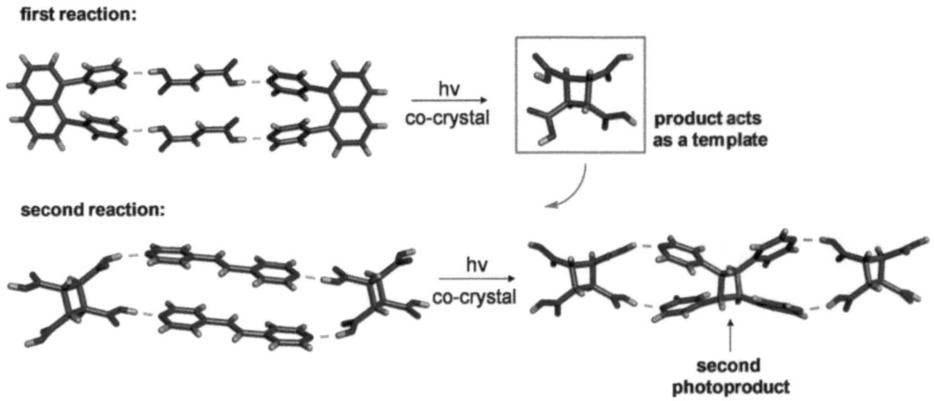

Fig. 8.19: Product of template photoreaction in first reaction, cbta, is subsequently used as a template to direct the formation of 4,4′-tpcb. *Yellow dashed lines* indicate hydrogen bonds.

1,8-dpn [48, 49]. Moreover, Saha and MacGillivray [59] showed that cbta can be subsequently used as a template to assemble 4,4′-bpe for photodimerisation in 2(4,4′-bpe) · (cbta) that afforded 4,4′-tpcb in quantitative yield (Figure 8.19). In the first reaction, the template was a hydrogen bond acceptor, whereas in the second reaction the code was reversed and the template was a hydrogen bond donor. A photoactive polymorph of 2(4,4′-bpe) · (cbta) was also reported [59].

8.3 Co-crystals for thermal expansion

Materials typically respond to temperature changes by undergoing TE. When heated, most materials usually expand, i.e. positive thermal expansion (PTE), but in rarer instances, contraction with applied heat, i.e. negative thermal expansion (NTE), or no effective expansion, i.e. zero thermal expansion (ZTE), have been observed. Materials that exhibit TE properties, including linear (anisotropic) and volumetric (isotropic) expansions that range from large to zero, have experienced significant applications in industry, cookware and technology [60–67]. In recent years, discussions of TE behaviour have attributed reasons for anomalous expansion to be found the molecular level [68]. In this regard, studies of inorganic and metal-organic compounds are numerous [69–72], whereas purely organic systems have been much less frequently explored [60, 73, 74]. Organic systems offer a wide range of functionalities, interactions and crystal packing, which may afford unique TE properties. Moreover, it has been documented that stronger and weaker interactions are less and more affected by changes in temperature, which can give rise to smaller and larger expansions, respectively [75–77]. Gaining an understanding of the types of intermolecular interactions that give rise to and control TE in materials aids in designing materials with the desired TE properties.

In this context, recent work by Saha's group [68] reported a hydrogen-bonded co-crystal involving benzene-1,3,5tricarboxylic acid (bta), 4,4′-methylenebis(2,6-dimethylaniline) (mbda) and methanol (MeOH) to form two polymorphs based on a 1D assembly or 3D network (Figure 8.20). The 3D network featured strong hydrogen bonds in all three directions, whereas the 1D assembly exhibited strong hydrogen bonds in only one direction. Owing to the presence of the weaker intermolecular interactions in the 1D form, the anisotropic TE was expected to be larger than the 3D form. However, the 3D form underwent greater structural deformations and was shown to exhibit weaker packing of guest molecules (MeOH) that collectively manifested as a larger anisotropic TE. The two polymorphs exhibited similar volumetric expansions, which was attributed to the greater crystal packing efficacy of the 1D form. The study illustrated the importance of examining crystal structures and intermolecular interactions as a means of understanding TE properties.

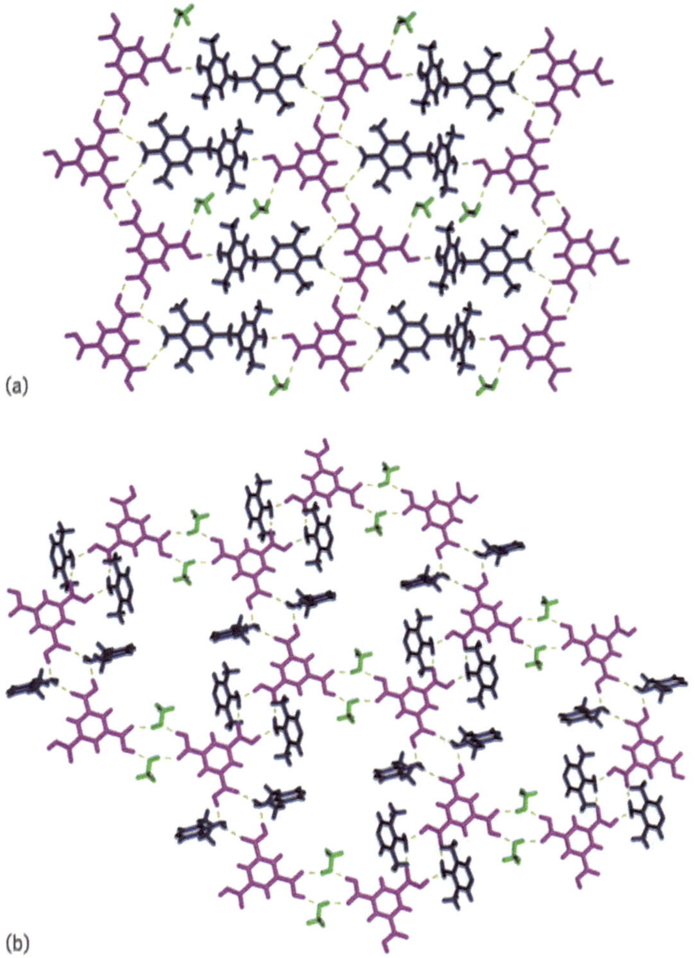

Fig. 8.20: Hydrogen-bonded co-crystals (bta) · (mbda) · (MeOH): (a) 1D form and (b) 3D form (one half of mbda molecule shown for clarity). *Yellow dashed lines* indicate hydrogen bonds. Colour coding: bta (*pink*), mbda (*blue*) and MeOH (*green*).

8.3.1 Thermal expansion involving co-crystals based on resorcinol

The work of Saha et al. [78–80] suggested to us that modifying the non-covalent interactions and components that comprise a co-crystal might provide a method of systematically altering TE properties. Co-crystallisations [81, 82] offer opportunities to modify properties of materials, even when the property may be absent from one of the components, e.g. photoreactivity [83–85]. Moreover, systematically changing one component in a two-component solid can result in different properties manifesting within a series [86, 87]. The use of co-crystals to enable systematic modification of molecular solids as related to TE was therefore expected to provide an opportunity to control TE properties by making subtle changes [79] to one component [88] and determining the resulting structure/property relationships [89–93].

As control of TE properties in molecular solids was at a nascent stage, we hypothesised that incorporation of a functional group known to undergo molecular movement when exposed to an external stimulus, e.g. temperature change, might provide an opportunity to achieve movement and, possibly, TE in a co-crystal. The azo moiety was known to exhibit molecular pedal motion in the crystalline state [94–96]. The process is analogous to the pedalling motion of a bicycle (Figure 8.21) [88], where two aromatic rings serve as the pedals that rotate around the azo core, the azo group undergoing molecular rotation.

Our successes in using res as a template to direct solid-state reactivity led us to further hypothesise that we might be able to utilise res-based CCFs to achieve molecular motion and expansion properties in co-crystals. We could then incorporate a hydrogen bond acceptor molecule that comprised a functional group capable of molecular motion. A series of co-crystals based on hydrogen bonding could be synthesised by modifying the CCF. The resulting differences in crystal packing and intermolecular interactions would provide an insight into structural effects on TE properties. Azo groups

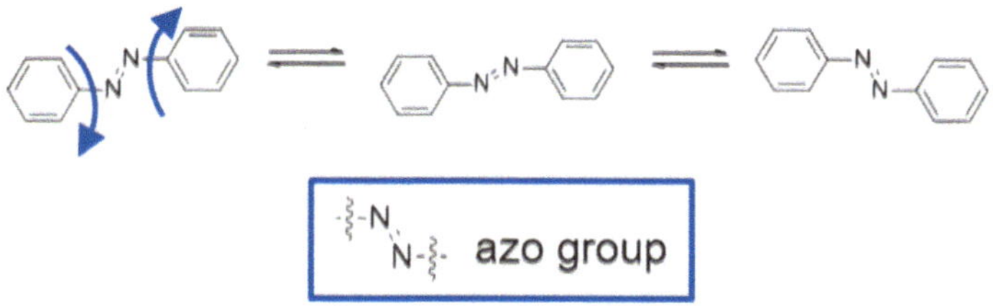

Fig. 8.21: Pedal motion of azo groups (azobenzene shown as an example).

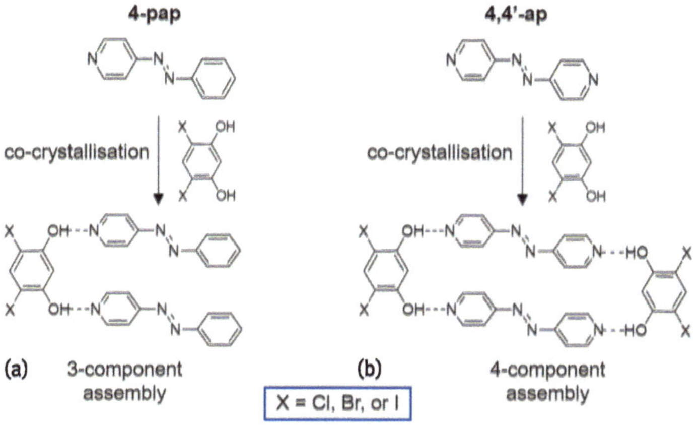

Fig. 8.22: Co-crystallisation approach for thermal expansion (TE) involving res-based co-crystal formers (CCFs) and azo-based compounds: (a) unsymmetrical 4-pap and (b) symmetrical 4,4′-ap.

in unsymmetrical and symmetrical 4-phenylazopyridine (4-pap) and 4,4′-azopyridine (4,4′-ap) respectively were targeted (Figure 8.22). A series of isosteric [97] res-based CCFs that act as ditopic hydrogen bond donors were employed.

It was determined that 4-pap and 4,4′-ap, as pure components, do not undergo molecular pedal motion and exhibit minimal TE in the solid state. Therefore, a co-crystallisation strategy was applied to afford hydrogen-bonded architectures able to support molecular motion [33, 85, 98].

In our initial work, co-crystallisation of unsymmetrical 4-pap with either 4,6-dichlororesorcinol (4,6-diCl res) or 4,6-dibromoresorcinol (4,6-diBr res) yielded isostructural co-crystals, namely 2(4-pap)·(4,6-diCl res) and 2(4-pap)·(4,6-diBr res) [99]. Both co-crystals were defined by three-component assemblies sustained by two O–H···N hydrogen bonds (Figure 8.23a, b), and the extended packings were based on tongue-and-groove structures [100, 101]. Upon cooling, molecular pedal motion in the azo groups of 4-pap occurred in both 2(4-pap)·(4,6-diCl res) and 2(4-pap)·(4,6-diBr res). Twisting of the aromatic rings of 4-pap was accompanied by pedal motion during

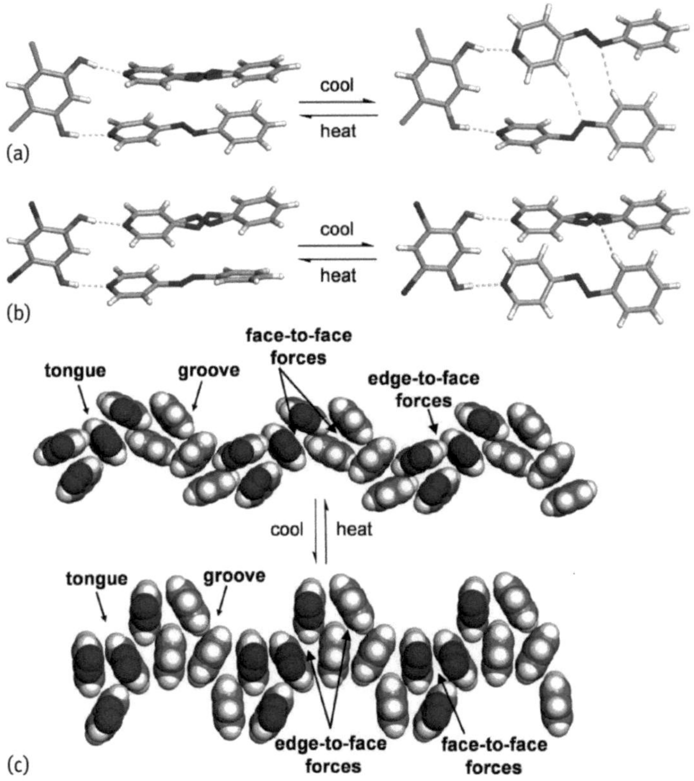

Fig. 8.23: Structural changes during TE: (a) 2(4-pap)·(4,6-diCl res) assembly, (b) 2(4-pap)·(4,6-diBr res) assembly and (c) tongue-and-groove of 2(4-pap)·(4,6-diCl res) (*blue* = azo group). *Yellow dashed lines* indicate hydrogen bonds.

cooling, which resulted in dramatic and widespread interconversion of face-to-face and edge-to-face aromatic stacking in the tongue-and-groove structure (Figure 8.23c).

Significant changes (approximately 3–18%) to all three crystallographic axes occurred as a result of the motion and twisting within the co-crystals. Both solids 2(4-pap)·(4,6-diCl res) and 2(4-pap)·(4,6-diBr res) experienced "colossal" [62] negative and positive expansions [99], commensurate with the largest reported TE coefficients [102] for organic compounds [60, 78, 79, 93, 103–106]. The TE coefficients for 2(4-pap)·(4,6-diCl res) were larger than 2(4-pap)·(4,6-diBr res), which was attributed to the presence of Br···Br interactions [107, 108] that inhibit full aromatic ring twisting [99]. Analogous halogen···halogen interactions were not present in 2(4-pap)·(4,6-diCl res).

Co-crystallisation of 4-pap [97] with 4,6-diiodoresorcinol (4,6-diI res) afforded 2(4-pap)·(4,6-diI res) (Figure 8.24). Upon cooling, molecular pedal motion in the azo groups occurred [88], but dramatic changes in unit cell dimensions were not observed. When the CCF was switched to 4,6-diI res, the packing was altered to such an extent that aromatic ring twisting was not favoured.

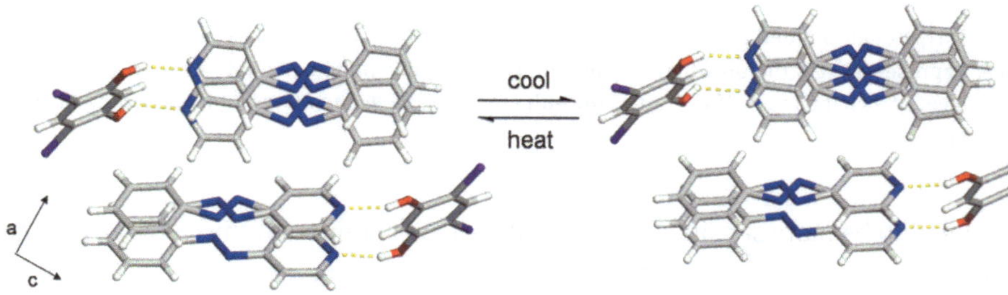

Fig. 8.24: Moderate structural changes within the asymmetric unit of 2(4-pap)·(4,6-diI res) during TE. *Yellow dashed lines* indicate hydrogen bonds.

The symmetrical azo derivative 4,4′-ap (Figure 8.22), was studied in follow-up work. Co-crystallisation of 4,4′-ap with isosteric 4,6-diCl res, 4,6-diBr res and 4,6-diI res afforded three isostructural co-crystals [109]. Each co-crystal was dominated by four-component assemblies in 2(4,4′-ap)·2(4,6-diX res) (where X = Cl, Br, or I) sustained by four O–H···N hydrogen bonds (Figure 8.25a–c). Extended crystal packing was defined by layers held together by a combination of O···X and C–H···X halogen bonds and C–H···π forces (Figure 8.25d, e) [109].

In contrast to the cases involving 4-pap, molecular pedal motion of the azo groups and twisting were not observed upon cooling 2(4,4′-ap)·2(4,6-diX res) [109]. However, each solid exhibited "colossal" [80, 110] TE in a direction approximately orthogonal to the stacked 2D layers (Figure 8.25e). The hydrogen and halogen bonds served to limit expansion of the 2D layers, whereas the weaker interlayer interactions, which can be

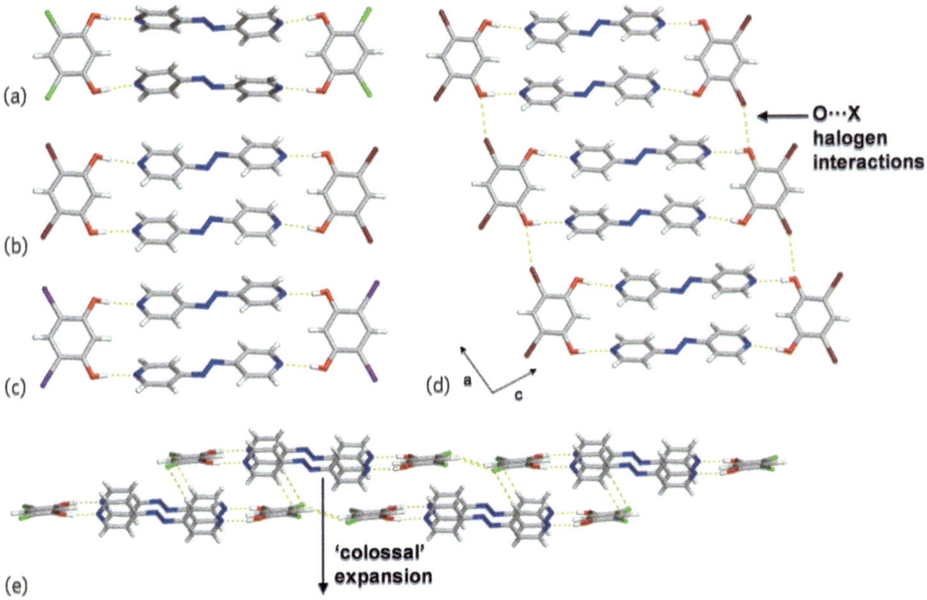

Fig. 8.25: Hydrogen-bonded assemblies of 4,4′-ap with (a) 4,6-diCl res, (b) 4,6-diBr res and (c) 4,6-diI res. Representative structures highlighting: (d) O···X halogen bonds in 2(4,4′-ap) · 2(4,6-diBr res) and (e) layered assemblies in 2(4,4′-ap) · 2(4,6-diCl res). *Yellow dashed lines* indicate hydrogen and halogen bonds.

considered to be more sensitive to temperature [106], supported the colossal TE. The TE properties were only achieved using the co-crystal approach.

8.4 Conclusions

Applications of co-crystals have been presented to direct [2+2] photodimerisations and to affect TE properties in the organic solid state. It has been demonstrated how co-crystals based on small-molecule templates can effectively enforce chemical reactivity to construct molecular targets in the solid state that are otherwise difficult to obtain in solution. Small-molecule CCFs were also used to furnish environments that support molecular motion and corresponding TE. Developments are envisaged in supramolecular chemistry and crystal engineering that will further enable co-crystallisation strategies to address increasingly complex challenges in synthetic and materials chemistry and therefore allow the effects of crystal packing to be confronted with appreciable success.

Acknowledgement: L.R.M. gratefully acknowledges support from the National Science Foundation (DMR-1408834). R.H.G. gratefully acknowledges support from Webster University from both a Faculty Research Grant and Faculty Development Fund.

Bibliography

[1] Lehn J-M. Supramolecular chemistry – scope and perspectives molecules – supermolecules – molecular devices. World Scientific Publishing Co.: Singapore, 1992.

[2] Wolf KL, Wolff R. Übermolekeln. Angew. Chem. 1949, 61, 191–201.

[3] Phillip D, Stoddart JF. Self-assembly in natural and unnatural systems. Angew. Chem. Int. Ed. 1996, 35, 1155–1196.

[4] Hof F, Craig SL, Nuckolls C, Rebek J. Molecular encapsulation. Angew. Chem. Int. Ed. 2002, 41, 1488–1508.

[5] Lehn J-M. Cryptates – chemistry of macropolycyclic inclusion complexes. Acc. Chem. Res. 1978, 11, 49–57.

[6] Pedersen CJ. Cyclic polyethers and their complexes with metal salts. J. Am. Chem. Soc. 1967, 89, 7017–7036.

[7] El Essawi M, Tebbe KF. Studies on polyhalides, XXXIV – Cesium(18-crown-6)dichloroiodate, $[Cs(C_{12}H_{24}O_6)]ICl_2$. Z. Naturforsch., B: Chem. Sci. 1998, 53, 263–265.

[8] Ward DL, Huang RH, Dye JL. Structures of alkalides and electrides. I. Structure of potassium cryptand[2.2.2] electride. Acta Crystallogr. C 1988, 44, 1374–1376.

[9] Bernstein J, Davis RE, Shimoni L, Chang NL. Patterns in hydrogen bonding – functionality and graph set analysis in crystals. Angew. Chem. Int. Ed. 1995, 34, 1555–1573.

[10] Desiraju GR, Gavezzotti A. From molecular to crystal-structure; polynuclear aromatic hydrocarbons. J. Chem. Soc., Chem. Commun. 1989, 621–623.

[11] Hunter CA, Sanders JKM. The nature of pi–pi interactions. J. Am. Chem. Soc. 1990, 112, 5525–5534.

[12] Metrangolo P, Meyer F, Pilati T, Resnati G, Terraneo G. Halogen bonding in supramolecular chemistry. Angew. Chem. Int. Ed. 2008, 47, 6114–6127.

[13] Moulton B, Zaworotko MJ. From molecules to crystal engineering: supramolecular isomerism and polymorphism in network solids. Chem. Rev. 2001, 101, 1629–1658.

[14] Dunitz JD. From crystal statics towards molecular dynamics. Trans. Am. Crystallogr. Assoc. 1984, 20, 1–14.

[15] Schmidt GMJ. Photodimerization in the solid state. Pure Appl. Chem. 1971, 27, 647–678.

[16] MacGillivray LR, Reid JL, Ripmeester JA. Supramolecular control of reactivity in the solid state using linear molecular templates. J. Am. Chem. Soc. 2000, 122, 7817–7818.

[17] Eddleston MD, Sivachelvam S, Jones W. Screening for polymorphs of cocrystals: a case study. CrystEngComm 2013, 15, 175–181.

[18] Das D, Jacobs T, Pietraszko A, Barbour LJ. Anomalous thermal expansion of an organic crystal – implications for elucidating the mechanism of an enantiotropic phase transformation. Chem. Commun. 2011, 47, 6009–6011.

[19] Cohen MD, Schmidt GMJ, Sonntag FI. Topochemistry. II. Photochemistry of trans-cinnamic acids. J. Chem. Soc. 1964, 2000–2013.

[20] Schmidt, GMJ. Topochemistry. Part III. The crystal chemistry of some trans-cinnamic acids. J. Chem. Soc. 1964, 2014–2021.

[21] Tanaka K, Toda F. In: Toda F (ed). Organic solid-state reactions. Springer Netherlands: Dordrecht, 2002.

[22] Stoermer R, Laage E. (Truxillic acids.) M. Natural and artificial truxinic and truxinic acids. Berichte der Dtsch. Chem. Gesellschaft Abteilung B Abhandlungen 1921, 54B, 77–85.

[23] Stoermer R, Laage, E. (Truxillic acid.) V. The seventh acid of the truxillic acid group, neo-truxinic acid. Berichte der Dtsch. Chem. Gesellschaft Abteilung B Abhandlungen 1921, 54B, 96–101.

[24] Ladell J, Mcdonald TRR, Schmidt GMJ. The crystal structure of alpha-trans-cinnamic acid. Acta Crystallogr 1956, 9, 195–195.

[25] Bryan RF, Freyberg DP. Crystal-structures of alpha-trans-methoxy-cinnamic and para-methoxy-cinnamic acids and their relation to thermal mesomorphism. J. Chem. Soc., Perkin Trans. 2 1975, 1835–1840.

[26] Desiraju GR. Crystal engineering: a holistic view. Angew. Chem. Int. Ed. 2007, 46, 8342–8356.

[27] Xu R, Schweizer B, Frauenrath H. Soluble poly(diacetylene)s using the perfluorophenyl-phenyl motif as a supermolecule synthon. J. Am. Chem. Soc. 2008, 130, 11437–11445.

[28] Coates GW, Dunn AR, Henling LM, Ziller JW, Lobkovsky EB, Grubbs RH. Phenyl-perfluorophenyl stacking interactions: topochemical[2+2] photodimerization and photopolymerization of olefinic compounds. J. Am. Chem. Soc. 1998, 120, 3641–3649.

[29] Sharma CVK, Panneerselvam K, Shimoni L, Katz H, Carrell HL, Desiraju GR. 3-(3′,5′-Dinitrophenyl)-4-(2′,5′-dimethoxyphenyl)cyclobutane-1,2-dicarboxylic acid: engineered topochemical synthesis and molecular and supramolecular properties. Chem. Mater. 1994, 6, 1282–1292.

[30] Aoyama Y, Endo K, Anzai T, Yamaguchi Y, Sawaki T, Kobayashi K, Kanehisa N, Hashimoto H, Kai Y, Masuda, H. Crystal engineering of stacked aromatic columns. Three-dimensional control of the alignment of orthogonal aromatic triads and guest quinones via self-assembly of hydrogen-bonded networks. J. Am. Chem. Soc. 1996, 118, 5562–5571.

[31] Ito Y, Borecka B, Trotter J, Scheffer JR. Control of solid-state photodimerization of trans-cinnamic acid by double salt formation with diamines. Tetrahedron Lett. 1995, 36, 6083–6086.

[32] Feldman KS, Campbell RF. Efficient stereo- and regiocontrolled alkene photodimerization through hydrogen bond enforced preorganization in the solid state. J. Org. Chem. 1995, 60, 1924–1925.

[33] MacGillivray LR, Papaefstathiou GS, Friščić T, Hamilton TD, Bučar D-K, Chu Q, Varshney DB, Georgiev IG. Supramolecular control of reactivity in the solid state: from templates to ladderanes to metal-organic frameworks. Acc. Chem. Res. 2008, 41, 280–291.

[34] MacGillivray LR, Reid JL, Ripmeester JA, Papaefstathiou GS. Toward a reactant library in template-directed solid-state organic synthesis: reactivity involving a monofunctional reactant based on a stilbazole. Ind. Eng. Chem. Res. 2002, 41, 4494–4497.

[35] Hutchins KM, Sumrak JC, MacGillivray LR. Resorcinol-templated head-to-head photodimerization of a thiophene in the solid state and unusual edge-to-face stacking in a discrete hydrogen-bonded assembly. Org. Lett. 2014, 16, 1052–1055.

[36] Bučar D-K, Sen A, Mariappan SVS, MacGillivray LR. A [2+2] cross-photodimerisation of photostable olefins via a three-component cocrystal solid solution. Chem. Commun. 2012, 48, 1790–1792.

[37] Gao XC, Friščić T, MacGillivray LR. Supramolecular construction of molecular ladders in the solid state. Angew. Chem. Int. Ed. 2004, 43, 232–236.

[38] Mascitti V, Corey EJ. Total synthesis of (+/-)-pentacycloanammoxic acid. J. Am. Chem. Soc. 2004, 126, 15664–15665.

[39] Friščić T, MacGillivray LR. 'Template-switching': a supramolecular strategy for the quantitative, gram-scale construction of a molecular target in the solid state. Chem. Commun. 2003, 1306–1307.

[40] Friščić T, MacGillivray LR. A step towards a [2.2]paracyclophane: a single crystal to single crystal reaction involving a hydrogen-bonded molecular assembly with multiple reaction centres. Aust. J. Chem. 2006, 59, 613–616.

[41] Santra R, Biradha K. Stepwise dimerization of double [2 + 2] reaction in the co-crystals of 1,5-bis(4-pyridyl)-1,4-pentadiene-3-one and phloroglucinol: a single-crystal to single-crystal transformation. CrystEngComm 2008, 10, 1524–1526.

[42] Meng F, Min J, Wang C, Wang L. Preparation of cocrystals of 1,4-diimidazolylbutadiene with a template molecule and dependence of the cocrystal photochemical products on wavelength. Eur. J. Org. Chem. 2016, 2016, 2220–2225.

[43] Papaefstathiou GS, Kipp AJ, MacGillivray LR. Exploiting modularity in template-controlled synthesis: a new linear template to direct reactivity within discrete hydrogen bonded molecular assemblies in the solid state. Chem. Commun. 2001, 2462–2463.

[44] Hutchins KM, Sumrak JC, Swenson DC, MacGillivray LR. Head-to-tail photodimerization of a thiophene in a co-crystal and a rare adipic acid dimer in the presence of a heterosynthon. CrystEngComm 2014, 16, 5762–5764.

[45] Shan N, Jones W. Identification of supramolecular templates: design of solid-state photoreactivity using structural similarity. Tetrahedron Lett. 2003, 44, 3687–3689.

[46] Lee E, Ju H, Lee SS, Park K-M. Tubular arrangement of 1,3-alternate calix[4]arene with bipyridines: a platform for [2 + 2] photocyclization. Cryst. Growth Des. 2013, 13, 992–995.

[47] Varshney DB, Gao XC, Friščić T, MacGillivray LR. Heteroditopic Rebek's imide directs the reactivity of homoditopic olefins within desolvated quaternary assemblies in the solid state. Angew. Chem. Int. Ed. 2006, 45, 646–650.

[48] Friščić T, MacGillivray LR. Reversing the code of a template-directed solid-state synthesis: a bipyridine template that directs a single-crystal-to-single-crystal [2 + 2] photodimerisation of a dicarboxylic acid. Chem. Commun. 2005, 5748–5750.

[49] Mei XF, Liu SL, Wolf C. Template-controlled face-to-face stacking of olefinic and aromatic carboxylic acids in the solid state. Org. Lett. 2007, 9, 2729–2732.

[50] Papaefstathiou GS, Duncan AJE, MacGillivray LR. Two act as one: unexpected dimers of catechol direct a solid-state [2+2] photodimerization in a six-component hydrogen-bonded assembly. Chem. Commun. 2014, 50, 15960–15962.

[51] Stojaković J, Whitis AM, MacGillivray LR. Discrete double-to-quadruple aromatic stacks: stepwise integration of face-to-face geometries in cocrystals based on indolocarbazole. Angew. Chem. Int. Ed. 2013, 52, 12127–12130.

[52] Gilday LC, Robinson SW, Barendt TA, Langton MJ, Mullaney BR, Beer PD. Halogen bonding in supramolecular chemistry. Chem. Rev. 2015, 115, 7118–7195.

[53] Mukherjee A, Tothadi S, Desiraju GR. Halogen bonds in crystal engineering: like hydrogen bonds yet different. Acc. Chem. Res. 2014, 47, 2514–2524.

[54] Aakeröy CB, Baldrighi M, Desper J, Metrangolo P, Resnati G. Supramolecular hierarchy among halogen-bond donors. Chem. Eur. J. 2013, 19, 16240–16247.

[55] Caronna T, Liantonio R, Logothetis TA, Metrangolo P, Pilati T, Resnati G. Halogen bonding and $\pi \cdots \pi$ stacking control reactivity in the solid state. J. Am. Chem. Soc. 2004, 126, 4500–4501.

[56] Zhu B, Wang J-R, Zhang Q, Mei X. Greener solid-state synthesis: stereo-selective [2 + 2] photodimerization of vitamin K_3 controlled by halogen bonding. CrystEngComm 2016, 18, 6327–6330.

[57] Sinnwell MA, MacGillivray LR. Halogen-bond-templated [2+2] photodimerization in the solid state: directed synthesis and rare self-inclusion of a halogenated product. Angew. Chem. Int. Ed. 2016, 55, 3477–3480.
[58] Berg JM, Tymoczko JL, Stryer L. Biochemistry, 5th edition; W H Freeman: New York, 2002.
[59] Bhattacharya S, Stojaković J, Saha BK, MacGillivray LR. A product of a templated solid-state photodimerization acts as a template: single-crystal reactivity in a single polymorph of a cocrystal. Org. Lett. 2013, 15, 744–747.
[60] Das D, Jacobs T, Barbour LJ. Exceptionally large positive and negative anisotropic thermal expansion of an organic crystalline material. Nat. Mater. 2010, 9, 36–39.
[61] Khuong TAV, Nuñez JE, Godinez CE, Garcia-Garibay MA. Crystalline molecular machines: a quest toward solid-state dynamics and function. Acc. Chem. Res. 2006, 39, 413–422.
[62] Goodwin AL, Calleja M, Conterio MJ, Dove MT, Evans JSO, Keen DA, Peters L, Tucker MG. Colossal positive and negative thermal expansion in the framework material $Ag_3[Co(CN)_6]$. Science 2008, 319, 794–797.
[63] Grobler I, Smith VJ, Bhatt PM, Herbert SA, Barbour LJ. Tunable anisotropic thermal expansion of a porous zinc(II) metal-organic framework. J. Am. Chem. Soc. 2013, 135, 6411–6414.
[64] Salvador JR, Gu F, Hogan T, Kanatzidis MG. Zero thermal expansion in YbGaGe due to an electronic valence transition. Nature 2003, 425, 702–705.
[65] Zhu Y, Corigliano A, Espinosa HD. A thermal actuator for nanoscale in situ microscopy testing: design and characterization. J. Micromech. Microeng. 2006, 16, 242–253.
[66] Roy R, Agrawal DK, Mckinstry HA. Very low thermal-expansion coefficient materials. Annu. Rev. Mater. Sci. 1989, 19, 59–81.
[67] Tran KD, Groshens TJ, Nelson JG. Fabrication of near-zero thermal expansion $(Fe_xSc_{1-x})2Mo_3O_{12}$-$MoO_3$ ceramic composite using the reaction sintering process. Mater. Sci. Eng. A 2001, 303, 234–240.
[68] Bhattacharya S, Saraswatula VG, Saha BK. Does higher-dimensional hydrogen bonding guarantee a smaller thermal expansion? A thermal expansion study of an interdigitated 1D and interpenetrated 3D hydrogen bonded colored dimorphic system. Cryst. Growth Des. 2016, 16, 277–284.
[69] Cliffe MJ, Hill JA, Murray CA, Coudert FX, Goodwin AL. Defect-dependent colossal negative thermal expansion in UiO-66(Hf) metal-organic framework. Phys. Chem. Chem. Phys. 2015, 17, 11586–11592.
[70] Hancock JC, Chapman KW, Halder GJ, Morelock CR, Karlan BS, Gallington LC, Bongiorno A, Han C, Zhou S, Wilkinson AP. Large negative thermal expansion and anomalous behavior on compression in cubic ReO_3-type A(II)B(IV)F(6): $CaZrF_6$ and $CaHfF_6$. Chem. Mater. 2015, 27, 3912–3918.
[71] Tallentire SE, Child F, Fall I, Vella-Zarb L, Evans IR, Tucker MG, Keen DA, Wilson C, Evans JSO. Systematic and controllable negative, zero, and positive thermal expansion in cubic $Zr_{1-x}Sn_xMo_2O_8$. J. Am. Chem. Soc. 2013, 135, 12849–12856.
[72] Yamada I, Shiro K, Etani H, Marukawa S, Hayashi N, Mizumaki M, Kusano Y, Ueda S, Abe H. Irifune, T. Valence transitions in negative thermal expansion material $SrCu_3Fe_4O_{12}$. Inorg. Chem. 2014, 53, 10563–10569.
[73] Saraswatula VG, Saha BK. Modulation of thermal expansion by guests and polymorphism in a hydrogen bonded host. Cryst. Growth Des. 2015, 15, 593–601.
[74] Saraswatula VG, Bhattacharya S, Saha BK. Can the thermal expansion be controlled by varying the hydrogen bond dimensionality in polymorphs? New J. Chem. 2015, 39, 3345–3348.
[75] Kitaigorodsky AI. Molecular Crystals and Molecules. Academic Press: New York, 1973.

[76] Weigel D, Beguemsi T, Garnier P, Berar JF. Evolution of thermal-expansion tensor as function of temperature. I. General Law of Evolution of Tensor Symmetry. J. Solid State Chem. 1978, 23, 241–251.
[77] Garnier P, Calvarin G, Weigel D. Determination of thermal dilation tensors by x-ray diffractions. II. Dilation of lead oxides and interpretation as function of their structure and their chemical bonds. J. Chim. Phys. Phys. Chim. Biol. 1972, 69, 1711–1718.
[78] Bhattacharya S, Saha BK. Uniaxial negative thermal expansion in an organic complex caused by sliding of layers. Cryst. Growth Des. 2012, 12, 4716–4719.
[79] Bhattacharya S, Saha BK. Interaction dependence and similarity in thermal expansion of a dimorphic 1D hydrogen-bonded organic complex. Cryst. Growth Des. 2013, 13, 3299–3302.
[80] Bhattacharya S, Saha BK. Steric guided anomalous thermal expansion in a dimorphic organic system. CrystEngComm 2014, 16, 2340–2343.
[81] Desiraju GR. Crystal engineering: a brief overview. J. Chem. Sci. 2010, 122, 667–675.
[82] Aakeröy CB, Salmon DJ. Building co-crystals with molecular sense and supramolecular sensibility. CrystEngComm 2005, 7, 439–448.
[83] Goud NR, Khan RA, Nangia A. Modulating the solubility of sulfacetamide by means of cocrystals. CrystEngComm 2014, 16, 5859–5869.
[84] Bushuyev OS, Corkery TC, Barrett CJ, Friščić T. Photo-mechanical azobenzene cocrystals and in situ X-ray diffraction monitoring of their optically-induced crystal-to-crystal isomerisation. Chem. Sci. 2014, 5, 3158–3164.
[85] Hutchins KM, Dutta S, Loren BP, MacGillivray LR. Co-crystals of a salicylideneaniline: photochromism involving planar dihedral angles. Chem. Mater. 2014, 26, 3042–3044.
[86] Aakeröy CB, Forbes S, Desper J. Altering physical properties of pharmaceutical co-crystals in a systematic manner. CrystEngComm 2014, 16, 5870–5877.
[87] Kapadia PP, Ditzler LR, Baltrusaitis J, Swenson DC, Tivanski AV, Pigge FC. Semiconducting organic assemblies prepared from tetraphenylethylene tetracarboxylic acid and bis(pyridine)s via charge-assisted hydrogen bonding. J. Am. Chem. Soc. 2011, 133, 8490–8493.
[88] Harada J, Ogawa K. What molecules are likely or unlikely to undergo pedal motions in crystals? Cryst. Growth Des. 2014, 14, 5182–5188.
[89] Bernstein J. Crystal-growth, polymorphism and structure–property relationships in organic-crystals. J. Phys. D: Appl. Phys. 1993, 26, B66–B76.
[90] Braga D, Grepioni F, Chelazzi L, Campana M, Confortini D, Viscomi GC. The structure-property relationship of four crystal forms of rifaximin. CrystEngComm 2012, 14, 6404–6411.
[91] Cinčić D, Friščić T, Jones W. Isostructural materials achieved by using structurally equivalent donors and acceptors in halogen-bonded cocrystals. Chem. Eur. J. 2008, 14, 747–753.
[92] Saraswatula VG, Saha BK. The effect of temperature on interhalogen interactions in a series of isostructural organic systems. New J. Chem. 2014, 38, 897–901.
[93] Bhattacharya S, Saraswatula VG, Saha BK. Thermal expansion in alkane diacids – another property showing alternation in an odd–even series. Cryst. Growth Des. 2013, 13, 3651–3656.
[94] Harada J, Ogawa K. Pedal motion in crystals. Chem. Soc. Rev. 2009, 38, 2244–2252.
[95] Harada J, Ogawa K, Tomoda S. Molecular motion and conformational interconversion of azobenzenes in crystals as studied by X-ray diffraction. Acta Crystallogr. B 1997, 53, 662–672.
[96] Singh R, Ahmad M, Bharadwaj PK. Coordination polymers of copper and zinc ions with a linear linker having imidazole at each end and an azo moiety in the middle: pedal motion, gas adsorption, and emission studies. Cryst. Growth Des. 2012, 12, 5025–5034.
[97] Dikundwar AG, Pete UD, Zade CM, Bendre RS, Row TNG. Tracing a common "origin" of phase transformation, polymorphism, disorder, isosterism, and isostructuralism in fluorobenzoylcarvacryl thiourea. Cryst. Growth Des. 2012, 12, 4530–4534.

[98] Elacqua E, Kaushik P, Groeneman RH, Sumrak JC, Bučar D-K, MacGillivray LR. A supramolecular protecting group strategy introduced to the organic solid state: enhanced reactivity through molecular pedal motion. Angew. Chem. Int. Ed. 2012, 51, 1037–1041.

[99] Hutchins KM, Groeneman RH, Reinheimer EW, Swenson DC, MacGillivray LR. Achieving dynamic behaviour nd thermal expansion in the organic solid state via co-crystallization. Chem. Sci. 2015, 6, 4717–4722.

[100] Carlucci L, Ciani G, Proserpio DM. Borromean links and other non-conventional links in 'polycatenated' coordination polymers: re-examination of some puzzling networks. CrystEngComm 2003, 5, 269–279.

[101] Kepert CJ, Rosseinsky MJ. Zeolite-like crystal structure of an empty microporous molecular framework. Chem. Commun. 1999, 375–376.

[102] Cliffe MJ, Goodwin AL. PASCal: a principal axis strain calculator for thermal expansion and compressibility determination. J. Appl. Crystallogr. 2012, 45, 1321–1329.

[103] Engel ER, Smith VJ, Bezuidenhout CX, Barbour LJ. Uniaxial negative thermal expansion facilitated by weak host-guest interactions. Chem. Commun. 2014, 50, 4238–4241.

[104] Negrier P, Tamarit JLl, Barrio M, Mondieig D. Polymorphism in halogen-ethane derivatives: CCl_3–CCl_3 and ClF_2C–CF_2Cl. Cryst. Growth Des. 2013, 13, 782–791.

[105] Fortes AD, Suard E, Knight KS. Negative linear compressibility and massive anisotropic thermal expansion in methanol monohydrate. Science 2011, 331, 742–746.

[106] Salud J, Barrio M, Lopez DO, Tamarit JL, Alcobe X. Anisotropy of intermolecular interactions from the study of the thermal-expansion tensor. J. Appl. Crystallogr. 1998, 31, 748–757.

[107] Medlycott EA, Udachin KA, Hanan GS. Non-covalent polymerisation in the solid state: halogen–halogen vs. methyl–methyl interactions in the complexes of 2,4-di(2-pyridyl)-1,3,5-triazine ligands. Dalton Trans. 2007, 430–438.

[108] Reddy CM, Kirchner MT, Gundakaram RC, Padmanabhan KA, Desiraju GR. Isostructurality, polymorphism and mechanical properties of some hexahalogenated benzenes: the nature of halogen···halogen interactions. Chem. Eur. J. 2006, 12, 2222–2234.

[109] Hutchins KM, Kummer KA, Groeneman RH, Reinheimer EW, Sinnwell MA, Swenson DC, MacGillivray LR. Thermal expansion properties of three isostructural co-crystals composed of isosteric components: interplay between halogen and hydrogen bonds. CrystEngComm 2016, 18, 8354–8357.

[110] Mahé N, Nicolaï B, Allouchi H, Barrio M, Do B, Ceolin R, Tamarit JL, Rietveld IB. Crystal structure and solid-state properties of 3,4-diaminopyridine dihydrogen phosphate and their comparison with other diaminopyridine salts. Cryst. Growth Des. 2013, 13, 708–715.

Tom Leyssens and Joop H. ter Horst

9 Solution co-crystallisation and its applications

9.1 Introduction

Even though the term co-crystal remains a topic of debate [1, 2], general consensus is evolving towards defining co-crystals as crystalline compounds constructed of two or more neutral components, which are, in their pure form, solid at room temperature when under atmospheric pressure. In most cases, this definition clearly distinguishes co-crystals from salts or solvates. For solvates one of the components in its pure form, i.e. the solvent, is a liquid. A co-crystal is different from a salt as it is constructed from two distinct neutral components without any charge transfer taking place between the components. Co-crystals can therefore be formed of components that lack ionisable functional groups. Figure 9.1 shows the structure of the 1 : 1 carbamazepine–saccharin co-crystal constructed from neutral carbamazepine and saccharin molecules. The co-formers carbamazepine and saccharin are solids at room temperature.

Co-crystals have become increasingly popular in the pharmaceutical industry as they can improve physicochemical properties of active pharmaceutical ingredients (APIs) [4] by enhancing drug bioavailability [5, 6], stability [7, 8] and even processability [9]. A pharmaceutical co-crystal is obtained when an API is combined with pharmaceutically acceptable co-formers, e.g. generally recognised as safe (GRAS) materials, amino acids, nutriments or even other drug molecules. To prepare a co-crystalline material, either a mechanochemical or solution crystallisation approach can be used. In the mechanochemical approach, the compound of interest is ground together with a potential co-former using amounts reflecting the expected co-crystal stoichiometry. In the solution crystallisation approach, a solution containing the correct amount of co-crystal components is supersaturated and co-crystallisation rather than pure component crystallisation occurs.

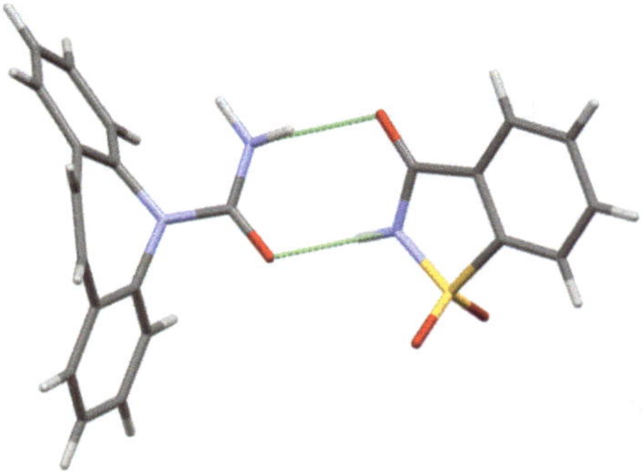

Fig. 9.1: A 1 : 1 co-crystal of carbamazepine and saccharin [3] showing a typical amide–amide hydrogen bond.

https://doi.org/10.1515/9783110464955-009

Both approaches have their advantages and disadvantages. The mechanochemical approach often requires less material and rapid screening is easy to implement. After grinding two components in a mortar or ball mill, the resulting powder can easily be analysed, e.g. X-ray powder diffraction, Raman and infrared (IR), to identify solid-state changes. The mechanochemical approach is often facilitated by the addition of a catalytic amount of solvent to the mixed solids (liquid assisted grinding; LAG) [10, 11]. The downside is that this approach can lead to polymorphic transformations, degradation of the compound, amorphous materials and lack of control over the homogeneity of the sample and process conditions, i.e. temperature and pressure.

Solution crystallisation, on the other hand, offers good control over process parameters and solution homogeneity, and is long proven as an industrial production technique [12]. In the discovery phase, novel forms such as co-crystal solvates, stoichiometrically diverse co-crystals [13] and co-crystal polymorphs [14] can furthermore be more easily identified, there is the advantage of crystallising under various conditions, for instance, changing the temperature profile, solvent composition and crystallisation technique. In addition, such small-scale solution co-crystallisations yield initial data for the future development of a large-scale industrial co-crystallisation process. Finally, solution co-crystallisation can lead to the formation of single crystals, which can give a definite confirmation of co-crystal formation through structural resolution.

The limited use of solution co-crystallisation compared with mechanochemistry can in part be ascribed to the increased complexity of the crystallising system, as the introduction of a solvent brings additional degrees of freedom to an already multi-component system. For instance, the solution composition required for co-crystallisation to occur does not have to coincide with the co-crystal composition: a 1 : 1 solution composition might result in the crystallisation of a pure component crystalline product rather than a 1 : 1 co-crystal. In this chapter, an attempt is made to alleviate the apparent complexity of solution co-crystallisation and convince the reader of its importance and relevance in co-crystal research and applications. The chapter therefore starts by giving an overview of solution co-crystallisation thermodynamics and kinetics and shows their importance for the different solution co-crystallisation methods. The second part illustrates the importance of solution co-crystallisation for enhanced drug product properties, but also highlights the opportunities co-crystallisation offers in the development of separation processes.

9.2 Solution co-crystallisation

Like all crystalline materials, co-crystals can be obtained from a supersaturated solution through the fundamental processes of nucleation and crystal growth. For co-crystallisation to be feasible, consistent and reliable, conditions must be established under which the suspension of co-crystalline material becomes the thermodynamically stable state. These conditions are graphically illustrated in phase diagrams that show, under equilibrium conditions, the stable phases and their compositions for a given set of conditions, such as temperature, pressure and overall composition. Phase diagrams thus allow the identification of the final suitable conditions for a robust co-crystallisation process. After discussing the thermodynamics of solution co-crystallisation, the discussion focuses upon solution co-crystallisation kinetics, as these define the route towards the final conditions of thermodynamic equilibrium. They are therefore of high importance in achieving full control over the solution co-crystallisation process.

9.2.1 Co-crystal solution thermodynamics

A phase diagram represents the thermodynamically stable phases for a given overall composition at a specific temperature and pressure. A binary melting temperature–composition diagram of a co-crystal system is shown in Figure 9.2 (left). Within this diagram, the three solid phases of pure components, A and B, and co-crystal, AB, each characterised by its melting temperature, and a melt, L, can be identified. For A, B and AB, the melting temperatures correspond respectively to $T_m(A)$, $T_m(B)$ and $T_m(AB)$ at mole fractions of A of 1, 0 and 0.5. Far below the melting temperature, any other overall composition corresponds to a physical bi-phasic mixture of either A and AB or B and AB. When temperatures are high enough to allow complete melting, a single liquid phase, L, is obtained. At intermediate temperatures, three bi-phasic situations occur (A + L; AB + L; B + L), which are separated by a liquidus line from the complete liquid phase and by the solidus line from the complete solid phase. The liquidus line indic-

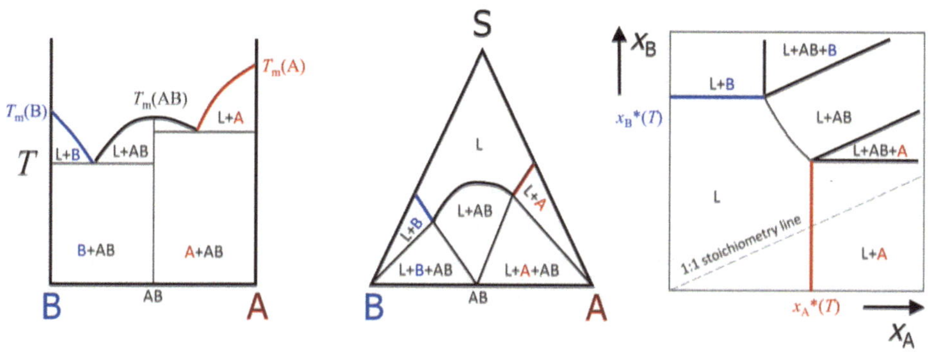

Fig. 9.2: *Left*: temperature–composition binary phase diagram of the AB co-crystal-forming components A and B. $T_m(A)$, $T_m(B)$ and $T_m(AB)$ represent the melting temperature of A and B and co-crystal AB respectively. *Middle*: ternary phase diagram at constant pressure and temperature of components A and B in solvent S. *Right*: an orthogonal ternary phase diagram of components A and B in a solvent. The mole fractions x_A^* and x_B^* represent the solubility of components A and B respectively.

ates the melting temperature of a specific composition and the solidus line indicates the eutectic temperature.

Within the binary phase diagram, two specific invariant points (eutectic points) correspond to the situation at which three phases co-exist in equilibrium, i.e. A+AB+L or B+AB+L. The eutectic point corresponds to the intercept point of two liquidus lines. A solid sample with an overall eutectic composition fully melts upon heating at the eutectic temperature. Any mixed sample showing an overall composition different from the eutectic composition partially melts upon reaching the eutectic temperature. Heating from the eutectic temperature to the liquidus line, more crystalline material is melted and enriched in the other component. At the liquidus line, the sample melts fully.

When performing co-crystallisation from solution, an additional degree of freedom is introduced, as not only the amount of both components, but also the amount of solvent can be varied. For the sake of clarity, and to remain with a 2D representation, phase diagrams are often given at constant temperature and pressure, limiting the variables to the relative amounts of all components. This typically leads to ternary phase diagrams, as presented in Figure 9.2 (middle and right) [15, 16]. To investigate temperature effects, multiple ternary diagrams can be easily superimposed. Using the representation in Figure 9.2 (middle), common practice is to place the solvent at the top of the triangle with the base representing the relative amounts of A and B. Adding solvent to a stoichiometric amount of A and B implies following a line leading from the middle of the base line to the top of the triangle. The orthogonal representation (Figure 9.2, right) is an alternative view of this diagram. Instead of representing the amounts of all components, the solvent is implicitly included by the concentration of components A and B on the horizontal and vertical axes respectively. Both phase diagram representations are thermodynamically identical and the choice of one or the other usually depends on personal preference.

The ternary phase diagrams (Figure 9.2) show one mono-phasic region (liquid L), three bi-phasic regions (A + L; B + L; AB + L) and two tri-phasic regions (A + AB + L; B + AB + L). The three bi-phasic regions, in which the solution and a solid are in equilibrium, are separated from the liquid region by solubility curves. The solubility curves of A (red line) and B (blue line) separate the bi-phasic regions A + L and B + L, respectively, from the liquid phase. As a co-crystal A_nB_m is a binary crystalline material; its solubility can be mathematically described through a solubility product K (Equation 9.1), with the solution mole fractions x_A and x_B of components A and B and the *, indicating that the solution has a composition for which the co-crystal material is in equilibrium with the solution [17, 18]:

$$(A_nB_m)_{solid} \rightleftarrows nA_{solution} + mB_{solution} \rightarrow K = (x_A^n.x_B^m)^* \tag{9.1}$$

This solubility product K is a function of temperature, but also depends on the stoichiometry in the solution; for instance, if solute complexation changes relative to the concentration of components A and B [19, 20].

In the bi-phasic regions, the equilibrium composition of the solution is given by the intersection between the solubility curve and the tie line, with the straight line starting at the pure solid composition with which the solution is in equilibrium and passing through the overall mixture composition. The composition of the liquid phase therefore varies along the solubility curves, according to the relative amounts of A and B present in the overall mixture and according to the solid phase, which is stable in suspension. The proportion of liquid versus solid phase can be determined using the lever rule. In the tri-phasic regions, A + AB + L and B + AB + L, the solution is in equilibrium with two solid phases. The solution composition in this region is indicated by the point at which the two solubility curves cross the eutectic point. This eutectic point is an invariant point and does not depend on the amount of solids present, as long as the overall composition lies in the three-phasic region. Two distinct eutectic points appear that correspond to the crossing of the co-crystal solubility curve with the solubility curve of A or B respectively.

Components A and B are different in nature and they show different pure component solubilities. These differences have an impact on the overall aspect of the ternary phase diagram. Generally, when both components show comparable solubility in a given solvent, a more symmetrical phase diagram is obtained (Figure 9.3, middle). Such diagrams often show a large bi-phasic region in which the co-crystal is stable in suspension. When the pure components show a significant difference in solubility, an asymmetrical ternary phase diagram usually results in the co-crystal bi-phasic region being skewed towards the more soluble compound (Figure 9.3, right and left). The greater the difference in solubility between the two components, the more pronounced this asymmetry becomes. According to the extent of the asymmetry, two cases are distinguished by evaluating at the stoichiometric line, the line showing all overall compositions having the same stoichiometric ratio as the co-crystal. In the first case, the 1 : 1 stoichiometric line in an AB co-crystal system, or the $n : m$ stoichiometric line in an A_nB_m co-crystal system crosses the co-crystal bi-phasic region. Then, the system is

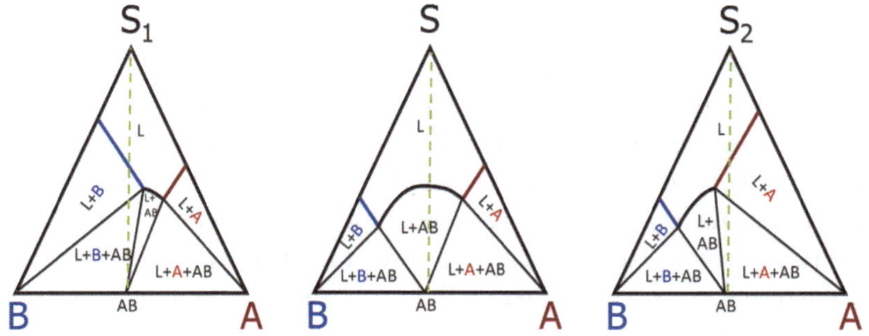

Fig. 9.3: The skewedness of the co-crystal region in the ternary phase diagram is determined by the pure component solubilities. The *dashed line* indicates the 1 : 1 stoichiometric line in this 1 : 1 co-crystal system.

said to behave congruently (Figure 9.3, middle): adding solvent to a co-crystal sample partially dissolves the co-crystals, creating a suspension of co-crystals without transformation to other solid phases. In this case, a robust co-crystallisation process can be developed starting from a stoichiometric component ratio. On the other hand, in a phase diagram with a highly skewed co-crystal bi-phasic region, the 1 : 1 stoichiometric line crosses the bi-phasic region of the less soluble component, region L+B in Figure 9.3 (left) or region L+A in Figure 9.3 (right), and the tri-phasic region. This tri-phasic region contains the co-crystal in addition to component A and B, in equilibrium with the solution which has the corresponding eutectic composition. Then, the system is said to behave incongruently: adding solvent to a co-crystal sample causes a (partial) transformation of the co-crystal phase to one of the pure component crystals. In this case, starting from a stoichiometric ratio of components to develop a co-crystallisation process is not the ideal choice. A system can behave incongruently in a given solvent, but does not necessarily do so in another solvent. The choice of solvent is therefore crucial when developing a co-crystallisation process, as it has a direct impact on the overall aspect of the phase diagram.

9.2.2 Solution co-crystallisation methods

To prepare co-crystals by solution crystallisation, supersaturation must be created for the co-crystalline material throughout the entire process. The method of supersaturation creation defines the crystallisation technique used. In addition to the conventional crystallisation techniques such as cooling and evaporative crystallisation, other techniques, such as precipitation and solvent-mediated phase transformations can be applied. As for single component systems, understanding the underlying thermodynamics is crucial. The various crystallisation techniques lead to different routes through the ternary phase diagrams shown in Figure 9.4, which are explained in the next subsections.

9.2.2.1 Evaporative co-crystallisation

During evaporative co-crystallisation, the supersaturation is created by evaporating the solvent, which increases the concentrations of the co-crystal-forming components in the solution. Usually, solvent evaporation takes place at a constant temperature and pressure. Industrially, the evaporation takes place at the boiling temperature of the solution, which is a function of the pressure. High evaporation rates can be achieved at the boiling point owing to good vapour removal. An increased evaporation rate leads to an increased rate of supersaturation generation, which speeds up the crystallisation process, but may also increase the chance of metastable formation.

The phase diagram in Figure 9.4a exemplifies evaporative co-crystallisation. Initially, the evaporative concentration at point 0, with composition (x_A^0, x_B^0), lies in the

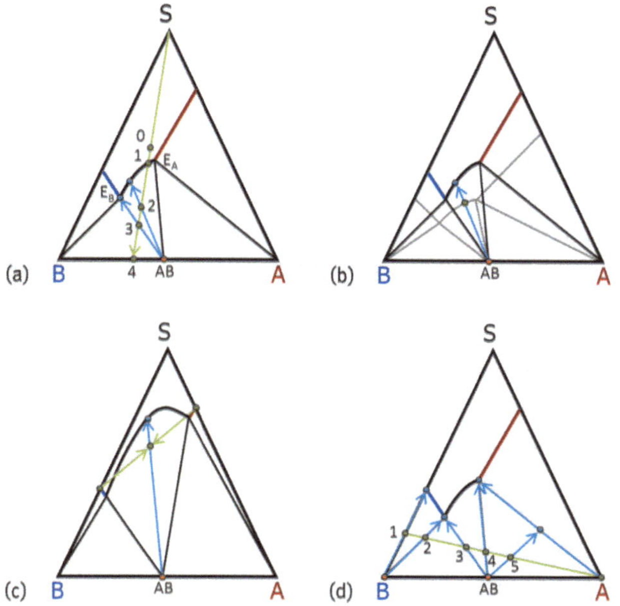

Fig. 9.4: Solution co-crystallisation methods and their relation to the ternary phase diagram. (a) Evaporative crystallisation; (b) cooling crystallisation; c: precipitation; (d) solvent-mediated transformation.

clear liquid region and represents an overall stoichiometric composition with a slight excess of B. Thus, in this example, the overall stoichiometry is unequal to that in the co-crystal. After evaporation, the solvent vapour is removed from the system and therefore, the location in the phase diagram of the overall composition changes. The removal of solvent increases the concentrations of the co-crystal-forming components while keeping the solution stoichiometry, the ratio of A versus B, constant (green line from points 0 to 1), which gives a relationship between the mole fractions x_A and x_B of components A and B in the solution during evaporation.

$$x_B = \frac{x_B^0}{x_A^0} x_A \tag{9.2}$$

This is an expression for the green line, the operating line (Figure 9.4a), showing the overall composition in the remaining solution or suspension during evaporation.

Based on the usually realistic assumption that only solvent evaporates and leaves the system, the mole fractions x_A and x_B for components A and B, which are the same as the solution concentrates as long as no crystals have formed, can be described using the initial concentrations x_A^0 and x_B^0 through the fraction x_e of solvent evaporated and removed.

$$x_A = \frac{x_A^0}{1 - x_e x_S^0}; \quad x_B = \frac{x_B^0}{1 - x_e x_S^0} \tag{9.3}$$

These equations describe the overall composition when the solvent is evaporated from the system as expressed by the green line in Figure 9.4a. In Equation 9.3 the initial solvent fraction, x_S^0, is used rather than $1 - x_A^0 - x_B^0$ to take into account any substantial impurities or additives in the system that would also be concentrated owing to solvent evaporation.

In point 1 of Figure 9.4a, the saturation concentration for the system is reached. At this saturation concentration, for an A_nB_m co-crystal, the solution composition product $(x_A)^n.(x_B)^m$ is equal to the solubility product K (Equation 9.1). For simplicity, the discussion here considers a 1 : 1 co-crystal, AB, with experimental solubility product $K = (x_A.x_B)^*$. The experimental solubility product K is represented by the thick, curved black line in Figure 9.4a. If the solubility product K is known from solubility measurements, the fraction x_e^* of the solvent that must be evaporated to reach saturation at point 1 in Figure 9.4a can be determined:

$$x_e^* = \frac{1 - \left[\frac{x_A^0 x_B^0}{K}\right]^{1/2}}{x_S^0} \tag{9.4}$$

Evaporating more solvent creates supersaturation towards the co-crystal phase and the overall composition enters the two-phase region. However, as no co-crystal material has yet formed, the stoichiometric ratio in the solution remains the same. A certain amount of supersaturation is needed to initiate the formation of the co-crystal phase owing to the energy barrier for crystal nucleation [21]. As long as no crystalline phase is present, the solution composition continues to develop along the green operating line in Figure 9.4a towards point 2.

At some point, the supersaturation for the co-crystalline phase is great enough to overcome the nucleation barrier and co-crystallisation commences. The solution A : B stoichiometry now starts to deviate from the overall ratio if the initial A : B stoichiometry is not the same as in the formed crystalline phase, such as in the case of Figure 9.4a. In the case of co-crystallisation in a solution with an overall composition equal to point 2, the solution concentrations of A and B start to decrease following the blue arrow pointing from the co-crystal composition on the AB axis to the point on the solubility line, indicating equilibrium between the two phases. The relationship between the actual solution concentrations x_A and x_B upon co-crystallisation can be described using the mole balance if the overall composition (x_A^{p2}, x_B^{p2}) in point 2 and the composition (x_A^{cr}, x_B^{cr}) in the co-crystalline material are known. For instance, the overall composition (x_A^{p2}, x_B^{p2}) is known from the fraction x_e of solvent evaporated and the composition (x_A^{cr}, x_B^{cr}) of a 1 : 1 AB co-crystal is (0.5, 0.5).

$$\frac{x_B^{cr} - x_B}{x_B^{cr} - x_B^{p2}} = \frac{x_A^{cr} - x_A}{x_A^{cr} - x_A^{p2}} \tag{9.5}$$

As co-crystallisation proceeds, the concentrations x_A and x_B continue to decrease following Equation 9.5 until, equilibrium is eventually reached. Equilibrium is achieved if the concentration product $x_A \cdot x_B$ of the concentrations following Equation 9.5 (Figure 9.4a, blue line), is equal to the solubility product K (Figure 9.4a, black line). Thus, the equilibrium solution composition can be derived from the intercept of the blue arrow (described by Equation 9.5) and the co-crystal solubility line represented by the solubility product K (described by Equation 9.1). Through equations 5 and 3, this gives an expression that can be solved for the composition (x_A^e, x_B^e) in equilibrium if the fraction x_e of solvent evaporated is known.

$$\frac{x_A^{cr} - x_A^e}{x_B^{cr} - \frac{K}{x_A^e}} = \frac{x_A^{cr} - \frac{x_A^0}{1-x_e x_S^0}}{x_B^{cr} - \frac{x_B^0}{1-x_e x_S^0}} \tag{9.6}$$

This equation can thus be used to determine the equilibrium concentrations x_A^e and $x_B^e = x_A^e/K$ for the fraction $x_e > x_e^*$ of solvent evaporated.

Evaporating solvent beyond point 2 creates more supersaturation and the co-crystallisation continues. As the overall composition of the suspension continues to develop along the line from pure solvent to point 3, more co-crystalline material is crystallising out and the solution composition develops from (close to) equilibrium in point 2 towards the eutectic point E_B (Figure 9.3). If the overall suspension composition arrives at point 3, the solution concentration in the suspension equilibrates towards the eutectic composition (x_A^{EB}, x_B^{EB}) of eutectic point E_B. Thus, there is also an upper limit x_e^{max} to the fraction of solvent evaporated beyond which an evaporative co-crystallisation results, thermodynamically, in a product contaminated with pure component B.

The eutectic composition at eutectic point E_B is an equilibrium composition towards the co-crystalline phase and therefore, the solubility product $K = x_A^{EB}. x_B^{EB}$ can be described using the eutectic composition (x_A^{EB}, x_B^{EB}). This enables the composition to be established in point 3 and therefore the fraction x_e^{max} of solvent needed to evaporate to reach an overall composition at point 3. Point 3 is positioned on the line going from co-crystal composition (x_A^{cr}, x_B^{cr}) to eutectic E_B composition (x_A^{EB}, x_B^{EB}), which is described by Equation 9.6 in the special case of $x_A^e = x_A^{EB}$ and $x_B^e = x_B^{EB}$. Point 3 is also positioned on the working line described by Equation 9.2. Point 3 is therefore determined by the intercept between these two lines, which leads to the following Equation 9.7, in which only the fraction x_e^{max} of solvent evaporated is unknown:

$$\frac{x_A^{cr} - x_A^{EB}}{x_B^{cr} - x_B^{EB}} = \frac{x_A^{cr} - \frac{x_A^0}{1-x_e^{max} x_S^0}}{x_B^{cr} - \frac{x_B^0}{1-x_e^{max} x_S^0}} \tag{9.7}$$

Therefore, the equilibrium solution concentrations x_A^e and x_B^e can be described when evaporating a sample using a fraction $x_e^* \le x_e \le x_e^{max}$ of solvent evaporated.

If the evaporation is continued beyond point 3, the overall suspension composition enters the tri-phasic region and the possibility exists that the third phase, crystals of pure component B, crystallise concomitantly with the co-crystals. When equilibrated, the suspension in the tri-phasic region comprises co-crystals, pure component B and a solution with the eutectic composition.

If all solvent is evaporated, the final mixture has the composition on the AB axis (point 4) and contains both co-crystal and B crystals. At the end of full evaporation, in particular, the concentrations can increase rapidly, even at low evaporation rate, and crystallisation may not be able to keep up with supersaturation reduction compared with the supersaturation increase due to evaporation. Therefore, in this part of the process, the chance of metastable formation and poor-quality crystals is substantial. For co-crystal discovery purposes, it is consequently better not to achieve full evaporation but to filter off obtained crystals from the suspension once sufficient evaporation has occurred.

The yield Y of a batch-wise industrial evaporative co-crystallisation can be defined as the fraction of co-crystallised component A:

$$Y = \frac{x_A^{p2} - x_A^*}{x_A^{p2}} = \frac{x_A^0 - x_A^*(1 - x_e x_S^0)}{x_A^0} \tag{9.8}$$

This is valid for $x_e^* \le x_e \le x_e^{max}$. Below x_e^* the yield is zero; above x_e^{max} the product is a mixture of co-crystals and pure component B. To control a batch-wise industrial evaporative co-crystallisation, it is recommended that the overall suspension composition should not be developed beyond point 3. Thus, the eutectic composition indicates the boundary for the process. If the eutectic composition is known, a co-crystallisation process can be designed that produces pure co-crystals in a stable process up to a fraction x_e^{max} of evaporated solvent achieving a final solution composition in the eutectic point. Usually, this is also the condition of maximum yield of a stable co-crystallisation.

For pharmaceutical component A, co-crystallisation should have an optimal yield with as much component A as possible co-crystallised during a stable operation.

If the stoichiometry in the right eutectic point is higher, e.g. 2 : 1, than that of the co-crystal, e.g. 1 : 1, and the stoichiometry in the left eutectic point is lower, e.g. 1 : 5, the system behaves congruently and it is possible to use stable evaporative co-crystallisation starting with an overall composition that is equal to that in the co-crystal, e.g. for the system in Figure 9.3 (middle). In principle, all solvent can be evaporated, which would transform all of component A into co-crystalline material. However, to maintain a stable operation towards the end of such a process is difficult owing to the large amount of crystalline mass present compared with the amount of solvent.

Another criterion to employ for choosing the end of an evaporative co-crystallisation is the suspension density, ρ. A stable operation can only be achieved up to a certain maximum crystal suspension density. Beyond that suspension density, the crystals are not sufficiently suspended and product quality may be compromised. The suspension density can, for instance, be determined by the molar amount $2(x_A^{p2} - x_A^*)$ of crystals produced per molar amount x_S of solvent present in equilibrium:

$$\rho = \frac{2\left(x_A^{p2} - x_A^*\right)}{x_S} \tag{9.9}$$

The yield and suspension density for the process in Figure 9.4a can be increased by changing the initial concentration towards slightly higher A:B stoichiometries. In this case, an initial composition can be chosen to enter the bi-phasic region at the other eutectic point on the right. Then, co-crystallisation could be performed until the solution composition is at the eutectic point on the left. This would maximise the yield under stable co-crystallisation conditions. Equally, if the suspension density appears too high for the process, slightly lower A : B stoichiometries can be employed. Also, the remaining solution after filtration could be reused, although higher levels of impurities can be expected to have an impact on the crystallisation process.

9.2.2.2 Cooling crystallisation

During a batch-wise cooling co-crystallisation, supersaturation is created by decreasing temperature, which has the effect of decreasing the solubility product of the co-crystal-forming components in the solution in addition to those of the pure component solubilities [22]. Various cooling profile strategies can be employed during the process [23]. Fast cooling rates should increase the rate at which the supersaturation is created, which may increase the probability of metastable formation.

Figure 9.4b exemplifies cooling co-crystallisation. In contrast to evaporative crystallisation, during cooling crystallisation the overall composition in the suspension does not change whereas the underlying phase diagram is manipulated by a reduction in temperature. In Figure 9.4b, the high-temperature phase diagram is shown in grey and the low temperature phase diagram is coloured. The composition is chosen in such a way that the initially clear solution is exactly saturated for the co-crystalline component and is therefore positioned on the co-crystal solubility line. The overall composition contains a slight excess of B. Thus, in this example, the solution stoichiometry is again unequal to that in the co-crystal, showing once more that the phase diagram rather than the co-crystal stoichiometry directs solution stoichiometry for co-crystal formation.

During cooling co-crystallisation, the decreasing temperature generates a decreasing solubility for the co-crystal. Once the co-crystalline component has nucleated, the co-crystals can precipitate to consume the supersaturation. It is also possible to seed the saturated solution with co-crystals and grow these during the process. If only co-crystals are crystallising, the solution composition follows Equation 9.5 until saturation is reached. This equilibrium composition in the solution is shown in

Figure 9.4b as the point on the low-temperature co-crystal solubility line. The arrow shows that the co-crystalline material AB is crystallised out of the solution until equilibrium is reached (the intercept between the arrow and the low-temperature co-crystal solubility line).

In Figure 9.4b, the overall composition is positioned in the two-phase region where equilibrium can be reached between the co-crystal and the solution. For a stable cooling co-crystallisation process, it is necessary to remain in this two-phase region throughout the process. In Figure 9.4b, the high-temperature compositional window of the two-phase co-crystal region does not perfectly overlap with that at the low temperature. This means that not all compositions in the high-temperature two-phase region can be used as the initial composition for a stable cooling crystallisation. Compositions with an excess of B can end up in the three-phase region at the end of the cooling crystallisation and the crystallisation product may contain pure B crystals along with the co-crystalline component. When too much A is present in the overall composition, the cooling crystallisation starts in the three-phase region, which may lead to concomitant crystallisation of the co-crystal and pure component A.

In the example of Figure 9.4b, the stoichiometry in the eutectic composition does not change with temperature as both the high- and low-temperature eutectic composition lie on a straight line starting at the solvent origin. However, in reality, temperature may induce large stoichiometric changes in the eutectic compositions. These eutectic composition shifts should be accounted for in the cooling co-crystallisation design, but they could also be exploited for separation purposes [24].

9.2.2.3 Co-crystal precipitation

During co-crystal precipitation, two streams of different compositions are mixed in such a way that the overall composition ends up in the two-phase region, as shown in Figure 9.4c. The final mixed solution is supersaturated because the concentration product $x_A.x_B$ is greater than the solubility product K. In this example, a solution saturated with A and a solution saturated with B are added together. Although the streams are mixed, the concentrations of A and B are decreased along the green arrows.

As the mixing of the streams generates supersaturation, the points where the feeds are mixed are the positions where high local supersaturations occur. Therefore, at these positions, nucleation is most likely to occur. The high local supersaturations tend to increase the chance of metastable formation. Therefore, a change in mixing strategy may coincide with a change in crystalline product outcome.

Various approaches of mixing strategies can be employed during the process and can be both continuous and batch-wise. For instance, a solution of A could be added to a vessel filled with a solution of B. In this case, at the inlet, a solution of A is mixed with the vessel solution, which contains increasing amounts of A and decreasing amounts of B. Another possibility is to add both solutions to a vessel at the same time. Such different mixing approaches create completely different local mixing conditions compared with the addition of solution B to a vessel filled with solution A and a completely different product quality may be obtained.

9.2.2.4 Co-crystal transformation

A transformation can be induced from a pure component suspension, such as a suspension of co-former crystals to a co-crystal suspension [25]. This is exemplified in Figure 9.4d. The process starts with the crystalliser containing a suspension of co-former B crystals. The overall composition is located in point 1 of Figure 9.4d. If the overall composition can be shifted to the two-phase region in which co-crystalline material and solution are in equilibrium, eventually, the co-crystal forms and starts crystallising towards equilibrium. For example, this can be achieved by adding component A crystals to the suspension of component B crystals. The resulting overall composition would then be positioned on the green line in Figure 9.4d.

Adding small amounts of A crystals from point 1 slightly increases the solution concentration of A, as all added crystals dissolve. The overall composition is positioned in the two-phase region of the solution and B crystals. For this system, in this two-phase region the length of the line from the overall composition to the co-crystal solubility line decreases with the amount of A present, going from points 1 to 2. Therefore, by adding A crystals, the B crystalline phase also partially dissolves.

Upon adding more A crystals, the overall composition enters the three-phase region to arrive at an overall composition point 2. In this region, equilibrium is established among pure B crystals, AB co-crystals and the solution, which has a eutectic composition. By adding more A crystals, the overall composition moves through the three-phase region where only the amounts of crystalline phases change and the solution composition remains unaffected to be the eutectic composition. Moving from points 2 to 3, the suspension changes from an all B crystal suspension at point 2 to an all co-crystal suspension at point 3.

If even more A is added, the overall composition enters the two-phase co-crystal region between points 3 and 4. Going towards point 4, the length of the line from the operating line to the co-crystal solubility line starts to increase. As a result, more and more co-crystalline material is formed until arriving at point 4, where the highest yield for co-crystalline material is achieved. The point of highest yield depends on how the operating and co-crystal solubility lines are positioned. For other systems, the point of maximum yield may well be located within the two-phase co-crystal region.

Again, beyond point 4, a three-phase region is entered. In this region, the suspension changes from all co-crystal in point 4 to all pure component A crystals in point 5. After point 5, a pure component A crystal suspension is present.

There are various approaches to inducing transformations. One interesting approach is to suspend pure component crystals A and B in pure solvent. If the overall composition is positioned in the co-crystal region, upon dissolution of both components in the solvent, supersaturation for the more stable co-crystal is eventually created and the co-crystal crystallises. If the pure component solubilities are known, an estimate of the location of the two-phase region can be made, making this approach interesting to use in co-crystal discovery [26, 27].

9.2.3 Co-crystallisation kinetics

Ternary phase diagrams as presented in Section 9.2.2 give the accessible thermodynamically stable state for a sample with a specific starting composition. However, a phase diagram identifies neither the kinetic route travelled, nor the time of travel, to that stable state. The formation of co-crystals in any crystalline form is initiated by nucleation followed by crystal growth, which both need a supersaturated solution to occur. Controlling and understanding the nucleation and growth process kinetics is as important as understanding the thermodynamics of the process. Kinetics determine the route towards the final stable state and taking a wrong turn can yield, within the timeframe of a given process, a very different outcome with metastable co-crystal polymorphs, solvates or pure component crystalline compounds crystallising out rather than the desired co-crystal compound. In this section, the importance of co-crystallisation kinetics in solution is illustrated by focusing on the competition among the various stable solid phases in a ternary phase diagram.

The phase diagrams shown up to now only show the solubility lines of the most stable crystalline phases in a particular compositional region. However, the solubilities of metastable compounds in phase diagram regions can be derived from their solubilities in the phase diagram regions in which they are stable. Based on the solubility prediction of the metastable pure components and co-crystals, Gagnière et al. [28] split the thermodynamic regions of the ternary phase diagram into kinetic regions. Figure 9.5 shows these kinetic regions for a schematic phase diagram in an orthogonal representation, with the dashed lines indicating the solubilities of the metastable compounds.

For instance, in the bi-phasic thermodynamic region 3, in which co-crystal and solution are the stable phases, four kinetic regions, 3A, 3B, 3C and 3D, can be identified. A kinetic region indicates a compositional region in which a specific crystalline form can spontaneously appear from a supersaturated solution. For example, any clear solution containing a concentration of $x_A > x_A^*$, which is in the kinetic regions 5A, 5B, 4A, 3B, 3D and 2B, can in principle lead to spontaneous crystallisation of A. Even though crystal compound A can crystallise spontaneously from supersaturated solutions in kinetic regions 3B, 3D and 2B, it is not the most thermodynamically stable compound in these kinetic regions. Similarly, B crystals can spontaneously crystallise from a supersaturated solution in kinetic regions 1A, 1B, 2A, 2B, 3C, 3D and 4B, although the crystalline phase B in the latter three kinetic regions is metastable. The extended co-crystal solubility line shows that spontaneous crystallisation of co-crystal

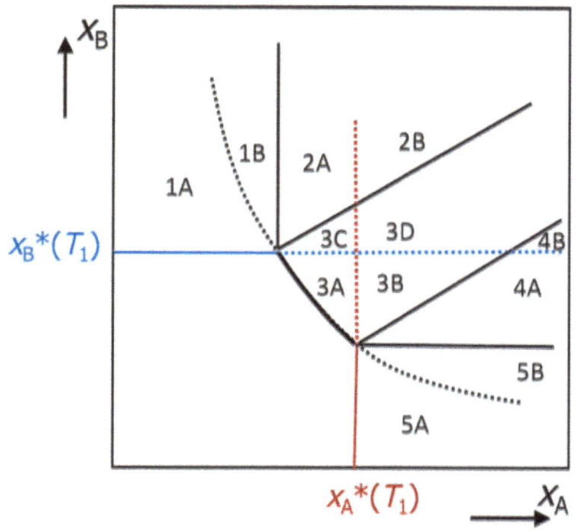

Fig. 9.5: Orthogonal co-crystal phase diagram at a given temperature and pressure, with kinetically accessible zones highlighted by *broken lines*; x_A^* and x_B^* represent the solubility of A and B respectively.

AB can take place from a supersaturated solution in kinetic regions 1B, 2A, 2B, 3A, 3B, 3C, 3D, 4A, 4B and 5B, whereas the co-crystal phase is only thermodynamically stable in thermodynamic regions 2, 3 and 4.

Thus, the thermodynamic regions 1 and 5 are both divided into two kinetic regions. In kinetic region 5A, the stable form A is the only crystalline phase that can form. Similarly, in kinetic region 1A, only the stable form B can crystallise spontaneously from a supersaturated solution. In kinetic regions 5B and 1B, a clear solution is supersaturated with respect to both A and AB or B and AB respectively, and both can therefore spontaneously crystallise from such a solution. Their relative crystallisation kinetics determine the initial outcome of crystallisation in that region. In kinetic region 5B, A can spontaneously nucleate and crystals grow, leading to a stable suspension of A crystals. Alternatively, the co-crystal phase AB can be the first to crystallise out, leading to a suspension of AB co-crystal materials in equilibrium with a solution that remains supersaturated with respect to A. This suspension is therefore metastable and, given enough time, A will nucleate and grow. Upon growth of A, the solution becomes under-saturated with respect to the co-crystal, which dissolves, leading to a final thermodynamically stable suspension of A crystals.

In the supersaturated solution in kinetic region 3A, the co-crystal AB is the only solid that can be formed. However, in supersaturated solutions in kinetic regions 3B and 3C respectively, A and B can also spontaneously form, next to the co-crystal. For a supersaturated solution in kinetic region 3D, formation of A, B and AB can occur. In these supersaturated solutions, crystals of A and B that kinetically form eventually dissolve once the AB co-crystal compound crystallises and the solution composition moves towards the co-crystal solubility line in region 3A.

The three-phase thermodynamic regions 2 and 4 are also divided into two kinetic regions. In solutions in region 4A, only the thermodynamically stable forms A and AB can spontaneously crystallise, with the same occurring in region 2A for forms B and AB. In solutions within kinetic region 4B, supersaturation with respect to A, B and AB is achieved; thus, both pure compounds and the co-crystal can spontaneously crystallise out. Ultimately, any compound B that crystallises spontaneously from a supersaturated solution in region 4B dissolves after formation of the co-crystal compound AB with the solution composition moving towards the eutectic point of region 4.

The importance of kinetics is illustrated by focusing on kinetic region 3D, in which a supersaturated solution can result in the spontaneous formation of all three solid forms, whereas only the co-crystal AB is the stable crystalline compound. Figure 9.6 (left) shows a case in which the thermodynamically stable co-crystal AB spontaneously crystallises out of a supersaturated solution with a composition at the black point. During the process, the solution concentrations of A and B drop stoichiometrically (green arrows) until the solubility line of the co-crystal is reached and the process terminates. The yield can be determined as described in Section 9.2 (Equation 9.5). At this stage, a thermodynamically stable suspension of co-crystals (green point) is obtained.

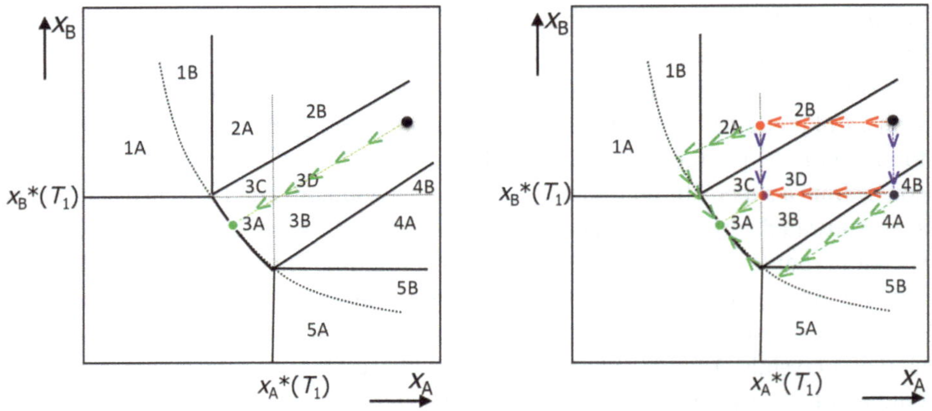

Fig. 9.6: Schematic phase diagrams at constant temperature and pressure in an orthogonal presentation showing the thermodynamic and kinetic domains. *Left*: solution composition during direct crystallisation of the thermodynamically stable co-crystal AB, starting from a supersaturated solution located in the *black dot*. *Right*: possible kinetic routes towards the same thermodynamically stable end point (*green dot*) starting from a supersaturated solution located in the *black dot*. *Green*, *red* and *purple arrows* indicate crystallisation of co-crystal AB, and pure A and B respectively.

Figure 9.6 (right) shows the kinetic route towards the same thermodynamic equilibrium if either metastable form A (red arrows) or B (purple arrows) crystallises from the supersaturated solution instead of the co-crystal. Upon spontaneous nucleation of pure component A, the solution concentration of A drops down to the solubility of A (red dot). At this stage, the solution is still supersaturated with respect to pure component B and the co-crystal AB. If eventually pure component B crystallises out, the solution is depleted of B down to the solubility of B (mixed red/purple dot). The solution now contains both A and B at their saturation concentrations, but remains supersaturated with respect to the co-crystal AB. Alternatively, this point can be reached by crystallising out B first (up to the purple dot) followed by crystallisation of A. Spontaneous crystallisation of the co-crystal eventually occurs followed by a co-crystal transformation process as discussed in Section 9.2.2.4.

Figure 9.6 (right) also shows an alternative route where crystallisation of A is not followed by crystallisation of B, but rather by crystallisation of the more stable co-crystal, AB. In the case represented here, rapid growth of the co-crystal is assumed, leading to a stoichiometric decrease in the concentrations of A and B until the solubility line of the co-crystal is reached. At this point, a suspension of A and co-crystal AB is obtained, but the solution has become under-saturated with respect to A. The crystals of A therefore dissolve, leading to an increase in the concentration of A in the solution, and further co-crystallisation. The solution composition follows the co-crystal solubility line, until the final thermodynamic point is reached (green point), with only the co-crystal suspended in solution. In practice, co-crystallisation rather than dissolution of A crystals is usually the rate limiting step and dissolution of A crystals commences directly after the formation of co-crystals.

The interplay of kinetics and thermodynamics in co-crystallisation was investigated experimentally by Gagnière et al. [29], who monitored carbamazepine and nicotinamide concentrations in solution. Using in situ IR spectroscopy, the solution concentrations of both species were monitored, allowing for the identification of the kinetic routes. A first series of experiments was conducted in kinetic region 3C, in which clear solutions were supersaturated with respect to both the metastable carbamazepine and the stable co-crystal phase. Spontaneous crystallisation from these supersaturated solutions led to the formation of the co-crystal. This was demonstrated by the stoichiometric consumption of both species until the co-crystal solubility curve was reached, similar to that shown in Figure 9.6 (left). A second series of experiments were this time conducted in kinetic region 4A, where clear solutions were supersaturated with respect to the stable phases of both carbamazepine and the co-crystal. Initially, co-crystal material formed, as shown by the stoichiometric decrease of both components. Before reaching the co-crystal solubility curve, a more substantial decrease in carbamazepine concentration than that of nicotinamide showed concomitant crystallisation of carbamazepine crystals. In the final stage, the co-crystals partially dissolved, with the carbamazepine crystallisation continuing until the eutectic point was reached. The solution concentration in close proximity to the eutectic point followed the co-crystal solubility curve, suggesting carbamazepine crystal growth to be rate limiting rather than co-crystal dissolution. This process corresponds to the green arrows in Figure 9.6 (right), starting in region 4A.

In summary, it is vital to determine location within the phase diagram, and what this predicts about the final outcome of the process. The route to this outcome is determined by kinetics and depends on solution nucleation and growth of the different forms, which in turn depend on the initial solution concentration and stoichiometry [30]. The kinetic route followed has a significant impact on product quality, i.e. size, shape and purity. In addition, as for classical, pure component crystallisation processes, using different process parameters such as seeding [26, 31], cooling profile and agitation, metastable crystal forms may be obtained. Finally, just as for single-component systems, co-crystals can also form hydrates, solvates or display polymorphism.

9.3 Co-crystallisation applications

As for any crystallisation process, solution co-crystallisation processes can be used for multiple purposes. In this section, solution co-crystallisation is discussed as a tool for improving dissolution behaviour and for achieving separation.

9.3.1 Improving dissolution behaviour of an active pharmaceutical ingredient

The biopharmaceutics classification system (BCS) [32] classifies orally administered drugs in four categories according to their dissolution and permeability behaviour. Class II drugs are termed as low solubility–high permeability drugs. an estimated 30%

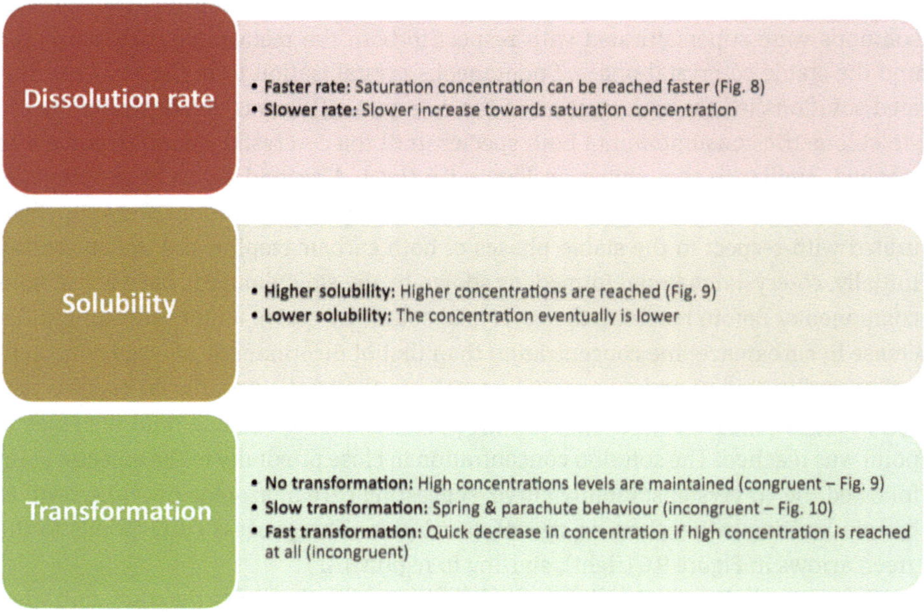

Fig. 9.7: Impact of co-crystals on dissolution behaviour of a drug compound. Co-crystallisation can have an impact on both dissolution rate and solubility. For incongruent systems, possible transformations to less-soluble compounds also need to be taken into account.

of drugs on the market belong to this category. Considering drugs under development, this percentage is predicted to show a significant increase in the future owing to an increasing molecular size and complexity of newly developed drugs [33]. The bioavailability of class II drugs is limited by their dissolution behaviour, with relationships often existing between their *in vitro* dissolution profile and their *in vivo* bioavailability.

Alternative solid forms such as co-crystals may to have an impact on the dissolution behaviour of a given class II drug form. Figure 9.7 gives an overview of the impact on dissolution behaviour of using co-crystal technology. Co-crystals have an effect on both the thermodynamic (solubility) and the kinetic (dissolution rate) aspects of drug dissolution. The dissolution profile of co-crystals is furthermore influenced by its congruent or incongruent behaviour in bodily fluids. Co-crystal dissolution can lead to a transformation of the co-crystal, reducing possible positive solubility and dissolution rate effects, which indicates incongruent dissolution behaviour. Therefore, the dissolution behaviour of co-crystals [34, 35] is discussed separately for congruent or incongruent systems, with relevant literature cited as examples in both cases.

9.3.1.1 Congruent co-crystal systems

When a pharmaceutical co-crystal is orally administered, dissolution occurs either in the stomach or in the intestines. This initially implies an increase in drug and co-former concentration in the stomach or intestinal fluids. For congruently dissolving systems, a typical dissolution profile is shown in Figure 9.8. When administering a co-crystal AB, both fluid concentrations of API A and co-former B are initially zero. Gradual dissolution of the co-crystal leads to an equimolar increase in both A and B in the fluid, which means that the overall concentration over time follows the 1 : 1 stoichiometric line (green arrows in Figure 9.8, right). When the co-crystal solubility curve characterised by the solubility product K is reached, the system is at equilibrium. As the concentrations of A and B follow the stoichiometric line, the final API concentration

$$x^*_{\mathrm{A(AB)}} = \sqrt{K}$$

is given by the solubility product. The dissolution profile levels off towards this value, which represents the highest amount of drug A that can be achieved upon orally administration of the co-crystal drug. As shown in Figure 9.8, this implies achieving a lower drug concentration in the case of administering a pharmaceutical co-crystal product compared with administering the pure form A. Despite the lower solubility, the dissolution rate can still be substantially faster, possibly realising higher *in vivo* concentrations.

Figure 9.8 shows an ideal system in which solution interactions between A and B do not play a determining role. In many cases, these solution interactions do come into play, with solution complexation equilibria acting on compound solubilities. Sometimes a decreasing solubility of A can be observed due to, for example, some sort of salting out effect. However, for these systems, an increase in the solubility of the tar-

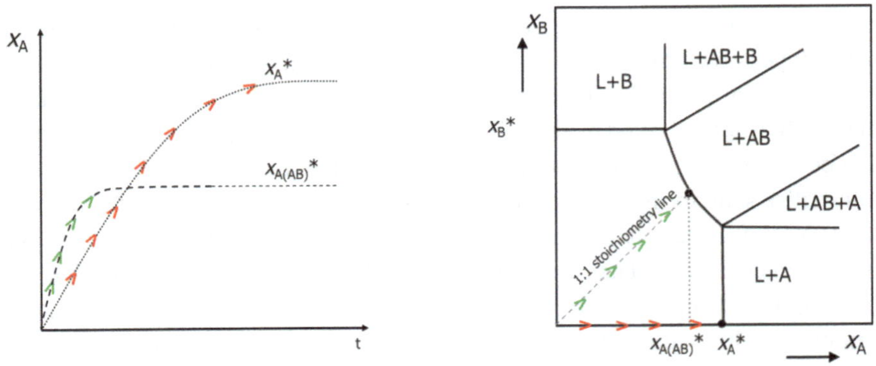

Fig. 9.8: Dissolution profile (*left*) and route through the phase diagram upon dissolution (*right*) of a pure drug A (*dotted line, red arrows*) and a congruent co-crystal system AB (*dashed line, green arrows*). The mole fractions x_A and x_B represent the solution concentrations of A and B respectively, and x_B^*, x_A^* and $x_{A(AB)}^*$ respectively represent the solubilities of components B and A, and the saturation concentration of A achieved when dissolving AB co-crystal material.

get compound A is typically noticed with increasing concentration of B in solution. This is illustrated in Figure 9.9 by a non-vertical solubility line of A with respect to the amount of B. As illustrated in Figure 9.9, dissolving pure co-crystal material, can lead to an increased concentration of A in the solution compared with dissolution in pure solvent. In a similar context, additives or excipients with a strong complexation with the API can also be used to increase the solubility of the pure component A crystals without the need to co-crystallise the components in the same solid.

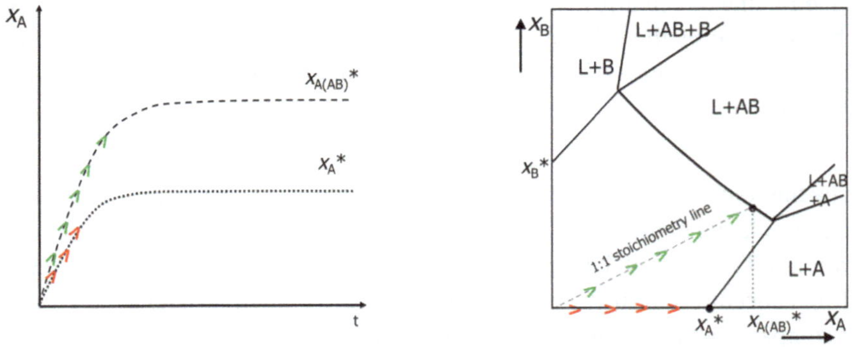

Fig. 9.9: Dissolution profile (*left*) and route through the phase diagram upon dissolution (*right*) of a pure drug A (*dotted line, red arrows*) and a congruent co-crystal system AB with strong solute interactions between components A and B. The mole fractions x_A and x_B represent the solution concentrations of A and B respectively, and x_B^*, x_A^* and $x_{A(AB)}^*$ respectively represent the solubilities of components B and A, and the saturation concentration of A achieved when dissolving AB co-crystal material.

However, this complexation behaviour makes it difficult to predict whether a system behaves congruently under conditions of drug administration. Therefore, the environment a drug is dissolved in should always be kept in mind. For instance, a drug that dissolves congruently in the stomach (low pH) does not necessarily do so in the intestine (high pH). Examples of congruently dissolving co-crystal systems are phenazopyridine:phthalimide [36], carbamazepine:cinnamic acid [37], carbamazepine:nicotinamide [38], acyclovir:fumaric acid [39], genistein:caffeine [40], ezetimibe:methylparaben [41] and adefovir:p-hydroxybenzoic acid [42].

9.3.1.2 Incongruent co-crystal systems

For incongruently dissolving co-crystal systems, a typical dissolution profile is shown in Figure 9.10 (left). Again, co-crystal dissolution leads to an equimolar increase in both A and B concentrations in solution, following the 1 : 1 stoichiometric line (Figure 9.10, right, green arrows), up until the solubility curve of the co-crystal is reached. However, at this stage, a meta-stable situation is achieved as the overall solution composition (black dot in Figure 9.10, right) corresponds to thermodynamic region 1 in the overall phase diagram (see Section 9.2) in which the free drug form A is the stable form in suspension. Often, this metastable concentration $x^*_{A(AB)}$ is referred to as a kinetic solubility [43] as at this stage, the solution is supersaturated with respect to the free form A, which eventually crystallises. After nucleation and crystal growth of A, a solvent-mediated transformation occurs [44], with the free form crystallising, and the co-crystal phase dissolving. When the co-crystal phase is depleted, further crystallisation of A leads to a final solution concentration equal to the thermodynamic saturation concentration x^*_A of the free form A. Alternatively, sufficient co-crystal phase may be

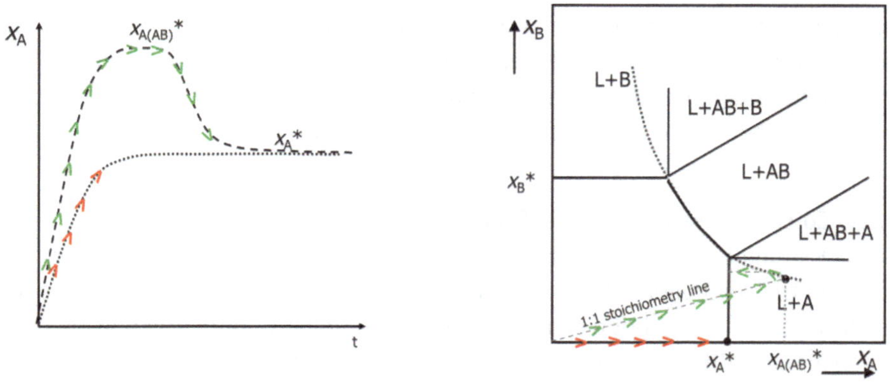

Fig. 9.10: Dissolution profile (*left*) and route through the phase diagram upon dissolution (*right*) of a pure drug A (*dotted line*, *red arrows*) and an incongruent co-crystal system AB. The mole fractions x_A and x_B represent the solution concentrations of A and B respectively, and x^*_B, x^*_A and $x^*_{A(AB)}$ respectively, represent the solubilities of components B and A, and the saturation concentration of A achieved when dissolving AB co-crystal material.

present, in which case the three-phase eutectic point is reached with both co-crystal and A crystals in suspension. It should be noted that the kinetic solubility cannot always be achieved, as nucleation of A may occur before co-crystal solubility has been reached.

Figure 9.10 shows a dissolution profile of an incongruent co-crystal system. This type of dissolution profile is of the "spring and parachute" type [45, 46]. The "spring" behaviour is characterised by a rapid increase in concentration A in solution due to an increased dissolution rate of the co-crystal. Furthermore, the amount of A dissolved in solution is higher for the co-crystal than for the free form A owing to an elevated co-crystal solubility product. Under these conditions, the co-crystal is not the stable form in suspension, with a solution-mediated transformation towards the free form A eventually occurring. As this solution transformation is often limited by the crystal growth kinetics of free form A, a "parachute"-like slow decrease in the concentration of A is observed in the solution. The rate at which the solution concentration of A decreases can be slowed down even further using additives and prolonged increased concentrations of A could be maintained. Examples of incongruent systems are described in the literature for drug candidates AMG517 [47], quercetin : proline [48], danazol–vannilin [49], myricetin–proline [50], megestrol acetate : saccharin [51] and dexlansoprazole:isonicotinamide [52].

9.3.2 Co-crystallisation as a separation technique

In addition to being a process for generating a particulate product, crystallisation is an economically viable and easy to implement industrial separation technology allowing efficient separation of a target compound from a solution containing multiple components, such as a reaction mixture. The main compound of interest is often the high concentration compound, whereas other compounds are often impurities. Similar to single-component crystallisation processes, multi-component crystallisation processes can be used as a form of separation technology [54].

9.3.2.1 Co-crystallisation instead of crystallisation

The most straightforward approach to using co-crystallisation as a separation tool is to target the component of interest A and removing it from the solution using co-crystallisation by the addition of an appropriate co-former B, rather than using pure component crystallisation. Targeting the main compound, excess co-former and impurities are removed with the mother liquor upon filtration. A co-crystallisation approach becomes particularly interesting when encountering difficulties with crystallisation of the pure compound A, for instance, owing to purity issues, amorphous compound formation or liquid–liquid de-mixing. Any method of co-crystallisation can then be used, as discussed in Section 9.2.2.

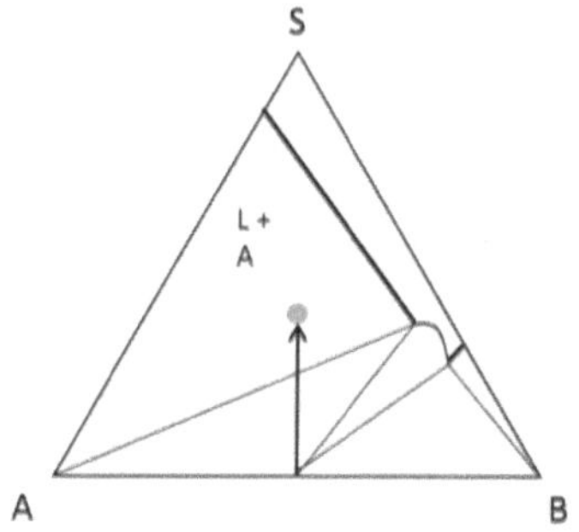

Fig. 9.11: To separate co-crystals into its components, a solvent is required in which the co-crystal dissolves incongruently. The pure component solubilities should preferably be very different, with a low solubility for the preferred compound, starting on the AB axis with co-crystal material and moving up towards the *green dot* as solvent is added.

However, using such a co-crystallisation approach results in a co-crystal product containing both the compound of interest and the co-former. As the product form of interest is not necessarily the co-crystal product, a second crystallisation step may have to be introduced to separate the co-crystal into its components. Once again, phase diagrams are crucial within this context. Pure co-crystal material can be obtained using any of the co-crystallisation processes described in Section 9.2.2. Next, suspending the co-crystal in a solvent in which the co-crystal dissolves incongruently, a transformation of the co-crystal to pure API A occurs for a given amount of solvent (Figure 9.11).

Such an approach was applied by Billot et al. [54] working on a drug candidate SAR1, an SRC kinase inhibitor with potential use as treatment for acute myeloid leukaemia. This component is obtained after three synthetic steps as a dry concentrate with a purity of 60%. Repeated crystallisation trials in various solvents led to 90% crystal purity at best. To improve this, a co-crystallisation approach was investigated, using benzoic acid as a co-former. The 1 : 1 SAR1 : benzoic acid co-crystal was formed from the solvent dichloromethane with a 60% yield and a purity of over 99.5%. Isopropanol was used as the solvent to separate the co-crystal into its components. The pure API has a very low solubility in this solvent, (< 1 g/L at room temperature) whereas that of benzoic acid is very high (> 300 g/L). Such high solubility differences often lead to incongruent co-crystal systems and the ternary phase diagram confirmed this to be the case. Mixing the co-crystal in isopropanol induced a transformation from the co-crystal to pure API crystals, with a final yield of 90%.

9.3.2.2 *In situ* product removal by co-crystallisation

Urbanus et al. [53, 55] used phase diagram knowledge to arrive at a new approach for the removal of cinnamic acid from its fermentation broth. In most bio-based processes, microbial production is inhibited at high fermentation product concentrations. *In situ* product removal is therefore required to keep the product concentration low during the fermentation process. The authors showed that cinnamic acid forms a co-crystal with 3-nitrobenzamide, at concentrations as low as seven times below the pure component solubility of cinnamic acid. Maximum *in situ* product removal capacity can be achieved by operating the process at the eutectic point, as shown in Figure 9.12.

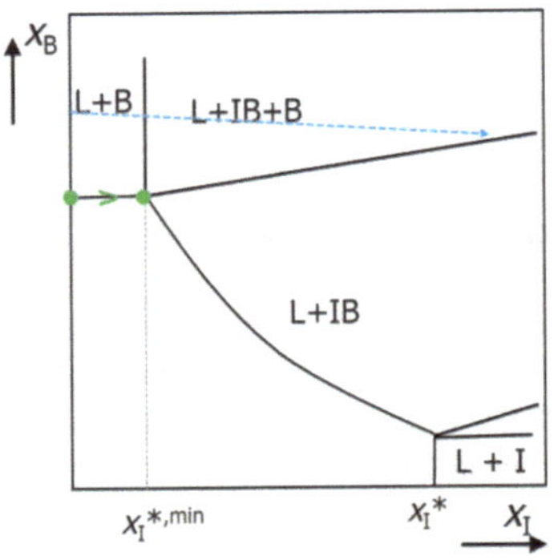

Fig. 9.12: *In situ* product removal through co-crystallisation. The *light-blue arrow* indicates the overall composition (operating line) indicating the increase in product concentration due to a reaction, whereas the *green arrow* indicates the solution composition. The impurity concentration in the solution can be substantially reduced through co-crystallisation. The mole fraction x_B represents the concentration of 3-nitrobenzamide, x_I represents the concentration of cinnamic acid, x_I^* is the solubility of cinnamic acid and $x_I^{*,min}$ is the apparent minimum solubility.

Starting off with an initial suspension of 3-nitrobenzamide, the overall composition follows the light-blue line, as the amount of cinnamic acid increases throughout fermentation. When crossing the apparent minimum solubility $x_1^{*,min}$, the co-crystal becomes a stable phase. At this stage, the composition of the solution (green line) equals that of the eutectic point, with both the co-crystal and 3-nitrobenzamide solid phases in suspension. Any cinnamic acid created at this stage through fermentation is removed by co-crystallisation with the co-former. The solution concentration remains unchanged owing to matching dissolution of 3-nitrobenzamide crystals, acting as a co-former buffer. This process can continue while in the tri-phasic equilibrium zone, which is achieved as long as the 3-nitrobenzamide buffer does not run out. This can easily be ensured by regularly adding more 3-nitrobenzamide.

9.3.2.3 Co-crystallisation for impurity removal

As shown by Hsi et al. working on ibuprofen/ketoprofen mixtures [56], impurity removal by co-crystallisation of the impurity with a co-former is possible. However, this leaves the final solution containing, along with a lower level of the impurity, some concentration of the co-former, which can be seen as an additional impurity. Furthermore, co-crystallisation remains challenging under such low impurity levels.

A more promising approach in this context was also suggested by Hsi et al. [57] building on the impurity co-crystal results. The idea behind their work was to prevent impurities from being incorporated into the crystal lattice of the target component. To do so, they started from the idea that a component co-crystallising with the impurity is also likely to interact strongly with the impurity in the solution, hence preventing it from influencing growth and entering the crystal lattice. They illustrated this approach using the benzamide/benzoic acid and cinnamamide/cinnamic acid systems, showing that impurity complexation can reduce the incorporation of impurities in the crystals of the main component by 20%.

This work was extended to amoxicillin β-lactam [59], for which all known synthetic pathways lead to a contamination with 4-hydroxyphenylglycine, a degradation product. Removal of this component through successive crystallisations is problematic because of the impurity incorporation in the amoxicillin β-lactam crystal. A co-crystal screen allowed the identification of 11 possible co-formers for the impurity, which were tested in 1 : 1 stoichiometric solution ratios. In the case of the co-formers L-lysine and L-leucine, this led to an impurity incorporation decrease of 85% in comparison with crystallisation performed without added complexing agents.

9.3.2.4 Co-crystallisation as a chiral separation technology

Resolution, the chiral separation of two different stereoisomers, is complex and challenging [59]. Chirality is of particular importance for pharmaceutically active compounds, as a given enantiomer can be active (eutomer), whereas the counter enantiomer (distomer) may be non-active, trigger an unwanted response by displaying unwelcome side-effects or, in worst case scenarios, be poisonous. For example, the eutomer of naproxen is used to treat arthritis whereas the distomer causes liver poisoning. Another example is thalidomide, where the eutomer is effective against morning sickness whereas the distomer is teratogenic. In spite of important advances in asymmetrical synthesis of the desired enantiomer, a prominent way of achieving enantiopure drugs still involves formation of a racemic mixture followed by chiral separation. As co-crystallisation can be used within the context of separation [60], it can also be used within the context of resolution. Again, phase diagrams are crucial for understanding the process and for developing a robust process. The challenge of chiral separation through co-crystallisation comes from the additional degrees of freedom due to the variable concentration of all components: *R*-enantiomer, *S*-enantiomer, solvent and (achiral or chiral) co-former. This can be described at a given temperature using a quaternary phase diagram.

Preferential crystallisation is a frequently used chiral resolution method as it is an economically interesting alternative to chiral chromatography that is easy to implement. The limitation of preferential crystallisation is the prerequisite for a conglomerate system in which enantiopure crystals form a physical racemic mixture. Co-crystallisation can be used within this context, as a racemic compound system can be turned into a conglomerate co-crystal system using a suitable co-former [61, 62]. As roughly 10% of the systems are conglomerates [63], on average, with trial and error, 1 in 10 co-formers should be expected to result in a conglomerate co-crystal, and therefore a potential 10-fold rise in conglomerate-based chiral resolution applications. However, in specific families of racemic compounds, it may be quite challenging to find a co-former that changes a very stable racemic compound into a conglomerate co-crystal system. This may be the reason why reports on co-crystal conglomerates are limited. Going from a racemic compound system to a conglomerate co-crystal system has an impact upon the ternary diagram, as shown in Figure 9.13.

The general idea behind preferential co-crystallisation is to start with an initial racemic solution, which is supersaturated with respect to both enantiomers. At this stage, co-crystals of a given enantiomer (e.g. A_RB) can be seeded. Co-crystallisation

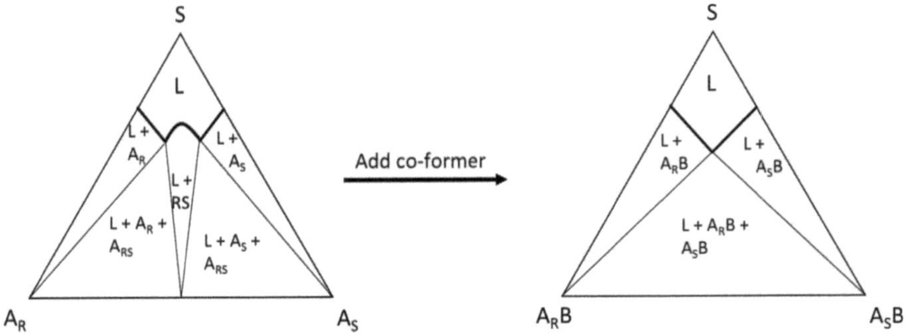

Fig. 9.13: Impact upon the ternary phase diagram of a racemic compound A, which forms a co-crystal conglomerate using co-former B. The representation of the co-crystal conglomerate is possible as a ternary phase diagram for overall 1 : 1 stoichiometry of B and the chiral compound.

depletes the solution in the *R*-enantiomer until the solubility line of the A_RB co-crystal is reached. Care should be taken that no spontaneous nucleation of the counter- enantiomer co-crystal occurs. Filtration yields the A_RB product and a liquid enriched in the *S*-enantiomer. Subsequently, the solution can be seeded with A_SB co-crystal seeds that grow and filtration yields A_SB crystals and a solution with a eutectic composition. Such a process was, for example, developed successfully for the resolution of a hybrid salt/co-crystal (±)trans-N,N′-dibenzyldiaminocyclohexane using 2,3-dichlorophenylacetic acid [64].

If the chiral compound can be racemised in solution, a conglomerate co-crystal of a racemic compound can furthermore be used in a de-racemisation process such as Viedma ripening [65, 66] or crystallisation-enabled dynamic kinetic resolution to obtain an enantiopure co-crystal product with high yield [67, 68]. Viedma ripening is an attrition-enhanced de-racemisation process and the operating principle is based on continuous grinding of a racemic suspension of crystals while racemising the chiral solute [69]. In this deracemisation process, the overall solid form transforms over time into an enantiopure end-product. A co-crystallisation-enabled dynamic kinetic resolution process is the preferential co-crystallisation in the presence of a solute racemisation. As the distomer solute is transferred into eutomer crystals, these de-racemisation processes in principle have substantially higher yields than processes conducted in the absence of solute racemisation.

For non-conglomerate forming compounds, co-crystallisation can be used in a manner similar to diastereomeric salt formation. A chiral resolving agent can form a co-crystal with both enantiomers of the target compound, or with only one of the target compounds, leading to diastereomeric co-crystal formation or enantio-specific co-crystal formation respectively [70]. The solid forms of each enantiomer has different physical properties such as solubility, which can be used for resolution.

Diastereomeric co-crystallisation has been used to resolve both enantiomers of the drug praziquantel [71], whereas (±)-trans-1,2-cyclohexanediol can be resolved using tartaric acid [72]. Enantiospecific co-crystallisation from solution has been developed by Springuel and Leyssens, resolving levetiracetam from a racemic mixture using mandelic acid [73]. The *S*-enantiomer (levetiracetam) forms a co-crystal with *S*-mandelic acid, whereas no co-crystal is formed between the R-enantiomer of the drug and *S*-mandelic acid. This led to the development of a resolution as shown in Figure 9.14. A racemic mixture of the drug was added to acetonitrile followed by the addition of *S*-mandelic acid. Levetiracetam : *S*-mandelic acid co-crystal crystallises out as the stable form in suspension, leaving the undesired *R*-enantiomer in solution. Up to 80% of the desired enantiomer could be removed using a single co-crystallisation step.

To identify the final suitable conditions for a robust resolution process through co-crystallisation, appropriate phase diagrams are required. Isothermal ternary phase diagrams no longer suffice as the amounts of both enantiomers, of resolving agent and solvent need to be taken into account. Quaternary diagrams such as those presented in Figure 9.14 need to be considered [74]. Just as for ternary phase diagrams, quaternary phase diagrams vary with the temperature and nature of the solvent. The complex interpretation of these diagrams can be reduced by working with well-selected cut planes of the overall quaternary diagram. As illustrated in Figure 9.14, while work-

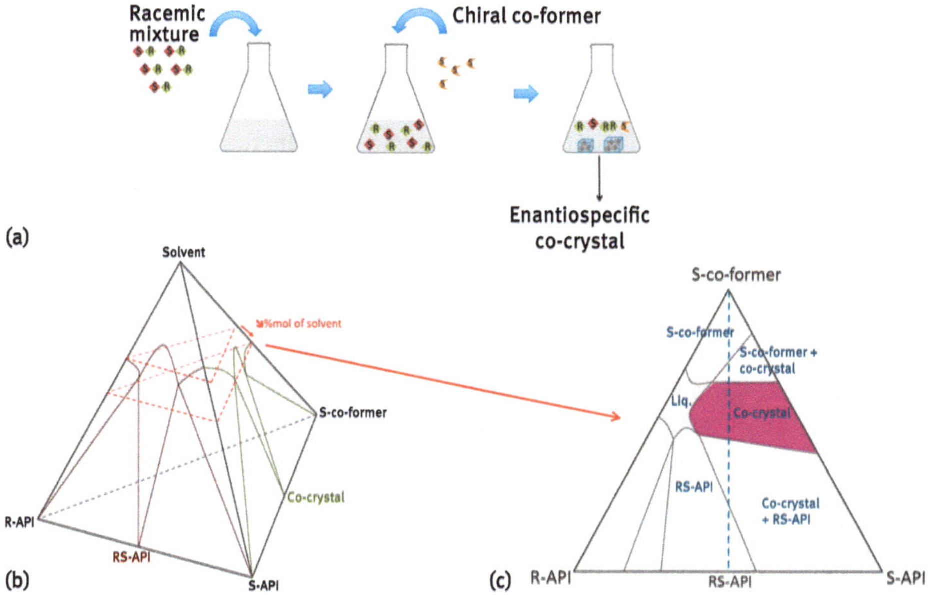

Fig. 9.14: Resolution from solution using enantiospecific co-crystals. (a) Racemic mixture is added to a solvent and fully dissolved. A chiral co-former is added that co-crystallises with only one of the two enantiomers. (b) Full isothermal quaternary phase diagram corresponding to a chiral resolution process through co-crystallisation. The apices represent pure *R*- and *S*-API, pure co-former and pure solvent. An example of a well-chosen cut plane for a given percentage of solvent in the quaternary co-crystal phase diagram showing the stability of the enantiospecific co-crystals in racemic solutions under specific conditions. *API* active pharmaceutical ingredient.

ing with a fixed molar percentage of solvent, a ternary cut can be made. Within the plane, the different solid states can be identified that are thermodynamically stable in suspension. Even though such a cut does not yield immediate information on the supernatant liquid composition, it will at least show whether resolution occurs, and for which amount of co-former added. Following the racemic line (broken blue line) in Figure 9.14, varying the amount of co-former leads to different situations. For low co-former concentrations, the RS-API is stable in suspension. By increasing the amount of co-former a suspension can be achieved for which both RS-API and co-crystal are stable in suspension. An even higher amount of co-former leads to a suspension where only the co-crystal is the stable solid phase, with even higher amounts of co-former leading to suspensions also containing co-former crystals. Resolution is ideally performed in the pure co-crystal zone, as only a single crystalline phase can be isolated.

9.4 Industrial co-crystallisation outlook

Co-crystals continue to gain interest in both scientific and industrial fields. Co-crystallisation allows the formation of various solid states of a specific pharmaceutical. As illustrated in this chapter, thermodynamics and kinetics have an impact upon co-crystal solubility and the dissolution rate, highlighting the direct pharmaceutical relevance. In addition, the industrial significance of solution co-crystallisation was demonstrated, as it can be used to develop novel applications for purification and separation purposes if basic thermodynamic and kinetics data are available. Thermodynamics, which is accessible through appropriate phase diagrams, yields the final process conditions in all circumstances, whereas kinetics is important for the control of the pathway towards these final conditions. Despite the complexity of solid form solubility and crystallisation kinetics in systems with multiple compounds, nowadays, knowledge, methods, approaches and techniques are available that enable the use of these for optimal product and process development.

Concerning the topic of binary co-crystals, the vast complexity of multi-component systems has only been touched upon. In the foreseeable future, it is expected that the demands of pharmaceutical properties will become more stringent as new pharmaceuticals increase in molecular size and complexity. The requirement for increasingly complex multi-component systems, their synthesis and their separation will become more challenging owing to the increasing number of components in the reaction mixtures. Complex, multi-component solid crystallisation from solution will not only demand a robust route towards these multicomponent products, but also offer new solution-based application possibilities. What remains clear is that the thermodynamics and kinetics complex multi-component crystallisation aspects will continue to be important for solution-based processes. Although understanding the thermodynamics and kinetics of these systems is challenging, it will also result in new opportunities in both the scientific and the industrial arenas.

Acknowledgement: We would like to acknowledge networking support by the COST Action CM1402 "Crystallize". JtH thanks the EPSRC Centre for Innovative Manufacturing in Continuous Manufacturing and Crystallisation (http://www.cmac.ac.uk) for supporting this work (EPSRC funding under grant reference: EP/I033459/1).

Bibliography

[1] Aitipamula S, Banerjee R, Bansal AK, et al. Polymorphs, salts, and co-crystals: what's in a name? Cryst. Growth Des. 2012, 12, 2147–2152.

[2] Grothe E, Meekes H, Vlieg E, ter Horst JH, de Gelder R. Solvates, walts, and co-crystals: a proposal for a feasible classification system. Cryst. Growth Des. 2016, 16, 3237–3243.

[3] Porter III WW, Elie SC, Matzger AJ. Polymorphism in carbamazepine co-crystals. Cryst. Growth Des. 2008, 8, 14–16.

[4] Schultheiss N, Newman A. Pharmaceutical co-crystals and their physicochemical properties. Cryst. Growth Des. 2009, 9, 2950–2967.

[5] Wang JR, Wang X, Yang Y, Chen X, Mei X. Solid state characterization of 17β-estradiol co-crystals presenting improved dissolution and bioavailability. CrystEngComm 2016, 18, 3498–3505.

[6] Zhang Z-H, Zhang Q, Zhang Q-Q, Chen C, He M-Y, Chen Q, Song G-Q, Xuan X-P, Huang X-F. From a binary salt to salt co-crystals of antibacterial agent lomefloxacin with improved solubility and bioavailability. Acta Crystallogr. B 71, 2015, 437–446.

[7] De Maere d'Aertrycke JB, Robeyns K, Willocq J, Leyssens T. Co-crystallization as a tool to solve deliquescence issues: the case of L-lactic acid. J. Cryst. Growth. 2017, doi:10.1016/j.jcrysgro.2017.02.025.

[8] Trask AV, Motherwell WDS, Jones W. Pharmaceutical co-crystallization: engineering a remedy for caffeine hydration. Cryst. Growth Des. 2005, 5, 1013–1021.

[9] Wouters J, Quéré L. Pharmaceutical salts and co-crystals. RSC Publishing. Cambridge, 2012.

[10] Friščic T, Jones W. Recent advances in understanding the mechanism of co-crystal formation via grinding. Cryst. Growth Des. 2009, 9, 1621–1637.

[11] Trask AV, Motherwell WDS, Jones W. Solvent-drop grinding: green polymorph control of co-crystallisation. Chem. Comm. 2004, 890–891.

[12] Ter Horst JH, Schmidt C, Ulrich J. Fundamentals of industrial crystallization. In: Nishinaga T, Rudolph P (eds). Handbook of crystal growth. 2015, Elsevier, Amsterdam, pp. 1317–1349.

[13] Leyssens T, Springuel G, Montis R, Candoni N, Veesler S. Importance of solvent selection for stoichiometrically diverse co-crystal systems: caffeine/maleic acid 1 : 1 and 2 : 1 co-crystals. Cryst. Growth Des. 2012, 12, 1520–1530.

[14] Aitipamula S, Chow PS, Tan RBH. Polymorphism in co-crystals: a review and assessment of its significance. CrystEngComm 2014, 16, 3451–3456.

[15] Chiarella RA, Davey RJ, Peterson ML. Making co-crystals: the utility of ternary phase diagrams. Cryst. Growth Des. 2007, 7, 1223–1226.

[16] Chadwick K, Davey R, Sadiq G, Cross W, Pritchard R. The utility of a ternary phase diagram in the discovery of new co-crystal forms. CrystEngComm 2009, 11, 412–414.

[17] Reus MA, van der Heijden AEDM, ter Horst JH. Solubility determination from clear points upon solvent addition. Org. Process Res. Dev. 2015, 19, 1004–1011.

[18] Thakuria R, Delori A, Jones W, Lipert MP, Roy L, Rodríguez-Hornedo N. Pharmaceutical co-crystals and poorly soluble drugs. Int. J. Pharm. 2013, 453,101–125.

[19] Alhalaweh A, George S, Boström D, Velaga SP. 1 : 1 and 2 : 1 Urea-succinic acid co-crystals: structural diversity, solution chemistry, and thermodynamic stability. Cryst. Growth Des. 2010, 10, 4847–4855.
[20] Jayasankar A, Reddy LS, Behune SJ, Rodriguez-Hornedo N. Role of co-crystal and solution chemistry on the formation and stability of co-crystals with different stoichiometry. Cryst. Growth Des. 2009, 9, 889–897.
[21] Yang H, ter Horst JH. Crystal nucleation of small organic molecules. In: Van Driessche AES, Kellermeier M, Benning LG, Gebauer D (eds). New perspectives on mineral nucleation and growth. Springer International Publishing, Cham, pp 317–337, 2017.
[22] Yu ZQ, Chow PS, Tan RBH. Operating regions in cooling co-crystallization of caffeine and glutaric Acid in acetonitrile. Cryst. Growth Des. 2010, 10, 2382–2387.
[23] Mullin JW. Crystallization. Butterworth Heinemann: Oxford, 2001.
[24] Lorenz H, Le Minh T, Kaemmerer H, Buchholz H, Seidel-Morgenstern A. Exploitation of shifts of eutectic compositions in crystallization-based enantioseparation. Chem. Eng. Res. Des. 2003, 91, 1890–1902.
[25] Croker DM, Rasmuson AC. Isothermal suspension conversion as a route to co-crystal production: one-pot scalable synthesis. Org. Process Res. Dev. 2014, 18, 941–946.
[26] Ter Horst JH, Cains PW. Co-crystal polymorphs from a solvent mediated transformation. Cryst. Growth Des. 2008, 8, 2537–2542.
[27] Leyssens T, Tumanova N, Robeyns K, Candoni N, Veesler S. Solution co-crystallization, an effective tool to explore the variety of co-crystal systems: caffeine/dicarboxylic acid co-crystals. CrystEngComm 2014, 16, 9603–9611.
[28] Gagnière E, Mangin D, Puel F, Rivoire A, Monnier O, Garcia E, Klein JP. Formation of co-crystals: kinetic and thermodynamic aspects. J. Cryst. Growth 2009, 311, 2689–2695.
[29] Gagnière E, Mangin D, Puel F, Bebon C, Klein JP, Garcia E. Co-crystal formation in solution: in situ solute concentration monitoring of the two components and kinetic pathways. Cryst. Growth Des. 2009, 9, 3376–3383.
[30] Boyd S, Back K, Chadwick K, Davey RJ, Seaton CC. Solubility, meta-stable zone width measurement and crystal growth of the 1 : 1 benzoic acid/isonicotinamide co-crystal in solutions of variable stoichiometry. J. Pharm. Sci. 2010, 99, 3779–3786.
[31] Gagnière E, Mangin D, Puel F, Valour JP, Klein JP, Monnier O. Co-crystal formation in solution: inducing phase transition by manipulating the amount of co-crystallizing agent. J. Cryst. Growth 2011, 316, 118–125.
[32] Amidon GL, Lennernas H, Shah VP, Crison JR. Theoretical basis for a biopharmaceutical drug classification: correlation of in vitro drug product dissolution and in vivo bioavailability. Pharm. Res. 1995, 12, 413–420.
[33] Thayer AM. Custom manufacturers take on drug solubility issues to help pharmaceutical firms move products through development. Chem. Eng. News 2010, 88, 13–18.
[34] Good DJ, Rodriguez-Hornedo N. Solubility advantage of pharmaceutical co-crystals. Cryst. Growth Des. 2009, 9, 2252–2254.
[35] Grossjohann C, Eccles KS, Maguire AR, Lawrence SE, Tajber L, Corrigan OI, Healy AM. Characterisation, solubility and intrinsic dissolution behavior of benzamide: dibenzyl sulfoxide co-crystal. Int. J. Pharm. 2012, 422, 24–32.
[36] Tao Q, Chen J-M,Ma L, Lu T-B. Phenazopyridine co-crystal and salts that exhibit enhanced solubility and stability. Cryst. Growth Des. 2012, 12, 3144–3152.
[37] Shayanfar A, Asadpour-Zeynali K, Jouyban A. Solubility and dissolution rate of carbamazepine-cinnamic acid co-crystal. J. Mol. Liq. 2013, 187, 171–176.
[38] Qiao N, Wang K, Schlindwein W, Davies A, Li M. In situ monitoring of carbamazepine-nicotinamide co-crystal intrinsic dissolution behaviour. Eur. J. Pharm. Biopharm. 2013, 83, 415–426.

[39] Bruni G, Maietta M, Maggi L, Mustarelli P, Ferrara C, Berbenni V, Milanese C, Girella A, Marini A. Preparation and physicochemical characterization of acyclovir co-crystals with improved dissolution properties. J. Pharm. Sc. 2013, 102, 4079–4086.
[40] Sowa M, Slepokura K, Matczak-Jon E. Solid-state characterization and solubility of a genistein-caffeine co-crystal. J. Mol. Struct. 2014, 1076, 80–88.
[41] Sugandha K, Kaity S, Mukherjee S, Isaac J, Ghosh A. Solubility enhancement of ezetimibe by a co-crystal engineering technique. Cryst. Growth Des. 2014, 14, 4475–4486.
[42] Zhang X, Sun F, Zhang T, Jia J, Su H, Wang C, Zhu G. Three pharmaceuticals co-crystals of adefovir: syntheses, structures and dissolution study. J. Mol. Struct. 2015, 110, 395–400.
[43] Brittain HG. Thermodynamic vs. kinetic solubility: knowing which is which. Am. Pharm. Rev. 2014, 17, issue 3.
[44] Greco K, Bogner R. Solution-mediated phase transformation: significance during dissolution and implications for bioavailability. J. Pharm Sc. 2012, 101, 2996–3018.
[45] Guzman HR, Tawa M, Zhang Z, Ratanabanangkoon P, Shaw P, Garnder CL, Chen H, Moreau JP, Almarsson O, Remenar JF. Combined use of crystalline salt forms and precipitation inhibitors to improve oral absorption of celecoxib form solid oral formulations. J. Pharm. Sci. 2007, 96, 2686–2702.
[46] Bavishi DD, Borkhataria CH. Spring and parachute: how co-crystals enhance solubility. Prog. Cryst. Growth Charact. Mater. 2016, 62, 1–8.
[47] Babu NJ, Nangia A. Solubility advantage of amorphous drugs and pharmaceutical co-crystals. Cryst. Growth Des. 2011, 11, 2662–2679.
[48] He H, Huang Y, Zhang Q, Wang JR, Mei X. Zwitterionic co-crystals of flavonoids and proline: solid-state characterization, pharmaceutical properties, and pharmacokinetic performance. Cryst. Growth Des. 2016, 16, 2348–2356.
[49] Childs SL, Kandi P, Lingireddy SR. Formulation of a danazol co-crystal with controlled supersaturation plays an essential role in improving bioavailability. Mol. Pharm. 2013, 10, 3112–3127.
[50] Liu M, Hong C, Yao Y, Shen H, Ji G, Li G, Xie Y. Development of a pharmaceutical co-crystal with solution crystallization technology: preparation, characterization, and evaluation of myricetin-proline co-crystals. Eur. J. Pharm Biopharm. 2016, 107, 151–159.
[51] Hisada N, Takano R, Takata N, Shiraki K, Ueto T, Tanida S, Kataoka M, Yamashita S. Characterizing the dissolution profiles of supersaturable salts, co-crystals and solvates to enhance in vivo oral absorption. Eur. J. Pharm. Biopharm. 2016, 103, 192–199.
[52] Li J, Wang L, Ye YQ, Fu X, Ren Q, Zhang H, Deng Z. Improving the solubility of dexlansoprazole by co-crystallization with isonicotinamide. Eur J. Pharm. Sci. 2016, 85, 47–52.
[53] Urbanus J, Roelands CPM, Verdoes D, Jansens PJ, ter Horst JH. Co-crystallization as a separation technology: controlling product concentrations by co-crystals. Cryst. Growth Des. 2010, 10, 1171–1179.
[54] Billot P, Hosek P, Perrin M-A. Efficient purification of an active pharmaceutical ingredient via co-crystallization: from thermodynamics to scale-up. Org. Process Res. Dev. 2013, 17, 505–511.
[55] Urbanus J, Roelands CPM, Mazurek J, Verdoes D, ter Horst JH. Electrochemically induced co-crystallization for product removal. CrystEngComm 2011, 13, 2817–2819.
[56] Hsi KHY, Chadwick K, Fried A, Kenny M, Myerson AS. Separation of impurities from solution by selective co-crystal formation. CrystEngComm 2012, 14, 2386–2388.
[57] Hsi KH, Kenny M, Simi A, Myerson AS. Purification of structurally similar compounds by the formation of impurity co-former complexes in solution. Cryst. Growth Des. 2013, 13, 1577–1582.
[58] Hsi KH, Concepción AJ, Kenny M, Magzoub AA, Myerson AS. Purification of amoxicillin trihydrate by impurity-coformer complexation in solution. CrystEngComm 2013, 15, 6776–6781.

[59] Lorenz H, Czapla F, Polenske D, Elsner MP, Seidel-Morgenstern A. Crystallization based separation of enantiomers. Journal of the University of Chemical Technology and Metallurgy 2007, 42, 5–16.

[60] Caira MR, Nassimbeni LR, Scott JL, Wildervanck AF. Resolution of optical isomers of 4-amino-p-chlorobutyric acid lactam by co-crystallization. J. Chem. Crystallogr. 1996, 26, 117–122.

[61] George F, Norberg B, Robeyns K, Wouters J, Leyssens T. Peculiar case of levetiracetam and etiracetam a-ketoglutaric acid co-crystals: obtaining a stable conglomerate of etiracetam. Cryst. Growth Des. 2016, 16, 5273–5282.

[62] George F, Tumanov N, Norberg B, Robeyns K, Filinchuk Y, Wouters J, Leyssens T. Does chirality influence the tendency toward co-crystal formation? Cryst. Growth Des. 2014, 14, 2880–2892.

[63] Jacques J, Collet A, Wilen SH. Enantiomers, racemates, and resolutions. J. Wiley &. Sons; New York, Chichester, Brisbane, Toronto. 1981.

[64] Mahieux J, Gonella S, Sanselme M, Coquerel G. Crystal structure of a hybrid salt co-crystal and its resolution by preferential crystallization: ((±)trans-*N*, *N′*-dibenzyldiaminocyclohexane)(2,3-dichlorophenylacetic acid. CrystEngComm 2012, 14, 103–111.

[65] Noorduin WL, Izumi T, Millemaggi A, Leeman M, Meekens H, van Enckevort WJP, Kellogg RM, Kaptein B, Vlieg E, Blackmond DG. Emergence of single solid chiral state from a nearly racemic amino acid derivative. J. Am. Chem. Soc. 2008, 130, 1158–1159.

[66] Kaptein B, Noorduin WL, Meekes H, van Enckevort WJP, Kellogg RM, Vlieg E. Attrition enhanced deracemization of an amino acid derivative that forms an epitaxial racemic conglomerate. Ang. Chem. Int. Ed. 2008, 47, 7226–7229.

[67] Yagishita F, Ishikawa H, Onuki T, Hachiya S, Mino T, Sakamoto M. Total spontaneous resolution by deracemization of isoindolinones. Angew. Chem. Int. Ed. 2012, 51, 13023–13025.

[68] Hachiya S, Kasashima Y, Yagishita F, Mino T, Masu H, Sakamoto M. Asymmetric transformation by dynamic crystallization of achiral succinimides. Chem. Commun. 2013, 49, 4776–4778.

[69] Sögütoglu LC, Steendam RRE, Meekes H, Vlieg E, Rutjes FPJT. Viedma ripening: a reliable crystallisation method to reach single chirality. Chem. Soc. Rev. 2015, 44, 6723–6732.

[70] Springuel G, Robeyns K, Norberg B, Wouters J, Leyssens T. Co-crystal formation between chiral compounds: how co-crystals differ from salts. Cryst. Growth Des. 2014, 14, 3996–4004.

[71] Sanchez-Guadarrama O, Mendoza-Navarro F, Cedillo-Cruz A, Jung-Cook H, Arenas-Garcia JI, Delgado-Diaz A, Herrera-Ruiz D, Morales-Rojas H, Höpfl H. Chiral resolution of RS-praziquantel via diastereomeric co-crystal pair formation with l-malic acid. Cryst. Growth Des. 2016, 16, 307–314.

[72] Molnar P, Thorey P, Bansaghi G, Szekely E, Poppe L, Tomin A, Kemeny S, Fogassy E, Simandi B. Resolution of racemic trans-1,2-cyclohexanediol with tartaric acid. Tetrahderon: Asymmetry 2008, 19, 1587–1592.

[73] Springuel G, Leyssens T. Innovative chiral resolution using enantiospecific co-crystallization in solution. Cryst. Growth. Des. 2012, 12, 3374–3378.

[74] Springuel G, Collard L, Leyssens T. Ternary and quaternary phase diagrams: key tools for chiral resolution through solution co-crystallization. CrystEngComm 2013, 15, 7951–7958.

Marc Fourmigué

10 The salt–co-crystal continuum in halogen-bonded systems

10.1 Introduction

The International Union of Pure and Applied Chemistry (IUPAC) [1] defines halogen bonding (XB) as occurring, "when there is evidence of a net attractive interaction between an electrophilic region associated with a halogen atom in a molecular entity and a nucleophilic region in another, or the same, molecular entity." It further elaborates by stating, "A typical halogen bond is denoted by the three dots in R–X···Y. R–X is the halogen bond donor, X is any halogen atom with an electrophilic (electron-poor) region, and R is a group covalently bound to X. In some cases, X may be covalently bound to more than one group". In other words, species Y is acting as a Lewis base. XB is therefore associated with the electronic polarisation of the halogen elements, particularly the heavier ones (X = Br, I), with a charge-depletion area in the prolongation of the R–X bond. This electrophilic region is called the σ-hole, as it is also associated with the anti-bonding σ* combination of the orbitals involved in the R–X bond. XB has already been the topic of several review papers [2, 3] and books [4, 5] and readers are invited to refer to these for further information. XB has been investigated in solution as in condensed matter, i.e. crystals, liquid crystals and gels. In most systems, the interaction strength is similar to that of weak to normal hydrogen bonds (HBs). In recent times, implications of XB in many domains have been revealed, such as in crystal engineering, anion recognition, conducting and magnetic materials, organocatalysis, biomolecules, drug design, photoactive materials, etc.

In crystal engineering, XB offers several attractive peculiarities, particularly a strong directionality with an essentially linear R–X···B motif. When combined with Lewis base acceptors, this linearity gives rise to high predictability, as illustrated by the numerous 1D, 2D or 3D networks purposely prepared not only from numerous ditopic XB donors (Figure 10.1a), but also from tritopic ones such as *sym*-triiodotrifluorobenzene [6, 7], or 1,3,5-trifluoro-2,4,6-tris(iodoethynyl)benzene (Figure 10.1b) [8, 9], or tetratopic ones, as in tetrabromomethane [10] or tetraiodoethylene (Figure 10.1c) [11, 12].

https://doi.org/10.1515/9783110464955-010

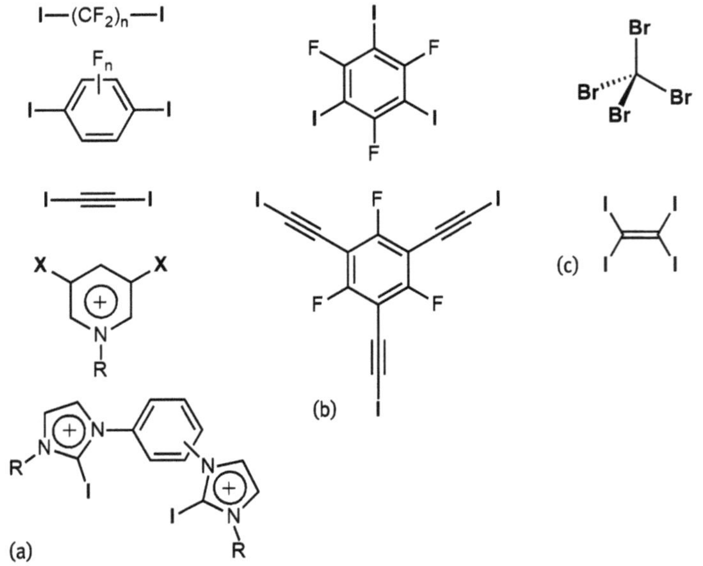

Fig. 10.1: Examples of (a) ditopic, (b) tritopic and (c) tetratopic halogen-bonded (XB) donors used in crystal engineering and anion recognition.

10.2 Neutral or ionic halogen-bonded co-crystals

10.2.1 Neutral or ionic? The example of hydrogen-bonded systems

The concept of charge-assisted HB is well established. This is the result of proton transfer between, for example, a carboxylic acid and an amine [13]. The $\Delta(\mathrm{p}K_a)$ values [14] have been widely applied to predict the actual state of a material: co-crystals where all components in the HB complex remain neutral, and salts where the molecules take ionic forms [15]. As charge-assisted HB gives rise to stronger HB interactions, it has been used successfully in the elaboration of robust supramolecular networks [16, 17]. One question then arises regarding the possible observation of analogous situations in halogen-bonded (XB) systems, where the exact position of the halogen atom would define the ionicity of the whole XB adduct, as illustrated in Figure 10.2.

Several conditions are needed to observe the situation described in Figure 10.2b. It implies first that the Y–I bond is labile enough to break, and the associated stabilisation/delocalisation of the negative charge in the resulting Y^- species. Simultaneously, the iodonium species illustrated in Figure 10.2b as a N-iodopyridinium cation, should also be accessible. Note also that the position of the iodine atom may well be intermediate where the determination of the degree of ionicity can be more difficult and require specific experimental and/or theoretical approaches. In the extreme ionic situation, the N-iodopyridinium now becomes the XB donor, which is associated through

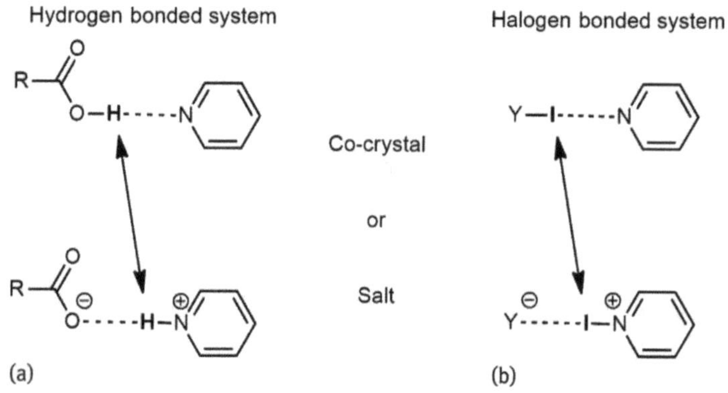

Fig. 10.2: (a) Co-crystal vs salt description in hydrogen and (b) XB systems.

XB with the Y^- species acting as an XB acceptor; a situation typical of charge-assisted XB, as detailed below.

10.2.2 Charge-assisted halogen bonding

Halogen bonding is particularly enhanced in adequately charged systems, with either positively charged molecules activating the halogen atom or negatively charged XB acceptors or even better, the association of both. For example, meta-substituted iodo- or bromopyridinium [18, 19], 2-halo-imidazolium [20–22], ethyl-4-halothiazolium [23, 24] and iodo-tetrathiafulvalenium salts [25] interact strongly through XB with neutral or anionic Lewis bases. Similarly, neutral XB donors such as perfluoroiodoalkanes, alkenes and arenes interact strongly with negatively charged species such as halide or pseudo-halide anions, oxo- and halometallate anions [26]. This charge assistance is also particularly strong in the N-iodonium derivatives, as most of them crystallise with a second pyridine moiety leading to symmetric XB systems, both in the solid state and in solution. Pyridine-N-oxides provide another example of this charge-assisted XB, particularly when engaged with N-iodoimide derivatives as XB donors. Indeed, in the latter, the strong polarisation of the N–I bond leading to a N^-–I^+ motif is perfectly fitted to interact with the opposite polarisation of the pyridine-N-oxide, which can be written as ^-O–N^+_{Py} to give extremely short I···O contacts, with a reduction ratio down to 0.66 [27].

The occurrence of charge-assisted XB has been nicely illustrated in isostructural systems of varying charge, such as associating iodinated tetrathiafulvalene (EDT-$TTFI_2$) with tetracyanoquinodimethane (TCNQ) derivatives of various electron acceptor capability, i.e. TCNQ, TCNQF and $TCNQF_2$ (Figure 10.3) [28]. A significantly shorter and stronger XB interaction between the iodine atom of EDT-$TTFI_2$ and the nitrile substituent of the $TCNQF_n$ molecules is indeed found in the ionic $(EDT\text{-}TTFI_2)_2^{+\bullet}$

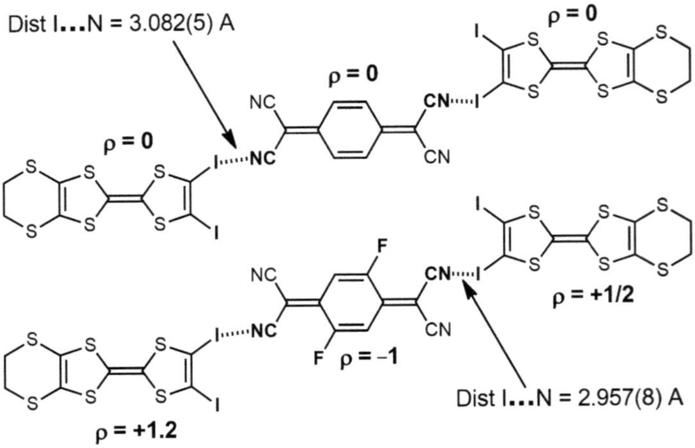

Fig. 10.3: Evolution of the I···N XB with the charge: neutral (with tetracyanoquinodimethane, TCNQ) vs ionic (with $TCNQF_2$) charge transfer complexes of iodinated tetrathiafulvalene ($EDT\text{-}TTFI_2$).

$(TCNQF_2)^{-\bullet}$ charge transfer salt than in the neutral $(EDT\text{-}TTFI_2)_2(TCNQ)$ charge transfer co-crystal. In the former, the building of charge in the two partners of the co-crystal association finds its origin in an electron transfer, made possible because of the careful redox adaptation of the two electron donor and acceptor moieties. Indeed, $TCNQF_4$ is a stronger oxidant (E_{red} = +0.30 V vs SCE) than TCNQ (E_{red} = +0.14 V vs SCE). An intermediate situation with a neutral–ionic conversion upon lowering the temperature was observed with TCNQF (E_{red} = +0.26 V vs SCE) [29]. This work was also extended to weaker electron donors (EDT-TTF–I) [30] or stronger electron acceptors (DDQ) [31].

10.2.3 Neutral or ionic? Halogen bonded systems

XB complexes between a Lewis base B and dihalogens or interhalogens (X–Y = I_2, Cl–F, I–Cl) can be formulated as B···X–Y. Legon [32] has shown that, in some instances, these systems were better described under an ionic form $[BX]^+\cdots Y^-$. Such systems are characterised by a large intermolecular stretching force that is constant compared with the so-called Mulliken outer complexes, B···X–Y. For example, a contribution of roughly 20% of the ionic form (Mulliken inner complex) was estimated for the ammonia/Cl–F complex, increasing up to 50–70% in the alkylated trimethylamine analogue, Me_3N/ClF [33, 34]. These studies also established an order of electrophilicity of the dihalogen molecules with $Cl_2 < Br_2 < BrCl < ClF < ICl$ [32]. In that respect, complexes involving dihalogens and interhalogens are most prone to exhibit ionic structures in the solid state, as has been comprehensively reviewed by Pennington et al. [35]. Examples of XB ionic (zwitterionic) structures, $[BX]^+\cdots Y^-$, are given with a specific reference to more recent examples in the following sections of this chapter.

10.3 Strong halogen bond donors

The propensity of XB systems B···X–Y to eventually ionise to $[BX]^+ \cdots Y^-$ relies on the possibility of breaking the X–Y bond. In XB systems, this has been observed with dihalogens and interhalogens, but also with N-haloimide derivatives, as detailed later in Section 10.3.2. The following focuses upon the evolutions observed in a series of compounds and the propensity to ultimately form ionic systems by increasing the electrophilic character of the XB donor and/or by increasing the Lewis base character of the XB acceptor.

10.3.1 Dihalogens and interhalogens

As already established by Pennington et al. [35], dihalogens and interhalogens can interact with Lewis bases at only one end, leading to the simple adduct B···X–Y, referred to as mode A in Figure 10.4. However, other possibilities are also observed and have been described as the amphoteric adduct (AA) mode, bridged amphoteric (BA) mode, bridged (B) mode (Figure 10.4). Hypervalent species described as "T-shaped" adducts are also known to evolve from the initially formed neutral adducts derived from thiones [36]. They are most often found with the weaker XB donors (I_2, Br_2) and are not discussed here to allow more attention to be given to the strongest, potentially ionic structures.

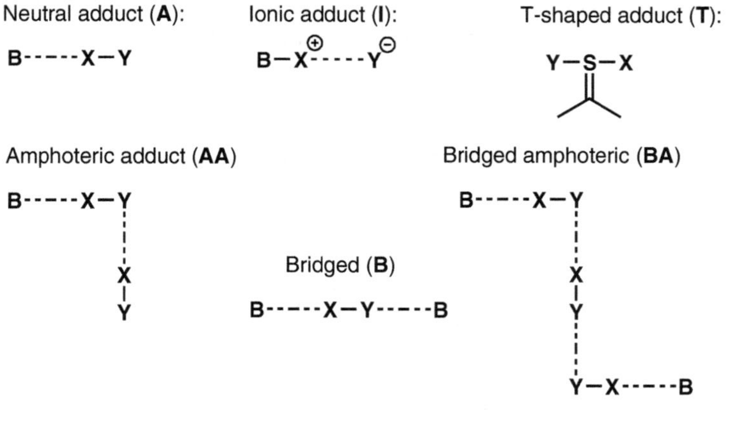

Fig. 10.4: Common geometries observed in dihalogens and interhalogen adducts.

10.3.1.1 Chalcogens as halogen bond acceptors

Chalcogen-containing molecules (S, Se) are particularly efficient XB acceptors. Many different XB adducts with dihalogens and interhalogens have been described [37], involving thio(seleno)ethers, thiones and thio(seleno)phosphines. In most cases, neut-

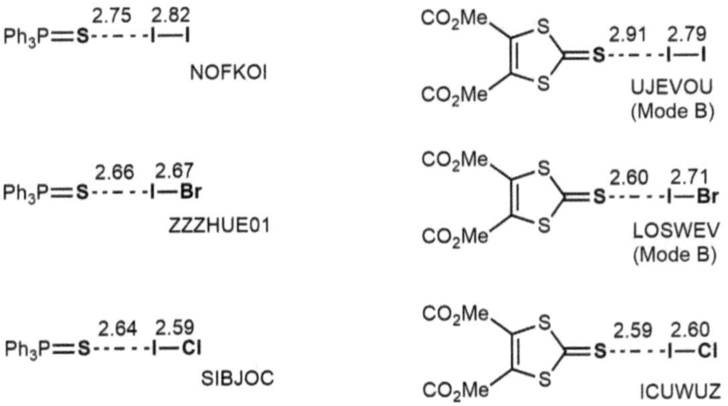

Fig. 10.5: Geometric characteristics of representative XB adducts (in Å) with their Cambridge Crystallographic Data Centre (CCDC) refcode.

ral adducts are observed, as illustrated in Figure 10.5, for the I_2, IBr and ICl adducts of (i) triphenylphosphine sulphide [38, 39] and (ii) 4,5-carboxymethyl-1,3-dithiole-2-thione [40]. As a general trend, much shorter XB interactions are observed with ICl as the XB donor. In these ICl adducts, the iodine atom is located essentially at the midpoint between sulphur and chlorine atoms, indicating an already relatively "ionic" structure.

The actual extent of charge transfer in such adducts was theoretically estimated by Koskinen et al. [41, 42] from quantum theory of atoms in molecules (QTAIM) calculations in a series of benzothiazole-2(3H)-thione and N-methylbenzothiazole-2-thione adducts with XY XB donors (XY = Br_2, I_2, IBr and ICl) (Figure 10.6).

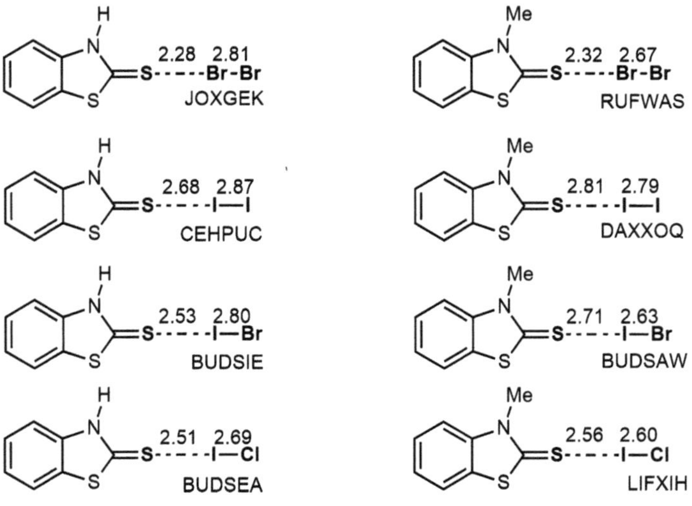

Fig. 10.6: Geometric characteristics of representative XB adducts (in Å) of benzothiazole-2(3H)-thione together with their CCDC refcode.

In the latter ICl adducts (BUDSEA, LIFXIH), the iodine atom is indeed notably closer to the sulphur than to the chlorine atom, which is another indication of a significant charge transfer. Both S···X and X···Y contacts were characterised (Table 10.1) by the interaction energy and by the ratio of the densities of potential energy V and kinetic energy G at bond critical points (BCPs), ($|V_{BCP}|/G_{BCP}$). The latter has values < 1 for non-covalent closed-shell (ionic) interactions and > 2 for covalent bonds, whereas values between 1 and 2 indicate a partial covalency. It is noted that in all systems, the V_{BCP}/G_{BCP} descriptor takes intermediate values between 1 and 2, indicative of a partial covalent character for both interactions. Furthermore, for the Br_2 and ICl adducts, this value is even greater on the S···X cf. the X···Y side, whereas in the ICl adduct, the charge on Cl exceeds −0.5, indicating that this adduct is now closer to the ionic side than to the neutral side.

It has been shown that the Lewis base character of the thione sulphur atom increases in the order dithiole-2-thione < thiazole-2-thione < imidazolidine-2-thione [37, 43]. It is therefore not surprising that the dihalogen adducts of imidazolidine-2-thione

Tab. 10.1: Properties of the electron density at selected bond critical points (BCPs) for adducts of N-methyl-benzothiazole-2(3H)-thione with XY halogen bond donors (XY = Br_2, I_2, IBr, ICl). See Figure 10.6 for refcode definitions.

Adduct	BCP	d (Å)	ρ (e Å^{-3})	E_{int}(kJ mol^{-1})	$\|V\|/G$	q(Y)	Reference
RUFWAS	S···Br	2.325	0.6362	−104	1.69	−0.457	[41]
	Br···Br	2.667	0.3745	−45	1.37		
DAXXOQ	S···I	2.808	0.2998	−39	1.31	−0.263	[42]
	I···I	2.791	0.4253	−52	1.73		
BUDSAW	S···I	2.706	0.3587	−50	1.41	−0.416	[42]
	I···Br	2.631	0.4645	−65	1.63		
LIFXIH	S···I	2.549	0.4757	−75	1.56		[42]
	I···Cl	2.617	0.4109	−58	1.49	−0.554	

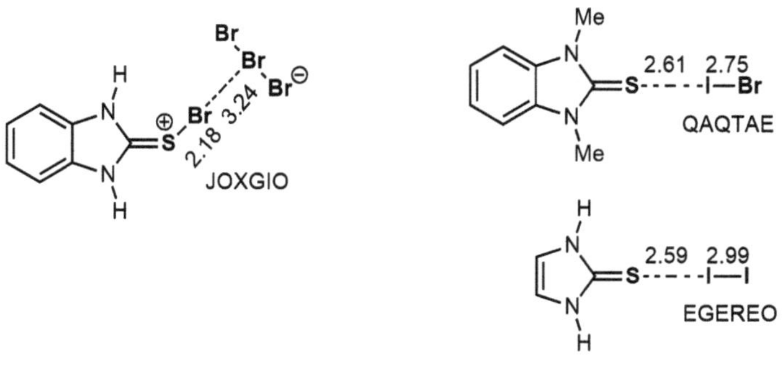

Fig. 10.7: Dihalogen adducts of imidazolidine-2-thione derivatives together with their CCDC refcode.

derivatives exhibit either fully ionic structures (as in JOXGIO, Figure 10.7) or regular adducts (QAQTAE, EGEREO in Figure 10.7), but with very short S···I distances and a reduction ratio below 0.688 with IBr or I_2. It is anticipated that ICl adducts would exhibit even shorter S···I distances.

10.3.1.2 Pyridines as halogen bond acceptors

Adducts of nitrogen compounds with dihalogens and interhalogens come second after sulphur-containing species for the number of reported crystal structures. Pyridine derivatives form Py–N···XY interactions and are particularly attractive owing to their potential for forming the ionic form $[Py–N–X]^+ \cdots [Y]^-$. They are highlighted via the bis(pyridine) iodine/bromine cations $[(Py–N)_2I(Br)]^+$, also known as bis-pyridine halonium cations, even if the actual charge on the iodine atom does not exceed +0.40 [44]. The latter compounds usually crystallise in symmetrical structures with equivalent N···X distances. This symmetry was also confirmed from nuclear magnetic resonance studies in solution [45, 46]. Structural characteristics of representative pyridine adducts with I_2, IBr and ICl systems have been collected in Table 10.2.

Interesting comparisons can be made with the stronger ICl adducts. The iodine atom "moves" towards the pyridyl nitrogen atom, going from the species with the electron-poor 3-bromopyridine (GANXAY) to that with the electron-rich 4-dimethylamino-pyridine (GANXOJ). Note that within these series, the total distance from the nitrogen to the outer chloride atom (N···Y in Table 10.2) is almost constant, only the position of the iodine atom varies. The comparison of both N···I and I···Cl distances

Tab. 10.2: Structural characteristics of halogen bonding in I_2, IBr or ICl adducts of pyridines.

Adduct	N_{Py}···I (Å)	I···Y(Å)	N···Y (Å)	N···I···Y (°)	CCDC	Reference
With ICl:						
(3-Br-Py)(ICl)	2.344 (5)	2.473 (5)	4.817	178.43 (5)	GANXAY	[47]
(2,2′-bipy)$(ICl)_2$[a]	2.321 (2)	2.4974 (7)	4.818	179.41 (5)	TAGNEV	[48]
	2.337 (2)	2.4878 (9)	4.825	175.33 (5)	TAGNEV01	[47]
	2.336 (7)	2.479 (11)	4.815	179.6 (3)		
	2.344 (7)	2.477 (11)	4.821	176.2 (2)		
(Py)(ICl)	2.29 (1)	2.510 (11)	4.800	178.7 (3)	PYRIIC10	[49]
(2,4,6-Me_3Py)(ICl)	2.294 (5)	2.531 (2)	4.825	179.05 (14)	GANXID	[47]
(4-Me_2N-Py)(ICl)	2.246 (2)	2.5615 (7)	4.807	179.24 (5)	GANXOJ	[47]
With IBr:						
(2,2′-bipy) $(IBr)_2$	2.46 (1)	2.577 (2)	5.037	175.9 (3)	TAGNIZ	[48]
(Py)(IBr)	2.26 (4)	2.66 (1)	4.920		PYIOBR	[50]
With I_2:						
(Py)(I_2)	2.425 (8)	2.8043 (9)	5.228	176.44 (18)	VUHDIN	[51]

[a] Two independent molecules in the asymmetric unit

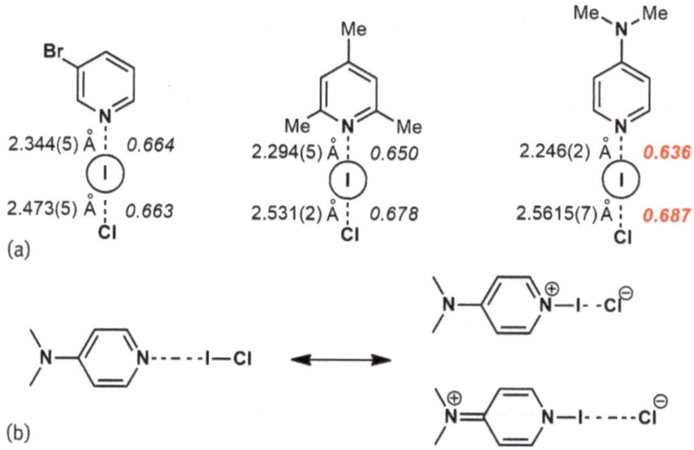

Fig. 10.8: (a) XB distances (Å) and reduction ration (in italics) for ICl adducts of pyridines and (b) resonance forms for the DMAP adduct.

vs the sum of the van der Waals radii (Figure 10.8a) indicates that in the latter 4-dimethylaminopyridine (DMAP) adduct, the iodine is indeed much closer to the nitrogen atom than to the chlorine, the signature of a more ionic structure and a consequence of the presence of the dimethylamino group in the para position, which stabilises this ionic form using the mesomeric effect (Figure 10.8b). QTAIM calculations would be highly desirable here to confirm these assumptions.

10.3.1.3 Phosphines as halogen bond acceptors

Dihalogen and interhalogen adducts of phosphorus derivatives provide an extensive series of compounds that can be classified into two groups: (i) simple adducts with a linear P–X–Y system (Figure 10.9a) and (ii) complex adducts where the cationic R_3P–X^+ moiety interacts with complex polyiodide salts (Figure 10.9b). For the latter, examples include $[Ph_3P–I]^+[I_3]^-$ [52], $[Ph_3P–I]^+[I_5]^-$ [53], $[^iPr_3P–I]^+[I_3^-]$ [54] and $[p\text{-}Tol_3P–I]^+[I_3]^-$ [55]. The simple adducts are described to adopt a “spoke” structure.

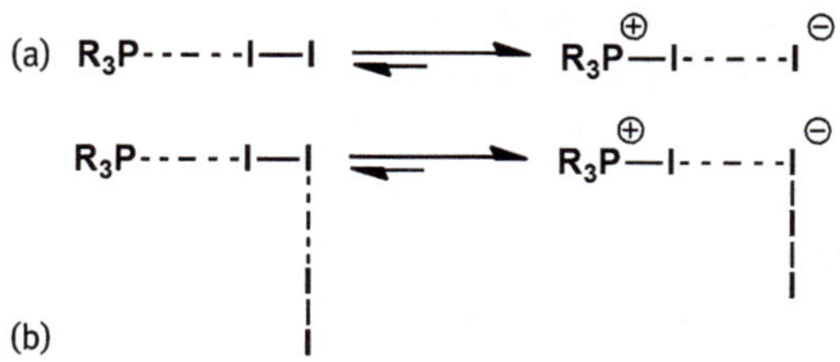

Fig. 10.9: Structures of diiodine phosphine adducts.

Tab. 10.3: Geometric features of selected simple diiodine phosphine adducts.

Adduct	P–I (Å)	I···I (Å)	P–I···I (°)	P(···I···)I (Å)	CCDC	Reference
(Mecarb)^{i}Pr$_2$P–I$_2$	2.5978 (14)	3.021 (1)	177.49 (3)	5.619	QAMZAF	[56]
(o-Tol)$_3$P–I$_2$	2.5523 (12)	3.0727 (4)	174.32 (3)	5.625	SEQKEF	[57]
(p-FC$_6$H$_4$)$_3$P–I$_2$^{a}	2.507 (3)	3.0807 (12)	177.76 (7)	5.588	RARXAL	[58]
	2.461 (3)	3.1529 (11)	173.07 (7)	5.614		
Ph$_3$P–I$_2$	2.481 (4)	3.161 (2)	178.23 (6)	5.642	JITSIO	[59]
Me$_2$PhP–I$_2$	2.410 (2)	3.408 (2)	177.01 (6)	5.518	ZEKMOR	[60]
[(Et$_2$N)$_3$P]–I$_2$	2.4730 (8)	3.6168 (4)	171.97 (2)	6.090	TIZBAG	[61]
[(^{n}Pr$_2$N)$_3$P]–I$_2$	2.455 (4)	3.6389 (14)	161.12 (9)	6.094	CALPEK	[61]

^{a} Two independent molecules in the asymmetric unit

These species differ significantly from the analogous nitrogen derivatives, which were shown above to adopt a "close to ionic" structure only in very specific conditions. Here, for example in the I_2 adducts collected in Table 10.3, the P–I distances are very short (2.43–2.60 Å), whereas the corresponding I···I contacts range from 3.0 to 3.6 Å. In other words, the simple diiodine adducts of phosphines uniformly have a significant ionic character. Only a combination of electron-withdrawing substituents and large cone angles decreases this ionic character. As a consequence, a strong evolution of the P(···I···)I distances between phosphorus and the outer iodine atom are observed (Table 10.3). Indeed, it amounts to 6.1 Å with the most electron-rich phosphines whereas it decreases to 5.6 Å with less electron-rich phosphines. This striking evolution is a consequence of the large ionic radius of the iodide anion (2.22 Å) involved in most ionic adducts, whereas covalent contributions play an important role in both P···I and I···I distances with the electron-poor phosphines, with a much smaller iodine covalent radius (1.33 Å).

10.3.2 N-iodo imides and analogues

Besides dihalogens and interhalogens, strong and highly polarisable XB donors can be found not only among organic N-iodo imides derivatives such as N-iodosuccimide (NIS), N-iodophthalimide (NIP) or N-iodosaccharin (NISac), and also *N, N′*-diiodohydantoin (Figure 10.10). These compounds are well known in synthetic chemistry as electrophilic iodination reagents and the corresponding imidate anions are also isolable. Therefore, they are attractive tools for conceiving Lewis base XB adducts with potentially a strong ionic character.

A recent review by Rissanen et al. [62] pointed out the strong XB donor character of these iodo derivatives (and their bromo analogues), towards themselves (with the carbonyl oxygen atom), towards neutral amines (DABCO [63], hexamethylenetetram-

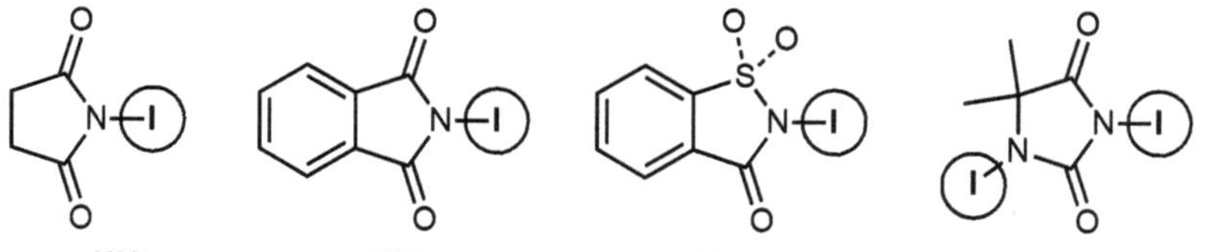

Fig. 10.10: Representative available N-iodo imide derivatives.

ine [63], pyridine [64], pyrazine [64]), and towards bromide anions [65]. In the reported examples with amines, the neutral character of the adduct was not questioned, as the N_{imide}–I bond is systematically and notably shorter than the I···$N_{pyridine}$ XB. One single exception, which was associated with a low-quality crystal structure, was reported for NISac···pyridine where the N_{imide}–I "covalent" bond is almost as short as the I···$N_{Pyridine}$ XB. To clarify this point and to determine the exact nature of the Ni-iodosaccharin/pyridine adducts (Figure 10.11), the investigation of a complete series of adducts of both NIS and NISac with para-substituted pyridines of varying Lewis base character through different substituents, R = H, Me and NMe_2, was undertaken [66].

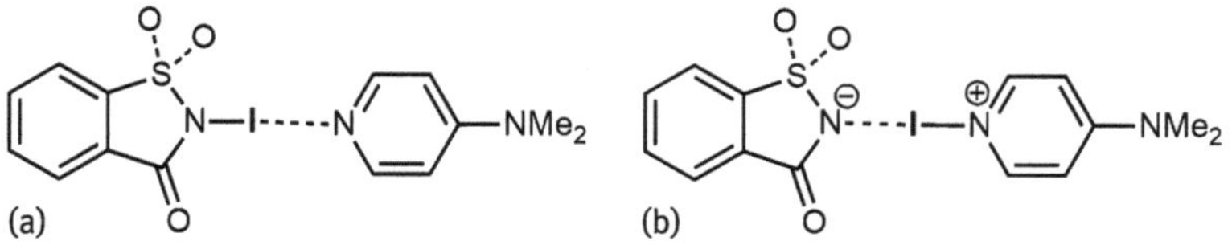

Fig. 10.11: The two extreme forms of the XB system, NISac · DMAP.

As shown in Table 10.4, the NIS adducts are less strongly bound than the NISac adducts. With the latter, the *para*-dimethylaminopyridine adduct, NISac · DMAP, is the only one where the iodine atom is closer to the pyridine than to the saccharin, indicating an important contribution of the ionic form shown in Figure 10.11b. This assumption was further confirmed by QTAIM calculations performed on isolated XB complexes extracted from the crystal phases. They showed that the NISac · DMAP complex can indeed be described as an ionic salt with a charge transfer of q = 0.68e, whereas NIS · DMAP should still be considered a co-crystal with q = 0.16e. Furthermore, the role of the crystal environment in the actual position of the iodine atom was also underlined, pointing out that the overall surroundings play a role that can be modelled by an external electric field.

Very recently, the ability of diiodohydantoins (DIHs, Figure 10.10) to act as a strong, ditopic XB donor was also investigated within a series of adducts with para-substituted pyridines of varying Lewis base character, with R = CN, CF_3, CO_2Me, H, Me, NMe_2 and pyrrolidinyl (Figure 10.12) [67]. Despite their preparation with a three-fold

Tab. 10.4: Room temperature geometric distances and angles (Å, °) of the N–I···$N'_{R\text{-}Py}$ halogen bond in the co-crystals with N-iodosuccimide (NIS) and NISac. d_1 = N–I distance, d_2 = I···N′ distance, $d_1 + d_2$ = N···N′ distance (the interaction is linear).

	d_1 (N–I)	d_2 (I···N′)	$d_1 - d_2$	N–I···N′	$d_1 + d_2$	Reference
NIS · Py[a]	2.116 (5) 2.144 (8)	2.493 (8) 2.430 (8)	−0.377 (9) −0.286 (11)	180180	4.609 (9) 4.574 (11)	[66]
NIS · Py–Me[a]	2.116 (4) 2.142 (4)	2.483 (4) 2.428 (4)	−0.367 (6) −0.286 (6)	180180	4.599 (6) 4.570 (6)	[66]
NIS · Py–NMe_2	2.146 (4)	2.407 (4)	−0.261 (6)	178.9 (1)	4.553 (6)	[66]
NISac · Py[b]	2.254 (11)	2.279 (11)	[b]	174.5 (4)	4.52 (2)	[64]
NISac · Py–Me	2.220 (3)	2.304 (3)	−0.084 (4)	178.2 (1)	4.523 (4)	[66]
NISac · Py–NMe_2	2.292 (1)	**2.228 (1)**	**+0.064 (1)**	178.8 (1)	4.520 (1)	[66]

[a] Two crystallographically independent NIS · Py or NIS · Py–Me complexes, both located on two-fold axes

[b] From Dolenc and Modec [64]: X-ray data quality is poor owing to the presence of superstructures on the diffraction images

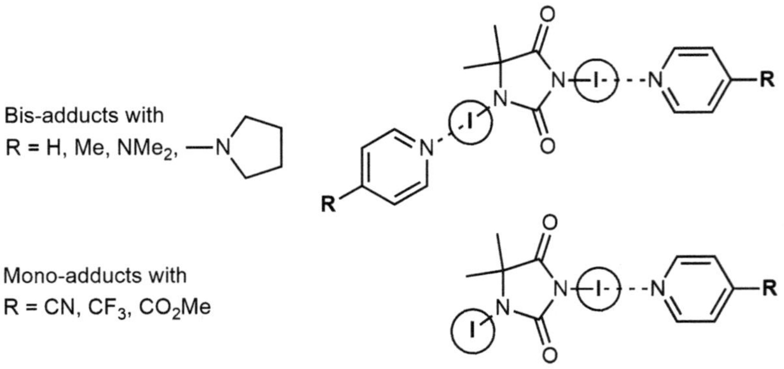

Fig. 10.12: Mono- and bis-adducts of diiodohydantoins with pyridines of varying Lewis base character.

excess of the different pyridines, the bis-substituted adducts involving both the imidic and amidic iodine atoms were only isolated with the most electron-rich pyridines (R = H, Me, NMe_2 and pyrrolidinyl) whereas the electron-poor pyridines (R = CN, CF_3, CO_2Me) favoured the formation of mono-adducts through XB with the imidic iodine atom only (Figure 10.12).

The analysis of the structural characteristics of these mono- and bis-adducts, combined with theoretical calculations, demonstrated that the formation of a first N_{DIH}–I···N_{Py} XB deeply modifies the XB (and associated σ-hole) potential of the second uncoordinated iodine atom. In other words, ditopic molecules such as DIHs are deactivated upon formation of a first XB and enter into a second only in the presence of the strongest XB acceptors. This indicates that the notable charge transfer from the pyridine to the DIH molecule, which takes place upon formation of a first "strong" XB, severely decreases the electrophilic character (σ-hole) of the second "uncoordinated" iodine atom. This charge effect is clearly identified in the structural characteristics of the mono- and bis-adducts of DIH which were isolated with the 4-pyrrolidinyl-pyridine (Figure 10.13). It is noted that the single XB interaction in the mono-adduct is much stronger than the two XB interactions in the bis-adduct.

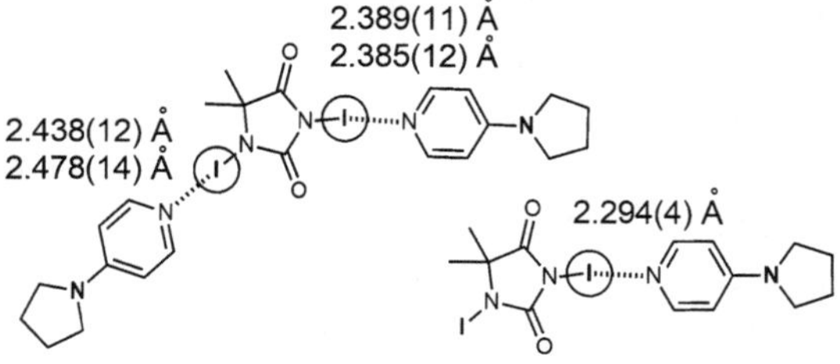

Fig. 10.13: Evolution of the XB distances in the mono- and bis adducts of DIH with 4-pyrrolidino-pyridine.

10.4 Discussion and perspectives

It has been noted above that organic N-iodo imides derivatives can act as very strong XB donors and, in some rare instances, are able to go beyond the "invisible" border between a neutral and an ionic description of the charge state. It should be stressed here that for all the adducts described above, with organic and inorganic (I_2, IBr, ICl) XB donors, such a border is only an image and that there is actually a smooth continuum from weak XB systems with little, if any, charge transfer to the much stronger interactions found for example in the phosphine adducts, with a full charge transfer and complete ionisation. The intermediate situations can be properly understood with the help of QTAIM calculations. This can provide excellent tools for evaluating the nature and strength of these interactions, which bear a notable covalent character. These theoretical analyses should be used in a more systematic way to rationalise the behaviour within a systematic series of structures. The example of the diiodine phosphine adducts reported in Table 10.3 is a case in point.

Most of the adducts described herein are based on the iodine atom as XB donor. This is not surprising as, in "normal" XB systems with little charge transfer, the ranking I > Br ≫ Cl is well established. On the other hand, there are some indications that in strong XB situations with a large covalent character, the one bromine member of the series might compare with or even exceed the iodine one in strength. For example, in the halonium salts of general formula $[(pyridine)X(pyridine)]^+$ (X = Br, I), the reduction ratio for the N···Br distance (0.63) is actually smaller than for the N···I distance (0.65) [68]. Similarly, with tertiary amines, the reduction ratio of 0.69 reported in the

NBS/DABCO adduct for the N···Br distance compares with that found in the NIS adduct of hexamethylenetetramine (0.72). A similar comparison also holds for the pyrazine adducts with NISac and NBSac, $(NISac)_2 \cdot$ pyrazine and $(NBSac)_2 \cdot$ pyrazine, with reduction of 0.69 for N···I and 0.71 for N···Br [62]. Again, a theoretical analysis of these evolutions would be highly profitable to rationalise these observations.

As seen above, the setting of an XB $B{\cdots}I_A{-}I_B$ between the Lewis base B and diiodine also leads to a lengthening of the $I_A{-}I_B$ bond, up to a point where it could even break to the $B{-}I^+, I^-$ species. This polarisation of the bond $I_A^{+\delta}{-}I_B^{-\delta}$ usually deactivates the dangling I_B atom from entering into a second XB as XB donor. Only recently, situations of very weak XB with halometallates as XB acceptors were reported by Ding et al. [69], where both ends of the I_2 molecule acted as XB donors. The example of the ditopic DIH molecule reported above in Section 10.3.2 has also shown the interdependence of the two iodine atoms upon XB formation. As a consequence of these polarisation effects, strong XB with a partial covalent character is limited to isolated molecular adducts, whereas the weaker normal XB obtained with di-, tri- or tetra-halogenated molecules (see Figure 10.1) are extensively used in crystal engineering for the elaboration of 1D, 2D or 3D networks. In this respect, the elaboration of extended networks based on charge-assisted XB with a partial covalent character would provide very robust systems with, for example, porosity.

Acknowledgement: Our own contribution to this topic owes a lot to the work of several colleagues in Rennes (F. Barrière, O. Jeannin, J. Lieffrig, O. Makhotkina and I. Nicolas) and in Nancy (E. Espinosa and E. Aubert). Their high competence and strong involvement are gratefully acknowledged. This work on halogen bonding was originally rooted in the development of molecular conductors involving halogen bonding interactions at the organic/inorganic interface, a work supported by ANR (France) under contract n° 08-BLAN-0091-02.

Bibliography

[1] Desiraju GR, Ho PS, Kloo L, Legon AC, Marquardt R, Metrangolo P, Politzer P, Resnati G, Rissanen K. Definition of the halogen bond (IUPAC Recommendations 2013). Pure Appl Chem 2013, 85, 1711–1713.

[2] Cavallo G, Metrangolo P, Milani R, Pilati T, Priimagi A, Resnati G, Terraneo G. The halogen bond. Chem Rev 2016, 116, 2478–2601.

[3] Fourmigué M. Halogen bonding: Recent advances. Curr Op Solid State Mater Sc 2009, 13, 36–45.

[4] Metrangolo P, Resnati G. Halogen bonding. Fundamentals and applications. Struct Bond vol. 126. Berlin, Heidelberg, Springer 2008.

[5] Metrangolo P, Resnati G. Halogen bonding I & II. Impact on materials chemistry and life sciences. Top Curr Chem, vols 358–359. Cham, Switzerland, Springer 2015.

[6] Triguero S, Llusar R, Polo V, Fourmigué M. Halogen bonding interactions of *sym*-triiodo-trifluorobenzene with halide anions, a combined structural and theoretical study. Cryst Growth Des, 2008, 8, 2241–2247.

[7] Metrangolo P, Pilati T, Terraneo G, Biella S, Resnati G. Anion coordination and anion-templated assembly under halogen bonding control. CrystEngComm 2009, 11, 1187–1196.

[8] Lieffrig J, Jeannin O, Fourmigué M. Expanded halogen-bonded anion organic networks with star-shaped iodoethynyl-substituted molecules: from corrugated 2D hexagonal lattices to pyrite-type two-fold interpenetrated cubic lattices. J Am Chem Soc 2013, 135, 6200–6210.

[9] Lieffrig J, Yamamoto HM, Kusamoto T, Cui H, Fourmigué M, Kato R. Halogen-bonded, eight-fold PtS-type interpenetrated supramolecular network. A study toward redundant and cross-bar supramolecular nanowire crystal. Cryst Growth Des 2011, 11, 4267–4271.

[10] Lindeman SV, Hecht J, Kochi JK. The charge-transfer motif in crystal engineering. Self-assembly of acentric (diamondoid) networks from halide salts and carbon tetrabromide as electron-donor/acceptor synthons. J Am Chem Soc 2003, 125, 11597–11606.

[11] Padgett CW, Walsh RD, Drake GW, Hanks TW, Pennington WT. New conformations and binding modes in halogen-bonded and ionic complexes of 2,3,5,6-Tetra(2′-pyridyl)pyrazine. Cryst Growth Des 2005, 5, 745–753.

[12] Walsh RB, Padgett CW, Metrangolo P, Resnati R, Hanks TW, Pennington WT. Crystal engineering through halogen bonding: complexes of nitrogen heterocycles with organic iodides. Cryst Growth Des 2001, 1, 165–175.

[13] Thomas LH, Jones AOF, Kallay AA, McIntyre GJ, Wilson CC. Engineering short, strong, charge-assisted hydrogen bonds in benzoic acid dimers through co-crystallization with proton sponge. Cryst. Growth Des. 2016, 16, 2112–2122.

[14] Lemmerer A, Govindraju S, Johnston M, Motloung X, Savig KL. Co-crystals and molecular salts of carboxylic acid/pyridine complexes: can calculated pK_a's predict proton transfer? A case study of nine complexes. CrystEngComm 2015, 17, 3591–3595.

[15] Childs SL, Stahly GP, Park A. The salt-co-crystal continuum: the influence of crystal structure on ionization state. Mol. Pharm 2007, 4, 323–338.

[16] Ward MD. Design of crystalline molecular networks with charge-assisted hydrogen bonds. Chem. Commun. 2005, 5838–5842.

[17] Ferlay S, Bulach V, Felix O, Hosseini MW, Planeix JM, Kyritsakas N. Molecular tectonics and supramolecular chirality: rational design of hybrid 1-D and 2-D H-bonded molecular networks based on bis-amidinium dication and metal cyanide anions. CrystEngComm 2002, 4, 447–453.

[18] Kosaka Y, Yamamoto HM, Nakao A, Tamura M, Kato R. Coexistence of conducting and magnetic electrons based on molecular π-electrons in the supramolecular conductor (Me-3,5-DIP)[Ni(dmit)$_2$]$_2$. J Am Chem Soc 2007, 129, 3054–3055.

[19] Kusamoto T, Yamamoto HM, Tajima N, Oshima Y, Yamashita S, Kato R. Bilayer Mott system based on Ni(dmit)$_2$ (dmit = 1,3-dithiole-2-thione-4,5-dithiolate) anion radicals: two isostructural salts exhibit contrasting magnetic behavior. Inorg Chem 2012, 51, 11645–11654.

[20] Serpell, CJ, Kilah, NL, Costa, PJ, Félix V, Beer, PD. Halogen bond anion templated assembly of an imidazolium pseudorotaxane. Angew Chem Int Ed 2010, 49, 5322–5326.

[21] Zapata F, Caballero A, White NG, Claridge TDW, Costa PJ, Félix V, Beer PD. Fluorescent charge-assisted halogen-bonding macrocyclic haloimidazolium receptors for anion recognition and sensing in aqueous media. J Am Chem Soc 2012, 134, 11533–11541.

[22] Walter SM, Kniep F, Rout L, Schmidtchen FP, Herdtweck E, Huber SM. Isothermal titration calorimetric titrations on charge-assisted halogen bonds: role of entropy, counterions, solvent, and temperature. J Am Chem Soc 2012, 134, 8507–8512.

[23] Kusamoto T, Yamamoto HM, Kato R. Utilization of σ-holes on sulfur and halogen atoms for supramolecular cation···anion interactions in bilayer Ni(dmit)$_2$ anion radical salts. Cryst Growth Des 2013, 13, 4533–4541.

[24] Kusamoto T, Yamamoto HM, Tajima N, Oshima Y, Yamashita S, Kato R. Bilayer Mott system with cation···anion supramolecular interactions based on a nickel dithiolene anion radical: coexistence of ferro- and antiferromagnetic anion layers and large negative magnetoresistance. Inorg Chem 2013, 52, 4759–4761.

[25] Fourmigué M, Batail P. Activation of hydrogen- and halogen-bonding interactions in tetrathiafulvalene-based crystalline molecular conductors. Chem Rev 2004, 104, 5379–5418.

[26] Cavallo G, Metrangolo P, Pilati T, Resnati G, Sansotera M, Terraneo G. Halogen bonding: a general route in anion recognition and coordination. Chem Soc Rev 2010, 39, 3772–3783.

[27] Puttreddy R, Jurček O, Bhowmik S, Mäkelä T, Rissanen K. Very strong $^{-}$N–X^{+}···$^{-}$O–N^{+} halogen bonds. Chem Commun 2016, 52, 2338–2341.

[28] Lieffrig J, Jeannin O, Frąckowiak A, Olejniczak I, Świetlik R, Dahaoui S, Aubert E, Espinosa E, Auban-Senzier P, Fourmigué M. Charge-assisted halogen bonding: donor-acceptor complexes with variable ionicity. Chem Eur J 2013, 19, 14804–14813.

[29] Frąckowiak A, Olejniczak I, Świetlik R, Jeannin O, Fourmigué M. Temperature induced neutral-ionic phase transition in the (EDT-TTF-I$_2$)$_2$TCNQF mixed-stack charge-transfer salt. J. Phys Chem C 2016, 120, 23740–23747.

[30] Lieffrig J, Jeannin O, Guizouarn T, Auban-Senzier P, Fourmigué M. Competition between C–H···N hydrogen bond and C–I···N halogen bond in TCNQF$_n$ (n = 0, 2, 4) salts with variable charge transfer. Cryst Growth Des 2012, 12, 4248–4257.

[31] Lieffrig J, Jeannin O, Shin KS, Auban-Senzier P, Fourmigué M. Halogen bonding interactions in DDQ charge transfer salts with iodinated TTFs. Crystals 2012, 2, 327–337.

[32] Legon AC. The interaction of dihalogens and hydrogen halides with Lewis bases in the gas phase: an experimental comparison of the halogen bond and the hydrogen bond. Struct Bond 2008, 126, 17–64.

[33] Bloemink HI, Evans CM, Holloway JH, Legon AC. Is the gas-phase complex of ammonia and chlorine monofluoride H$_3$N···ClF or [H$_3$NCl]$^+$···F$^-$? Evidence from rotational spectroscopy. Chem Phys Lett 1996, 248, 260–268.

[34] Bloemink HI, Holloway JH, Legon AC. Mulliken inner complexes [(CH$_3$)$_3$NCl]$^+$···F$^-$ in pre-reactive mixtures of trimethylamine and chlorine monofluoride: identification and characterisation by rotational spectroscopy. Chem Phys Lett 1996, 254, 59–68.

[35] Pennington WT, Hanks TW, Arman HD. Halogen bonding with dihalogens and interhalogens. Struct Bond 2008, 126, 65–104.

[36] Juárez-Pérez EJ, Aragoni MC, Arca M, Blake AJ, Devillanova FA, Garau A, Isaia F, Lippolis V, Núñez R, Pintus A, Wilson C. A unique case of oxidative addition of interhalogens IX (X=Cl, Br) to organodiselone ligands: nature of the chemical bonding in asymmetric I–Se–X polarised hypervalent systems. Chem Eur J 2011, 17, 11497–11514.

[37] Aragoni MC, Arca M, Devillanova FA, Garau A, Isaia F, Lippolis V, Verani G. Charge-transfer adducts between donors containing chalcogens (S and Se) and di-iodine: solution studies. Coord Chem Rev 1999, 184, 271–290.

[38] Apperley DC, Bricklebank N, Burns SL, Hibbs DE, Hursthouse MB, Malik KMA. Crystal structure of triphenylphosphine sulfide diiodine; the first crystallographically characterised 1~1 molecular charge-transfer complex of a tertiary phosphine sulfide with diiodine. J Chem Soc Dalton Trans 1998, 1289–1292.

[39] Arca I, Devillanova FA, Garau A, Isaia F, Lippolis V, Verani G, Demartin F. ^{31}P CP-MAS NMR, vibrational, and X-ray characterization of the adducts of triphenylphosphine sulfide with ICl and IBr. Z Anorg Allg Chem 1999, 624, 745–749.

[40] Skabara PJ, Berridge R, Bricklebank N, Lath H, Coles SJ, Horton PN. Self-assembly of halogen adducts of ester and carboxylic acid functionalised 1,3-dithiole-2-thiones. Polyhedron 2006, 25, 989–995.

[41] Koskinen L, Jääskeläinen S, Hirva P, Haukka M. Tunable interaction strength and nature of the S···Br halogen bonds in [(thione)Br_2] systems. Cryst Growth Des 2015, 15, 1160–1167.

[42] Koskinen L, Hirva P, Hasu A, Jääskeläinen S, Koivistoinen J, Petterson M, Haukka M. Modification of the supramolecular structure of [(thione)IY] (Y = Cl, Br) systems by cooperation of strong halogen bonds and hydrogen bonds. CrystEngComm 2015, 17, 2718–2727.

[43] Le Gal Y, Lorcy D, Jeannin J, Barrière F, Dorcet V, Lieffrig J, Fourmigué M. C=S···I halogen bonding interactions in crystalline iodinated dithiole-2-thiones and thiazole-2-thiones. CrystEngComm, 2016, 18, 5474–5481.

[44] Carlsson ACC, Veiga AX, Erdélyi M. Halogen bonding in solution. Top Curr Chem 2014, 359, 49–76.

[45] Carlsson ACC, Gräfenstein J, Budnjo A, Laurila JL, Bergquist J, Karim A, Kleinmaier R, Brath U, Erdélyi M. Symmetric halogen bonding is preferred in solution. J Am Chem Soc 2012, 134, 5706–5715.

[46] Carlsson ACC, Mehmeti K, Uhrbom M, Karim A, Bedin M, Puttreddy R, KleinmaierR, Neverov AA, Nekoueishahraki B, Gräfenstein J, Rissanen K, Erdélyi M. Substituent effects on the $[N–I–N]^+$ halogen bond. J Am Chem Soc 2016, 138, 9853–9863.

[47] Batsanov AS, Howard JAK, Lightfoot AP, Twiddle SJR, Whiting A. Stereoselective chloro-deboronation reactions induced by substituted pyridine-iodine chloride complexes. Eur J Org Chem 2005, 1876–1883.

[48] Soled S, Carpenter GB. The crystal structures of 2,2′-bipyridine.2ICl and 2,2′-bipyridine.2IBr. Acta Crystallogr B 1974, 30, 910–914

[49] Rømming C. Refinement of the crystal structure of the charge transfer compound pyridine–iodomonochloride. Acta Chem Scand 1972, 26, 1555–1560.

[50] Dahl T, Hassel O, Sky K. Structure of complexes formed by pyridine with cyanogen iodide and iodomonobromide. Acta Chem Scand 1967, 21, 592–593.

[51] Tuikka M, Haukka M. Crystal structure of the pyridine-diiodine (1/1) adduct. Acta Crystallogr 2015, E71, o463.

[52] Cotton FA, Kibala PA. Reactions of iodine with triphenylphosphine and triphenylarsine. J Am Chem Soc 1987, 109, 3308–3312.

[53] Pritchard RG, Moreland L. Solvent-free synthesis and crystal structure of $(Ph_3PI)I_5$, the third member in the series $Ph_3P(I_2)_n$ (n = 1, 2 and 3). Acta Crystallogr 2006, C62, o656-o658.

[54] Cross WI, Godfrey SM, McAuliffe CA, Pritchard RG, Sheffield JM, Thompson GM. The reaction of some tertiary phosphines with two mole equivalents of diiodine to produce the iodophosphonium triiodides $[R_3PI]I_3$; influence of R in causing subtle variations in solid state structures. J Chem Soc Dalton Trans 1999, 2795–2798.

[55] Alhanash FB, Barnes NA, Godfrey SM, Hurst PA, Hutchinson A, Khan RZ, Pritchard RG. Structural isomerism in tris-tolyl halo-phosphonium and halo-arsonium tri-halides, $[(CH_3C_6H_4)_3EX][X_3]$, (E = P, As; X = Br, I). Dalton Trans 2012, 41, 7708–7728.

[56] Teixidor F, Núñez R, Viñas C, Sillanpää R, Kivekäs R. The distinct effect of the o-carboranyl fragment: its influence on the I-I distance in R_3PI_2 complexes. Angew Chem Int Ed 2000, 39, 4290–4292.

[57] Barnes NA, Godfrey SM, Halton RTA, Mushtaq I, Pritchard RG. The reaction of tertiary phosphines with $(Ph_2Se_2I_2)_2$ – the influence of steric and electronic effects. Dalton Trans 2006, 4795–4804.

[58] Barnes NA, Godfrey SM, Khan RZ, Pierce A, Pritchard RG. A structural and spectroscopic study of tris-aryl substituted R_3PI_2 adducts. Polyhedron 2012, 35, 31–46.

[59] Godfrey SM, Kelly DG, McAuliffe CA, Mackie AG, Pritchard RG, Watson SM. The structure of triphenylphosphorus–diiodine, Ph_3PI_2, the first crystallographically characterised dihalogen derivative of a tertiary phosphine. J Chem Soc Chem Commun 1991, 1163–1164.
[60] Bricklebank N, Godfrey SM, Lane HP, McAuliffe CA, Pritchard RG, Moreno JM. The isolation from diethyl ether of ionic $[(Me_2N)_3PI]I$ and $[(CH_2{=}CHCH_2)_2PhPI]I$, and the crystallographically characterised molecular 'spoke' structure $PhMe_2PI_2$. J Chem Soc Dalton Trans 1995, 2421–2424.
[61] Barnes NA, Godfrey SM, Halton RTA, Mushtaq I, Pritchard RG. The reactions of alkylamino substituted phosphines with I_2 and $(Ph_2Se_2I_2)_2$: structural features of alkylamino phosphonium cations. Dalton Trans 2008, 1346–1354.
[62] Troff RW, Mäkelä T, Topić F, Valkonen A, Raatikainen K, Rissanen K. Alternative motifs for halogen bonding. Eur J Org. Chem 2013, 1617–1637.
[63] Raatikainen K, Rissanen K. Interaction between amines and N-haloimides: a new motif for unprecedentedly short Br···N and I···N halogen bonds. CrystEngComm 2011, 13, 6972–6977.
[64] Dolenc D, Modec B. EDA complexes of N-halosaccharins with N- and O-donor ligands. New J Chem 2009, 33, 2344–2349.
[65] Svensson C. Albertsson J. Eberson L. Structure of tetraethylammonium mono(N-bromosuccinimide) bromate(1–). Acta Chem Scand B1988, 42, 596–600.
[66] Makhotkina O, Lieffrig J, Jeannin O, Fourmigué M, Aubert E, Espinosa E. Cocrystal or salt: solid state-controlled iodine shift in crystalline halogen-bonded systems. Cryst Growth Des 2015, 15, 3464–3473.
[67] Nicolas I, Barrière F, Jeannin O, Fourmigué M. Sequential halogen bonding with ditopic donors: σ hole evolutions upon halogen bond formation. Cryst Growth Des 2016, 16, 2963–2971.
[68] Rissanen K, Haukka M. Halonium ions as halogen bond donors in the solid state $[XL_2]Y$ complexes. Top Curr Chem 2015, 359, 77–90.
[69] Ding X, Tuikka MJ, Hirva P, Kukushkin VY, Novikov AS, Haukka M. Fine-tuning halogen bonding properties of diiodine through halogen–halogen charge transfer–extended $[Ru(2,2'\text{-}bipyridine)(CO)_2X_2]\cdot I_2$ systems (X = Cl, Br, I). CrystEngComm 2016, 18, 1987–1995.

Dušan P. Malenov, Ivana S. Antonijević, and Snežana D. Zarić

11 Large horizontal displacements of benzene–benzene stacking interactions in co-crystals

11.1 Introduction

Aromatic non-covalent interactions are ubiquitous in chemical and biological systems, playing crucial roles in the processes of molecular recognition, catalysis and transport [1–9]. They are of great importance for the structure and the physical and chemical properties of the systems involved in these processes. Another important field that uses the knowledge of aromatic interactions is crystal engineering [10]. The formation and spatial distribution of interactions of aromatic moieties in co-crystals can lead to desirable properties [11–16], which enables their application in different areas [17, 18].

The prototype for studying aromatic interactions is the benzene dimer, which has two low-energy geometries [19–29]. In the first, the C–H group of one benzene is pointed towards the π-system of the other – the dimer with C–H/π interaction is often regarded as T-shaped (Figure 11.1) and has an interaction energy of −2.84 kcal/mol [24]. In the second geometry, two benzene molecules are parallel with their centres horizontally displaced at ~1.5 Å – the parallel interaction formed in this dimer is known as a stacking interaction (Figure 11.1) and has an energy −2.73 kcal/mol [24]. Stacking

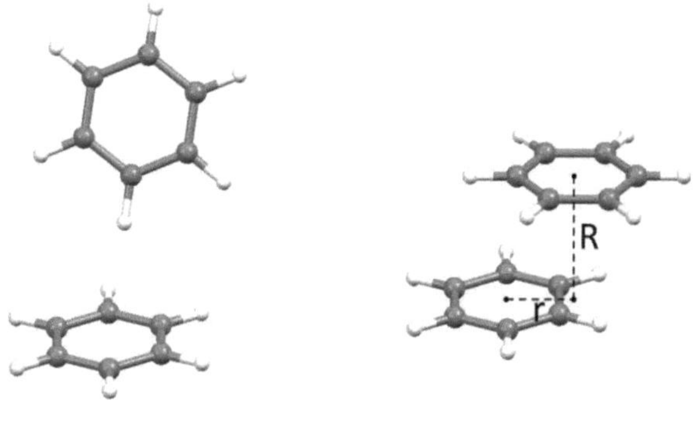

Fig. 11.1: Two low-energy geometries of the benzene dimer. Stacking interactions are characterised by normal distance R between the planes of interacting benzene molecules, and the horizontal displacement (offset) r, which represents the distance of the centre of one benzene to the projection of the centre of the other benzene onto the plane of the first benzene molecule.

https://doi.org/10.1515/9783110464955-011

interactions are of particular importance owing to their major roles in protein structure [30, 31], protein–ligand recognition [32, 33] and nucleic acid structure [34, 35].

Most studies of stacking interactions between aromatic moieties were, until recently, focused upon interactions that have small horizontal displacements [19–29]. However, in the past few years, several studies have been published showing substantial interaction energies of aromatic molecules at large horizontal displacements [28, 29, 36–39]. The topic of this chapter is parallel-stacking interactions between benzene molecules in co-crystals. The review deals with preferred geometries of stacking benzene interactions in co-crystal structures extracted from the Cambridge Structural Database (CSD), the energies of these interactions and their supramolecular assemblies.

11.2 Parallel-stacking interactions between benzenes in co-crystals from the CSD

11.2.1 Parallel-stacking interactions between non-coordinating benzenes

The CSD November 2015 release (version 5.37) [40] was subjected to a search for parallel benzene–benzene contacts [28]. The restrictions applied regarding the crystal structures included in the search were:

- Only co-crystal structures (i.e. structures containing at least two different chemical species) were included
- No polymer structures were included
- Structures determined from powder data were excluded
- The crystallographic R factor was less than 10%
- Only structures with error-free coordinates according to the CSD criteria were used
- H atom positions were normalised according to the default CSD lengths of X–H bonds

To ensure that the benzenes do not act as ligands to transition metals, each of their carbon atoms was set to be bonded exactly to three other atoms (two with carbon and one with hydrogen). Two such defined benzenes were considered to form a parallel-stacking interaction if the angle between their mean planes is less than 10° and their centres are within the ellipsoid defined by horizontal displacement of 7.5 Å and a normal distance of 4.0 Å (Figure 11.2). The search performed using these criteria yielded 1,294 stacking benzene–benzene interactions [28].

Relative to the benzene ring centre, the centres of mass of the carbon atoms are at distances of approximately 1.4 Å and centres of mass of hydrogen atoms are at distances of approximately 2.5 Å. According to these distances, benzene can be categorised into three regions – the ring region, the C–H bond region and the region beyond the C–H bonds (Figure 11.3a). However, if van der Waals radii are taken into ac-

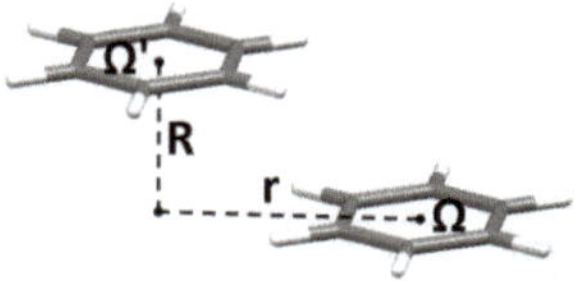

Fig. 11.2: The model system used for the search of parallel-stacking interactions between non-coordinating benzenes in co-crystals extracted from the Cambridge Structural Database (CSD); Ω and Ω′ are the centres (centroids) of interacting benzene molecules; r denotes horizontal displacement of these centres, whereas R denotes normal distance between the planes of the benzenes (see Figure 11.1).

count [41], which is a more relevant approach in the context of non-covalent interactions, both the ring and C–H bond regions are extended (Figure 11.3b), the latter providing a greater surface for potential overlapping of the molecules. If two parallel benzenes overlap with their ring regions, stacking interactions (at small horizontal displacement) are formed and the horizontal displacement is in the range r = 0.0–4.5 Å. If two parallel benzenes overlap only with their C–H bond regions, stacking interactions at large horizontal displacement (r = 4.5–7.5 Å) are formed.

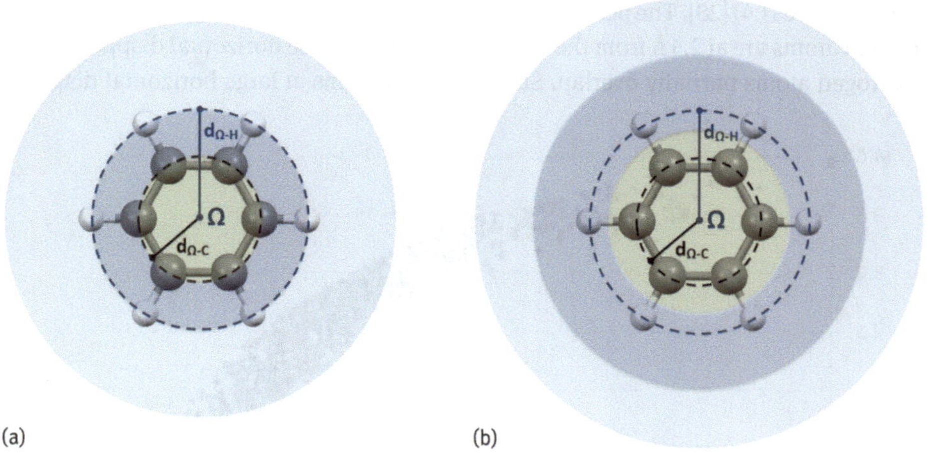

Fig. 11.3: Three regions of a benzene molecule – the ring region (*grey*), the C–H bond region (*blue*) and the region beyond the C–H bonds (*light blue*), taking into account only (a) the centres of the mass or (b) the van der Waals radii of atoms; Ω represents the ring centre; the marked distances are $d_{Ω\text{–}C} \approx 1.4$ Å and $d_{Ω\text{–}H} \approx 2.5$ Å.

The distribution of offset values for stacking contacts between two non-coordinating benzenes in co-crystals from the CSD shows that the vast majority of parallel interactions are at large horizontal displacements. Horizontal displacements greater than 4.5 Å are present in 1,097 of 1,294 contacts, which corresponds to 85% of all interac-

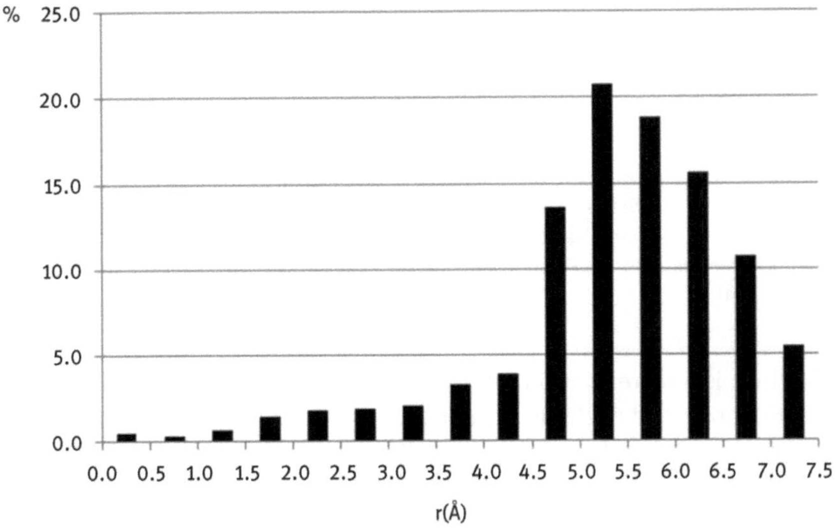

Fig. 11.4: The distribution of horizontal displacement (*r*) values for parallel-stacking benzene–benzene interactions found in co-crystals extracted from the CSD.

tions (Figure 11.4) [28]. The peak of this distribution is within the range 5.0–5.5 Å. As hydrogen atoms are at 2.5 Å from the benzene centre, at these horizontal displacements, hydrogen atoms partially overlap. Stacking interactions at large horizontal displace-

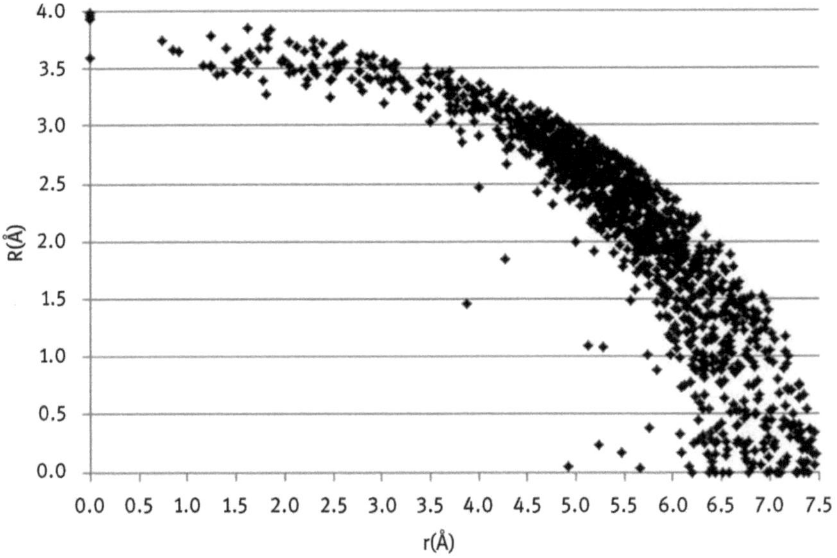

Fig. 11.5: Normal distance (*R*) versus horizontal displacement (*r*) plot for parallel-stacking benzene–benzene interactions in co-crystals from the CSD.

ments prevail [28]. Only 26 interactions (2% of all interactions) were found within the range 1.0–2.0 Å (Figure 11.4), which had been considered typical for stacking interactions [24] and features the geometry of one of the low-energy benzene dimers (Figure 11.1).

For offset values below 3.0 Å, the normal distances between interacting stacking benzenes are in the range 3.0–4.0 Å, with the majority being around 3.5 Å (Figure 11.5), which is typical for stacking interactions. However, at large horizontal displacements ($r > 4.5$ Å), normal distances are mostly below 3.0 Å, and decrease with the increase in horizontal displacements (Figure 11.5) [28].

11.2.2 Parallel-stacking interactions between coordinating benzenes

To study the effect of transition metal coordination upon parallel-stacking interactions of benzene, the CSD November 2015 (version 5.37) [40] was searched for contacts between two (η^6) coordinating benzene molecules with parallel orientation (Figure 11.6) [39]. Coordinating benzene was defined by setting all carbons to be bonded to exactly four atoms, three of them being within benzene (two carbons and one hydrogen) and one of them being transition metal. The crystal structure criteria and geometric criteria for this search were the same as for the previous search described in Section 11.2.1. Two coordinating benzenes were considered, forming a parallel-stacking interaction if the angle between their mean planes measures less than 10° and their centres are within the area corresponding to the ellipsoid defined by horizontal displacement $r = 7.5$ Å and normal distance $R = 4.0$ Å (Figure 11.6).

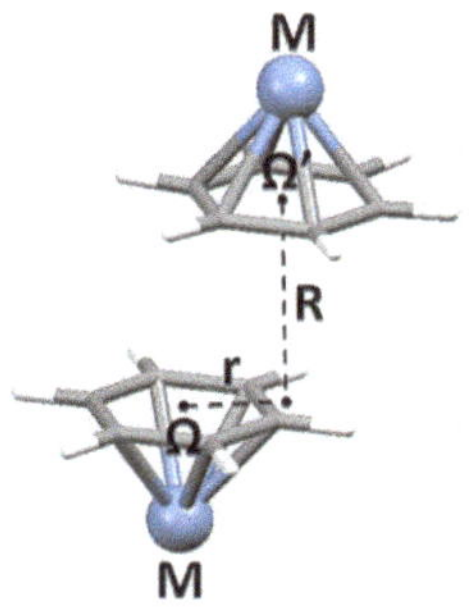

Fig. 11.6: The model system used for the search of parallel-stacking interactions between coordinating benzenes in co-crystals extracted from the CSD. *M* denotes transition metal coordinated by benzene, Ω and Ω′ are the centres of the interacting coordinating benzenes; *r* denotes horizontal displacement of these centres, whereas *R* denotes normal distance between the planes of the benzene molecules.

The search performed by these criteria yielded 150 parallel stacking interactions between coordinating benzenes [39]. Organometallic compounds containing benzene can be divided into sandwich compounds, which contain benzene and only one other planar ligand parallel to the benzene (Figure 11.7a), and half-sandwich compounds, which contain benzene and at least one other ligand not parallel to the benzene (Figure 11.7b). A total of 111 interactions were found between benzenes in half-sandwich compounds, whereas there were 39 parallel interactions between benzenes of sand-

Fig. 11.7: Molecular structures of (a) (η^6-benzene)-(η^6-methyl-2-methylbenzoate)-ruthenium [42], representing sandwich compounds, and (b) and (η^6-benzene)-bis(trichlorosilyl)-dihydrido-iron(IV) [43], representing half-sandwich compounds.

wich compounds [39]. The other ligand in almost all sandwich compounds studied in this search was also benzene.

The distribution of horizontal displacement values for sandwich–sandwich contacts shows that molecules prefer interactions at large horizontal displacements (Figure 11.8), as 72% of all contacts have offset values greater than 4.5 Å [39]. The percentage of parallel interactions at large horizontal displacements for half-sandwich–half-sandwich interactions is 50%, which is lower than both benzene–benzene and sandwich–sandwich contacts [28, 39]. The peak of the horizontal displacement distribution for coordinating benzenes is within the range 6.0–6.5 Å, both for sandwich and half-sandwich compounds (Figure 11.8), which is higher than for non-coordinating benzenes.

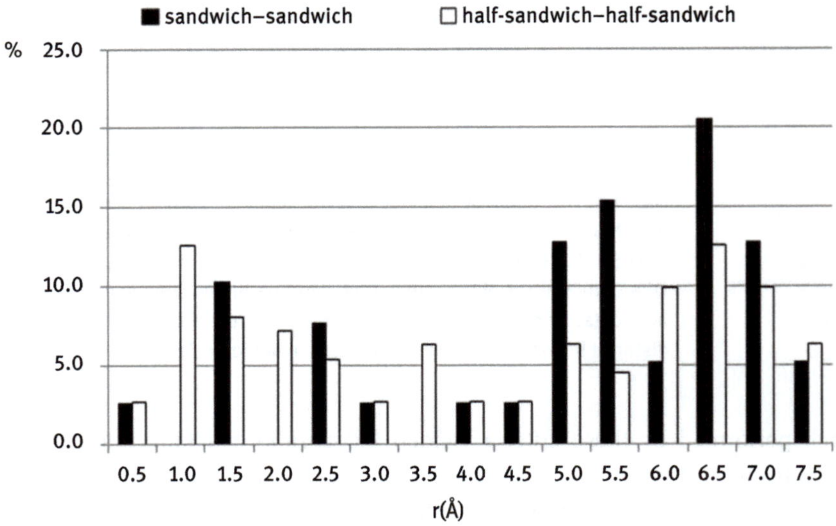

Fig. 11.8: The distribution of horizontal displacement (r) values for parallel-stacking interactions between coordinating benzenes found in co-crystals extracted from the CSD.

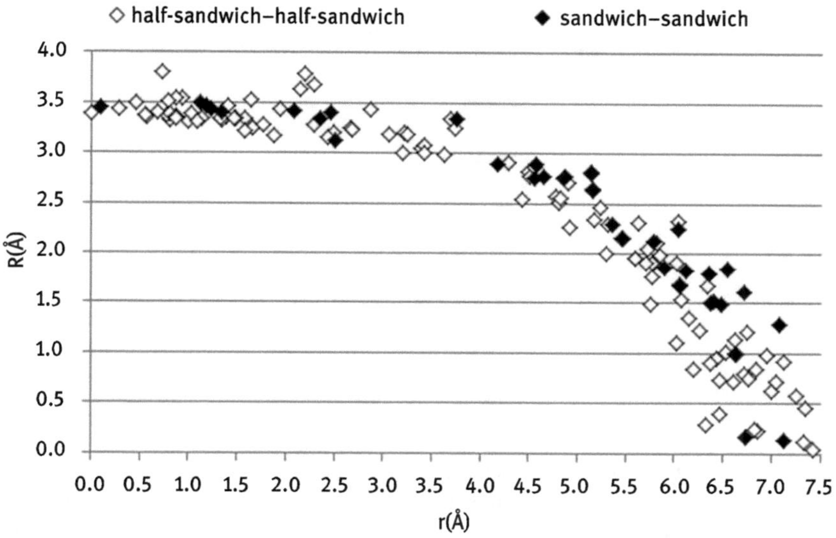

Fig. 11.9: Normal distance (*R*) versus horizontal displacement (*r*) plot for parallel-stacking interactions between coordinating benzenes in co-crystals extracted from the CSD.

The normal distance for the majority of parallel interactions between coordinating benzenes is below 3.5 Å (Figure 11.9). For all the contacts with horizontal displacements greater than 4.0 Å, normal distances are less than 3.0 Å, whereas for horizontal displacements greater than 5.5 Å, normal distances are mostly below 2.0 Å. The trend for coordinating benzenes is similar to that for non-coordinating benzenes – normal distances decrease with an increase in horizontal displacements (Figure 11.9) [39].

11.3 Energies of parallel-stacking interactions of non-coordinating and coordinating benzenes

To find interaction energies for benzene–benzene, half-sandwich–half-sandwich and sandwich–sandwich stacking interactions, calculations on model systems were performed [28, 39]. Along with the (non-coordinating) benzene molecule (Figure 11.10a), two model systems were employed for coordinating benzenes: staggered (benzene)tricarbonylchromium as an example of a half-sandwich compound (Figure 11.10b), and eclipsed, bis(benzene)chromium for sandwich compounds (Figure 11.10c). The model molecules were fully optimised by using the Tao–Perdew–Staroverov–Scuseria (TPSS) density functional [44] with a D2 dispersion correction [45] and the def2-TZVP (triple zeta valence polarized) basis set [46]. The potential surfaces for these interactions were calculated by keeping the monomer geometries rigid and by varying the normal distances for the series of offset values (0.0–6.0 Å) (Figure 11.10).

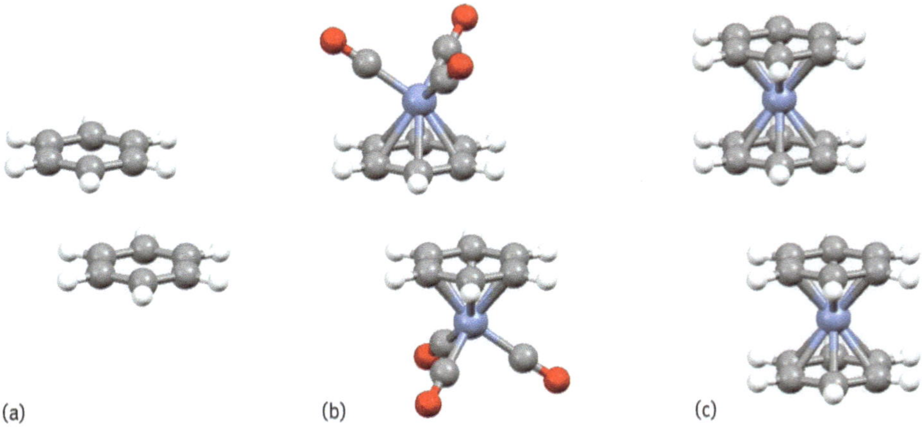

Fig. 11.10: Model systems for dispersion-corrected density functional theory calculations of stacking interaction energies between (a) non-coordinating benzenes, (b) coordinating benzenes of half-sandwich compounds and (c) coordinating benzenes of sandwich compounds. The model molecules are (a) benzene, (b) (benzene)tricarbonylchromium and (c) bis(benzene)chromium. The geometries with horizontal displacements of r = 1.5 Å are presented.

All calculations were performed using the ORCA program (version 2.9.1) [47]. The CCSD(T)/CBS method [48], which is considered the gold standard for quantum chemistry [49], is very demanding in terms of computational resources and costs, and thus cannot be used for calculations on large systems, particularly containing transition metals. It was shown that the TPSS density functional with a D2 dispersion correction, and the def2-TZVP basis set without BSSE correction (basis set superposition error) reproduces CCSD(T)/CBS interaction energies not only for benzene–benzene systems [50], but also for the benzene–half-sandwich systems [51]. Therefore, the TPSS-D2/def2-TZVP method was used for the calculation of potential energy surfaces for benzene–benzene, half-sandwich–half-sandwich and sandwich–sandwich interactions [39].

The calculations show that for all model systems the strongest interaction is formed at $r = 1.5$ Å (Figure 11.11a), which is a typical stacking interaction. The interaction between coordinating benzenes (−3.69 kcal/mol for sandwich–sandwich and −3.29 kcal/mol for half-sandwich–half-sandwich) is greater than the interaction between non-coordinating benzene molecules (−2.76 kcal/mol, Table 11.1) [39]. However, at greater horizontal displacements, interactions between non-coordinating benzenes are remarkably strong, with an interaction energy of −2.05 kcal/mol at $r = 5.0$ Å, which is 74% of the greatest interaction energy in the parallel benzene dimer system (Table 11.1) [28, 39]. Moreover, the potential energy curve is very flat in the benzene–benzene system in the range of $r = 3.5$–5.0 Å (Figure 11.11a) [28]. An even higher percentage of interaction energy is preserved in sandwich–sandwich systems, i.e. −3.03 kcal/mol at $r = 5.0$ Å [39]. This is not only a very large percentage

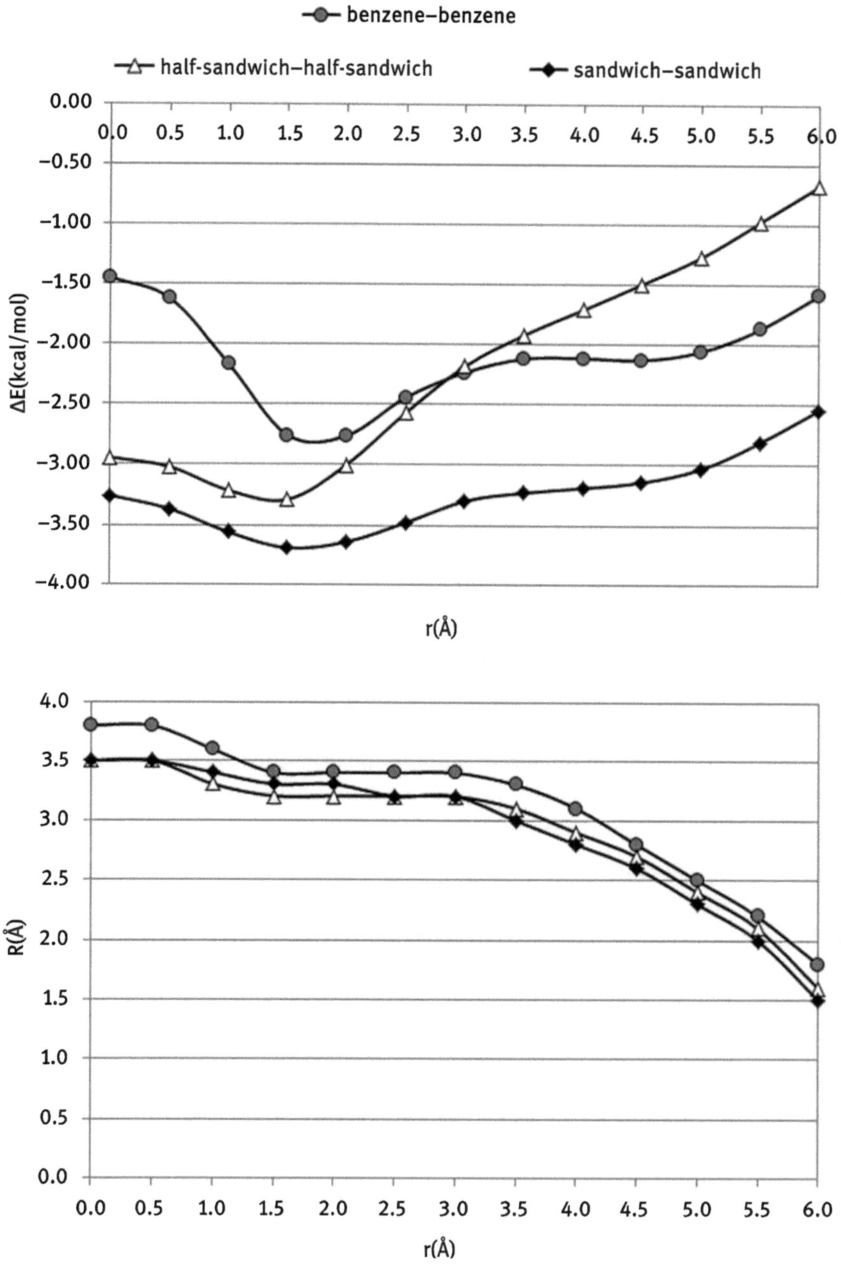

Fig. 11.11: (a) The potential energy curves and (b) the optimal normal distances at various offset values for benzene–benzene, half-sandwich–half-sandwich and sandwich–sandwich interactions calculated at the TPSS-D2/def2-TZVP level. The interaction energies for each offset value r were calculated by varying the normal distance (R) between two molecules in a series of single-point calculations. The greatest calculated energy for each offset value is presented.

of the strongest interaction energy (82%), but also a stronger interaction than the most stable stacking of non-coordinating benzenes (−2.76 kcal/mol, Table 11.1). In contrast to benzene–benzene and sandwich–sandwich systems, interactions at large horizontal displacements in half-sandwich–half-sandwich systems are not as strong, as from $r = 3.0$ Å the interaction energy decreases linearly (Figure 11.11a), being only −1.27 kcal/mol at $r = 5.0$ Å, which is 39% of the strongest interaction energy for this system (Table 11.1) [39].

Normal distances for all systems decrease with increasing horizontal displacements (Figure 11.11b). These distances are greatest for the benzene-benzene system at all offset values. At stacking offsets, normal distances are the smallest for half-sandwich–half-sandwich systems, whereas at large horizontal displacements the benzenes of two bis(benzene)chromium sandwich molecules are the closest [39].

11.4 Electrostatic potentials of coordinating and non-coordinating benzenes and their influence on parallel-stacking interactions

Electrostatic potential maps of benzene, (benzene)tricarbonylchromium and bis(benzene)chromium [52] can explain the peculiar energies of their parallel-stacking interactions at large horizontal displacements. Non-coordinating benzene and benzene of bis(benzene)chromium have very similar electrostatic potentials, which shift from negative at the face (above the ring) to positive at the edges (around hydrogen atoms), forming an electrostatic potential gradient (Figure 11.12a and c) [39]. At large horizontal displacements, two molecules are positioned so that these gradients are overlapping by their opposite ends so that areas of positive potential interact with areas of negative potential, forming remarkably strong interactions (Figure 11.11a and Table 11.1). The electrostatic potential around the benzene of (benzene)tricarbonylchromium is entirely positive (Figure 11.12b) [39], as electron density is withdrawn by chromium and the carbonyl ligands. Therefore, at large horizontal displacements two molecules interact via their positive potentials and the interaction is thus not as strong as in the other systems (Figure 11.11a and Table 11.1). This

Tab. 11.1: Energies of parallel-stacking interactions between benzenes calculated at the TPSS-D2/def2-TZVP level. The percentage values indicate the strongest interaction energy (at $r = 1.5$ Å).

	$r = 1.5$ Å ΔE (kcal/mol)	$r = 5.0$ Å ΔE (kcal/mol)	Percentage
Benzene–benzene	−2.76	−2.05	74
Half-sandwich–half-sandwich	−3.29	−1.27	39
Sandwich–sandwich	−3.69	−3.03	82

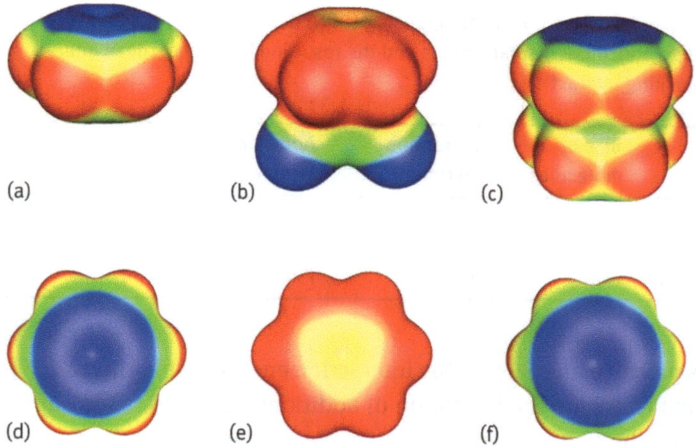

Fig. 11.12: Two views of electrostatic potential maps of (a) benzene, (b) (benzene)tricarbonylchromium and (c) bis(benzene)chromium calculated at the TPSS-D2/def2-TZVP level and plotted at the outer contour of electron density. This has the value 0.001 a.u. The electrostatic potential increases from *blue* and *green* to *yellow* and *red*.

is supported by the calculation of the electrostatic energy at offsets, $r = 5.0$ Å, performed at the MP2/6-311G(d,2p) level, which is attractive for sandwich–sandwich (−0.95 kcal/mol) and benzene–benzene (−0.81 kcal/mol) systems, and repulsive for the half-sandwich–half-sandwich system (+0.32 kcal/mol) [39].

11.5 Correlation of interaction geometries in co-crystals and the effect of supramolecular structures upon calculated interaction geometries and energies

The results of DFT calculations showed that both coordinating and non-coordinating benzenes form the strongest stacking interactions at small horizontal displacements ($r = 1.5$ Å) (Section 11.3) [28, 39]. However, stacking interactions at small horizontal displacements between benzenes in co-crystals are only significant if benzene is part of a half-sandwich compound (Figure 11.8, Section 11.2.2) [39]. Non-coordinating benzenes and benzenes of sandwich compounds prefer to form parallel-stacking interactions at large horizontal displacements in co-crystals (Figures 11.4 and 11.8, Section 11.2), which are weaker than stacking interactions at small displacements (Table 11.1) [28, 39]. One of the reasons for this observation is the relative strength of these interactions in comparison to stacking interactions at small offsets (Table 11.2). Namely, by shifting from small ($r = 1.5$ Å) to large horizontal displacements ($r = 5.0$ Å), parallel benzenes of sandwich compounds preserve 82% of interaction energy, non-coordinating benzenes preserve 74%, whereas benzenes of half-

Tab. 11.2: A summary of factors favouring (*PRO*) and opposing (*CON*) benzene–benzene parallel-stacking interactions at large horizontal displacements and the resulting fraction of these interactions among all parallel-stacking interactions.

System	Size of molecules	Interaction energies	Additional interactions	Percentage of all interactions
Benzene–benzene	PRO	PRO	PRO	85
Sandwich–sandwich	CON	PRO	PRO	72
Half-sandwich–half-sandwich	CON	CON	PRO	50

sandwich compounds preserve only 39% of the interaction energy (Table 11.1) [28, 39]. The other two defining factors are the formation of additional interactions in supramolecular structures and the size of interacting molecules (Table 11.2).

The possibilities for additional interactions can be analysed in the crystal structures containing parallel-stacking interactions. Both benzene molecules forming stacking interaction at small horizontal displacement have one of their faces involved in this interaction, thus preventing the formation of any additional interactions (Figure 11.1). However, if an interaction at a large horizontal displacement is formed, both faces can simultaneously form additional interactions with other constituents of the co-crystal (Figure 11.13). An example of this phenomenon is found in the co-crystal formed by syncarpurea and benzene [53], where benzenes forming parallel-stacking interactions at large horizontal displacements form additional interactions with benzene and syncarpurea molecules through C–H/π interactions (Figure 11.13).

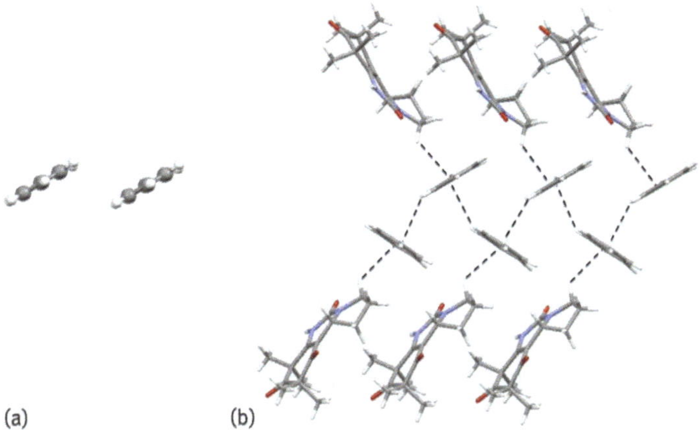

(a) (b)

Fig. 11.13: (a) Benzene–benzene stacking interaction with a horizontal displacement of r = 5.10 Å in the co-crystal of benzene and syncarpurea and (b) extended structure showing that, owing to large horizontal displacement of stacked benzenes, one benzene forms an additional aromatic C–H/π interaction with a neighbouring benzene molecule, whereas the other forms an additional C–H/π interaction with one of the syncarpurea C–H groups.

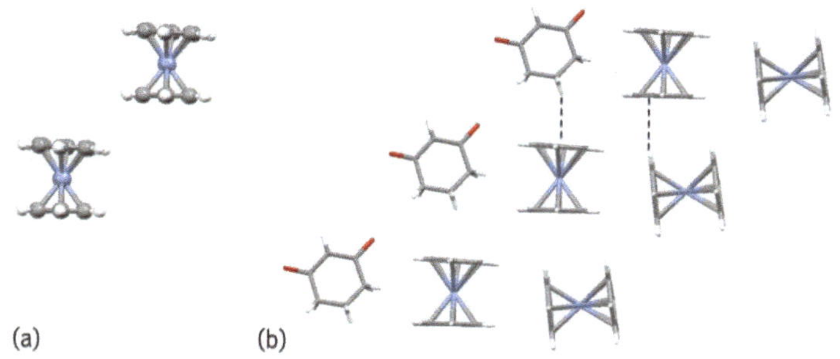

Fig. 11.14: (a) Parallel-stacking interactions between coordinating benzenes at horizontal displacement of $r = 5.083$ Å in the co-crystal of bis(benzene)chromium and 1,3-cyclohexanedionate and (b) voluminous molecules in the co-crystal fill the cavity created by voluminous sandwich compounds and form additional interactions that provide the molecular packing with further stabilisation.

Hence, forming a parallel-stacking interaction at large instead of small offset causes the decrease in interaction energy by 26% (Table 11.1), but enables the formation of additional interactions that provide additional stabilisation to the supramolecular structure of the co-crystal (Table 11.2).

A similar effect arising from additional interactions to those found for non-coordinating benzene can be expected for benzenes of sandwich and half-sandwich compounds (Table 11.2). However, these compounds are significantly more voluminous than non-coordinating benzene, a factor that can influence the packing and the formation of certain interactions. It is for this reason that, in spite of a large fraction of interaction energy being preserved for interactions of benzenes of sandwich compounds (82%) in comparison with non-coordinating benzenes (74%), stacking interactions at large offsets occur less frequently in case of benzenes of sandwich compounds (Table 11.2). In the co-crystal containing bis(benzene)chromium and 1,3-cyclohexanedionate [54], parallel-stacking interactions at large horizontal displacements are formed, leaving non-coordinating faces of both benzenes available for the formation of additional interactions (Figure 11.14a). The large cavities formed by this arrangement can only be filled by large molecules such as other bis(benzene)chromium or 1,3-cyclohexanedionate molecules (Figure 11.14b), which additionally stabilise the system by forming additional interactions. However, if the other molecule in the co-crystal is not as voluminous as a sandwich/half-sandwich compound, the cavity created by two voluminous sandwich/half-sandwich compounds at large horizontal displacement cannot be filled to ensure efficient packing. Therefore, in the co-crystal containing bis(benzene)chromium and methyl-chromate, a co-former less voluminous than 1,3-cyclohexanedionate, stacking interactions are formed at small instead of large horizontal displacements (Figure 11.15) [55].

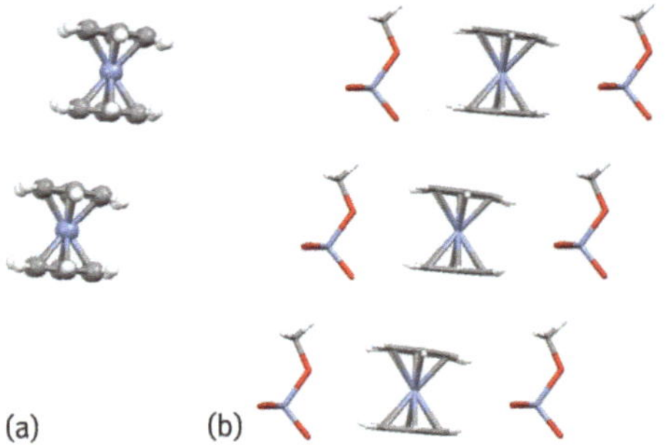

Fig. 11.15: (a) Parallel-stacking interactions between coordinating benzenes at a horizontal displacement of $r = 1.19$ Å in the co-crystal of bis(benzene)chromium and methyl-chromate and (b) when the cavity is formed by voluminous sandwich compounds at large horizontal displacement, methyl-chromate which is not so voluminous cannot fill it, so the stacking interaction at small horizontal displacement is formed instead.

11.6 Summary

Should we expect small or large stacking offsets between benzenes in co-crystals? The answer depends on whether benzene coordinates transition metal (and how) or not. Among the factors influencing parallel-stacking interactions, the only one strongly favouring large horizontal displacements, regardless of the type of benzene, is the possibility of forming additional supramolecular interactions. Non-coordinating benzenes, benzenes of half-sandwich and sandwich compounds can all form simultaneous interactions if their faces are available. Quantum chemical calculations indicate that stacking interactions at large horizontal displacements are strong for non-coordinating benzenes (74% of the strongest interaction energy) and benzenes of sandwich compounds (82%), but they are not strong for benzenes of half-sandwich compounds (only 39%). In addition, very large sandwich and half-sandwich compounds must have accordingly large fellow co-formers in their co-crystals to achieve efficient packing and to form parallel-stacking interactions at large horizontal displacements. All of the named factors are in favour of stacking interactions at large horizontal displacements between non-coordinating benzene molecules, and these interactions constitute a remarkable 85% of all stacking interactions between non-coordinating benzenes in co-crystals. The share of stacking interactions at large horizontal displacements is smaller for benzenes of sandwich compounds (72%), in spite of favourable interaction energy (82% of the strongest energy) and the possibility for additional interactions, as molecular size can oppose their formation. Finally, the lowest share of these interactions is present in contacts between benzenes of half-sandwich compounds (50%), as the only factor favouring them is the possibility of forming additional interactions, whereas energies (only 39% of the strongest energy) and size are not favourable.

The observed preferences for parallel-stacking interactions at large horizontal displacements in co-crystals, calculated interaction energies and rich supramolecular structures emerging from them all show that these interactions must be seriously considered along with classical stacking interactions at small horizontal displacements in understanding the organisation of systems containing aromatic moieties. These interactions are of particular interest in the field of crystal engineering and their properties can be exploited in the area of materials design.

Acknowledgement: This work was supported by the Ministry of Education, Science and Technological Development of the Republic of Serbia (grant no. 172065).

Bibliography

[1] Salonen LM, Ellermann M, Diederich F. Aromatic rings in chemical and biological recognition: energetics and structures. Angew Chem Int Ed 2011, 50, 4808–4842.

[2] Persch E, Dumele O, Diederich F. Molecular recognition in chemical and biological systems. Angew Chem Int Ed 2015, 54, 3290–3327.

[3] Bissantz C, Kuhn B, Stahl M. A medicinal chemist's guide to molecular interactions. J Med Chem 2010, 53, 5061–5084.

[4] Nlung M, Kuang Y, Gao Y, Zhang Y, et al. Aromatic-aromatic interactions induce the self-assembly of pentapeptidic derivatives in water to form nanofibers and supramolecular hydrogels. J Am Chem Soc 2010, 132, 2719–2728.

[5] Schneider HJ. Binding mechanisms in supramolecular complexes. Angew Chem Int Ed 2009, 48, 3924–3977.

[6] Motherwell WB, Moïse J, Aliev AE, Nič M, et al. Noncovalent functional-group-arene interactions. Angew Chem Int Ed 2007, 46, 7823–7826.

[7] Kang YK, Iovine PM, Therien MJ. Electron transfer reactions of rigid, cofacially compressed, π-stacked porphyrin-bridge-quinone systems. Coord Chem Rev 2011, 255, 804–824.

[8] Tsuzuki S. In: Intermolecular forces and clusters I, vol. 115, Springer, Berlin/Heidelberg 2005, pp. 149–193.

[9] Robson H, Fraaije JGEM, Kros A. Introducing quadrupole interactions into the peptide design toolkit. Angew Chem Int Ed 2010, 49, 8570–8572.

[10] Desiraju GR. Crystal engineering: a holistic view. Angew Chem Int Ed 2007, 46, 8342–8356.

[11] Ghosh S, Reddy CM. Elastic and bendable caffeine co-crystals: implications for the design of flexible organic materials. Angew Chem Int Ed 2012, 51, 10319–10323.

[12] Reddy CM, Padmanabhan KA, Desiraju GR. Structure-property correlations in bending and brittle organic crystals. Cryst Growth Des 2006, 6, 2720–2731.

[13] Reddy CM, Gundakaram RC, Basavoju S, Kirchner MT, et al. Structural basis for bending of organic crystals. Chem Commun 2005, pp. 3945–3947.

[14] Ghosh S, Malla Reddy C. Co-crystals of caffeine with substituted nitroanilines and nitrobenzoic acids: structure–mechanical property and thermal studies. CrystEngComm 2012, 14, 2444.

[15] Reddy CM, Rama Krishna G, Ghosh S. Mechanical properties of molecular crystals – applications to crystal engineering. CrystEngComm 2010, 12, 2296.

[16] Reddy CM, Kirchner MT, Gundakaram RC, Padmanabhan KA, Desiraju GR. Isostructurality, polymorphism and mechanical properties of some hexahalogenated benzenes: The nature of halogen···halogen interactions. Chem Eur J 2006, 12, 2222–2234.

[17] Terao F, Morimoto M, Irie M. Light-driven molecular-crystal actuators: rapid and reversible bending of rodlike mixed crystals of diarylethene derivatives. Angew Chem Int Ed 2012, 51, 901–904.

[18] Naumov P, Kowalik J, Solntsev KM, Baldridge A, et al. Topochemistry and photomechanical effects in crystals of green fluorescent protein-like chromophores: effects of hydrogen bonding and crystal packing. J Am Chem Soc 2010, 132, 5845–5857.

[19] Pitoňák M, Neogrády P, Řezáč J, Jurečka P, et al. Benzene dimer: high-level wave function and density functional theory calculations. J Chem Theory Comput 2008, 4, 1829–1834.
[20] Bludský O, Rubeš M, Soldán P, Nachtigall P. Investigation of the benzene-dimer potential energy surface: DFT/CCSD(T) correction scheme. J Chem Phys 2008, 128, art no 114102.
[21] Janowski T, Pulay P. High accuracy benchmark calculations on the benzene dimer potential energy surface. Chem Phys Lett 2007, 447, 27–32.
[22] Bettinger HF, Kar T, Sánchez-García E. Borazine and benzene homo- and heterodimers. J Phys Chem A 2009, 113, 3353–3359.
[23] Sherrill DC, Takatani T, Hohenstein EG. An assessment of theoretical methods for nonbonded interactions: comparison to complete basis set limit coupled-cluster potential energy curves for the benzene dimer, the methane dimer, benzene-methane, and benzene-H_2S. J Phys Chem A 2009, 113, 10146–10159.
[24] Lee EC, Kim D, Jurečka P, Tarakeshwar P, et al. Understanding of assembly phenomena by aromatic-aromatic interactions: Benzene dimer and the substituted systems. J Phys Chem A 2007, 111, 3446–3457.
[25] Headen TF, Howard CA, Skipper NT, Wilkinson MA, et al. Structure of π–π interactions in aromatic liquids. J Am Chem Soc 2010, 132, 5735–5742.
[26] Schweizer WB, Dunitz JD. Quantum mechanical calculations for benzene dimer energies: present problems and future challenges. J Chem Theory Comput 2006, 2, 288–291.
[27] Rezáč J, Hobza P. Benzene dimer: dynamic structure and thermodynamics derived from on-the-fly ab initio DFT-D molecular dynamic simulations. J Chem Theory Comput 2008, 4, 1835–1840.
[28] Ninković DB, Janjić GV, Veljković DŽ, Sredojević DN, Zarić SD. What are the preferred horizontal displacements in parallel aromatic-aromatic interactions? Significant interactions at large displacements. ChemPhysChem 2011, 12, 3511–3514.
[29] Ninković DB, Andrić JM, Malkov SN, Zarić SD. What are the preferred horizontal displacements of aromatic-aromatic interactions in proteins? Comparison with the calculated benzene-benzene potential energy surface. Phys Chem Chem Phys 2014, 16, 11173–11177.
[30] Burley SK, Petsko GA. Aromatic-aromatic interaction: a mechanism of protein structure stabilization. Science 1985, 229, 23–28.
[31] Chakrabarti P, Samanta U. CH/pi interaction in the packing of the adenine ring in protein structures. J Mol Biol 1995, 251, 9–14.
[32] Samanta U, Chakrabarti P. Assessing the role of tryptophan residues in the binding site. Protein Eng 2001, 14, 7–15.
[33] Chelli R, Gervasio FL, Procacci P, Schettino V. Stacking and T-shape competition in aromatic-aromatic amino acid interactions. J Am Chem Soc 2002, 124, 33–43.
[34] Hunter CA, Sanders JKM. The nature of .pi.-.pi. interactions. J Am Chem Soc 1990, 112, 5525–5534.
[35] Malinovskii VL, Samain F, Häner R. Helical arrangement of interstrand stacked pyrenes in a DNA framework. Angew Chem Int Ed 2007, 46, 4464–4467.
[36] Ninković DB, Andrić JM, Zarić SD. Parallel interactions at large horizontal displacement in pyridine-pyridine and benzene-pyridine dimers. ChemPhysChem 2013, 14, 237–243.
[37] Ninković DB, Janjić GV, Zarić SD. Crystallographic and ab initio study of pyridine stacking interactions. Local nature of hydrogen bond effect in stacking interactions. Cryst Growth Des 2012, 12, 1060–1063.
[38] Janjić GV, Veljković DŽ, Zarić SD. Water/aromatic parallel alignment interactions. Significant interactions at large horizontal displacements. Cryst Growth Des 2011, 11, 2680–2683.
[39] Malenov DP, Dragelj JL, Janjić GV, Zarić SD. Coordinating benzenes stack stronger than noncoordinating benzenes, even at large horizontal displacements. Cryst Growth Des 2016, 16, 4169–4172.

[40] Groom CR, Allen FH. The Cambridge Structural Database in retrospect and prospect. Angew Chem Int Ed 2014, 53, 662–671.

[41] Bondi A. Van der Waals volumes and radii. J Phys Chem 1964, 68, 441–451.

[42] Braga D, Abati A, Scaccianoce L, Johnson BFG, Grepioni F. Organometallic polymorphism. Synthesis and structural characterization of two forms of $[Ru(\eta^6\text{-}C_6H_6)(\eta^6\text{-}C_6H_4(CH_3)COOCH_3)][BF_4]_2$ and the phase transition in $[Ru(\eta^5\text{-}C_5H_5)(\eta^6\text{-}C_6H_5OH)][PF_6]$. Solid State Sci 2001, 3, 783–788.

[43] Yao Z, Klabunde KJ, Asirvatham AS. Unusual iron(IV) complexes: metal atom syntheses and studies of $(\eta^6\text{-arene})Fe(H)_2(SiCl_3)_2$ (arene = benzene, toluene, p-xylene). Inorg Chem 1995, 34, 5289–5294.

[44] Tao J, Perdew JP, Staroverov VN, Scuseria GE. Climbing the density functional ladder: non-empirical meta-generalized gradient approximation designed for molecules and solids. Phys Rev Lett 2003, 91, art no146401.

[45] Grimme S. Semiempirical GGA-type density functional constructed with a long-range dispersion correction. J Comput Chem 2006, 27, 1787–1799.

[46] Weigend F, Ahlrichs R. Balanced basis sets of split valence, triple zeta valence and quadruple zeta valence quality for H to Rn: design and assessment of accuracy. Phys Chem Chem Phys 2005, 7, 3297–3305.

[47] Neese F. The ORCA program system. WIRE Comput Mol Sci 2012, 2, 73–78.

[48] Pople JA, Head-Gordon M, Raghavachari K. Quadratic configuration interaction. A general technique for determining electron correlation energies. J Chem Phys 1987, 87, 5968–5975.

[49] Sinnokrot MO, Valeev EF, Sherrill CD. Estimates of the ab initio limit for π–π interactions: the benzene dimer. J Am Chem Soc 2002, 124, 10887–10893.

[50] Malenov DP, Ninković DB, Sredojević DN, Zarić SD. Stacking of benzene with metal chelates: calculated CCSD(T)/CBS interaction energies and potential-energy curves. ChemPhysChem 2014, 15, 2458–2461.

[51] Mutter ST, Platts JA. Modulation of stacking interactions by transition-metal coordination: ab initio benchmark studies. Chem Eur J 2010, 16, 5391–5399.

[52] Bulat FA, Toro-Labbé A, Brinck T, Murray JS, Politzer P. Quantitative analysis of molecular surfaces: areas, volumes, electrostatic potentials and average local ionization energies. J Mol Model 2010, 16, 1679–1691.

[53] Hufford CD, Oguntimein B, Martin M, Clardy J. Syncarpurea, a novel metabolite from uvaria afzelii. Tetrahedron Lett 1984, 25, 371–374.

[54] Braga D, Costa AL, Grepioni F, Scaccianoce L, Tagliavini E. Organic-organometallic crystal synthesis. 1. Hosting paramagnetic [(h6-arene)2Cr]+ (arene = benzene, toluene) in organic anion frameworks via O-H···O and C–H···O hydrogen bonds. Organometallics 1997, 16, 2070–2079.

[55] Braga D, Grepioni F, Tagliavini E, Novoa JJ, Mota F. C–H···O hydrogen bonds in the mixed-valence salt $[(\eta^6\text{-}C_6H_6)_2Cr]+[CrO_3(OCH_3)]-$ and the breakdown of the length/strength analogy. New J Chem 1998, 22, 755–757.

Daniel A. Decato and Orion B. Berryman

12 Simultaneous halogen and hydrogen bonding to carbonyl and thiocarbonyl functionality

12.1 Introduction

The importance of non-covalent interactions is undisputable and understanding these forces is the crux of numerous research programs. For over 100 years the hydrogen bond (HB) has captivated the interests of scientists, resulting in an extensive understanding of the interaction [1–11]. The HB is primarily an electrostatic interaction between an electropositive hydrogen atom (HB donor) and an electronegative species (HB acceptor) with energies ranging from 0.5 to 40 kcal/mol [1]. The directional preference of the interaction is predictable and enables geometric control at the molecular level. In particular, the field of crystal engineering has employed HBs to produce predictable supramolecular synthons [12]. However, in more complicated systems such as ternary and quaternary co-crystals, unfavourable/uncontrolled competition between multiple HB donors and acceptors has been envisioned [13]. To address these challenges, other non-covalent interactions have been studied. To demonstrate, the halogen bond (XB) has been found to exhibit characteristics similar to the HB, and has emerged as an alternative tool for constructing new materials. The XB, as defined by the International Union of Pure and Applied Chemistry (IUPAC), "occurs when there is evidence of a net attractive interaction between an electrophilic region associated with a halogen atom in a molecular entity and a nucleophilic region in another, or the same, molecular entity" [14]. The efforts of many have accelerated our understanding of this non-covalent interaction, prompting its widespread use [15–26]. Although comparable to the HB with regard to strength, the XB is considerably more directional (the R-X···B angle tends to be close to 180°, where X is a halogen, R a covalently bound group, and B the halogen bond acceptor). The similarities, along with the comprehensive knowledge of the HB, have motivated studies comparing the two in the gas [27–31], solution [15, 17, 18, 32, 33] and solid phases [26, 34, 35]. In addition, there are studies examining their simultaneous use [36–41]. In particular, Aakeröy's foundational work regarding XB and HB tandem application to crystal engineering has provided techniques to avoid "synthon crossover" [13, 42–46]. These studies have revealed a complex relationship between the two interactions, with observations of both competition and cooperation. Understanding how these similar interactions behave when proximal to each other is of great interest to many fields, especially because the recent ascendancy of the XB field leaves much to be discovered.

The HB and the XB can interact with Lewis bases. For example, the carbonyl accepts both XBs and HBs, with typical donors directed at the lone-pair electrons of oxy-

https://doi.org/10.1515/9783110464955-012

gen. In fact, early crystallographic evidence of halogens interacting with carbonyls was presented by Hassel in 1959, i.e. between Br_2 and acetone [47]. Here, the distal ends of the bromine atoms were directed at the lone-pairs of the acetone resulting in infinite 1D chains. Since then, it has been shown that the carbonyl is a capable XB acceptor and can act as a monodentate or bidentate acceptor [48, 49]. The carbonyl acceptor offers a unique opportunity to examine the XB and HB together in the same system. In this setting, the forces can compete for, or share the two classical electron-rich regions. The relationship with carbonyls has gained interest since Voth et al. [50] presented an examination of the Protein Data Bank (PDB), investigating simultaneous XB and HB to carbonyls in protein–ligand complexes. They reported that the XB and HB are geometrically orthogonal in protein–ligand complexes. Complementary computational work to this PDB search revealed that the XB is also energetically independent of the HB [50]. Only recently has this orthogonality concept been purposefully engineered in small molecule co-crystals, and some reports include a succinct Cambridge Structural Database (CSD) search regarding the relationship of the two interactions [51, 52].

Herein, we present a detailed examination of the CSD, where XB and HB share a carbonyl or thiocarbonyl acceptor. The following survey of the CSD parallels the findings of the PDB search, indicating that the geometric orthogonality between the interactions is a robust and genuine phenomenon. However, similar previous CSD examinations of carbonyls with non-covalent interactions revealed that the spatial relationship is not limited to XB–HB situations, suggesting that the orthogonal relationship might be largely dictated by the carbonyl acceptor. This hypothesis, when applied to the thiocarbonyl, is less conclusive owing to the small data set and diffuse nature of the acceptor. The data indicate that in some situations the spatial relationship is comparable, yet structurally the angle between donors is more acute.

12.2 Structural database search parameters

To understand the interplay between XB and HBs, the CSD was examined for bifurcated carbonyl acceptors that accept a XB and a HB. ConQuest (version 1.18) was used to collect and measure contact data from the CSD (version 5.37 updates through May 2016) [53]. XB distances and angles were restricted to $\leq \sum$ vdW radii and 150–180°. HB contacts were restricted to rather inclusive parameters outlined by Steiner [1], which include hydrogen-acceptor distances from 0.0 to 3.0 Å and angles of 110.0–180.0°. Parent HB donor atoms were restricted to N, O, and S atoms, and non-traditional C–H HBs were not included because of the liberal HB geometric parameters outlined. The data were limited to strictly organic structures to eliminate any possible influence by well-defined metal coordination geometries. No restrictions on disorder or R value were implemented to augment the small data set, particularly for the thiocarbonyl compounds. Standard deviations are provided in parenthesis after the measured values presented. The search results from carbonyl acceptors and thiocarbonyl receptors are discussed separately below.

12.3 Carbonyl search results

A total of 473 structures, resulting in a total of 595 interactions, were identified from the carbonyl search. To evaluate the geometric orthogonality of HB and XB donors the X···O···H angles were measured. A histogram of the X···O···H angles (Figure 12.1) displays a significant bias for the contacts to be orthogonal, with a mean value of 89.2 (18.2)°. This is in agreement with the previously mentioned PDB search, which reported an average X···O···H angle of 88.2 (14.1)° from 76 protein–ligand interactions [50]. This also parallels a histogram presented by Vasylyeva et al. [52] in a 2014 CSD search. The persistence of the orthogonal orientation in both small molecules and bio-macromolecules suggests that the relationship might not be an anomaly. The foundation of the phenomenon is unknown, and is undoubtedly complex given the many factors influencing solid-state arrangements. However, a few justifications for the spatial arrangement have emerged.

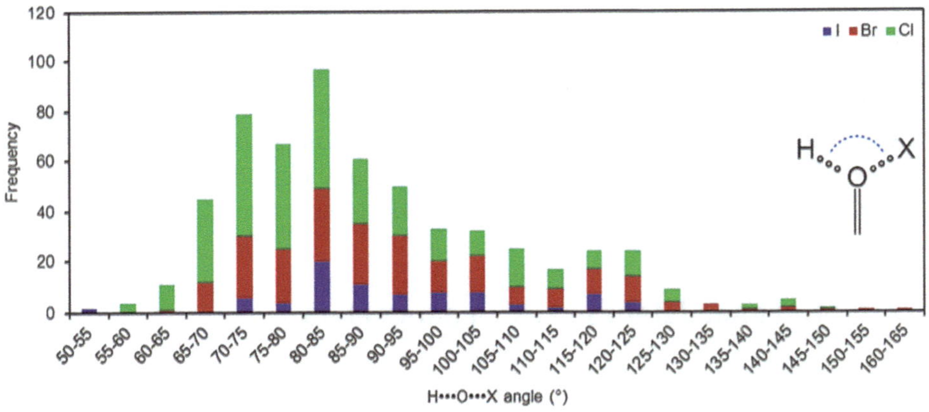

Fig. 12.1: Histogram of the X···O···H angles (X = Cl, Br, I) observed in the Cambridge Structural Database (CSD).

One partial explanation may be Etter's [54] "best donor–best acceptor" guideline for predicting HBs in the solid state. If both lone-pair sites are occupied by one interaction, and assuming there are no other acceptors present, the next best acceptor positions are orthogonal to the lone-pair electrons. An examination of the data set reveals that, of the 473 structures, only 65 demonstrate two HBs and one XB to the carbonyl. Additionally, of the 473 structures, only seven structures exhibit two XBs and one HB to the carbonyl. This suggests that some instances of orthogonal XB–HB contacts may result from the HB, preferentially interacting with both lone-pairs. This also indicates that some XB–HB orthogonality is due to both classical lone-pair sites being occupied. However, this does not rule out situations of a lone-pair site being sterically blocked, resulting in a similar forced orthogonal condition. The application of forced orthogonality via steric hindrance may be a viable strategy for assembling new materials, but it does not explain all the near-orthogonal relationships found within the data. Therefore, the data were further examined.

A steric argument could also be made as halogen atoms are considerably larger than hydrogen atoms. This larger size may result in less favourable XB contacts with the classical lone-pair sites because of steric effects with groups adjacent to the carbonyl. Under these conditions, a perpendicular approach to the carbonyl plane may be more energetically favourable for the XB. The steric influence of adjacent groups on the arrangement of the atoms involved in these two forces has been previously discussed and supported through computational studies by Voth et al. [50]. This steric argument was augmented by highlighting the electrostatic potential surface of an amide oxygen, which displays in-plane and out-of-plane negative electrostatic potentials. Thus, the current rationale for XB–HB orthogonality in protein–ligand complexes is founded on both steric factors and electronics of the carbonyl acceptor. The spatially orthogonal phenomenon in small molecules is also likely to be a combination of many other factors, owing to the numerous influences on the solid-state packing of small molecules.

Given the largely electrostatic natures of the XB and HB, it stands to reason that the geometric orthogonality is also driven by the electronics of the carbonyl acceptor in small molecules. If the orthogonal relationship is rooted in the electrostatic potential surface of the carbonyl oxygen, then this observation should persist under similar conditions. For instance, in the case of two HBs to a carbonyl, the spatial orthogonality should remain because of the electrostatic nature of the interaction. A report by Simón and Goodman [55] analysed carbonyls that accept two HBs. The histogram of the H···O···H angles measured from the data they collected from the CSD clearly indicates an orthogonal relationship. This observation implies that the unique XB–HB spatial orthogonality originates from the acceptor. The hypothesis was tested further by analysing instances where two XBs interact with a carbonyl. Here the search reveals a similar bias for an orthogonal relationship with an average X···O···X angle of 91.4 (18.5)° from a total of 92 structures. Based on these findings it can be concluded that the spatially orthogonal relationship is heavily influenced by the electronics of the carbonyl acceptor and is not exclusive to small molecule crystal structures that contain simultaneous XB–HB interactions. The repeated observation of geometrically orthogonal non-covalent contacts to bifurcated carbonyl acceptors is striking and should be examined further. Examination of other geometric parameters may reveal further nuances to the XB–HB carbonyl relationship.

Additional search parameters were employed to obtain further insight into the orthogonal XB–HB relationship of carbonyls. For example, the X···O···H angle distribution suggests that, in the majority of contacts, either only one of the donors interacts with the electron-dense region of the oxygen carbonyl or neither does. A dihedral angle analysis of Y–C=O···X (where X = H, or Cl, Br or I and Y = C, N or O) establishes whether the contact is in the plane of the carbonyl (dihedral of ≈ 0°) or orthogonal to the plane (dihedral angle ≈ 90°). A histogram of the HB dihedral angle indicates a clear bias for the HB to be in the plane of the carbonyl (dihedral angle ≈ 0°) (Figure 12.2). This is comparable with a 2012 CSD search, which investigated two HBs sharing a carbonyl acceptor [55]. Here, it is shown that HBs tend to interact with the oxygen in the plane of the carbonyl (dihedral angle ≈ 0°). Other solid-state examinations of HB to sp^2-hybridized carbonyl oxygen atoms have also displayed this trend [1, 56–58]. By contrast, a PDB search presented by Simón and Goodman [55, 59] highlights that two simultaneous HBs prefer to be orthogonal to the carbonyl double-bond (dihedral angle ≈ 90°). However, this search was limited to catalytic oxyanion holes, where it was suggested that the unique arrangement relates to the catalytic properties of these enzymes, and the distribution across the PDB may not reflect this observation.

An evaluation of dihedral angles does not reveal where HB donors are directed with respect to the oxygen lone-pairs. For example, the HB could be in the plane of the carbonyl (dihedral angle ≈ 0°) yet directed at the distal/polar end of the oxygen. A scatter plot of the C=O···H angle vs the Y–C=O···H (Y = C, N, O) dihedral angle reveals the location of the HB with respect to the carbonyl lone-pairs (Figure 12.3). The HBs generally have a small dihedral angle (< 15°) and a C=O···H angle of approximately 120°,

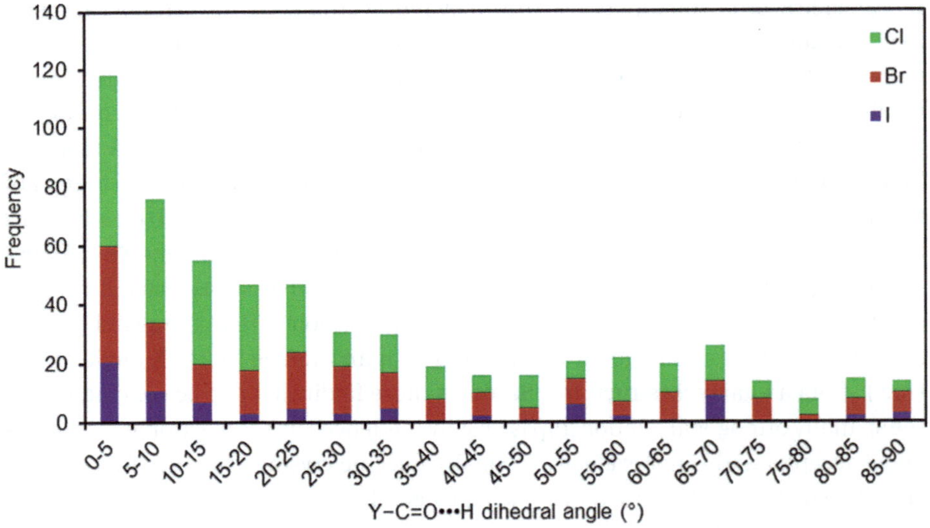

Fig. 12.2: Histogram of HB dihedral angles Y–C=O···H (Y = C, N, O). The dihedral values are displayed as ≤90°. This is achieved by displaying the supplementary angle for angles greater than 90°. This value retains the information concerning the location of the interaction in relation to the carbonyl plane.

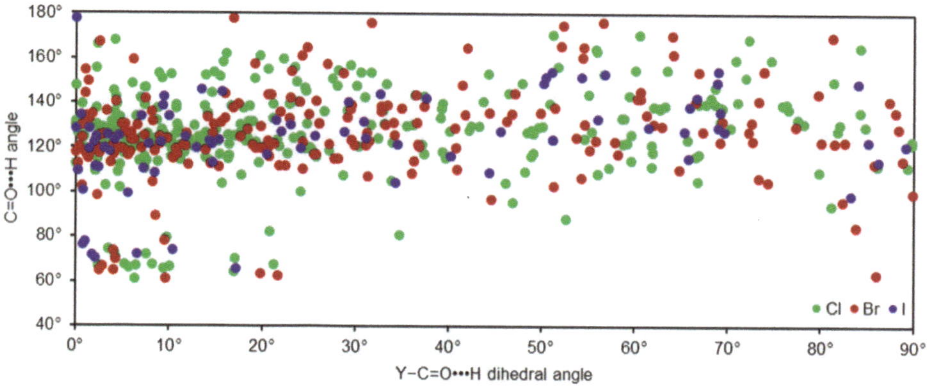

Fig. 12.3: Scatterplot of C=O···H angle vs the Y–C=O···H dihedral angle (Y = C, N, O). The HB prefers to be directed at the carbonyl lone-pairs of electrons.

directed at the putative location of the lone-pair electrons. This analysis highlights that the HB prefers to interact with at least one of the carbonyl oxygen lone-pairs. This conclusion predicts that the XB would have to interact with the remaining lone-pair electrons or at less electron-dense sites (orthogonal to the lone-pairs). However, the X···O···H angles suggests that the XB donor might interact with the less electron-rich sites.

Definitive trends in the XB data are less clear, which may be a result of the greater number of structures containing the weaker XB donors, chlorine and bromine. A histogram of the XB dihedral angle Y–C=O···X (Y = C, N and O; X = Cl, Br and I) appears to show a preference for the XB to approach orthogonal to the carbonyl plane, which is once again analogous to the PDB findings of Voth et al. (Figure 12.4) [50]. A closer examination reveals that stronger iodine XB donors do not show a dihedral preference, possibly indicating that they are more competitive with HB donors. Here, the average X···O···H angle remains orthogonal with a mean value of 92.2 (15.4)°. This supports the notion that the spatial relationship is dictated by the carbonyl acceptor electronics. However, the small iodine data set (82 interactions) may be biasing this interpretation. A scatterplot of the C=O···X angle vs the Y–C=O···X dihedral angle shows a densely populated area of contacts with a dihedral angle of 70–90° and a C=O···X angle of approximately 90°, which again is similar to the interactions found in the PDB (Figure 12.5). The deviation from this dihedral region (70–90°) results in C=O···X angles that tend to be > 120°(especially at dihedral angles < 45°). The orthogonal relationship at these smaller XB dihedral angles persists displaying an average X···O···H angle of 90.5 (18.4)° for structures with Y–C=O···X dihedral angles < 45°. This observation is analogous to that found for α-helical carbonyls in proteins containing XB and HB. The HB is often displaced from the plane of the carbonyl, yet the two interactions remain orthogonal to each other [50]. Despite the direction of the donor approach, this con-

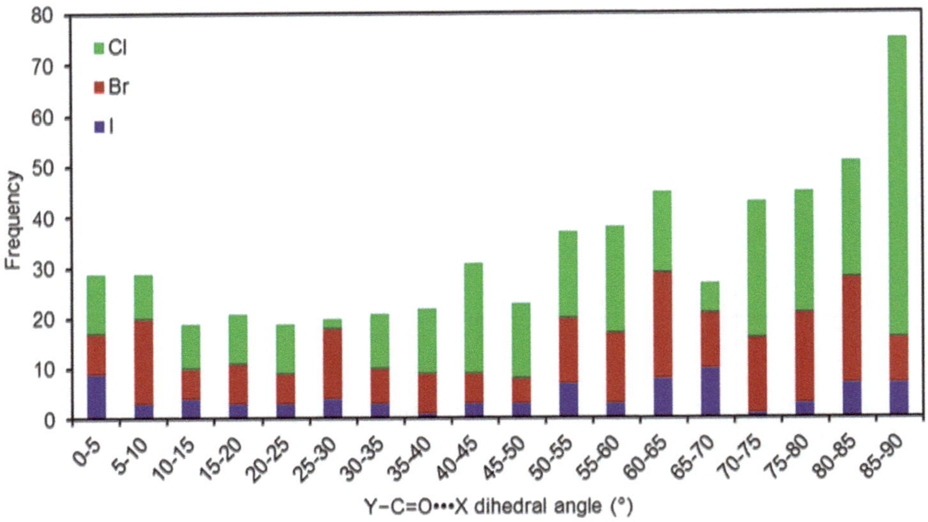

Fig. 12.4: Histogram of the XB dihedral angle Y–C=O···X (Y = C, N, O and X = Cl, Br, I).

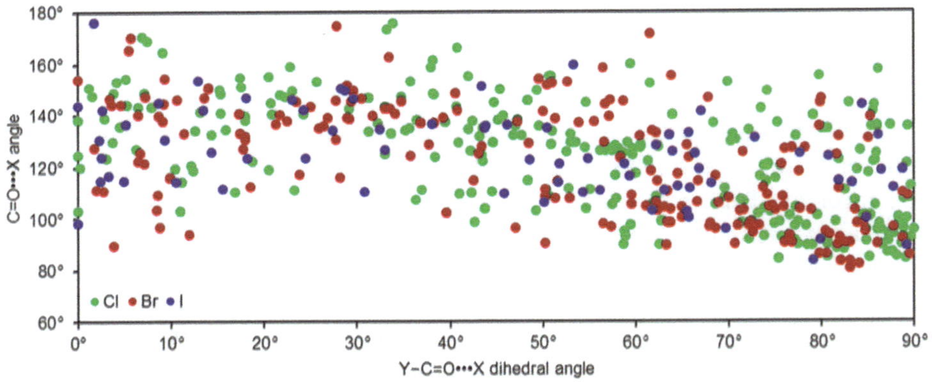

Fig. 12.5: Scatterplot of C=O···X angle vs the Y–C=O···X dihedral angle (Y = C, N, O and X = Cl, Br, I).

sistency in both small molecules and bio-macromolecules provides further evidence of the orthogonal relationship.

To further probe the XB–HB relationship, a scatterplot of the XB dihedral vs the HB dihedral angle was produced revealing where the interactions occur with respect to the carbonyl plane (Figure 12.6). First, there is a significant clustering of hits where the HB is in the plane of the double-bond and the XB is normal to the double-bond (HB dihedral angle ≈ 0° and XB dihedral angle ≈ 90°). There are also occurrences where both XB and HB are in the plane of the carbonyl (HB and XB dihedral angle ≈ 0°). The lack of HB contacts, where HB dihedral angle ≈ 90°, is evidence that the HB effectively competes with the XB to be in the plane of the carbonyl to occupy the most

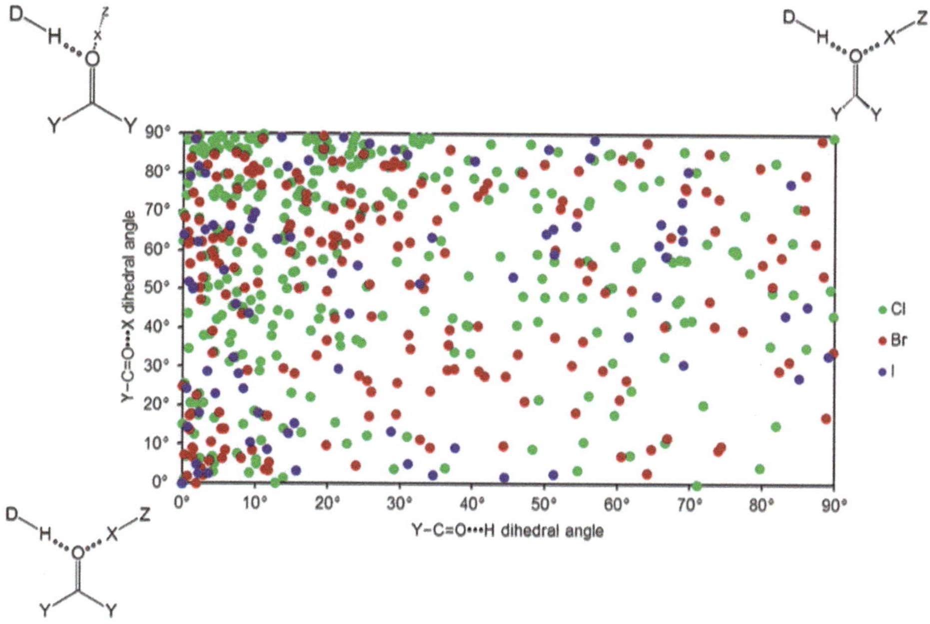

Fig. 12.6: Scatterplot of Y–C=O···X dihedral angle vs the Y–C=O···H dihedral angle (Y = C, N, O and X = Cl, Br, I).

favourable interaction site. The predilection of carbonyl functional groups forming multipoint contacts (acid dimers, amide ribbons, etc.) with lone-pair electrons may also help to account for the lack of HB contacts perpendicular to the carbonyl [13]. Another likely contributor is the abundance of data from the weaker bromine and chlorine XB donors. These weaker XB donors are less likely to displace the HB from the most electron-rich sites than iodine donors. The influence of weaker XB donors in designed XB–HB orthogonal co-crystals is currently unclear. Takemura et al. [51] produced orthogonal assemblies with iodine derivatives, yet no such features were observed with bromine derivatives. By contrast, Vasylyeva et al. [52] were able to produce the orthogonality with both iodine and bromine donors within their small molecule peptide bond models. The contrasting results highlight not only the difficulty of predicting solid-state structures, but also the need to further examine the orthogonal relationship.

The present CSD search has demonstrated that the spatially orthogonal relationship between XB and HB donors interacting with a carbonyl is robust and persists in small molecule crystal structures. The data indicate that this relationship originates from the carbonyl acceptor, as it is not exclusive to the XB–HB combination. Examples of XB–XB and HB–HB with carbonyls also demonstrate a perpendicular preference. Further evaluation of this geometric relationship should be pursued further in silico and in solution as the spatial orthogonality displayed in the solid-state reveals nothing tangible about the energetics of the interaction.

12.4 Thiocarbonyl search

The HB contacts with sulphur acceptors are weaker and less directional than with oxygen counterparts [57]. Compared with oxygen, the sulphur atom is more diffuse, larger, and less electronegative and also participates in charge-transfer to a greater extent [60–63]. These characteristics make sulphur derivatives intriguing and challenging targets. Interestingly, XB interactions exhibit a greater component of charge-transfer than HB, suggesting complementarity with sulphur derivatives. Yet evaluations of sulphur acceptors and XB donors have largely been limited to computational and solid-state investigations [64–69]. Solution studies in this area are very limited [70] and comparative studies of oxygen and sulphur acceptors with XB donors would elucidate the differences between these two forces.

It is this paucity of studies that motivated the following CSD search of thiocarbonyls accepting a HB and XB. The search parameters parallel the above evaluation of the carbonyl, examining the geometric parameters of the contacts around the thiocarbonyl. The orthogonality of the interactions was inspected and additional trends were discussed. Using the same parameters described above, the search returned a total of 52 structures and 88 interactions of simultaneous XB and HB to a thiocarbonyl. Given this relatively small number of contacts, and the electronic nuances of thiocarbonyls (which are discussed later), conclusions drawn from this search are subject to further refinement.

An analysis of the X···S···H angle data reveals no singular bias, yet there appears to be a bimodal distribution, centred around 75° and about 120° (Figure 12.7). The angles around 75° indicate that a spatial relationship similar to the carbonyl is preferred, yet is more geometrically acute. The 120° relationship suggests that both the XB and HB might be directed at locations reminiscent of classically Sp2 hybridized lone pairs. This clustering of data around 120° could be the result of the small data set. Alternatively, this could also be a result of the longer thiocarbonyl bond (1.61 Å for thiocarbonyl vs 1.21 Å for carbonyl). This subtle difference may relieve some of the unfavourable steric factors with adjacent groups that were observed in calculations provided by Voth et al. [50] for the carbonyl functionality. The lack of a distinct preference is conceivably due to the small data set and the diffuse nature of sulphur, which highlights this area as deserving of more attention.

The histogram of the Y–C=S···H dihedral angles highlights a preference for the HB to be in the plane of the thiocarbonyl (Figure 12.8). A scatterplot of the C=S···H angle vs the Y–C=S···H dihedral angle (Figure 12.9) shows that the C=S···H angle is not dependent on the dihedral, with a mean value of 108.9 (7.0)°. This angle is consistent with other HB studies with thiocarbonyl acceptors (< 120°) [1], and is also in agreement with

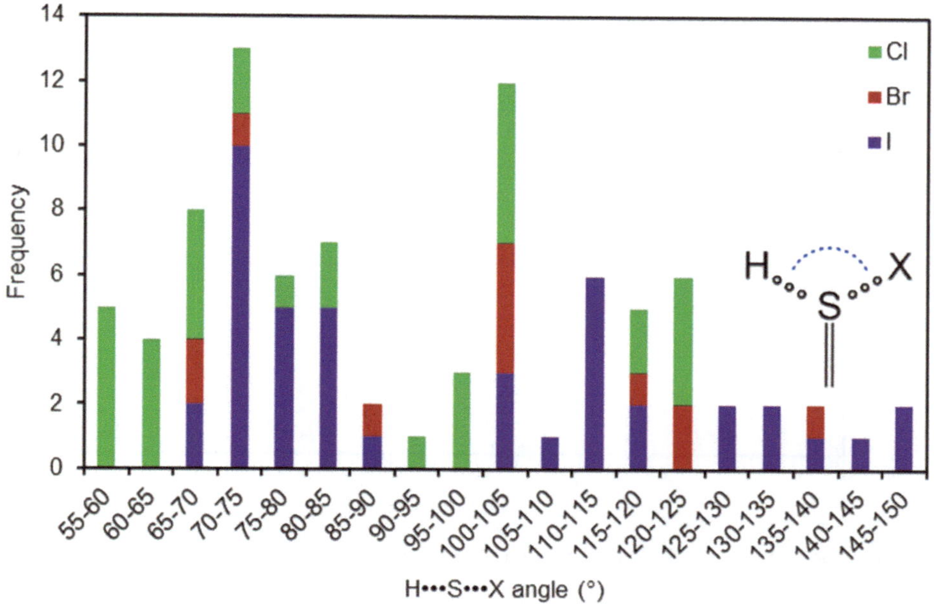

Fig. 12.7: Histogram of the X···S···H angles (X = Cl, Br, I) observed in the CSD.

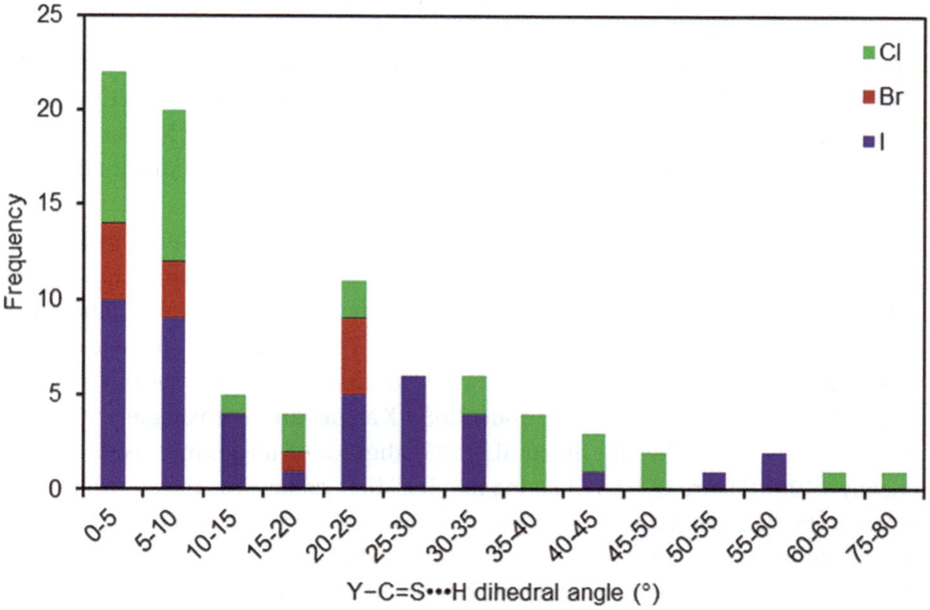

Fig. 12.8: Histogram of HB dihedral angles Y–C=S···H (Y = C, N, O).

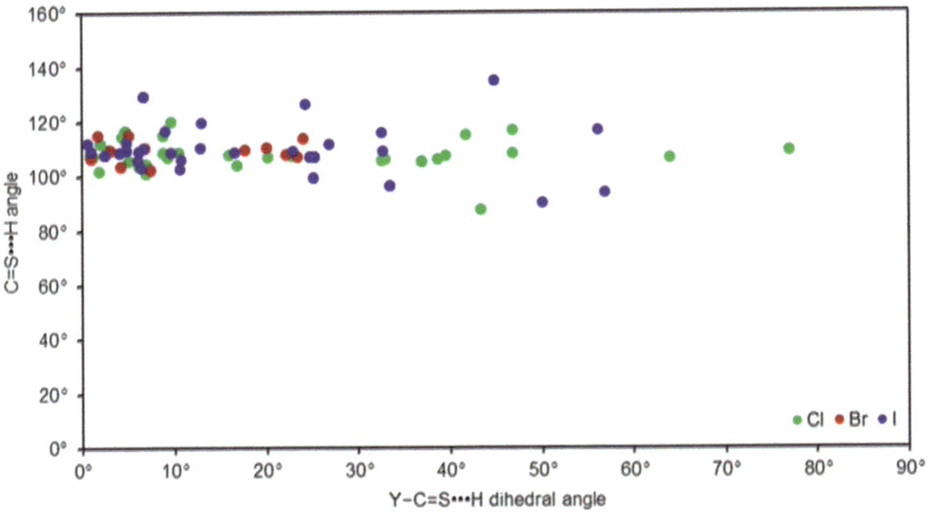

Fig. 12.9: Scatterplot of C=S···H angle vs the Y–C=S···H dihedral angle (Y = C, N, O).

Allen and co-workers [71], who described thiocarbonyls receiving multiple HB donors (up to six) and reported a bias for C=S···H angles of 105–115°. Therefore, the electronic nature of the thiocarbonyl-sulphur is highlighted, as it displays a belt of electron-density normal to the carbon sulphur double-bond in the region of the classical Ione pair site. Overall, when in the presence of an XB, the HB geometries are reminiscent of the HB findings by Allen and co-workers [71] indicating that the XB does not influence how the HB approaches the sulphur acceptor. This finding provides further information regarding the nature of the thiocarbonyl acceptor and the relationship between the HB and XB.

An additional analysis of the XB contacts with thiocarbonyl-sulphur began with the Y–C=S···X dihedral angle. A histogram from this search shows that there is no discernible trend; however, the data may slightly favour the out-of-plane contact (Figure 12.10). A scatterplot of the C=S···X angle vs the Y–C=S···X dihedral angle data reveals that the stronger iodine donor tends to form contacts closer to the electron-rich belt of the sulphur atom. The iodine donor C=S···X angles are approximately 90° when the dihedral is > 45°. When the dihedral is < 45°, the C=S···X angles increase to approximately > 100° (Figure 12.11), indicating that it is directed towards the electron-dense regions. On the other hand, no clear trend is observed for the weaker donors – chlorine and bromine. A scatterplot of Y–C=S···X dihedral vs Y–C=S···H dihedral angles displays similar features to the carbonyl plot where the majority of interactions have the HB donor near the plane of the thiocarbonyl and the XB donor out of plane (Figure 12.12).

The lack of thiocarbonyl data available for this search makes it difficult to draw conclusions about the XB–HB orthogonality. The electronic nature of the sulphur

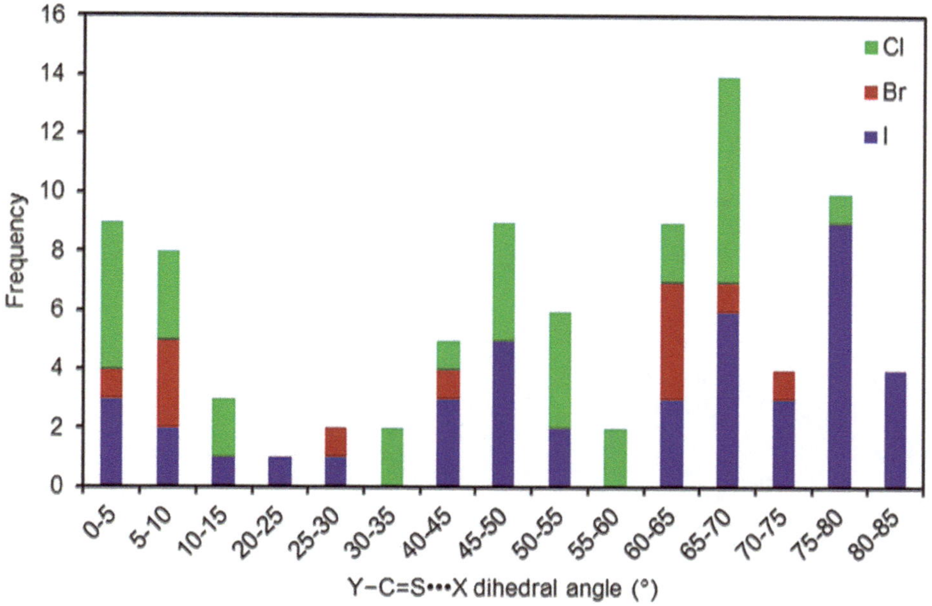

Fig. 12.10: Histogram of XB dihedral angles Y–C=S···X (Y = C, N, O and X = Cl, Br, I).

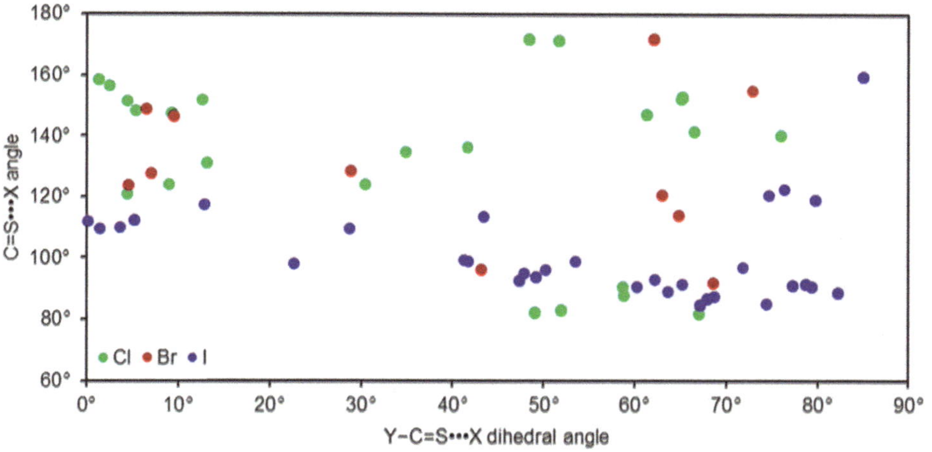

Fig. 12.11: Scatterplot of C=S···X angle vs the Y–C=S···X dihedral angle (Y = C, N, O and X = Cl, Br, I).

atom further complicates the analysis. Calculations of XB to a thiocarbonyl depict a shallow energy surface when varying both the XB torsional angle and C=S···I angle [72]. This causes the directionality of the XB contact in relation to the acceptor to be less predictable, as the penalty for drifting is minimal. Computations from Le Questel et al. [73] also indicate that an HB to the thiocarbonyl of thioureas produces a decrease in the electron density surrounding the sulphur, resulting in a weaker

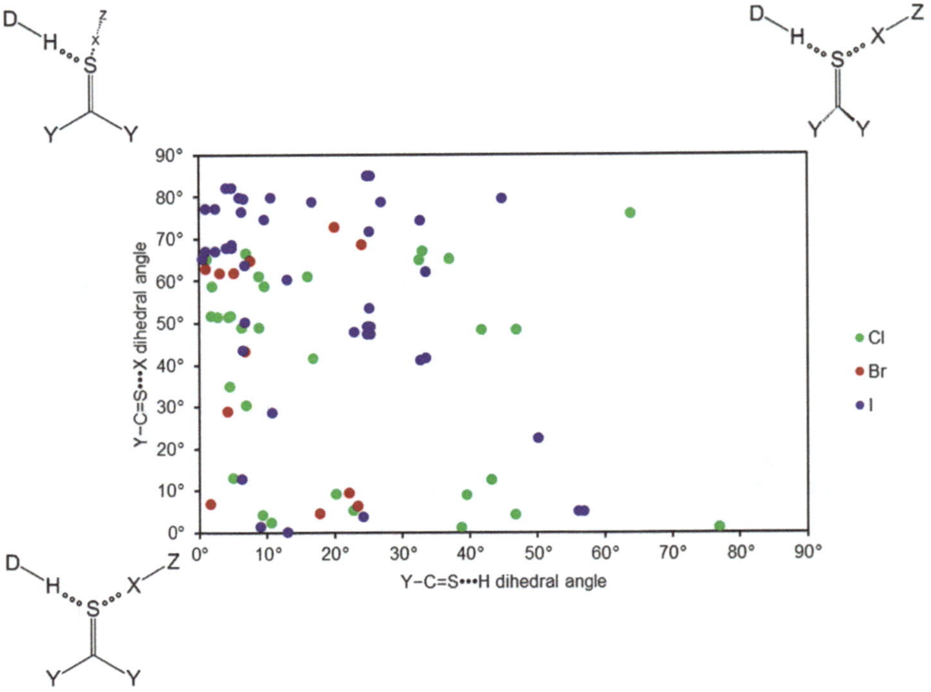

Fig. 12.12: Scatterplot of Y–C=S···X dihedral angle vs the Y–C=S···H dihedral angle (Y = C, N, O and X = Cl, Br, I).

XB interaction. By contrast, Eccles et al. [64] did not observe this phenomenon with thioamides. They suggested that the adjacent groups to the thiocarbonyl might have a more significant impact on the XB, where electron-donating groups led to shorter contacts. Although these findings are similar to those of a previous CSD investigation of HBs to thiocarbonyls, there is still much to be investigated. The relationship of XB and HB to thiocarbonyls is an area that requires more attention. A greater understanding of the XB and HB interplay and the influence of the acceptor will have a widespread impact on the design of materials based on these synthons.

12.5 Conclusions

The presented CSD search provides further evidence of a spatial orthogonal relationship between XBs and HBs when simultaneously interacting with carbonyl derivatives of small molecules. This arrangement was shown to occur outside of instances where one interaction dictates an orthogonal approach via steric occupation of both lone-pair sites. The observation of XB–HB spatial orthogonality in both biomacromolecules and small molecules justifies exploring the fundamental basis of the interaction and examining its ability to persist outside the solid state. More importantly, the observations presented here alongside previous CSD searches reveal that the XB–HB geometric relationship is not exclusive to this situation. The spatial orthogonality between two non-covalent interactions also appears in bifurcated carbonyl acceptors of either two HBs or two XBs, indicating that the observed feature is a result of the acceptor. The repeated observation of an orthogonal arrangement around carbonyls by various combinations of XBs and HBs demands attention outside the solid state. The design and examination of molecules with orthogonal donors to investigate thermodynamic impacts would be beneficial to numerous fields.

The first CSD analysis of XBs and HBs interacting with a common thiocarbonyl highlights a deficiency of studies involving the thiocarbonyl functionality. However, from the available data, some early conclusions can be made. The HB geometries mimic previous HB findings, demonstrating that the presence of an XB does not perturb the HB contact geometry. However, this is not unsurprising given the diffuse nature of the sulphur. The deficiency of thiocarbonyl structures highlights the importance of further fundamental investigations, providing an auspicious opportunity for solid, solution, and computational studies.

Acknowledgement: This work was funded by the National Science Foundation (NSF CAREER CHE-1555324), the Centre for Biomolecular Structure and Dynamics COBRE (NIH NIGMS grant P20GM103546), the Montana University System (MREDI 51030-MUSRI2015-02) and the University of Montana.

Bibliography

[1] Steiner T. The hydrogen bond in the solid state. Angew. Chem. Int. Ed. Engl. 2002, 41, 48–76.
[2] Scheiner S. Hydrogen bonding: a theoretical perspective. Oxford University Press 1997.
[3] Jeffrey GA, Saenger W. Hydrogen bonding in biological structures: Springer Berlin Heidelberg; 1991.
[4] Jeffrey GA. An introduction to hydrogen bonding: Oxford University Press, New York; 1997.
[5] Grabowski SJ. What is the covalency of hydrogen bonding? Chem. Rev. 2011, 111, 2597–2625.
[6] Desiraju GR. The C-H···O hydrogen bond in crystals: what is it? Acc. Chem. Res. 1991, 24, 290–296.
[7] Perrin CL, Nielson JB. "Strong" hydrogen bonds in chemistry and biology. Ann. Rev. Phys. Chem. 1997, 48, 511–544.
[8] Pimentel GC, McClellan AL. The hydrogen bond. San Francisco: WH Freeman; 1960.
[9] Pauling L. The nature of the chemical bond and the structure of molecules and crystals: an introduction to modern structural chemistry. Cornell University Press; 1960.
[10] Latimer WM, Rodebush WH. Polarity and ionization from the standpoint of the Lewis theory of valence. J. Am. Chem. Soc. 1920, 42, 1419–1433.
[11] Arunan E, Desiraju GR, Klein RA, Sadlej J, Scheiner S, Alkorta I, Clary DC, Crabtree RH, Dannenberg JJ, Hobza P. Definition of the hydrogen bond (IUPAC recommendations 2011). Pure Appl. Chem. 2011, 83, 1637–1641.

[12] Desiraju GR. Supramolecular synthons in crystal engineering – a new organic synthesis. Angew. Chem. Int. Ed. Engl. 1995, 34, 2311–2327.
[13] Aakeröy CB, Chopade PD, Desper J. Avoiding "synthon crossover" in crystal engineering with halogen bonds and hydrogen bonds. Cryst. Growth Des. 2011, 11, 5333–5336.
[14] Desiraju GR, Ho PS, Kloo L, Legon AC, Marquardt R, Metrangolo P, Politzer P, Resnati G, Rissanen K. Definition of the halogen bond (IUPAC recommendations 2013). Pure Appl. Chem. 2013, 85, 1711–1713.
[15] Gilday LC, Robinson SW, Barendt TA, Langton MJ, Mullaney BR, Beer PD. Halogen bonding in supramolecular chemistry. Chem. Rev. 2015, 115, 7118–7195.
[16] Cavallo G, Metrangolo P, Milani R, Pilati T, Priimagi A, Resnati G, Terraneo G. The halogen bond. Chem. Rev. 2016, 116, 2478–2601.
[17] Beale TM, Chudzinski MG, Sarwar MG, Taylor MS. Halogen bonding in solution: thermodynamics and applications. Chem. Soc. Rev. 2013, 42, 1667–1680.
[18] Erdelyi M. Halogen bonding in solution. Chem. Soc. Rev. 2012, 41, 3547–3557.
[19] Ford MC, Ho PS. Computational tools to model halogen bonds in medicinal chemistry. J. Med. Chem. 2016, 59,1655–1670.
[20] Fourmigué M. Halogen bonding: recent advances. Curr. Opin. Solid State Mater. Sci. 2009, 13, 36–45.
[21] Kolář MH, Hobza P. Computer modeling of halogen bonds and other σ-hole interactions. Chem. Rev. 2016, 116, 5155–5187.
[22] Legon AC. The Halogen bond: an interim perspective. Phys. Chem. Chem. Phys. 2010, 12, 7736–7747.
[23] Politzer P, Murray JS, Clark T. Halogen bonding and other σ-hole interactions: a perspective. Phys. Chem. Chem. Phys. 2013, 15, 11178–11189.
[24] Vargas JA. Applications of halogen bonding in solution. Pure Appl. Chem. 2015, 15.
[25] Wilcken R, Zimmermann MO, Lange A, Joerger AC, Boeckler FM. Principles and applications of halogen bonding in medicinal chemistry and chemical biology. J. Med. Chem. 2013, 56, 1363–1388.
[26] Mukherjee A, Tothadi S, Desiraju GR. Halogen bonds in crystal engineering: like hydrogen bonds yet different. Acc. Chem. Res. 2014, 47, 2514–2524.
[27] Wolters LP, Bickelhaupt FM. Halogen bonding versus hydrogen bonding: a molecular orbital perspective. ChemistryOpen 2012, 1, 96–105.
[28] Nepal B, Scheiner S. Competitive halide binding by halogen versus hydrogen bonding: bis-triazole pyridinium. Chem. Eur. J. 2015, 21, 13330–13335.
[29] Legon AC. The interaction of dihalogens and hydrogen halides with Lewis bases in the gas phase: an experimental comparison of the halogen bond and the hydrogen bond. In: Metrangolo P, Resnati G, eds. Halogen bonding: fundamentals and applications. Berlin, Heidelberg: Springer Berlin Heidelberg; 2008, 17–64.
[30] Hill GJ, Legon AC. On the directionality and non-linearity of halogen and hydrogen bonds. Phys. Chem. Chem. Phys. 2015, 17, 858–867.
[31] Grabowski SJ. Hydrogen and halogen bonds are ruled by the same mechanisms. Phys. Chem. Chem. Phys. 2013, 15, 7249–7259.
[32] Wageling NB, Neuhaus GF, Rose AM, Decato DA, Berryman OB. Advantages of organic halogen bonding for halide recognition. Supramol. Chem. 2016, 28, 665–672.
[33] Tepper R, Schulze B, Jäger M, Friebe C, Scharf DH, Görls H, Schubert US. Anion receptors based on halogen bonding with halo-1,2,3-triazoliums. J. Org. Chem. 2015, 80, 3139–3150.
[34] Corradi E, Meille SV, Messina MT, Metrangolo P, Resnati G. Halogen bonding versus hydrogen bonding in driving self-assembly processes. Angew. Chem. Int. Ed. Engl. 2000, 39, 1782–1786.

[35] Portalone G, Moilanen JO, Tuononen HM, Rissanen K. Role of weak hydrogen bonds and halogen bonds in 5-halo-1,3-dimethyluracils and their cocrystals – a combined experimental and computational study. Cryst. Growth Des. 2016, 16, 2631–2639.
[36] Mullaney BR, Partridge BE, Beer PD. A halogen-bonding bis-triazolium rotaxane for halide-selective anion recognition. Chem. Eur. J. 2015, 21, 1660–1665.
[37] Kilah NL, Wise MD, Serpell CJ, Thompson AL, White NG, Christensen KE, Beer PD. Enhancement of anion recognition exhibited by a halogen-bonding rotaxane host system. J. Am. Chem. Soc. 2010, 132, 11893–11895.
[38] Chudzinski MG, McClary CA, Taylor MS. Anion receptors composed of hydrogen and halogen-bond donor groups: modulating selectivity with combinations of distinct noncovalent interactions. J. Am. Chem. Soc. 2011, 133, 10559–10567.
[39] Beyeh NK, Pan F, Rissanen K. A halogen-bonded dimeric resorcinarene capsule. Angew. Chem., Int. Ed. Engl. 2015, 54, 7303–7307.
[40] Marti-Rujas J, Colombo L, Lu J, Dey A, Terraneo G, Metrangolo P, Pilati T, Resnati G. Hydrogen and halogen bonding drive the orthogonal self-assembly of an organic framework possessing 2D channels. ChemComm 2012, 48, 8207–8209.
[41] Raatikainen K, Rissanen K. Modulation of N···I and $^{+}$N-H···Cl^{-}···I halogen bonding: folding, inclusion, and self-assembly of tri- and tetraamino piperazine cyclophanes. Cryst. Growth Des. 2010, 10, 3638–3646.
[42] Aakeröy CB, Spartz CL, Dembowski S, Dwyre S, Desper J. A systematic structural study of halogen bonding versus hydrogen bonding within competitive supramolecular systems. IUCrJ 2015, 2, 498–510.
[43] Aakeröy CB, Schultheiss NC, Rajbanshi A, Desper J, Moore C. Supramolecular synthesis based on a combination of hydrogen and halogen bonds. Cryst. Growth Des. 2009, 9, 432–441.
[44] Aakeröy CB, Fasulo M, Schultheiss N, Desper J, Moore C. Structural competition between hydrogen bonds and halogen bonds. J. Am. Chem. Soc. 2007, 129, 13772–13773.
[45] Aakeröy CB, Chopade PD, Ganser C, Desper J. Facile synthesis and supramolecular chemistry of hydrogen bond/halogen bond-driven multi-tasking tectons. ChemComm 2011, 47, 4688–4690.
[46] Aakeröy CB, Beatty AM, Lorimer KR. Charge-assisted hydrogen bonds and halogen-halogen interactions in organic salts: benzylammonium benzoates and pentafluorobenzoates. Struct. Chem. 1999, 10, 229–242.
[47] Hassel O, Strømme K. Crystal structure of the 1 : 1 addition compound formed by acetone and bromine. Acta Chem Scand 1959;13.
[48] Syssa-Magalé J-L, Boubekeur K, Schöllhorn B. First molecular self-assembly of 1,4-diiodo-tetrafluoro-benzene and a ketone Via (O···I) non-covalent halogen bonds. J. Mol. Struct. 2005, 737, 103–107.
[49] Pigge FC, Vangala VR, Swenson DC. Relative importance of X···O=C vs. X···X halogen bonding as structural determinants in 4-halotriaroylbenzenes. Chem Commun. 2006, 2123–2125.
[50] Voth AR, Khuu P, Oishi K, Ho PS. Halogen bonds as orthogonal molecular interactions to hydrogen bonds. Nat. Chem. 2009, 1, 74–79.
[51] Takemura A, McAllister LJ, Hart S, Pridmore NE, Karadakov PB, Whitwood AC, Bruce DW. Halogen- and hydrogen-bonded salts and co-crystals formed from 4-halo-2,3,5,6-tetrafluorophenol and cyclic secondary and tertiary amines: orthogonal and non-orthogonal halogen and hydrogen bonding, and synthetic analogues of halogen-bonded biological Systems. Chem. Eur. J. 2014, 20, 6721–6732.
[52] Vasylyeva V, Nayak SK, Terraneo G, Cavallo G, Metrangolo P, Resnati G. Orthogonal halogen and hydrogen bonds involving a peptide bond model. CrystEngComm 2014, 16, 8102–8105.
[53] Groom CR, Bruno IJ, Lightfoot MP, Ward SC. The Cambridge Structural Database. Acta Crystallogr. B 2016, 72, 171–179.

[54] Etter MC. Encoding and decoding hydrogen-bond patterns of organic compounds. Acc. Chem. Res. 1990, 23, 120–126.
[55] Simón L, Goodman JM. Enzyme catalysis by hydrogen bonds: the balance between transition state binding and substrate binding in oxyanion holes. J. Org. Chem. 2010, 75, 1831–1840.
[56] Taylor R, Kennard O. Hydrogen-bond geometry in organic crystals. Acc. Chem. Res. 1984, 17, 320–326.
[57] Platts JA, Howard ST, Bracke BRF. Directionality of hydrogen bonds to sulfur and oxygen. J. Am. Chem. Soc. 1996, 118, 2726–2733.
[58] Murray-Rust P, Glusker JP. Directional hydrogen bonding to sp^2- and sp^3- hybridized oxygen atoms and its relevance to ligand-macromolecule interactions. J. Am. Chem. Soc. 1984,106, 1018–1025.
[59] Simon L, Goodman JM. Hydrogen-bond stabilization in oxyanion holes: grand jeté to three dimensions. Org. Biomol. Chem. 2012, 10, 1905–1913.
[60] Novaković SB, Fraisse B, Bogdanović GA, Spasojević-de Biré A. Experimental charge density evidence for the existence of high polarizability of the electron density of the free electron pairs on the sulfur atom of the thioureido group, NH-C=S-NH_2, induced by N-H···S and C-H···S interactions. Cryst. Growth Des. 2007, 7, 191–195.
[61] Howard DL, Kjaergaard HG. Hydrogen bonding to divalent sulfur. Phys. Chem. Chem. Phys. 2008,10, 4113–4118.
[62] Biswal HS, Shirhatti PR, Wategaonkar S. O-H···O versus O-H···S hydrogen bonding. 2. Alcohols and thiols as hydrogen bond acceptors. J. Phys. Chem. A 2010, 114, 6944–6955.
[63] Kaur D, Aulakh D, Khanna S, Singh H. Theoretical study on the Nature of S···H and O···H hydrogen bonds. J. Sulfur Chem. 2014, 35, 290–303.
[64] Eccles KS, Morrison RE, Sinha AS, Maguire AR, Lawrence SE. Investigating C=S···I halogen bonding for cocrystallization with primary thioamides. Cryst. Growth Des. 2015, 15, 3442–3451.
[65] Grant Hill J. The halogen bond in thiirane···ClF: an example of a Mulliken inner complex. Phys. Chem. Chem. Phys. 2014, 16, 19137–19140.
[66] Le Gal Y, Lorcy D, Jeannin O, Barriere F, Dorcet V, Lieffrig J, Fourmigue M. C=S···I halogen bonding interactions in crystalline iodinated dithiole-2-thiones and thiazole-2-thiones. CrystEngComm 2016, 18, 5474–5481.
[67] Nagels N, Geboes Y, Pinter B, De Proft F, Herrebout WA. Tuning the halogen/hydrogen bond competition: a spectroscopic and conceptual DFT study of some model complexes involving CHF_2I. Chemistry 2014, 20, 8433–8443.
[68] Topić F, Rissanen K. Systematic construction of ternary cocrystals by orthogonal and robust hydrogen and halogen bonds. J. Am. Chem. Soc. 2016, 138, 6610–6616.
[69] Jay JI, Padgett CW, Walsh RDB, Hanks TW, Pennington WT. Noncovalent interactions in 2-mercapto-1-methylimidazole complexes with organic iodides. Cryst. Growth Des. 2001, 1, 501–507.
[70] Robertson CC, Perutz RN, Brammer L, Hunter CA. A solvent-resistant halogen bond. Chem. Sci. 2014, 5, 4179–4183.
[71] Allen FH, Bird CM, Rowland RS, Raithby PR. Resonance-induced hydrogen bonding at sulfur acceptors in $R_1R_2C=S$ and $R1CS_2^-$ systems. Acta Crystallogr. B 1997, 53, 680–695.
[72] Arman HD, Gieseking RL, Hanks TW, Pennington WT. Complementary halogen and hydrogen bonding: sulfur···iodine interactions and thioamide ribbons. Chem Commun 2010, 46, 1854–1856.
[73] Le Questel J-Y, Laurence C, Graton J. Halogen-bond interactions: a crystallographic basicity scale towards iodoorganic compounds. CrystEngComm 2013, 15, 3212–3221.

Edward R.T. Tiekink

13 Crystal chemistry of the isomeric N,N′-bis(pyridin-*n*-ylmethyl)-ethanediamides, $n = 2, 3$ or 4

13.1 Introduction

A working definition for a "multi-component crystal" is a crystal comprising more than one molecular entity with all components present in stoichiometric quantities. This volume, "*Aspects of Multi-Component Crystals: Synthesis, Concepts and Function*", contains many such examples, e.g. solvates, salts and neutral species. Without dispute, co-crystals are attracting greatest attention in the general area of multi-component crystals, comprising molecules that are normally solids under ambient conditions with the co-crystal co-formers (the individual species comprising the multi-component crystal) being in their neutral states. With the latter definition in mind, there are many varied motivations for studying organic co-crystals. These encompass concepts such as modifying luminescence properties, developing new non-linear optical materials, chiral resolution, promoting crystallisation of otherwise uncrystallisable species and the stabilisation of unusual/unstable molecules. In addition to these considerations, the area undoubtedly attracting the most interest is the development of new pharmaceuticals based on co-crystal technology. Recently published reviews provide excellent introductions to the field [1–3].

Molecules that featured in early systematic co-crystallisation investigations form the focus of this chapter; namely the isomeric N,N′-bis(pyridin-*n*-ylmethyl)ethanediamides, $n = 2, 3$ or 4, which are abbreviated as $^2LH_2$, $^3LH_2$ and $^4LH_2$, respectively. As seen from Figure 13.1, these molecules feature diamide moieties, located centrally, flanked by terminal pyridyl rings differing in the location of the pyridyl-N atoms. The presence of these functional groups makes them ideally suited to form co-crystals stabilised by hydrogen and/or halogen bonding. The purpose of this overview is to demonstrate the variety and complexity of co-crystals formed by these compounds. This is complemented by a brief overview of the complexing ability of $^2LH_2$, $^3LH_2$ and $^4LH_2$ and some protonated/deprotonated forms of these towards metal centres, both transition metals and main group elements. This allows both correlations with all-organic analogues and provides clues to the capabilities and versatility of these multi-functional molecules/ions.

The ensuing discussion is broken down as follows: First, a description of available structures of parent molecules is provided. An overview of the co-crystals formed by $^2LH_2$ is then presented, then any salts, followed by complexation patterns to metals in

https://doi.org/10.1515/9783110464955-013

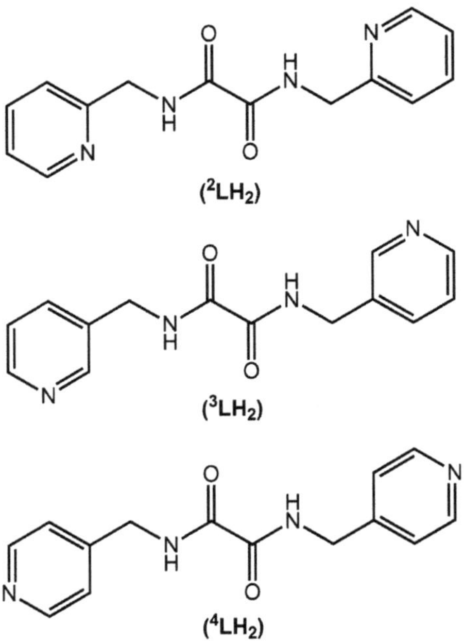

Fig. 13.1: Chemical diagrams for isomeric N,N′-bis(pyridin-*n*-ylmethyl)ethanediamides, n = 2, 3 and 4, showing the abbreviations employed for each isomer.

neutral form and ionic forms. The subsequent sections cover the structural chemistry of the higher homologues, $^3LH_2$ and $^4LH_2$.

Data, in the form of crystallographic information files (CIFs), were extracted from the Cambridge Crystallographic Database [4] and interrogated using PLATON [5]. Diagrams were drawn using DIAMOND [6]. Structures with unresolvable disorders were not included. However, if the disorder was associated with "spectator" species, e.g. counter-ions, solvent, etc., and an analysis of the intermolecular interactions/coordinating ability of $^nLH_2$ was possible, the structure was retained in this review. No other restrictions were applied.

In the case of all-organic species, the focus of the discussion is upon the nature of interactions involving the heteroatoms and of their metal complexes upon the co-ordinating sites, meaning that complete descriptions of the overall molecular packing for most species is beyond the scope of this survey. Rather, the focus of the discussion for metal-containing species is upon the supramolecular association of $^nLH_2$ and derived ionic species via hydrogen bonding interactions.

13.2 Crystal chemistry of $^2LH_2$, $^3LH_2$ and $^4LH_2$

Although the crystal structure of $^2LH_2$ has yet to be reported, two polymorphs are known for each of $^3LH_2$ [7] and $^4LH_2$ [8, 9]. The dithiodiamide analogue of $^2LH_2$ has been described in terms of both experimental and theoretical (DFT) structures [10]. It was suggested that an all planar conformation located about a centre of inversion is most stable owing to the presence of intramolecular amide-N–H···N(pyridyl) hydrogen bonding [10]. With these observations in mind, it is quite possible that a similar conformation, i.e. centrosymmetric and planar applied to $^2LH_2$.

The two polymorphs of $^3LH_2$ have only just recently been described [7], and each crystallises in the monoclinic space group $P2_1/c$. In form I, one entire molecule comprises the asymmetric unit, i.e. $Z' = 1$, whereas in form II there are two independent molecules, each located about a centre of inversion, i.e. $Z' = 2 \times 0.5$. An overlay of the three molecules is shown in Figure 13.2, from which it is apparent that although the central cores are coincident, the relative orientations of the pyridyl rings differ significantly.

Although the emphasis of this chapter is supramolecular chemistry, here, a brief description of the salient features of the central core is given, as these are common to the all-organic structures described herein. For example, their coordination to metals often results in significant differences as highlighted in the relevant sections. The diamide moiety is usually strictly planar with rarely only minor deviations apparent. The arrangement is always anti-periplanar, facilitating the formation of intramolecular amide-N–H···O(amide) hydrogen bonds. Calculations (DFT) show that alternative arrangements are more than 6 kcal mol^{-1} less stable [7]. Finally, being attached to two electronegative substituents, the amide-C–C(amide) bond is usually longer than anticipated for an sp^2-C–C(sp^2) bond [5]. The striking difference between the polymorphs of $^3LH_2$ relates to the disposition of the pyridyl rings, being syn-periplanar in form I and anti-periplanar in both independent molecules in form II. Furthermore, in form I, the pyridyl-N atoms are orientated to the same side of the molecule. However, the observed conformational flexibility does not have great energy significance, as computational chemistry (DFT) shows the differences in energy for the different conformations to be less than 1–2 kcal mol^{-1}, thereby allowing for the adoption of different molecular packing arrangements, which can be quite distinct.

The intermolecular interactions involving the heteroatoms in form I of $^3LH_2$ leads to a supramolecular layer Figure 13.3a, with the corrugated topology related to the

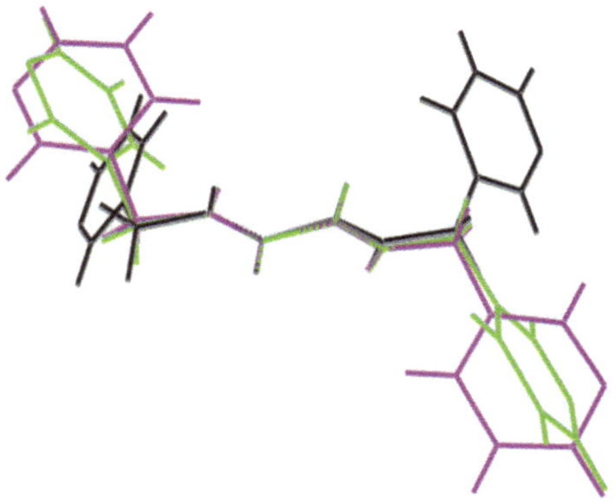

Fig. 13.2: An overlay diagram for the three independent molecules for $^2LH_2$: form I (*black image*) and form II (*green and pink*). The molecules have been overlapped so that the central chromophores are coincident.

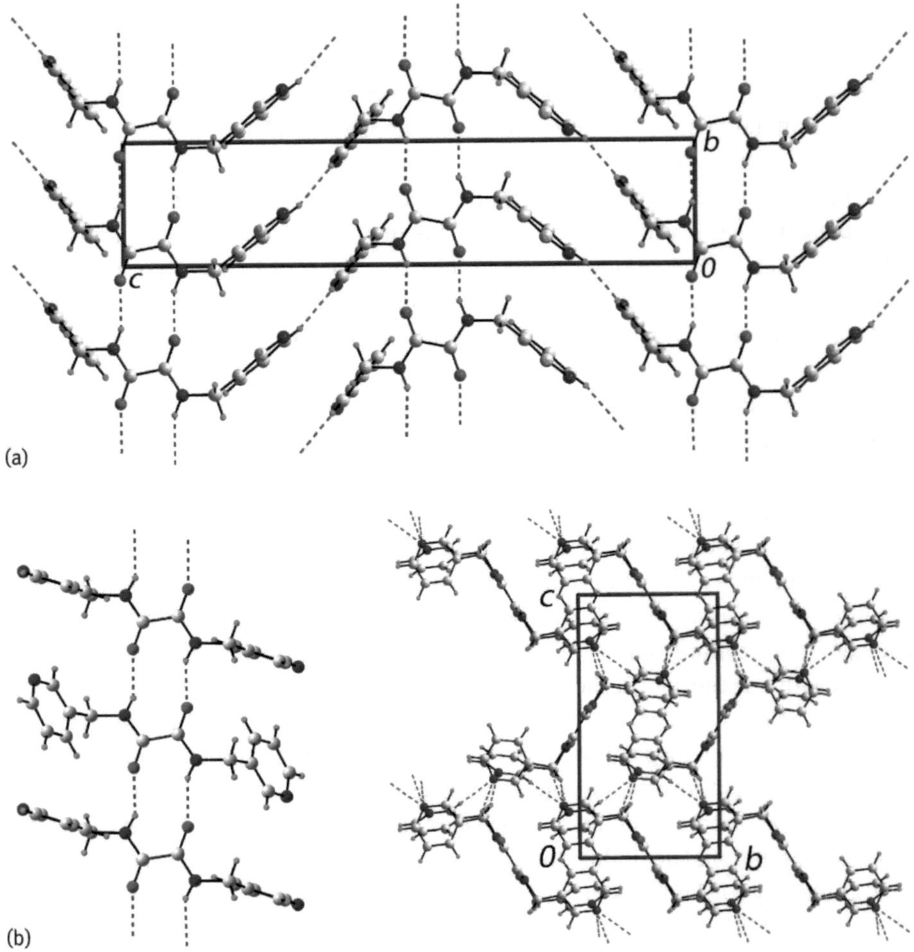

Fig. 13.3: (a) Molecular packing for $^{3}LH_2$, form I, showing the supramolecular layer in the bc-plane, and (b) and (c) molecular packing for $^{2}LH_2$, form II, showing the supramolecular tape and view of the unit cell contents in projection down the a-axis respectively. The amide-N–H···O(amide) hydrogen bonds are shown as *blue dashed lines* whereas the C–H···N(pyridyl) interactions are shown as *pink dashed lines*.

syn-periplanar disposition of the pyridyl rings. The main feature of the molecular packing is the formation of supramolecular tapes involving both amide moieties and strong amide-N–H···O(amide) hydrogen bonds leading to 10-membered $\{\cdots HNC_2O\}_2$ synthons. The tapes are connected into a two-dimensional arrangement by pyridyl-C–H···N(pyridyl) interactions, involving two pyridyl rings but one pyridyl-N atom only as the other pyridyl-N atom does not form an intermolecular interaction within the standard distance criteria assumed in PLATON [5]. Supramolecular tapes, akin to those of form I, are found in the crystal of form II (Figure 13.3b, left), and comprise alternating independent molecules. The tapes are consolidated into the three-dimensional packing by pyridyl- and methylene-C–H···N(pyridyl) interactions involving both molecules (Figure 13.3b, right). Two pyridyl-N atoms do not form significant intermolecular interactions.

As mentioned above, two polymorphs are found for $^4LH_2$ with, in a sense, the opposite characteristics of forms I and II of $^3LH_2$. That is, although both forms of $^4LH_2$ are monoclinic, in form I of $^4LH_2$ [8], two independent molecules, neither with symmetry, comprise the asymmetric unit, i.e. $Z' = 2$, whereas in form II [9], which is also monoclinic, the molecule lies about a centre of inversion, i.e. $Z' = 0.5$. A common feature of each molecule is the anti-periplanar disposition of the pyridyl rings, but, as seen in the overlay diagram of Figure 13.4, in each of the molecules the pyridyl rings have distinct orientations with respect to the central diamide moiety. In the second independent molecule of form I, the twist about the central amide-C–C(amide) is significantly greater, at about 5º, than is normally observed in non-coordinating $^nLH_2$ molecules discussed herein (see green image in Figure 13.4).

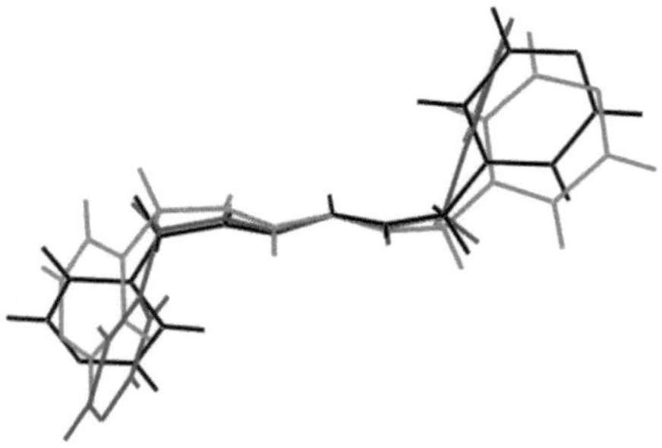

Fig. 13.4: An overlay diagram for the three independent molecules for $^2LH_2$: form I (black and green images) and form II (pink). The molecules have been overlapped so that the central chromophores are coincident.

The molecular packing in the two polymorphic forms of $^4LH_2$ differ greatly. In form I [8], the two independent molecules associate into dimeric aggregates via the 10-membered amide synthon, $\{\ldots HNC_2O\}_2$. The exocyclic amide groups associate with pyridyl-N, i.e. via amide-N–H···N(pyridyl) hydrogen bonds, so that a supramolecular layer is formed (Figure 13.5a). The topology is cross-linked. In this hydrogen bonding scheme there are no apparent roles for the amide-O and half the available pyridyl-N atoms. The presence of pyridyl-C–H···O(amide) interactions, with each amide-O atom accepting two such interactions, link the layers into the three-dimensional crystal (Figure 13.5b).

Supramolecular layers sustained by hydrogen bonding are also evident in the crystal of form II of $^4LH_2$ [9] (Figure 13.6a). However these, with a sinusoidal topology, are sustained exclusively by amide-N–H···N(pyridyl) hydrogen bonding. The layers are linked by pyridyl-C–H···O(amide) contacts (Figure 13.6b). Clearly, in the two forms of $^4LH_2$, there is a trade-off between utilising half or all available pyridyl-N atoms in the hydrogen bonding scheme, forming amide $\{\ldots HNC_2O\}_2$ synthons and compensating C–H···O(amide) interactions.

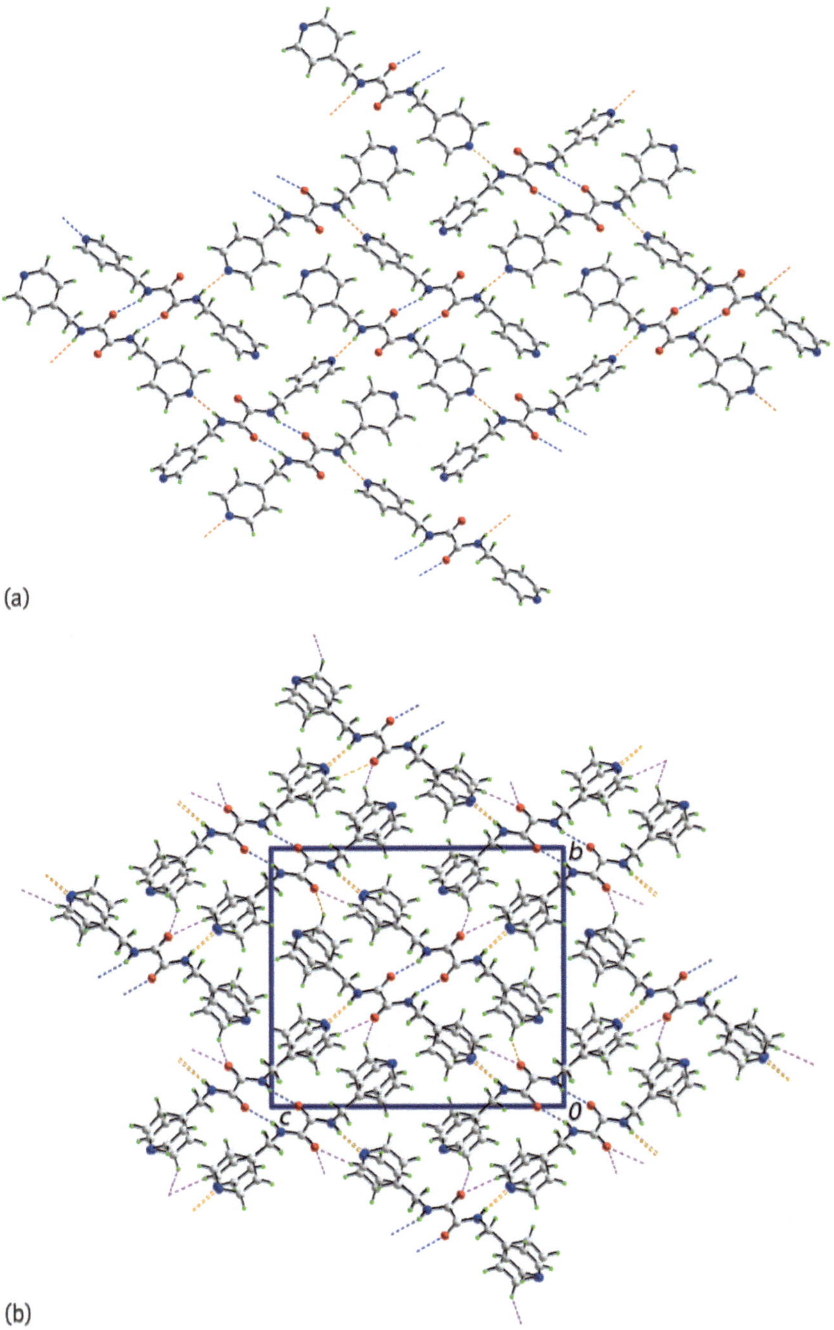

(a)

(b)

Fig. 13.5: Molecular packing for ${}^{4}LH_2$, form I, showing the (a) supramolecular layer in the bc-plane, and (b) view of the unit cell contents in projection down the a-axis. The amide-N–H···O(amide) and amide-N–H···N(pyridyl) interactions are shown as *blue and orange dashed lines*, respectively, whereas the pyridyl-C–H···O(amide) interactions are shown as *pink dashed lines*.

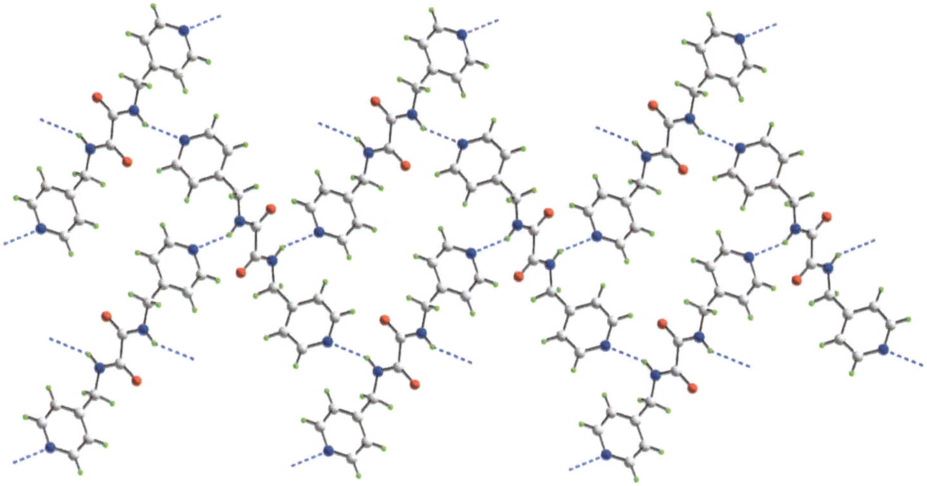

(a)

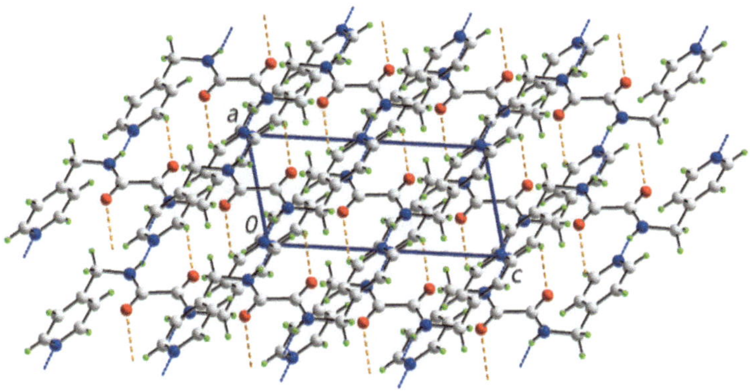

(b)

Fig. 13.6: Molecular packing for $^4LH_2$, form II, showing (a) the supramolecular layer in the bc-plane, and (b) the view of the unit cell contents in projection down the b-axis. The amide-N–H···N(pyridyl) and C–H···O(amide) interactions are shown as *blue and pink dashed lines*, respectively.

13.3 Multi-component crystals of $^2LH_2$ and derivatives

In keeping with the lack of structural data for $^2LH_2$, there is a paucity of structural data for co-crystals/salts involving this isomer. By contrast, there are well over ten coordination compounds containing $^2LH_2$ and ions derived from this. A thorough analysis of the hydrogen bonding patterns of the all-organic species is presented. However, only a summary of key coordination modes and, when applicable, hydrogen bonding involving $^2LH_2$/ions derived from $^2LH_2$ is included for the metal-containing species.

13.3.1 Co-crystal

There is a sole example of a co-crystal involving $^2LH_2$, namely a 1 : 2 co-crystal with 2-aminobenzoic acid and $^2LH_2$ [11]. As $^2LH_2$ lies about a centre of inversion, there is only one unique acid molecule. The key feature of the supramolecular association is the formation of a layer, whereby the pyridyl-N atoms are connected to carboxylic acids via hydroxyl-O–H···N(pyridyl) hydrogen bonds and a seven-membered {C(=O)OH. . . NCH} heterosynthon. This brings about a pyridyl-C–H···O(carbonyl) interaction, which is well known for being robust in co-crystal technology [12]. The amino group is pivotal in the molecular packing by donating, via amino-N–H···O(amide), and accepting, via amide-N–H···N(amino), hydrogen bonds leading to 14-membered $\{\cdots NH\cdots OC_2NH\}_2$ synthons (Figure 13.7). The layers are connected by pyridyl-C–H···O(amide, carbonyl) interactions to consolidate the crystal.

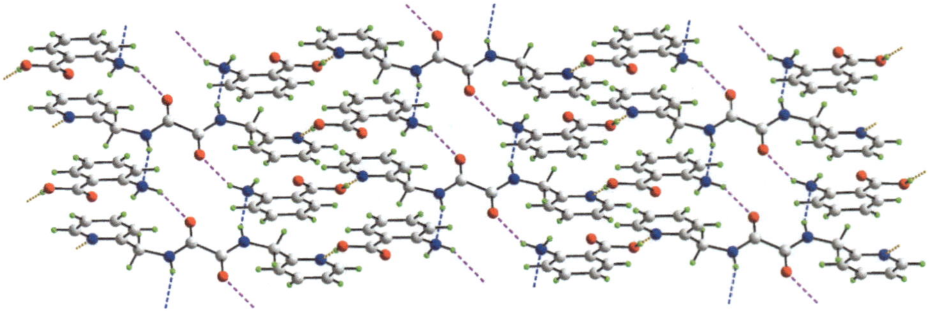

Fig. 13.7: Supramolecular layer in the 1 : 2 co-crystal formed between $^2LH_2$ and 2-aminobenzic acid, with hydroxyl-O–H···N(pyridyl), amide-N–H···O(amide) and amino-N–H···N(amino) hydrogen bonds shown as *orange*, *blue* and *pink dashed lines*, respectively.

13.3.2 Salts

There are two salts in the crystallographic literature with each anion derived from benzoic acid derivatives. The first salt to be described is a 1 : 2 salt between $[^2LH_4]^{2+}$ and 3,5-dicarboxybenzoate [13], where the centrosymmetric cation is doubly protonated at the pyridyl-N atoms, i.e. $[^2LH_4]^{2+}$ is a dipyridinium di-cation. The di-cations associate into a supramolecular tape with translational symmetry via a sequence of amide $\{\ldots HNC_2O\}_2$ synthons akin to that shown in Figure 13.3a for form I of $^3LH_2$ [7]. As illustrated in Figure 13.8a, the tapes are connected into a three-dimensional architecture by charge-assisted pyridinium-N–H···O(carboxylate) and hydroxyl-O–H···O(carboxylate) hydrogen bonds, in addition to eight-membered carboxylic acid dimer $\{\ldots HOCO\}_2$ synthons.

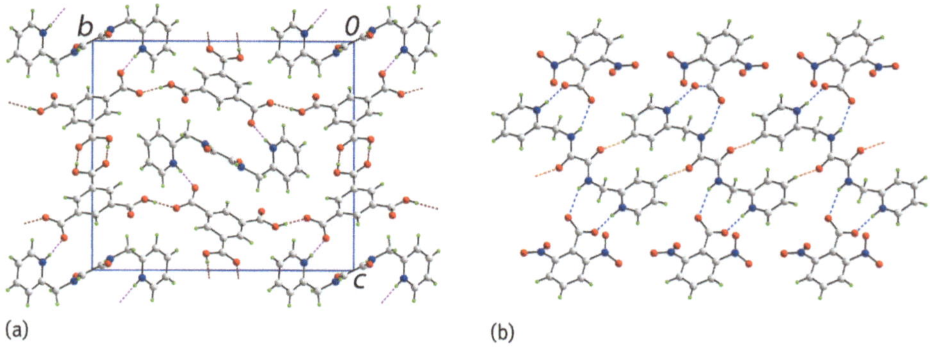

Fig. 13.8: (a) A view of the unit cell contents in projection down the a-axis in the 1 : 2 salt: $[^2LH_4]^{2+}$ and 3,5-dicarboxbenzoate. The amide tapes along the a-axis are connected by hydroxyl-O–H···O(carboxylate) and pyridinium-N–H···O(carboxylate) hydrogen bonds shown as *brown and pink dashed lines*, respectively, and (b) supramolecular layer in the 1 : 2 salt: $[^2LH_4]^{2+}$ and 2,6-dinitobenzoate. The amide- and pyridinium-N–H···O(carboxylate) and pyridyl-C–H···O(amide) interactions are shown as *blue and orange dashed lines*, respectively.

The second salt also has a 1 : 2 stoichiometry and is found between $[^2LH_4]^{2+}$ and 2,6-dinitrobenzoate and again, owing to the di-cation being centrosymmetric, there is only one independent anion in the asymmetric unit [14]. Charge-assisted hydrogen bonding also features in the crystal, whereby the carboxylate groups span adjacent pyridinium-N–H and amide-N–H groups to form a nine-membered heterosynthon {. . . HNC_2NH···OCO}. The resulting three-ion but electrically neutral aggregate associates with symmetry-related neighbours via pyridyl-C–H···O(amide) interactions to form a supramolecular layer, as shown in Figure 13.8b.

13.3.3 Metal-containing species

There are three examples in the literature where $^2LH_2$ coordinates in the neutral mode and, curiously, each displays a distinctive mode of coordination. In the centrosymmetric six-coordinate dimethylplatinum(IV) di-cationic complex [15], where the platinum is further coordinated by a 2,2′-bipyridyl-type ligand and a 4-carboxybenzyl group, the $^2LH_2$ molecule is bidentate, bridging via the pyridyl-N atoms (Figure 13.9a). As the diamide chromophore is flanked on either side by bulky substituents, it does not participate in hydrogen bonding interactions. The terminal carboxylic acid groups associate via $\{\ldots HOCO\}_2$ synthons to form a supramolecular chain (not illustrated). In the mononuclear compound $HgI_2(^2LH_2)_2$ [16], both $^2LH_2$ molecules coordinate in the monodentate mode (Figure 13.9b), with the pendent pyridyl-N atoms participating in neither coordination nor hydrogen bonding interactions. However, each of the independent $^2LH_2$ molecules self-associates to form amide tapes, via $\{\ldots HNC_2O\}_2$ synthons, with the result that a supramolecular tube is con-

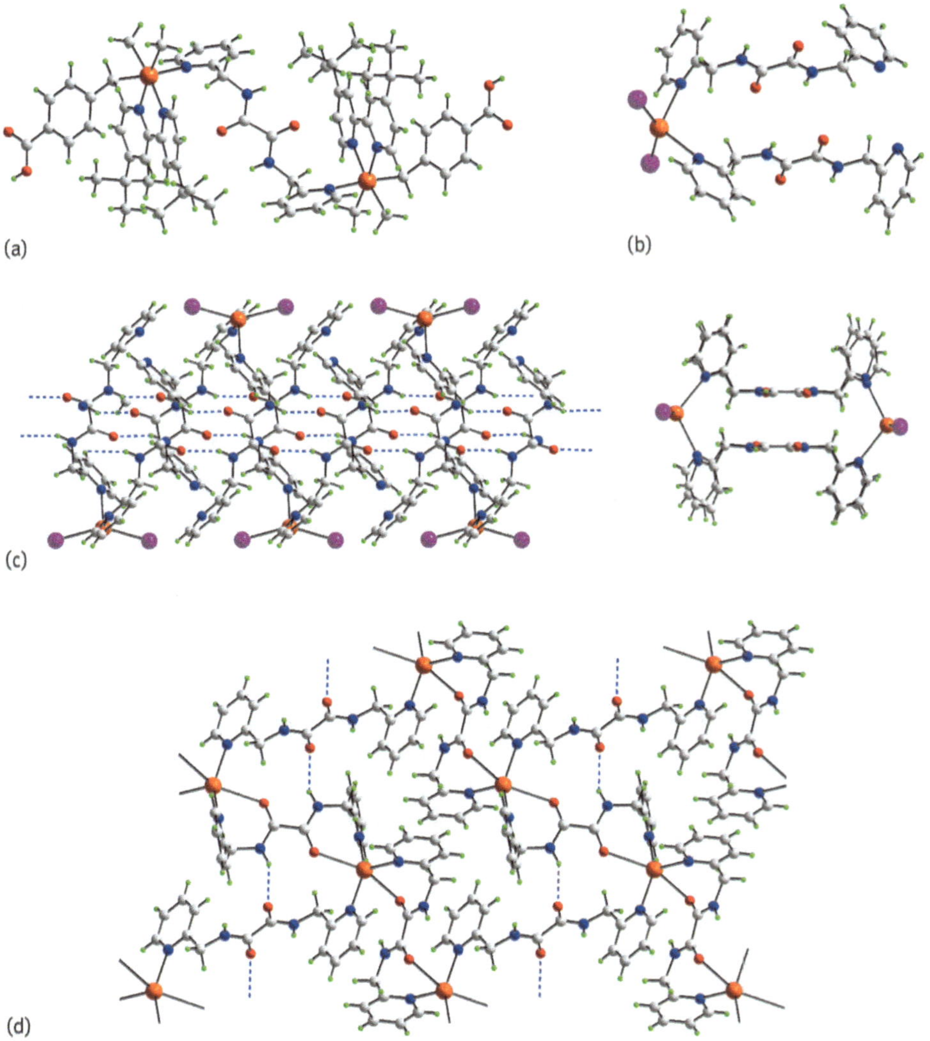

Fig. 13.9: Neutral coordination mode for $^2LH_2$. (a) A bridging mode via pyridyl-N atoms in a dicationic dimethylplatinum(IV) complex, (b) the asymmetric unit in $HgI_2(^2LH_2)_2$ shown as a monodentate mode of coordination via a single pyridyl-N atom, (c) side-on and end-on views of the supramolecular tube in $HgI_2(^2LH_2)_2$ mediated by amide-N–H···O(amide) hydrogen bonding (*blue dashed lines*), and (d) both bidentate (pyridyl-N) and tetradentate (pyridyl-N and amide-O) bridging modes in sheets of $[Ag_2(^2LH_2)_3^{2+}]_n$.

structed (as shown in the side- and end-on views of Figure 13.9c). The third structure is the most interesting of all three and, in fact, displays two distinct coordination modes. In $\{Ag_2(^2LH_2)_3(\text{trifluoromethanesulfonate})_2]_n$ [17] and referring to Figure 13.8d, the silver ions are connected into a two-dimensional grid via one bidentate bridging $^2LH_2$ molecule, coordinating via both pyridyl-N atoms, and two tetradentate $^2LH_2$ molecules, which employ all pyridyl-N and amide-O atoms in coordination. The tetradentate coordinate mode results in the formation of seven-membered Ag–N–C–C–N–C–O chelate rings. Further stability to the array is provided by amide-N–H···O(amide) hydrogen bonds between bi- and tetra-dentate coordinating ligands. The acidic sites not involved in the construction of the aforementioned layer are connected to the trifluoromethanesulfonate anions via amide-N–N···O(trifluoromethanesulfonate) hydrogen bonding to contribute to the stabilisation between layers.

The remaining six structures available in the literature feature the $[^2L]^{2-}$ di-anion [18–21]. In the mononuclear and neutral complex Pd(^{2}L) (Figure 13.10a), the $[^2L]^{2-}$ di-anion wraps around the palladium(II) centre to define a square planar geometry defined by the four nitrogen atoms [18]. In the di-cation of $[(DMF)_2Cu(^2L)Cu(DMF)_2](ClO_4)_2$ (Figure 13.10b), $[^2L]^{2-}$ is hexadentate, employing all six heteroatoms in co-

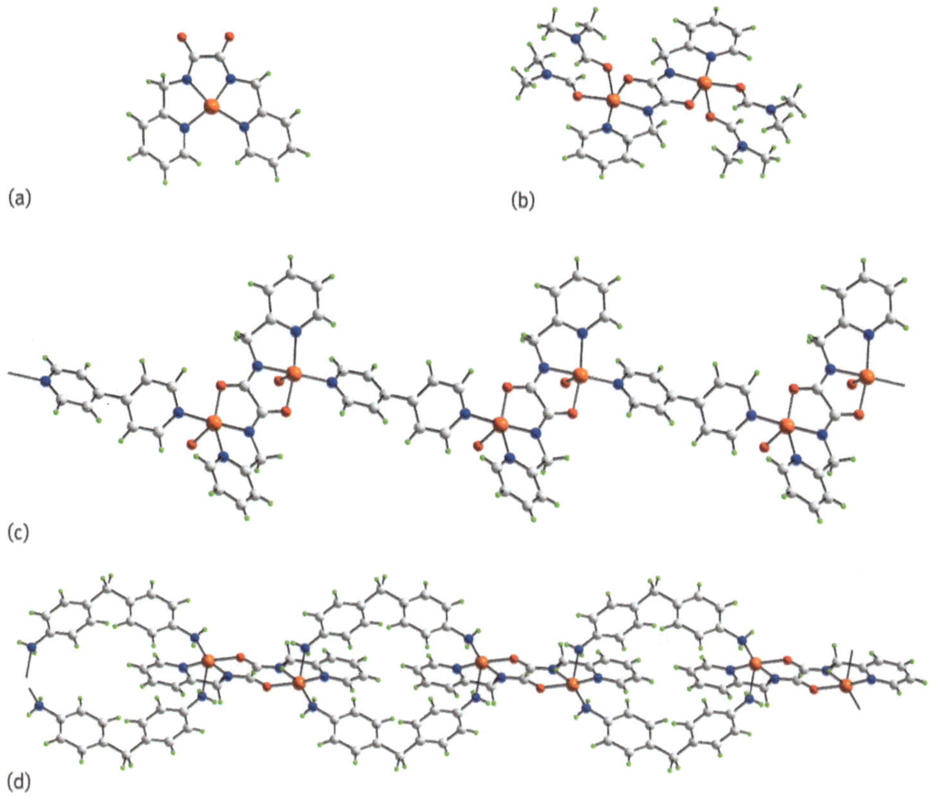

Fig. 13.10: (a) A tetradentate-N_4 coordination mode for $[^2L]^{2-}$ in Pd(^{2}L), (b) a hexadentate N_4O_2 coordination mode for $[^2L]^{2-}$ in $[(DMF)_2Cu(^2L)Cu(DMF)_2]^{2+}$, (c) and (d) hexadentate N_4O_2 coordination modes for $[^2L]^{2-}$ leading to a coordination polymers in each of (c) $\{(H_2O)Cu(^2L)Cu(OH_2)(4,4'\text{–bipyridyl})^{2+}\}_n$ and (d) $\{Cu(^2L)Cu[\text{bis(4–aminophenyl)methyl}]_2^{2+}\}_n$.

ordination with two copper(II) atoms [19]; DMF is *N,N*-dimethylformamide. When DMF is displaced as in $\{[H_2OCu(^2L)Cu(OH_2)](NO_3)2.8H_2O\}_n$, a coordination polymer ensues [20]. A similar coordination mode is also found in each of the next two structures whereby similar binuclear units of the $[Cu(^2L)Cu]$ di-cation, described above, are linked into coordination polymers by bridging 4,4′-bipyridyl ligands, $\{[(H_2O)Cu(^2L)Cu(OH_2)(4,4'\text{-bipyridyl})](NO_3)_2\}_n$ (Figure 13.10c), or two bis(4-aminophenyl)methyl ligands in $[\{Cu(^2L)Cu[\text{bis(4-aminophenyl)methyl}]_2\}(NO_3)_2]_n$ (Figure 13.10d), giving a chain of loops, whereby the coordinated water molecules of $\{[(H_2O)Cu(^2L)Cu(OH_2)(4,4'\text{-bipyridyl})](NO_3)_2\}_n$ are displaced [21].

13.4 Multi-component crystals of $^3LH_2$ and derivatives

Significantly more examples of $^3LH_2$/ions in the forms of co-crystals, salts and metal-containing species have been reported compared with the $^2LH_2$ isomer described in the previous section. Noteworthy is the observation of co-crystals stabilised by halogen bonding, a mono-protonated form, $[^3LH_3]^+$, in addition to a tautomeric form of the $^3LH_2$ structure shown in Figure 13.1.

13.4.1 Co-crystals

Co-crystals involving $^3LH_2$ fall into two classes, namely involving carboxylic acids and halogens, with the former described first. Reflecting the propensity of carboxylic acids to associate with pyridyl-N, the 1 : 2 co-crystal comprising $^3LH_2$ and 4-nitrobenzoic assemble into a three-molecule aggregate. $^3LH_2$ is located about a centre of inversion [22]. As shown in Figure 13.11a, the aggregates are linked into a supramolecular ladder as the amide-N–H atom forms a hydrogen bond with a nitro-O atom. The role of the amide-O atom is to participate in three C–H···O interactions, one hydrogen being derived from a phenyl ring and two from pyridyl rings. These sustain a two-dimensional double layer (Figure 13.11a, lower image). The $^3LH_2$ molecule in its 1 : 2 co-crystal with 2-aminobenzoic acid accepts two hydroxyl-O–H···N(pyridyl) and two amino-N–H···O(amide) hydrogen bonds. At the same time, it donates two amide-N–H···O(carbonyl) hydrogen bonds and is thereby surrounded by six 2-aminobenzoic acid molecules [11], indicating that $^3LH_2$ is saturated in terms of its hydrogen bonding capacity (Figure 13.11b). The hydrogen bonding extends in three dimensions to generate the supramolecular architecture (Figure 13.11c).

The next co-crystal to be described is the 1 : 1 co-crystal between $^3LH_2$ and the bi-functional carboxylic acid, 2,2′-disulfanediyldibenzoic acid [23]. As is evident from Figure 13.11d, a supramolecular tape of linked 10-membered $\{\ldots HNC_2O\}_2$ synthons is formed with the pendant pyridyl-N atoms bridged by the carboxylic acid groups via hydroxyl-O–H···N(pyridyl) hydrogen bonds.

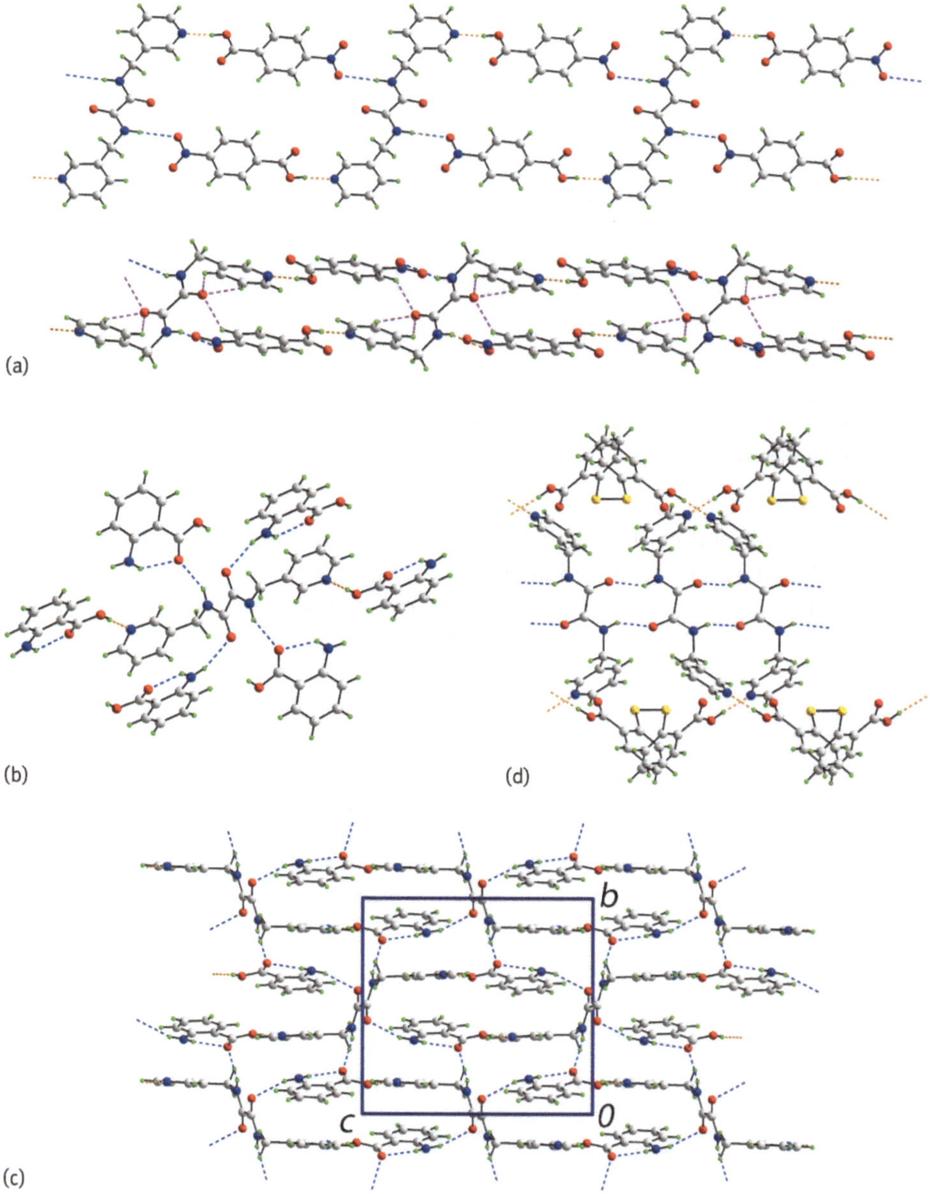

Fig. 13.11: (a) Supramolecular aggregation in the 1 : 2 co-crystal formed between $^{3}LH_2$ and 4-nitrobenzoic acid: ladder and double layer, (b) hydrogen-bonding interactions involving $^{3}LH_2$ and 2-aminobenzoic acid in the 1 : 2 co-crystal, showing the immediate environment about one $^{3}LH_2$ molecule and (c) showing the overall three-dimensional architecture, and (d) a supramolecular chain in the 1 : 1 co-crystal comprising $^{3}LH_2$ and 2,2′-disulfanediyldibenzoic acid. The hydroxyl-O–H···N(pyridyl) and amide/amine-N–H···O hydrogen bonds are shown as *orange and blue dashed lines*, respectively.

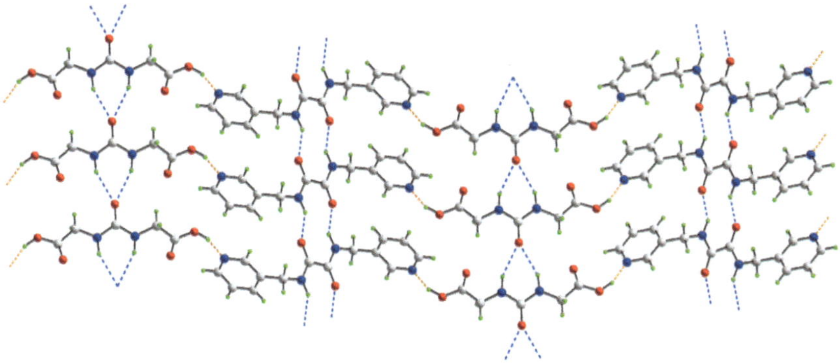

Fig. 13.12: Supramolecular layer in the 1 : 1 co-crystal of $^3LH_2$ with N,N′-dicarboxymethylurea. The hydroxyl-O–H···N(pyridyl) and amide-N–H···O hydrogen bonds are shown as *orange and blue dashed lines*, respectively.

In the 1 : 1 co-crystal of $^3LH_2$ with N,N′-dicarboxymethylurea [24], supramolecular tapes mediated by ten-membered amide synthons, $\{\ldots HNC_2O\}_2$, are formed by $^3LH_2$. Related tapes are formed by the N,N′-dicarboxymethylurea molecules as a result

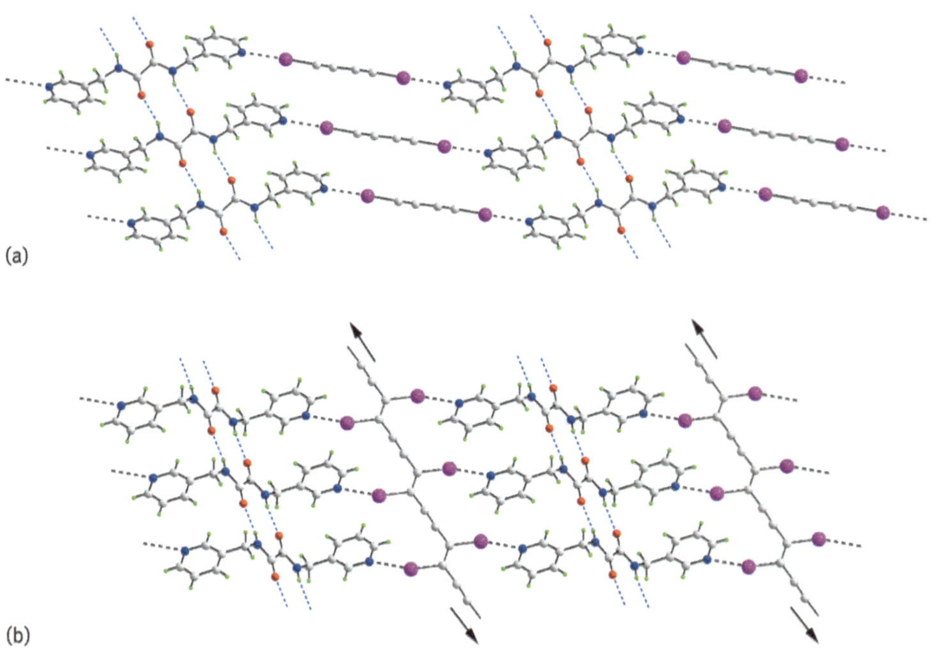

Fig. 13.13: Combination of hydrogen and N···I halogen bonding leading to two-dimensional arrays in 1 : 1 co-crystals of $^3LH_2$ with (a) 1,4-di-iodobuta-1,3-diyne and (b) polymeric hemikis(iododiacetylene). The amide-N–H···O(amide) hydrogen bonding and halogen bonding are shown as *blue and black dashed lines*, respectively.

of urea-N–H···O(carbonyl) hydrogen bonds via six-membered {...O···HNCNH} synthons as the carbonyl-O atom accepts two hydrogen bonds (Figure 13.12). Hydrogen bond links between the chains are hydroxyl-O–H···N(pyridyl), resulting in a two-dimensional array. A similar array is found in the related 1 : 1 co-crystal with N,N′-diglycine oxamide, whereby chains of amide tapes formed by each of the co-crystal co-formers are linked by hydroxyl-O–H···N(pyridyl) hydrogen bonds [24].

As mentioned in the preamble to this chapter, isomeric $^{n}LH_2$ molecules featured in early co-crystal investigations were based on halogen bonding. In this context, there are five co-crystals of $^{3}LH_2$ in the literature featuring halogen bonding [25–28] and four of these adopt similar two-dimensional arrays. One such array is illustrated for the 1 : 1 co-crystal formed between $^{3}LH_2$ and 1,4-di-iodobuta-1,3-diyne [25], where $^{3}LH_2$ molecules are linked into supramolecular tapes via 10-membered $\{...HNC_2O\}_2$ synthons. These, in turn, are linked laterally by N···I halogen bonding, as illustrated in Figure 13.13a. More recently, a 1 : 1 co-crystal of $^{3}LH_2$ with a polymer was described, namely with hemikis(iododiacetylene) (Figure 13.13b) [26]. As mentioned above in the Introduction (Section 13.1), unstable molecules can be stabilised by co-crystal formation and this is realised in the 1 : 1 co-crystal formed between $^{3}LH_2$ and 1,4-dibromobuta-1,3-diyne [27].

In contrast, supramolecular aggregation in the 1 : 1 co-crystal formed between $^{3}LH_2$ and 2,3,5,6-tetrafluoro-1,4-di-iodobenzene is quite distinct from the co-crystal structures just described, with each co-former being centrosymmetric [28]. This struc-

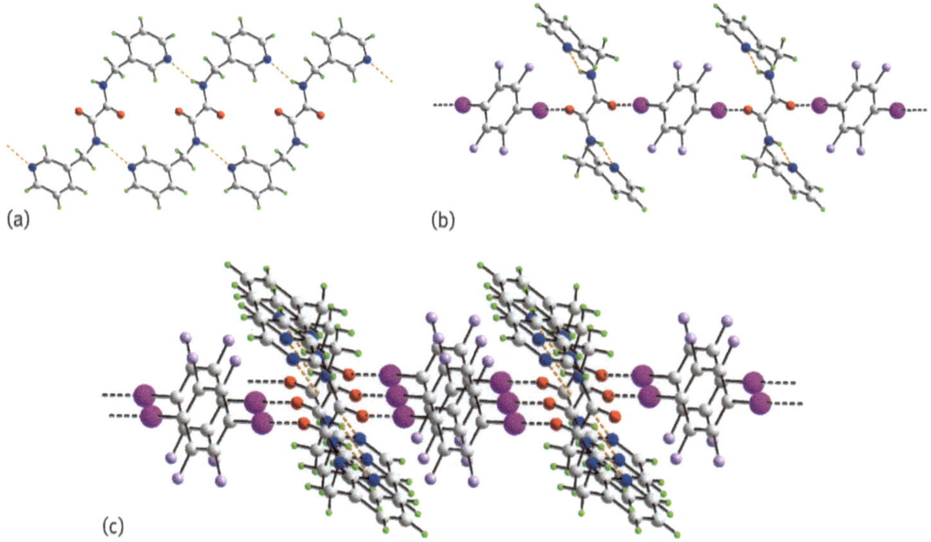

Fig. 13.14: Supramolecular aggregation in the 1 : 1 co-crystal of $^{3}LH_2$ with 2,3,5,6-tetrafluoro-1,4-di-iodobenzene: (a) supramolecular tape sustained by amide-N–H···N(pyridyl) hydrogen bonds, (b) chain sustained by O···I halogen bonding and (c) two-dimensional array. The amide-N–H···N(pyridyl) hydrogen bonding and halogen bonding are shown as *orange and black dashed lines*, respectively.

ture is unusual in that, while translationally related molecules of $^3LH_2$ self-assemble into linear supramolecular chains, the association is via amide-N–H⋯N(pyridyl) hydrogen bonds (Figure 13.14a) resulting in 18-membered {…HNC$_2$NC$_3$N}$_2$ synthons. This arrangement frees up the amide-O atoms to form O⋯I halogen bonds, leading to supramolecular chains (Figure 13.14b). The net result is a two-dimensional array, as shown in Figure 13.14c.

13.4.2 Salts

A single example of a salt derived from $^3LH_2$ is known, namely between $[^3LH_4]^{2+}$ and two equivalents of 2,6-dinitrobenzoate [14]. As the di-cation lacks symmetry, two independent anions are found in the asymmetric unit and each forms quite distinct interactions with the di-cation. The di-cation is protonated at each pyridyl-N atom. One carboxylate group bridges the amide-N–H and pyridinium-N–H atoms derived from two molecules and through the application of a centre of inversion, 20-membered {…OCO⋯HNC$_3$NH…}$_2$ synthons arise from these interactions. The second carboxylate group bridges the equivalent atoms as just described, but employs a single oxygen atom only, with the result that smaller 16-membered {…O⋯HNC$_3$NH…}$_2$ synthons are formed. The supramolecular assembly arising from the hydrogen bonding is a chain as illustrated in Figure 13.15. The benzoate atom not participating in the hydrogen bonding scheme just described participates in C–H⋯O interactions, which, along with other C–H⋯O(nitro) contacts, consolidates the molecular packing in three dimensions.

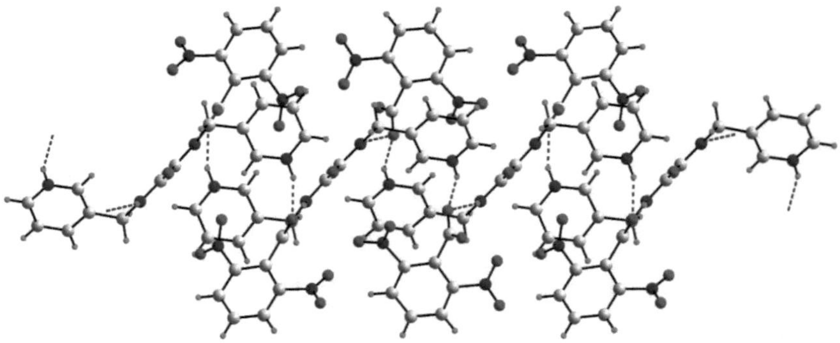

Fig. 13.15: Supramolecular chain in the 1 : 2 salt formed between $[^3LH_4]^{2+}$ and 2,6-dinitrobenzoate. The amide-N–H⋯O(carboxylate) and charge-assisted pyridinium-N–H⋯O(carboxylate) hydrogen bonds are shown as *blue dashed lines*.

13.4.3 Metal-containing species

A number of metal complexes of $^3LH_2$ and derivatives are available. Several of these feature a bidentate bridging mode exemplified in Figure 13.16a, whereby two $Zn[S_2CN(Me)CH_2CH_2OH]_2$ entities are linked by $^3LH_2$, utilising both pyridyl-N atoms [29]. In the molecular packing, the $Zn[S_2CN(Me)CH_2CH_2OH]_2$ residues associate about a centre of inversion via hydroxyl-O–H···O(hydroxyl) hydrogen bonds, leading to supramolecular chains and open 28-membered $\{\ldots HOC_2NCSZnSCNC_2O\}_2$ synthons. The voids defined by the aforementioned synthons are sufficiently large to accommodate centrosymmetrically related chains whereby the $^3LH_2$ ligands of one chain thread through the cavities of the other to form a doubly interpenetrated chain, as shown in Figure 13.16b. Similar bridging modes for $^3LH_2$ are found in binuclear complexes where the pendant groups are Zn(phthalocyaninato) complexes with the bulky nature of the terminal residues precluding supramolecular association involving the diamide functionality [30]. In another binuclear complex, two $^3LH_2$ molecules link two palladium(II) centres in the centrosymmetric, tetra-positive cation $[(dppm)Pd(^3LH_2)_2Pd(dppm)]^{4+}$ [31]; dppm is $Ph_2PCH_2PPh_2$. Within the bridging region, there are intramolecular amide-N–H···O(amide) hydrogen bonds, leading to a 10-membered $\{\ldots HNC_2O\}_2$ synthon. The exocyclic amide-N–H atoms form hydrogen bonds to oxygen atoms of trifluoromethanesulfonate anions. In adducts of binuclear $[Co(oxalate)]_2$, i.e. $\{[Co(oxalate)(^3LH_2)]_2\}_n$, each octahedral cobalt centre forms two Co–N(pyridyl) bonds derived from two bidentate, bridging $^3LH_2$ ligands to form a two-dimensional array [32].

In an unprecedented example of a tautomeric form of $^3LH_2$, the dimine-dihydroxyl tautomer, i.e. $3\text{-}NC_5H_4CH_2N{=}C(OH){-}C(OH){=}NCH_2C_5H_4N$, is observed in its 1 : 2 adduct with $Zn[S_2CN(Me)CH_2CH_2OH]_2$, i.e. a tautomer of the structure shown in Figure 13.16a [29]. As shown in Figure 13.16c, the overall molecular conformations of the two structures are comparable. Indeed, similar inter-woven double chains are again found, but this time mediated by hydroxyl-O–H···N(imine) hydrogen bonds. In a variation on the above, the $[^3LH_4]^{2+}$ di-cation can also function as a bidentate, bridging ligand. Thus, in the supramolecular chain of $\{[^3LH_2][EuCl_3(OH_2)]\}_n$ [33] (Figure 13.16d), $[^3LH_4]^{2+}$ employs the amide-O atoms in linking successive europium atoms. The charge balance is provided by non-coordinating chloride anions.

The final three structures to be described in this section also feature bidentate, bridging $^3LH_2$ ligands [16, 34]. A view of one of these, i.e. $\{[Ag(^3LH_2)_2]BF_4\}_n$ [34], is illustrated in Figure 13.17. Each $^3LH_2$ ligand is bidentate, bridging with the resulting N_4 donor set for silver(I), defining a tetrahedral geometry. The molecules are assembled into a three-dimensional array by linear tapes of interlinked $\{\ldots HNC_2O\}_2$ synthons (not evident in Figure 13.17). This arrangement leads to cross-like channels in which reside the BF_4^- anions. A very similar structure is found in the nitrate analogue [34]. To a first approximation, a similar three-dimensional architecture is found in $\{[Cu(^3LH_2)_2Br].Br.H_2O\}_n$ [16]. Each of the $^3LH_2$ and coordinated bromide ligands

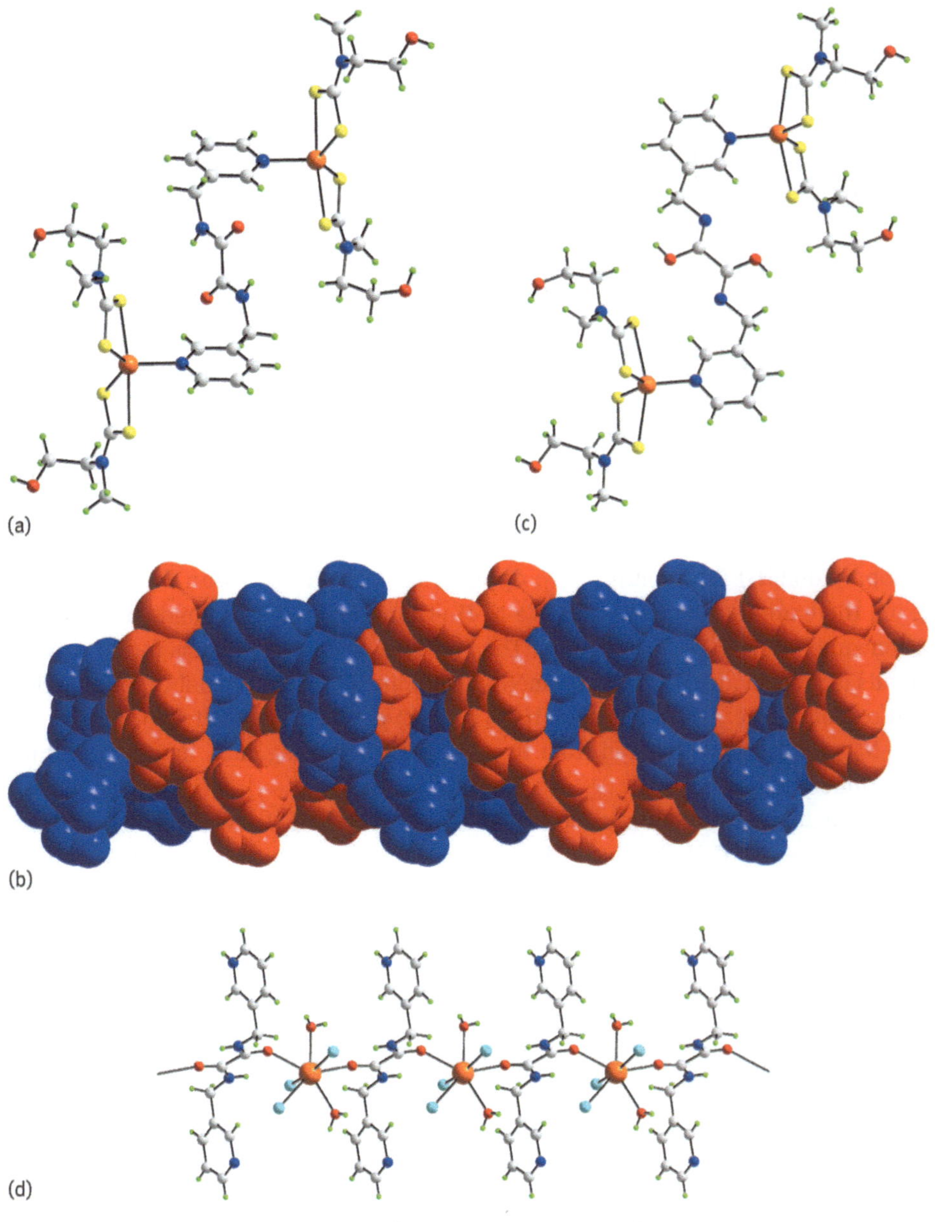

Fig. 13.16: Bidentate, bridging modes for [$^3LH_2$] and derivatives: (a) [$^3LH_2$] bridging two $Zn[S_2CN(Me)CH_2CH_2OH]_2$ entities, (b) interwoven double chains from the supramolecular chain formed by the molecule shown in (a), (c) imine-hydroxyl tautomeric form of [$^3LH_2$], bridging two $Zn[S_2CN(Me)CH_2CH_2OH]_2$ entities, and (d) [$^3LH_4$]$^{2+}$ bridging $[EuCl_3(OH_2)]^-$ anions within a chain.

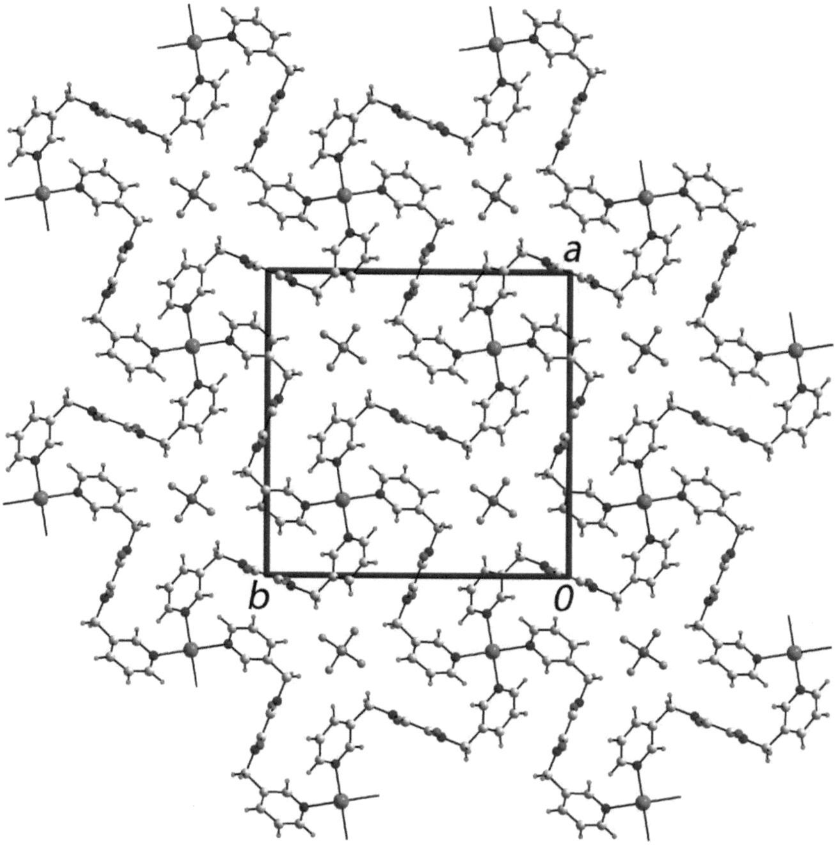

Fig. 13.17: A view of the three-dimensional architecture showing in projection down the c-axis in the structure of $\{[Ag(^3LH_2)_2]BF_4\}_n$.

is bidentate-bridging leading to trans-Br_2N_4 coordination geometries for copper(II). The formation of diamide tapes, mediated by $\{...HNC_2O\}_2$ synthons, reinforces the Cu–Br–Cu bridges.

13.5 Multi-component crystals of $^4LH_2$ and derivatives

The most frequently studied of the isomeric $^nLH_2$ molecules by far is the one with $n = 4$. As with the previous section, co-crystals are described first followed by salts and then metal-containing species. The discussion on co-crystals begins with those involving carboxylic acids, then hydroxyl-containing molecules and finally, those that form halogen-bonding interactions.

13.5.1 Co-crystals

There are five co-crystals in the literature involving $^4LH_2$ and carboxylic acids, one of which is a monofunctional carboxylic acid [35] and the other four are bifunctional, having two carboxylic acid residues [24, 36, 37]. In the former, $^4LH_2$ is located about a centre of inversion and so in this 1 : 2 co-crystal, there is one independent molecule of (4-nitrophenyl)acetic acid [35]. As might now be anticipated based on related structures described earlier, a three-molecule aggregate is formed, which is stabilised by hydroxyl-O–H⋯N(pyridyl) hydrogen bonding. These aggregates are linked into a linear supramolecular chain via amide-N–H⋯O(hydroxyl) hydrogen bonds leading to the formation of 14-membered $\{\ldots HNC_4N\}_2$ synthons (Figure 13.18). The amide-O forms a weak C–H⋯O interaction within the chain.

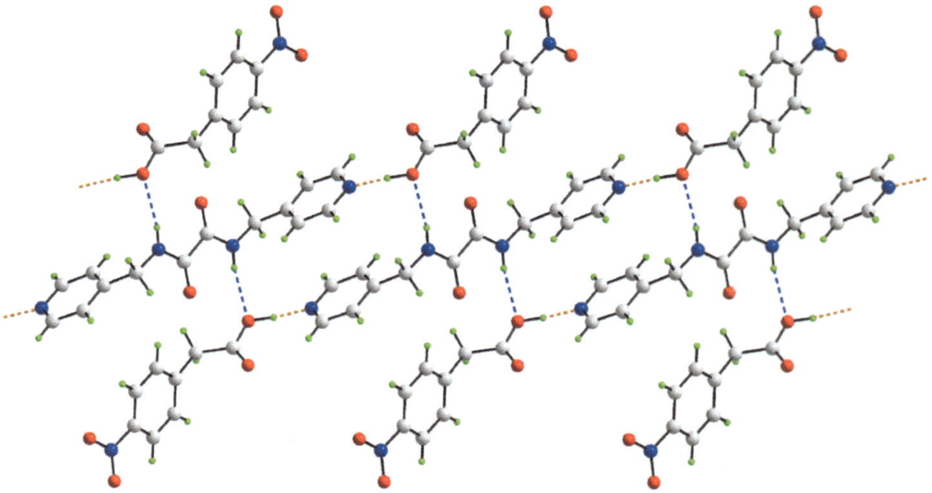

Fig. 13.18: Supramolecular chain in the 2 : 1 co-crystal of [$^4LH_2$] with (4-nitrophenyl)acetic acid. The hydroxyl-O–H⋯N(pyridyl) and amide-N–H⋯O(hydroxyl) hydrogen bonds are shown as *orange and blue dashed lines*, respectively.

The common features of the remaining 1 : 1 co-crystals containing bifunctional carboxylic acids are the formation of amide tapes, via 10-membered $\{\ldots HNC_2O\}_2$ synthons, and the formation of hydroxyl-O–H⋯N(pyridyl) hydrogen bonding, which links the tapes into two-dimensional arrays. In the co-crystal with N,N′-bis(carboxymethyl) urea [24], the urea molecules are connected by urea-N–H⋯O(carbonyl) hydrogen bonds (Figure 13.19a), whereas in the co-crystals with N,N′-bis(2-hydroxymethyl)oxalamide (Figure 13.19b) and N,N′-bis(2-hydroxyethyl)oxalamide [36], the links between each of these co-formers are also of the type amide-N–H⋯O(amide). In an interesting variation, a polymer, namely poly(1,2-bis(2-carboxyethyl)tetra-1-en-3-yn-1,4-diyl, provides the bridges between amide tapes (Figure 13.19c) [37].

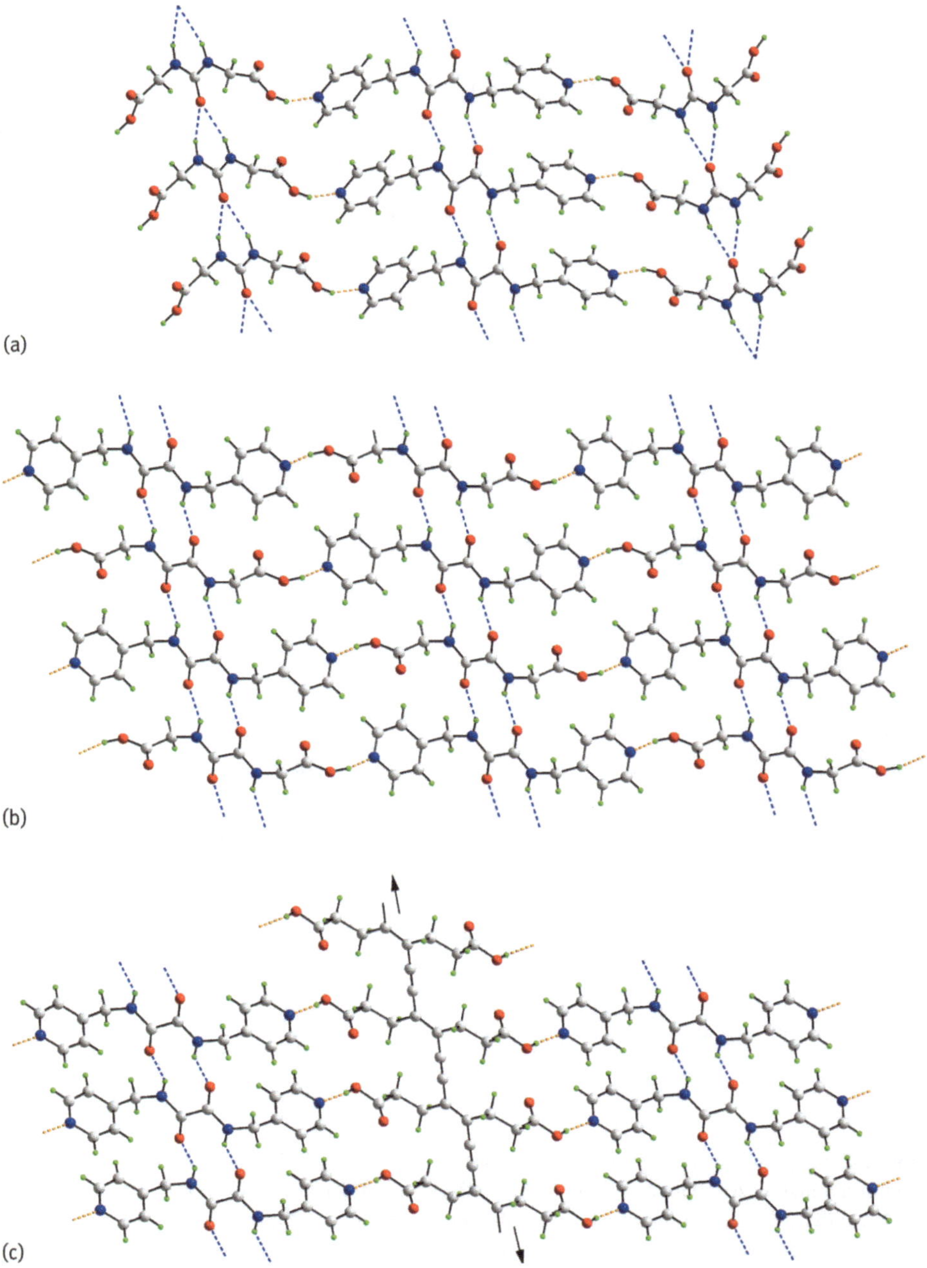

(a)

(b)

(c)

Fig. 13.19: Two-dimensional supramolecular arrays of 1 : 1 co-crystals of [$^4LH_2$] with (a) N,N′-bis(carboxymethyl)urea, (b) N,N′-bis(2-hydroxymethyl)oxalamide and (c) poly(1,2-bis(2-carboxyethyl)tetra-1-en-3-yn-1,4-diyl. The hydroxyl-O–H···N(pyridyl) and amide-N–H···O(amide) hydrogen bonds are shown as *orange and blue dashed lines*, respectively. In (a), the amide-N–H···O(carbonyl) hydrogen bonds are shown as *blue dashed lines*.

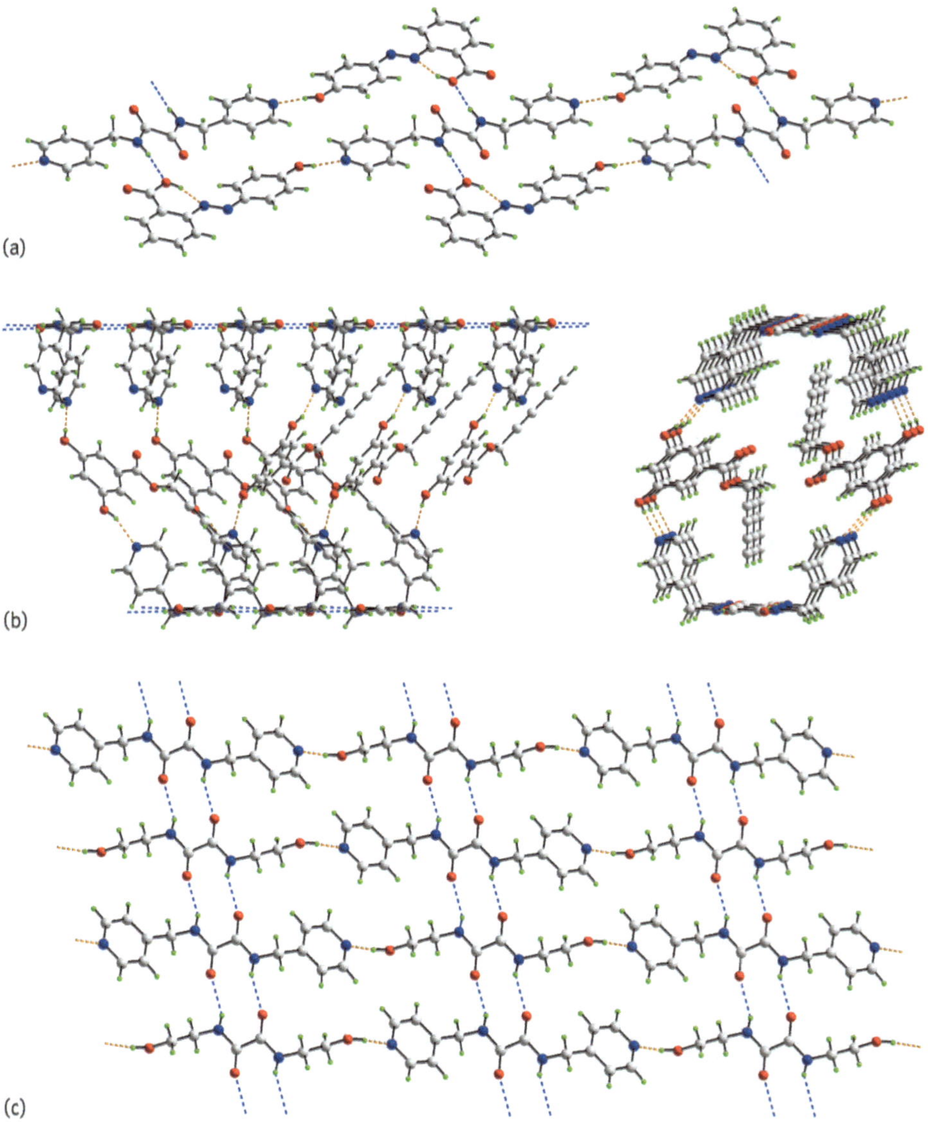

Fig. 13.20: Supramolecular association in co-crystals of [$^4LH_2$]: (a) 1 : 2 with bis(2-((4-hydroxyphenyl)diazenyl)benzoic acid), (b) 1 : 1 with penta-2,4-diynyl 3,5-dihydroxybenzoate and (c) 1 : 1 with N,N′-bis(2-hydroxyethyl)oxalamide. The hydroxyl-O–H···N(pyridyl) and hydroxyl-O–H···N(azo) hydrogen bonds are shown as *orange dashed lines* and the amide-N-H···O(hydroxyl) and amide-N–H···O(amide) hydrogen bonding are shown as *blue dashed lines*.

There are four structures where the co-former with $^4LH_2$ has hydroxyl substituents and these are discussed next as a group. The 1 : 2 co-crystal formed with 2-[(4-hydroxyphenyl)diazenyl]benzoic acid, both hydroxyl and carboxylic acid residues, are present in the co-former [38], thereby providing a nice segue between the previously described co-crystal structures and those under consideration here. This structure is particularly notable for the absence of the hydroxyl-O–H···N(pyridyl) hydrogen bonding normally observed in the carboxylic acid co-crystal structures described herein. Instead, an intramolecular hydroxyl-O–H···N(azo) hydrogen bond is formed to close a six-membered loop. Links between the co-formers are hydrogen bonds of the type hydroxyl-O–H···N(pyridyl) and amide-N–H···O(hydroxyl), so that comparatively large 40-membered $\{\ldots HNC_2NC_4N\cdots HOC_4N_2\ldots HO\}_2$ synthons are formed (Figure 13.20a). The supramolecular assembly arising from the specified hydrogen bonding is a linear supramolecular chain; the amide-O atom does not form a significant intermolecular contact. Two very similar architectures are found in 1 : 1 co-crystals with penta-2,4-diynyl 3,5-dihydroxybenzoate (Figure 13.20b), and with the polymer, poly(penta-2,4-diynyl) [39]. In the crystal of the former, the pyridyl rings adopt, to a first approximation, a syn-periplanar conformation and associate into amide tapes via $\{\ldots HNC_2O\}_2$ synthons. Two such tapes are connected by hydroxyl-O–H···N(pyridyl) hydrogen bonds to generate a supramolecular tube, propagated by a 2_1 screw axis, as shown in the two views of Figure 13.20b. Linear, supramolecular amide tapes mediated by $\{\ldots HNC_2O\}_2$ synthons are also found in the fourth co-crystal in this sub-category; namely the 1 : 1 co-crystal between $^{4}LH_2$ and N,N′-bis(2-hydroxyethyl)oxalamide [36].

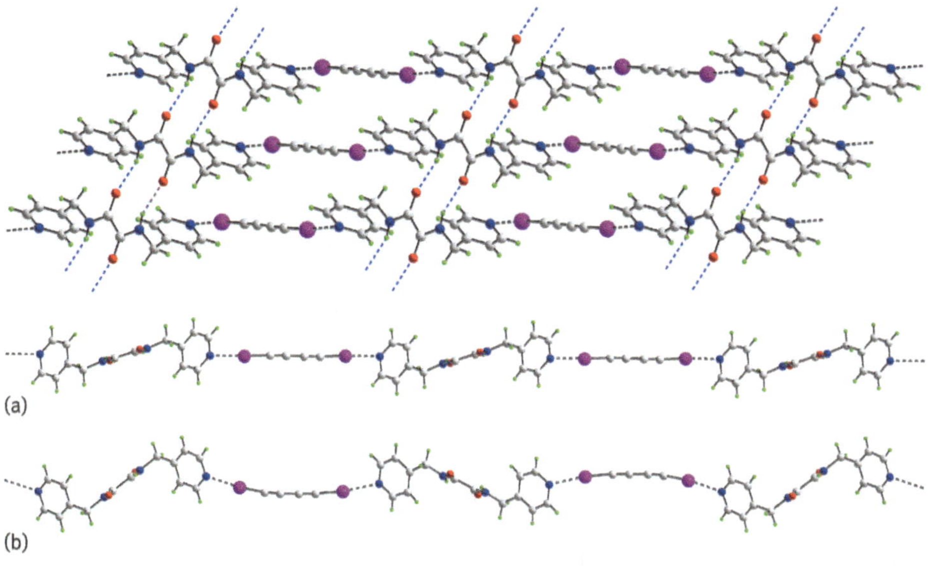

Fig. 13.21: Supramolecular aggregation in the 1 : 1 co-crystals of $^{4}LH_2$ with 1,4-di-iodobuta-1,3-diyne: (a) planar and side-on views of the two-dimensional array sustained by amide-N–H···O(amide) hydrogen bonds and I···N halogen bonds on the co-crystal determined at 3 GPa, and (b) side-on view of the same co-crystal determined under ambient conditions. The amide-N–H···O(amide) hydrogen bonding and halogen bonding are shown as *blue and black dashed lines*, respectively.

For the closely related di-carboxylic acid analogues described above, the tapes are connected into two-dimensional arrays by hydroxyl-O–H···N(pyridyl) hydrogen bonding, with each tape mediated by amide-N–H···O(amide) hydrogen bonding and $\{\ldots HNCO\}_2$ synthons, as shown in Figure 13.20c.

Two 1 : 1 co-crystals sustained by both hydrogen bonding and halogen bonding have been described above, i.e. with 1,4-di-iodobuta-1,3-diyne [25, 40] and with 1,6-di-iodohexa-1,3,5-triyne [25]. These closely resemble the supramolecular association in analogues formed with $^3LH_2$. The former is of particular interest as the structure has been determined under ambient and high pressure (3 GPa) conditions. Two views of the supramolecular aggregation are shown in Figure 13.21a, from which it can be seen that the two-dimensional array is clearly flatter in the high-pressure form than that determined under normal pressure conditions (Figure 13.21b) [25].

13.5.2 Salts

Two salts, both containing $[^4LH_4]^{2+}$ di-cations, have been reported [10, 14]. In the di-nitrate salt (Figure 13.22a) where the di-cations are centrosymmetric, charge-assisted pyridinium-N–H···O(nitrate) and amide-N–H···O(nitrate) hydrogen bonds are evident [10]. In essence, the nitrate anion employs one oxygen atom to bridge symmetry-related molecules so that a three-dimensional architecture ensues. Although there is no formal role for the amide-O atoms in this scheme, it is noted that these form close contacts with pyridinium-C–H hydrogen atoms. In the second salt, the $[^4LH_4]^{2+}$ di-cations are also located about a centre of inversion and similar charge-assisted hydrogen bonding is observed to that just described, but involving a single benzoate-oxygen atom only, i.e. pyridinium-N–H···O(benzoate) and amide-N–H···O(benzoate) hydrogen bonds, leading to 22-membered $\{\ldots HNC_2NC_4NH\}_2$ synthons [14]. This contrasts with the more open arrangement in $[^4LH_4][NO_3]_2$ [10], resulting in a linear supramolecular chain that is formed instead, with a flat topology (Figure 13.22b). The amide-O atoms form very weak contacts with pyridinium-C–H hydrogen atoms.

13.5.3 Metal-containing species

The coordination chemistry of $^4LH_2$ is relatively simple compared with those exhibited by the $^2LH_2$ and $^3LH_2$ isomers in that it is restricted, with one exception, to bidentate bridging coordination modes. This bridging mode is clearly illustrated in Figure 13.23a for the di-cation in $[Ph_3PAu(^4LH_2)AuPPh_3][ClO_4]_2$ [41]. The presence of charge-assisted amide-N–H···O(perchlorate) hydrogen bonding restricts further supramolecular association between $^4LH_2$ entities. In the same way, steric effects in a congested zinc porphyrinate complex restrict supramolecular association involving $^4LH_2$ [30]. In a binuclear six-coordinate dimethylplatinum(IV) di-cationic

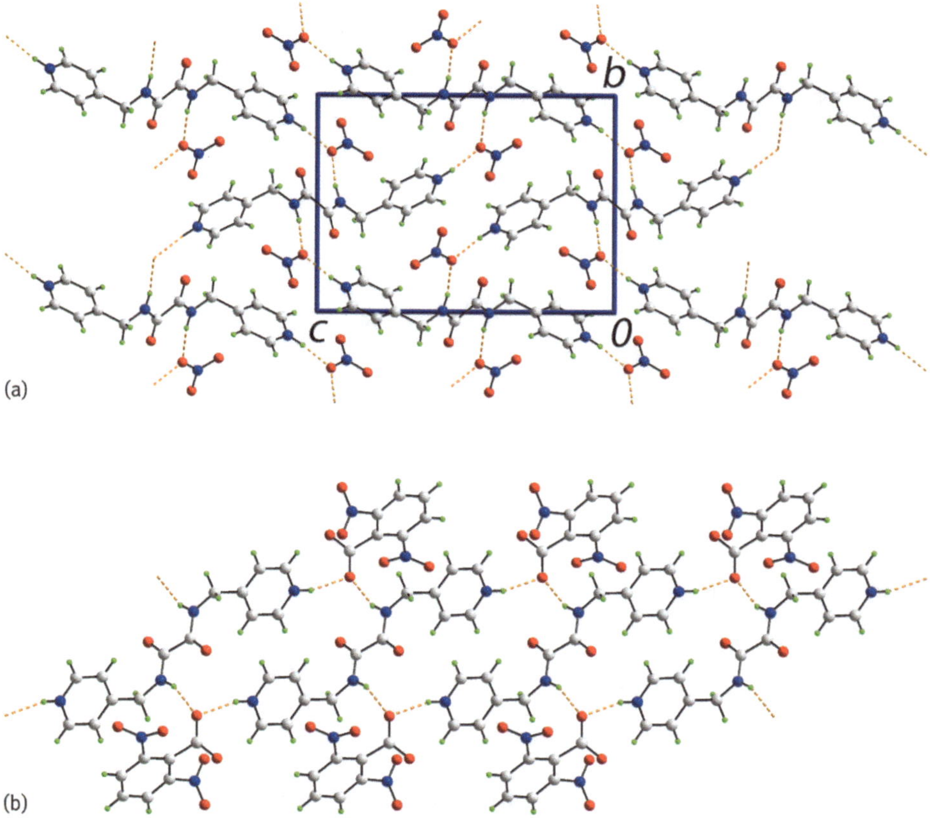

Fig. 13.22: Supramolecular association in salts of $[^{4}LH_4]^{2+}$. (a) 1 : 2 salt with nitrate, and (b) 1 : 2 salt with 2,6-dinitrobenzoate. The charge-assisted N–H···O hydrogen bonds are shown as *orange dashed lines*.

complex [15], the platinum atoms are bridged by a $^{4}LH_2$ molecule (Figure 13.23b). Each platinum centre is also coordinated by a 2,2′-bipyridyl-type ligand and a 4-caroxybenzyl group. In this case, the diamide is accessible for hydrogen bonding and forms a $\{\ldots HNC_2O\}_2$ synthon via hydrogen bonds. The terminal 4-carboxybenzyl groups also self-associate via the carboxylic acid synthon, $\{\ldots HOC_2O\}_2$, with the result that a linear supramolecular chain is formed (Figure 13.23c). In the centrosymmetric, tetranuclear cation, $[(dppp)Au_2(^{4}LH_2)_2Au_2(dppp)]^{4+}$, two $^{4}LH_2$ bridges are present and these are connected by a 10-membered $\{\ldots HNC_2O\}_2$ synthon within the cavity [41]; dppp is $Ph_2P(CH_2)_3PPh_2$. Further self-association of $^{4}LH_2$ is precluded by the formation of charge-assisted amide-N–H···O(perchlorate) hydrogen bonds (Figure 13.23d).

The final zero-dimensional structure to be described is that of $[I_2Zn(^{4}LH_2)ZnI_2]$ [41]. As shown in Figure 13.23e, an oligomeric structure is formed. Owing to severe disorder associated with the co-crystallised methanol molecule, a detailed analysis of

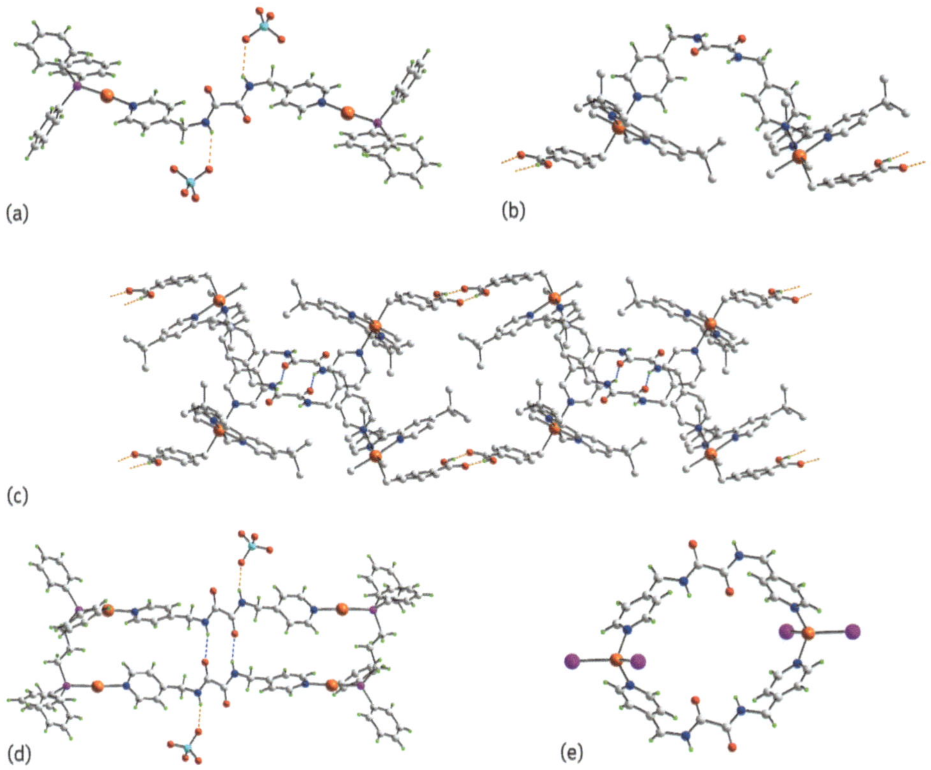

Fig. 13.23: Bidentate bridging modes for [$^{4}LH_2$] in: (a) {[$Ph_3PAu(^{4}LH_2)AuPPh_3$][ClO_4]$_2$}, (b) a binuclear platinum(IV) complex (non-$^{4}LH_2$ hydrogen atoms have been removed), (c) a supramolecular chain whereby the cations in (c) are connected by amide–N···O(amide) and carboxylic acid-O–H···O(carboxylic acid) hydrogen bonds, shown as *blue and orange dashed lines*, respectively; non-acidic hydrogen atoms have been removed, (d) {[(dppp)$Au_2(^{4}LH_2)_2Au_2$(dppp)][ClO_4]$_2$}$^{2+}$ and (e) [$I_2Zn(^{4}LH_2)ZnI_2$].

the supramolecular association in the crystal is precluded. However, the structure is included here as it varies significantly from the equivalent formulations with either chloride or bromide rather than iodide.

The supramolecular aggregation patterns for molecules of the general formula [$X_2Zn(^{4}LH_2)ZnX_2$] is dependent on the nature of X [42]. When X = I, an oligomeric structure is formed (Figure 13.23e). When X = Cl or Br (Figure 13.24a), a zigzag chain is formed with no supramolecular association between chains mediated by $^{4}LH_2$. A similar zigzag chain is found in [$CoCl_2(^{4}LH_2)$].½H_2O, where the chain is decorated by the water molecules of crystallisation, which participate in water-O–H···O(amide) hydrogen bonds [8]. In [$Hg_2I_4(^{4}LH_2)$]$_n$ (Figure 13.24b), $I_2Hg(^{4}LH_2)HgI_2$ units are connected into a supramolecular chain via tapes mediated by 10-membered {…HNC_2O}$_2$ synthons [16]. The standout coordination mode for $^{4}LH_2$ occurs in the

(a)

(b)

(c)

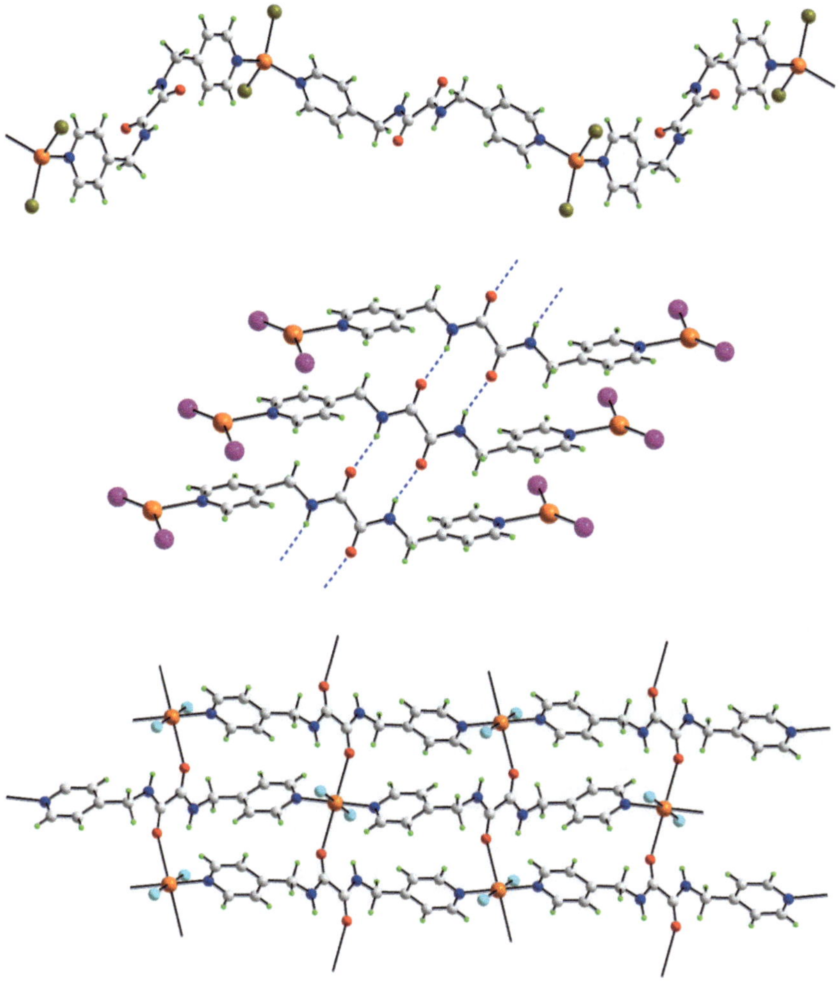

Fig. 13.24: Higher-dimensional aggregation mediated by [$^4LH_2$] in: (a) [$Br_2Zn(^4LH_2)ZnBr_2$], (b) $[Hg_2I_4(^4LH_2)]_n$ in a supramolecular chain connected by amide-N···O(amide) hydrogen bonds shown as *blue dashed lines*, and (c) two-dimensional $[CuCl_2(^4LH_2)]_n$.

two-dimensional structure of $[CuCl_2(^4LH_2)]_n$ [43]. As seen from Figure 13.24c, $^4LH_2$ is bidentate bridging, employing pyridyl-N atoms to generate chains that are linked via Cu–O(amide) bonds indicating a neutral N_2O_2 coordination mode for $^4LH_2$.

The final structure to be described is that of $[Cd_2(NO_3)_4(^4LH_2)_3]_n$ [43]. Each of the three independent and centrosymmetric $^4LH_2$ molecules adopts a bidentate bridging mode of coordination via the pyridyl-N atoms leading to a two-dimensional array shown in Figure 13.25a. The overall architecture is based on mutual orthogonal interpenetration of layers (Figure 13.25b), with the links between the layers being of the type amide-N–H···O(amide) and amide-N–H···O(nitrate) hydrogen bonding.

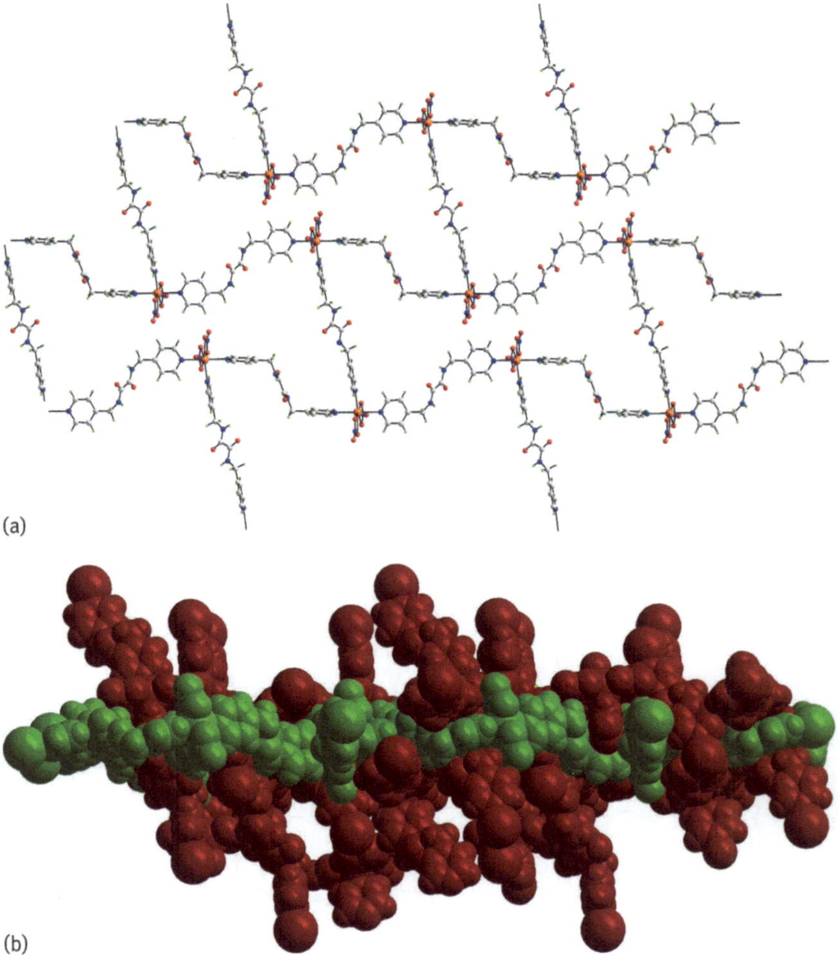

(a)

(b)

Fig. 13.25: Interpenetration in the structure of $[Cd_2(NO_3)_4(^4LH_2)_3]_n$. (a) View of one two-dimensional array, and (b) orthogonal interpenetration of two arrays, shown in space-filling mode.

13.6 Conclusions and outlook

Herein, it has been demonstrated that $^nLH_2$ and ions derived from these have a high degree of conformational flexibility, engendering these molecules/ions to form disparate supramolecular architectures. In their co-crystals, reliable supramolecular synthons are often formed by $^nLH_2$, e.g. hydroxyl-O–H···N(pyridyl) and amide-N–H···O(amide), with the latter leading to $\{\ldots HNC_2O\}_2$ synthons and often supramolecular tapes. This suggests that targeted co-crystal formation might be feasible for this class of compound. Furthermore, a rich variety of coordination modes are noted for neutral for $^nLH_2$, including monodentate, via pyridyl-N, the common bidentate bridging mode, via two pyridyl-N atoms, and even tetradentate modes via two pyridyl-N and two amide-O atoms. For ions derived from these, up to hexadentate coordination modes are possible. This suggests that the relatively unexplored coordination chemistry of $^{n}LH_2$ and their ions are sure to produce exciting discoveries. This all indicates that further investigation of isomeric $^{n}LH_2$ in the context of both multi-component crystals and coordination chemistry is clearly warranted.

Acknowledgement: The support of Sunway University for studies in co-crystals, through Grant No. INT-FST-RCCM-2016-01, is gratefully acknowledged.

Bibliography

[1] Aakeröy C. Is there any point in making co-crystals? Acta Crystallogr B 2015, 71, 387–391.

[2] Duggiral NK, Perry ML, Almarsson Ö, Zaworotko MJ. Pharmaceutical co-crystals: along the path to improved medicines. Chem Commun 2016, 52, 640–655.

[3] Boll G, Nangia A. Pharmaceutical cocrystals: walking the talk. Chem Commun 2016, 52, 8342–8360.

[4] Groom CR, Bruno IJ, Lightfoot MP, Ward SC. The Cambridge Structural Database. Acta Crystallogr B 2016, 72, 171–179.

[5] Spek AL. Single-crystal structure validation with the program PLATON. J Appl Crystallogr 2003, 36, 7–13.

[6] Brandenburg K. DIAMOND. Crystal Impact GbR, Bonn, Germany, 2006.

[7] Jotani MM, Zukerman-Schpector J, Sousa Madureira L, Poplaukhin P, Arman HD, Miller T, Tiekink, ERT. Structural, Hirshfeld surface and theoretical analysis of two conformational polymorphs of N,N′-bis(pyridin-3-ylmethyl)oxalamide. Z Kristallogr 2016, 231, 415–425.

[8] Lee G-H, Wang, HT. Hydrogen-bonded supramolecule of N,N′-bis(4-pyridylmethyl)oxalamide and a zigzag chain structure of catena-poly[[[dichloridocobalt(II)]-μ-N,N′-bis(4-pyridylmethyl)oxalamide-κ^2N4:N4′] hemihydrate]. Acta Crystallogr C 2007, 63, m216–219.

[9] Lee G-H. Hydrogen -bonded supramolecular networks of N,N′-bis(4-pyridylmethyl)oxalamide and 4,4′-{[oxalylbis(azanediyl)]dimethylene}dipyridinium dinitrate. Acta Crystallogr C 2010, 66, o241–244.

[10] Zukerman-Schpector J, Sousa Madureira L, Poplaukhin P, Arman HD, Miller T, Tiekink ERT. Conformational preferences for isomeric N,N′-bis(pyridin-n-ylmethyl)ethanedithiodiamides, $n = 2, 3$ and 4: a combined crystallographic and DFT study. Z Kristallogr 2015, 230, 531–541.

[11] Arman HD, Miller T, Tiekink ERT. The robust {C(=O)OH···N(py)} heterosynthon persists in co-crystals formed between anthranilic acid and molecules with amide/pyridyl functionality. Z Kristallogr 2012, 227, 825–830.

[12] Shattock T, Arora KK, Vishweshwar P, Zaworotko, MJ. Hierarchy of supramolecular synthons: persistent carboxylic acid... pyridine hydrogen bonds in co-crystals that also contain a hydroxyl moiety. Cryst Growth Des 2008, 8, 4533–4545.

[13] Jotani M.M, Syed S, Halim SNA, Tiekink ERT. 2-({[(Pyridin-1-ium-2-ylmethyl)carbamoyl]formamido}methyl)pyridin-1-ium bis(3,5-dicarboxybenzoate): crystal structure and Hirshfeld surface analysis. Acta Crystallogr E 2016, 72, 241–248.

[14] Arman HD, Miller T, Poplaukhin P, Tiekink ERT. 2,6-Dinitrobenzoic acid and its 2 : 1 salts formed with isomeric n-({[(pyridin-1-ium-n-ylmethyl)carbamoyl]formamido}-methyl)pyridin-1-ium, n = 2, 3 and 4. Z Kristallogr 2013, 228, 295–303.

[15] Fraser CSA, Eisler DJ, Jennings MC, Puddephatt RJ. Self-assembly of an organometallic side-by-side double helix. Chem Commun 2002, 1224–1225.

[16] Zeng Q, Li M, Wu D, Lei S, Liu C, Piao L, Yang Y, An S, Wang C. Organic-inorganic hybrid aligned by the ligand-ligand hydrogen bonds by using pyridyl-substituted oxalamides as the building blocks. Cryst Growth Des 2008, 8, 869–876.

[17] Arman HD, Miller T, Poplaukhin P, Tiekink ERT. Poly[[bis[μ_2–*N*,*N*′-bis(2-pyridylmethyl)oxalamide-$\kappa^4N,O{:}N',O'$][μ_2–*N*,*N*′-bis(2-pyridylmethyl)oxalamide-$\kappa^2N{:}N'$]disilver(I)] bis(trifluoromethanesulfonate)]. Acta Crystallogr E 2010, 66, m1167–1168.

[18] Reger DL, Smith DMC, Shimizu KD, Smith MD. [N,N′-Bis(2-pyridylmethyl)oxamidato]-palladium(II) monohydrate chloroform hemisolvate. Acta Crystallogr E 2003, 59, m652–654.

[19] Xu L-H, Wang H-X, Zhu, L-N. A new oxamido-bridged dinuclear copper(II) complex with large antiferromagnetic exchange interactions. J Coord Chem 2012, 65, 1051–1061.

[20] Lloret F, Julve M, Faus J, Journaux Y, Philoche-Levisalles M, Jeannin Y. Oxamidato complexes. 1. A study of the formation of complexes between copper(II) and H_2pmox (H_2pmox = N,N′-bis(2-pyridylmethyl)oxamide). Synthesis, crystal structure, and magnetic properties of the alternating-chain compound $[Cu_2(pmox)(H_2pmox)(H_2O)_2](NO_3)_2.8\,H_2O$. Inorg Chem 1989, 28, 3702–3706.

[21] Zhang H-X, Kang B-S, Zhou Z-Y, Chan ASC, Chen Zhong-Ning, Ren C. 1-D Structures of assemblies containing oxamidato dicopper building blocks controlled by ditopic N-donor spacers. J Chem Soc Dalton Trans 2001, 1664–1669.

[22] Syed S, Halim SNA, Jotani MM, Tiekink ERT. A 2 : 1 co-crystal of p-nitrobenzoic acid and N,N′-bis(pyridin-3-ylmethyl)ethanediamide: crystal structure and Hirshfeld surface analysis. Acta Crystallogr E 2016, 72, 76–82.

[23] Arman HD, Miller T, Poplaukhin P, Tiekink ERT. 2,2′-(Disulfanediyl)dibenzoic acid-N,N′-bis(3-pyridylmethyl)ethanediamide (1/1). Acta Crystallogr E 2010, 66, o2590–2591.

[24] Nguyen TL, Fowler FW, Lauher JW. Commensurate and incommensurate hydrogen bonds. An exercise in crystal engineering. J Am Chem Soc 2001, 123, 11057–11064.

[25] Goroff NS, Curtis SM, Webb JA, Fowler FW, Lauher JW. Designed co-crystals based on the pyridine-iodoalkyne halogen bond. Org Lett 2005, 7, 1891–1893.

[26] Jin H, Plonka AM, Parise JB, Goroff NS. Pressure induced topochemical polymerization of diiodobutadiyne: a single-crystal-to-single-crystal transformation. CrystEngComm 2013, 15, 3106–3110.

[27] Jin H, Young CN, Halada GP, Phillips BL, Goroff NS. Synthesis of the stable ordered conjugated polymer poly(dibromodiacetylene) from an explosive monomer. Angew Chem Int Ed 2015, 54, 14690–14695.

[28] Hursthouse MB, Gelbrich T, Plater MJ. Private communication to the CSD, 2003. CSD Refcode: IPOSIP.

[29] Poplaukhin P, Tiekink ERT. Interwoven coordination polymers sustained by tautomeric forms of the bridging ligand. CrystEngComm 2010, 12, 1302–1306.

[30] Li X, He X, Chen Y, Fan X, Zeng, Q. Solid-state supramolecular chemistry of zinc tetraphenylporphyrin and zinc phthalocyanine with bis(pyridyl) ligands. J Mol Struct 2011, 1002, 145–150.

[31] Qin Z, Jennings MC, Puddephatt RJ. Self-assembly in palladium(II) and platinum(II) chemistry: The biomimetic approach. Inorg Chem 2003, 42, 1956–1965.

[32] Zou H, Qi Y. Crystal structure of poly[{μ-N,N′-bis[(pyridin-4-yl)methyl]oxalamide}-μ-oxalato-cobalt(II)]. Acta Crystallogr E 2014, 70, m307–308.

[33] Reger DL, Smith DMC, Shimizu KD, Smith MD. Syntheses and solid state structures of europium and terbium complexes of N,N′-bis(2-pyridylmethyl)urea and N,N′-bis(3-pyridylmethyl)oxalamide. Polyhedron 2004, 23, 711–717.

[34] Schauer CL, Matwey E, Fowler FW, Lauher JW. Controlled spacing of metal atoms via ligand hydrogen bonds. J Am Chem Soc 1997, 119, 10245–10246.

[35] Arman HD, Kaulgud T, Miller T, Poplaukhin P, Tiekink ERT. Crystal and molecular structures of the 2 : 1 co-crystal of 4-nitrophenylacetic acid and N,N′-bis(pyridin-3-ylmethyl)oxalamide, and with the thioxalamide analogue. J Chem Cryst 2012, 42, 673–679.

[36] Nguyen TL, Scott A, Dinkelmeyer B, Fowler FW, Lauher JW. Design of molecular solids: utility of the hydroxyl functionality as a predictable design element. New J Chem 1998, 22, 129–135.

[37] Curtis SM, Le N, Nguyen T, Xi O, Tran T, Fowler FW, Lauher JW. What have we learned about topochemical diacetylene polymerizations? Supramol Chem 2005, 17, 31–36.

[38] Arman HD, Miller T, Poplaukhin P, Tiekink ERT. 2-[(4-Hydroxyphenyl)diazenyl]benzoic acid–N,N′-bis(4-pyridylmethyl)oxamide (2/1). Acta Crystallogr E 2009, 65, o3178–3179.

[39] Ouyang X, Fowler FW, Lauher JW. Single-crystal-to-single-crystal topochemical polymerizations of a terminal diacetylene: two remarkable transformations give the same conjugated polymer. J Am Chem Soc 2003, 125, 12400–12401.

[40] Wilhelm C, Boyd SA, Chawda S, Fowler FW, Goroff NS, Halada GP, Grey CP, Lauher JW, Luo L, Martin CD, Parise JB, Tarabrella C, Webb JA. Pressure-induced polymerization of diiodobutadiyne in assembled cocrystals. J Am Chem Soc 2008, 130, 4415–4420.

[41] Tzeng C-C, Yeh H-T, Wu Y-L, Kuo J-H, Lee G-H, Peng S-M. Supramolecular assembly of gold(I) complexes of diphosphines and N,N′-bis-4-methylpyridyl oxalamide, Inorg Chem 2006, 45, 591–598.

[42] Tzeng B-C, Chen B-S, Lee S-Y, Liu W-H, Lee G-H, Peng S-M. Anion-directed assembly of supramolecular zinc(II) halides with N,N′-bis-4-methyl-pyridyl oxalamide. New J Chem 2005, 29, 1254–1257.

[43] Tzeng B-C, Huang Y-C, Chen B-S, Wu W-M, Lee S-Y, Lee G-H, Peng S-M. Crystal-engineering studies of coordination polymers and a molecular-looped complex containing dipyridyl-amide ligands. Inorg Chem 2007, 46, 186–195.

Edward R. T. Tiekink and Julio Zukerman-Schpector

14 Solute–solvent interactions mediated by main group element(lone-pair)···π(aryl) interactions

14.1 Introduction

Since the disclosure of tellurium(II, IV)···π(aryl) interactions in their crystal structures [1], there has been increasing interest in ascertaining the importance of this type of supramolecular synthon, i.e. main group element(lp)···π(arene), in the structural chemistry of main group element chemistry [2–10]. It is noted in passing that analogous interactions have been revealed in structural biology and are known to provide important stabilising contacts [11, 12]. In molecular structural chemistry, the emphasis thus far has been upon seeking “stand-alone” element(lp)···π(arene) contacts and determining the supramolecular architectures they sustain [2–10]. Usually these architectures are zero- and one-dimensional; however, examples of two- and even three-dimensional aggregation patterns are known [9, 10]. With the restriction that the element(lp)···π(arene) interaction operates in a given dimension within a crystal, in isolation from other supramolecular synthons, these interactions occur in a maximum of 14% of thallium(I) structures. This is reduced to 9% in bismuth(III) structures and down to a minimum of 2–3% for tin(II) and lead(II) compounds [9, 10]. In terms of bonding, it is likely that the lone-pair···π interaction has similar features to halogen bonding [13] in that there is an asymmetric distribution of electron density in the lone-pair of electrons. This leaves an electropositive region at the tip of the lone-pair that interacts with the π-electrons of the ring. This electron-deficient region is termed variously a σ-hole or polar cap, and the energy of stabilisation imparted by these interactions are upward of several kcal mol^{-1}. In keeping with the theme of the present volume, “*Aspects of Multi-Component Crystals: Synthesis, Concepts and Function*”, a description herein of the supramolecular architectures sustained by main group element(lp)···π(aryl) interactions is given where the π-system is a solvent molecule such as benzene.

14.2 Procedures

Standard protocols were utilised to identify structures for the current survey [10]. The Cambridge Structural Database (CSD: version 5.37 + 2 updates) [14] was searched using CONQUEST (version 1.18) [15]. There were two key geometric restraints, as illustrated in Figure 14.1a. The first restraint is the distance d, between the metal centre and the ring centroid, labelled Cg, which was set at 4.0 Å. This distance is shown to be suffi-

https://doi.org/10.1515/9783110464955-014

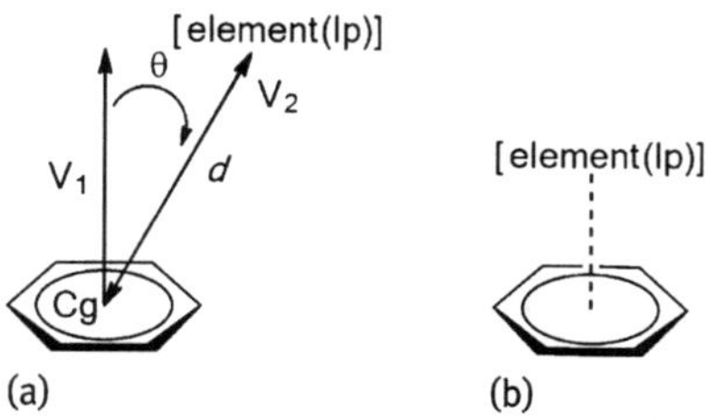

Fig. 14.1: (a) The geometric restrictions employed to identify element(lp)···π(arene) interactions. The distance *d* (Å) is the separation between the ring centroid, Cg, and the element. The angle θ (°) is between the normal to the arene ring, $\boldsymbol{V}_1$, and the vector, $\boldsymbol{V}_2$, between the ring centroid and the element. (b) A representation of a delocalised element(lp)···π(arene) interaction where the element's lone-pair interacts approximately equally with all six carbon atoms of the aryl ring.

cient to identify relevant contacts [10]. The second restraint is the angle θ, which is the angle between the normal to the plane through the aryl ring ($\boldsymbol{V}_1$) and the vector from Cg to the main group element ($\boldsymbol{V}_2$). This angle indicates the relative location of the element above the C_6 plane and was restricted to ≤ 20°, so that only "delocalised" [16] element(lp)···π(aryl) interactions were retrieved. The consequence of this restriction is that the lp···C separations are approximately equal (Figure 14.1b). This restriction removes "localised" interactions where the lone-pair is orientated towards a single atom of the aryl ring only as well as "semi-localised" interactions where the lone-pair is engaged with only one of the bonds of the ring [16]. The only other restrictions that were applied were those structures with unresolvable disorders, which were omitted. Furthermore, structures determined from powder data were not included. This resulted in 37 "hits" satisfying the specified criteria. Each data set, in the form of a Crystallographic Information File (CIF), was manually scrutinised and analysed quantitatively using PLATON [17]. Diagrams are original and were drawn using DIAMOND [18].

14.3 Discussion

Structures are discussed starting from the gallium-group elements with benzene, followed by examining the substituted forms of the C_6 ring. The tin group elements are described, etc. Within each group, mononuclear species are discussed before binuclear species, etc. Finally, for each sub-category, lower nuclearity aggregates are described before higher nuclearity aggregates.

The chemical diagrams were generated using ChemDraw® and show only the interacting species. As such, other species in the crystal, such as counter ions, are not illustrated. In the diagrams showing aggregation patterns, all hydrogen atoms have been omitted for reasons of clarity. The arene rings forming the element(lp)···π(arene) interactions are highlighted in purple. The interaction is indicated by a dashed purple line in each case.

14.3.1 Group III

Of the Group III elements, only thallium(I) compounds have been found to form Tl(lp)···π(arene) interactions where the arene ring is a solvent molecule. There are six examples, (1)–(6), and their chemical diagrams are shown in Figure 14.2.

The first three structures in this category are mononuclear thallium(I) compounds. In (1), a molecule of mononuclear {bis[3-(9-triptycyl)pyrazolyl]hydroborato} thallium(I) [19] associates with a solvent toluene molecule via a Tl(lp)···π(toluene) interaction [$d = 3.22$ Å and $\theta = 3.7°$] to form a zero-dimensional aggregate in which the thallium(I) atom increases its coordination number from 2 (N_2 donor set) to 3 within a distorted trigonal planar geometry (Figure 14.3a). A similar aggregation pattern is

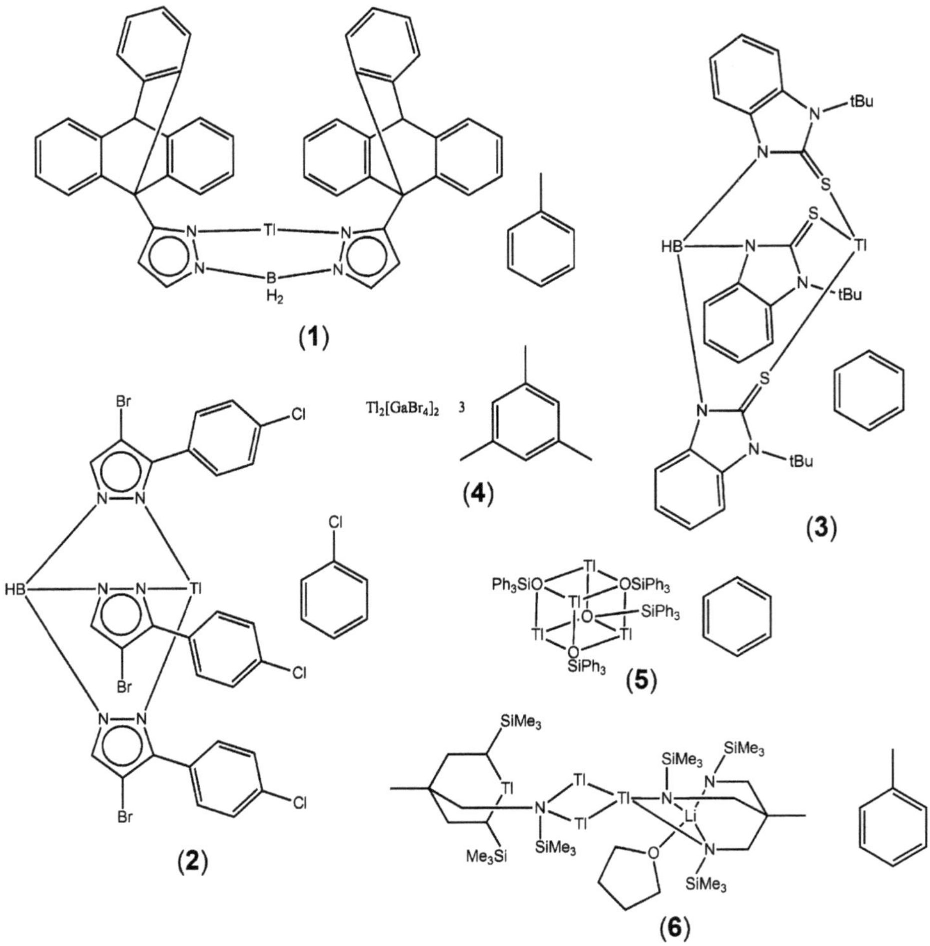

Fig. 14.2: Chemical diagrams for (1)–(6). In this and subsequent figures showing chemical diagrams, only species participating in element(lp)...π(arene) interactions are shown.

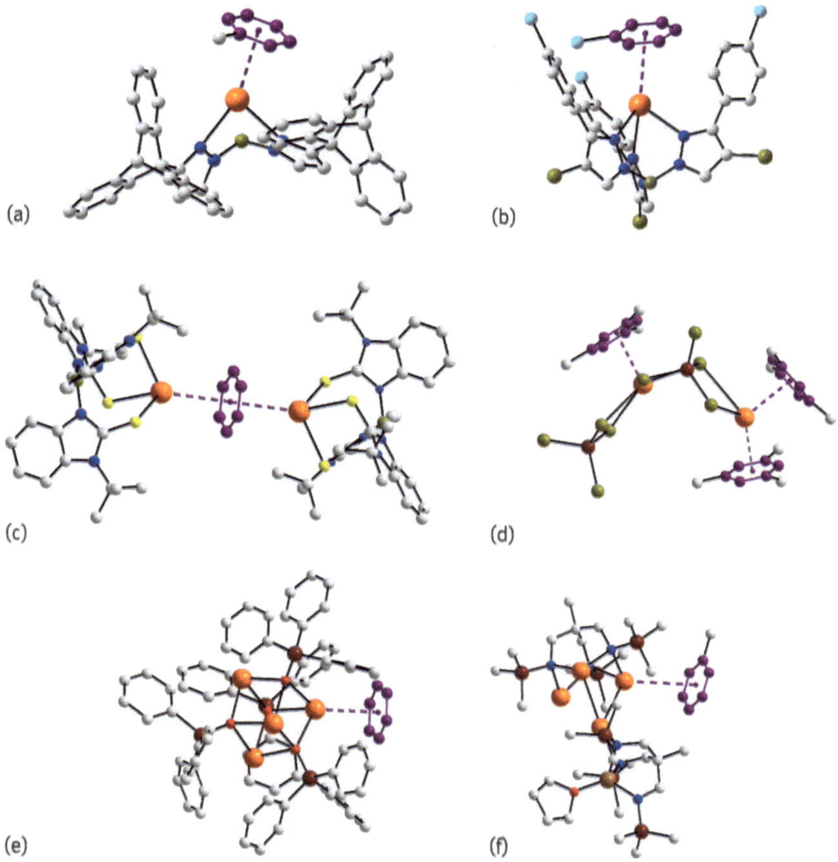

Fig. 14.3: Supramolecular aggregation via Tl(lp)···π(arene) interactions in the crystal structures of (1)–(6) shown in (a)–(f), respectively. Colour code: Tl (or other main group element), *orange*; bromide, *olive-green*; chloride, *cyan*; sulphur, *yellow*; silicon, *brown*; oxygen, *red*; nitrogen, *blue*; boron, *grey-green*; and lithium, *beige*.

found in (2), whereby a single Tl(lp)···π(chlorobenzene) interaction [d = 3.69 Å and θ = 14.5°] is found in (hydrogen tris(4-bromo-3-(p-chlorophenyl)pyrazolyl)borato) thallium(I) chlorobenzene solvate [20] to give a two-molecule, zero-dimensional aggregate (Figure 14.3b). The thallium(I) centre in bis(hydrogen tris(2-thioxo-3-t-butylbenzimidazol-1-yl)borate)thallium(I) (3) [21] is located on a three-fold axis of symmetry and was characterised crystallographically as its hemi-benzene solvate. From symmetry, the benzene ring accepts two Tl(lp)···π(toluene) interactions [d = 3.39 Å and θ = 0.0°] to form a binuclear three-molecule aggregate (Figure 14.3c). The original coordination geometry for each thallium atom in (2) and (3) was pyramidal, defined by N_3 and S_3 donor sets, respectively. The coordination geometry in each is expanded to four with the interaction of the benzene ring included.

In binuclear di-thallium(I) bis[tetrabromo-gallium(III)] tris(1,3,5-trimethylbenzene) solvate (4) [22] one $[GaBr_4]^-$ anion bridges two thallium(I) centres, whereas the other is terminally bound, giving two independent thallium centres, each of which is associated with 1,3,5-trimethylbenzene molecules. The thallium forming four Tl···Br bonds is connected to one 1,3,5-trimethylbenzene molecule [d = 2.94 Å, α = 2.9°] whereas the other coordinatively unsaturated thallium(I) atom is connected to two 1,3,5-trimethylbenzene molecules [d = 3.01 Å, α = 5.4°; d = 3.03 Å, α = 3.1°]. The result is a binuclear four-molecule, zero-dimensional aggregate (Figure 14.3d). In tetranuclear tetrakis(μ_3-triphenylsilanolato)-tetra-thallium(I) benzene solvate (5) [23] only one of the four thallium(I) centres forms a single Tl(lp)...π(benzene) interaction [d = 3.20 Å and θ = 2.7°] to form a tetranuclear, two-molecule and zero-dimensional aggregate (Figure 14.3e). In the same way, a single Tl(lp)···π(toluene) interaction [d = 3.20 Å and θ = 2.7°] is found in the structure of tetranuclear (μ_3-1,1,1-tris(trimethylsilylamido)ethane)(μ_3-1,1-bis(trimethylsilylamido)-1-(trimethylsilylamino)ethane)-tetrahydrofuran-tetra-thallium(I)-lithium toluene solvate (6) (Figure 14.3f) [24].

14.3.2 Group IV

14.3.2.1 Tin(II)

Three tin(II) structures, (7)–(9), have been revealed to feature Sn(lp)···π(arene) interactions in their crystal structures. Chemical diagrams for these are shown in Figure 14.4.

In the binuclear structure of bis(μ_2-2,6-dimethylphenolato)-bis(2,6-dimethylphenolato)-di-tin(II).2(toluene) (7) [25], two centrosymmetrically related molecules associate via Sn(lp)···π(aryl) interactions [d = 3.36 Å and θ = 10.8°] to form a dimeric, zero-dimensional aggregate. Toluene molecules interact with the terminal tin atoms [d = 3.59 Å and θ = 7.6°] so that each Sn(II) atom in (7) forms a single Sn(lp)···π(aryl) interaction and adopts a pseudo-tetrahedral geometry. In mononuclear 2,3-bis(trimethylsilyl)-1-stanna-2,3-dicarba-closoheptaborane(4) hemi-benzene solvate (8) [26], the tin(II) centre is coordinated by three boron and two carbon atoms, but in a way that leaves the tin(II) atom exposed. This enables the formation of a Sn(lp)···π(benzene) contact [d = 3.73 Å and θ = 13.6°] and, as the benzene molecule of solvation is located about a centre of inversion, a binuclear three-molecule aggregate ensues. The tetranuclear cluster molecule, bis(μ_3-oxo)-octakis(μ-trifluoroacetato)-di-tin(II)-di-tin(IV), is disposed about a centre of inversion and is crystallised as a benzene solvate, also located about a centre of inversion, with the ratio between molecules being 1 : 1 (9) [27]. The exocyclic tin atoms, i.e. being tin(II) as opposed to the endocyclic tin(IV) atoms and having square-pyramidal geometries defined by five oxygen atoms, form Sn(lp)···π(benzene) contacts [d = 3.32 Å and θ = 6.1°] to generate a supramolecular chain with a flattened step topology.

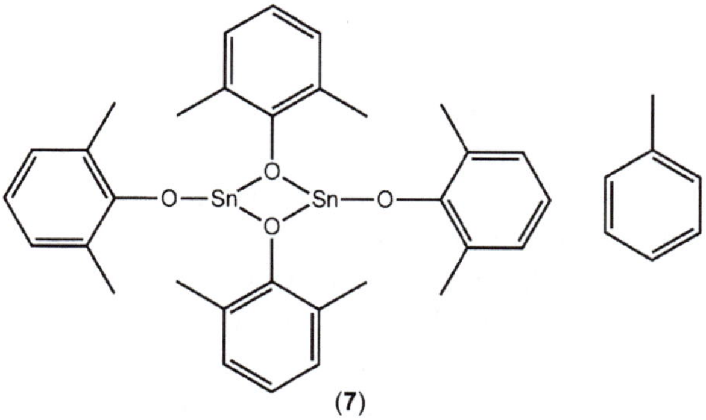

(**7**)

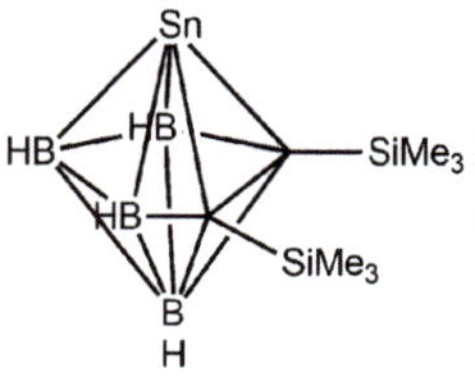

(**8**)

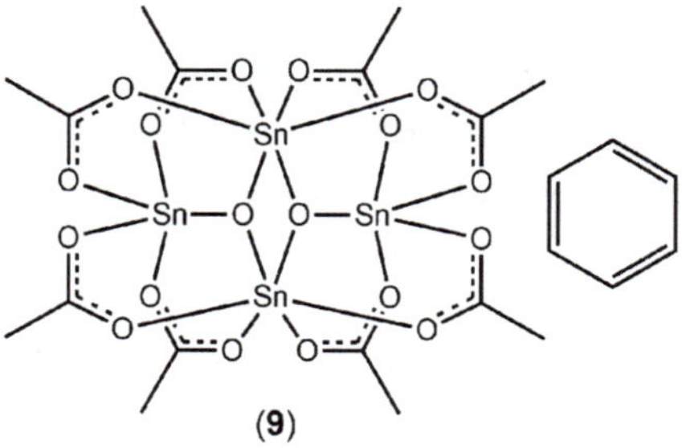

(**9**)

Fig. 14.4: Chemical diagrams for (7)–(9).

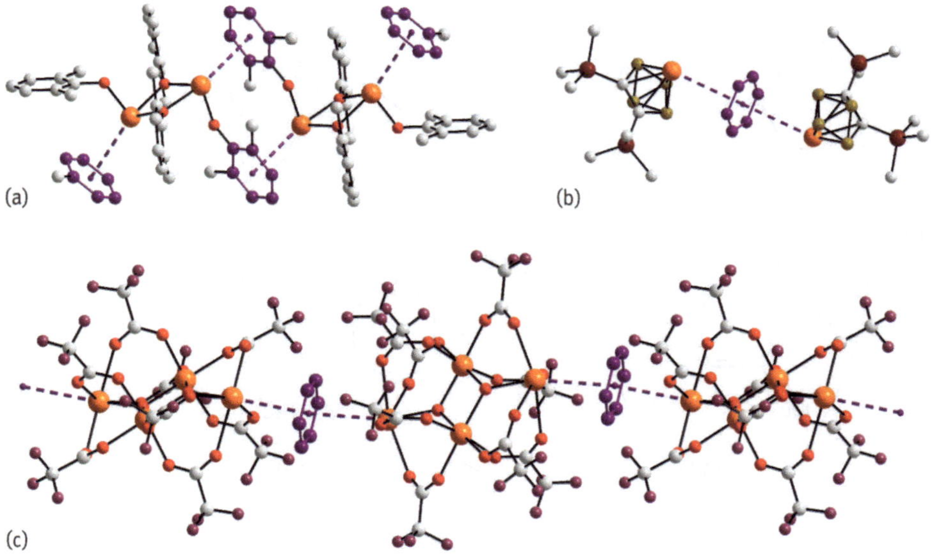

Fig. 14.5: Supramolecular aggregation via Sn(lp)... π(arene) interactions in the crystal structures of (7)–(9), shown in (a)–(c), respectively. Additional colour code: fluoride, *violet-red*.

14.3.2.2 Lead(II)

Five lead(II) structures, (10)–(14), feature at least one Pb(lp)···π(arene) interaction in their crystal structures. See Figure 14.6 for chemical diagrams.

The first three structures in this category are mononuclear, with each lead(II) centre forming a single Pb(lp)···π(arene) contact. Thus, the lead(II) atom in (10), i.e. bis(diphenylpentane-2,4-diiminato)lead(II) toluene solvate [28], is coordinated by four nitrogen atoms, but there is a large void above the lead(II) atom (Figure 14.7a), which is occupied by the toluene molecule [d = 3.66 Å and θ = 8.6°]. A relatively high coordination number is also found in (11), bis(2,4,6,8-tetra-t-butyl-1-oxo-1H-phenoxazin-9-olato)lead(II) benzene di-solvate [29], where the lead(II) is in a N_2O_4 donor set. However, once again, there is a large void to accommodate one of the two benzene molecules co-crystallised with the lead compound via a Pb(II)···π(benzene) interaction [d = 3.77 Å and θ = 17.9°] (Figure 14.7b). The mononuclear structure of 2,3-bis(trimethylsilyl)-2,3-dicarba-1-plumba-closo-heptaborane(6) benzene solvate

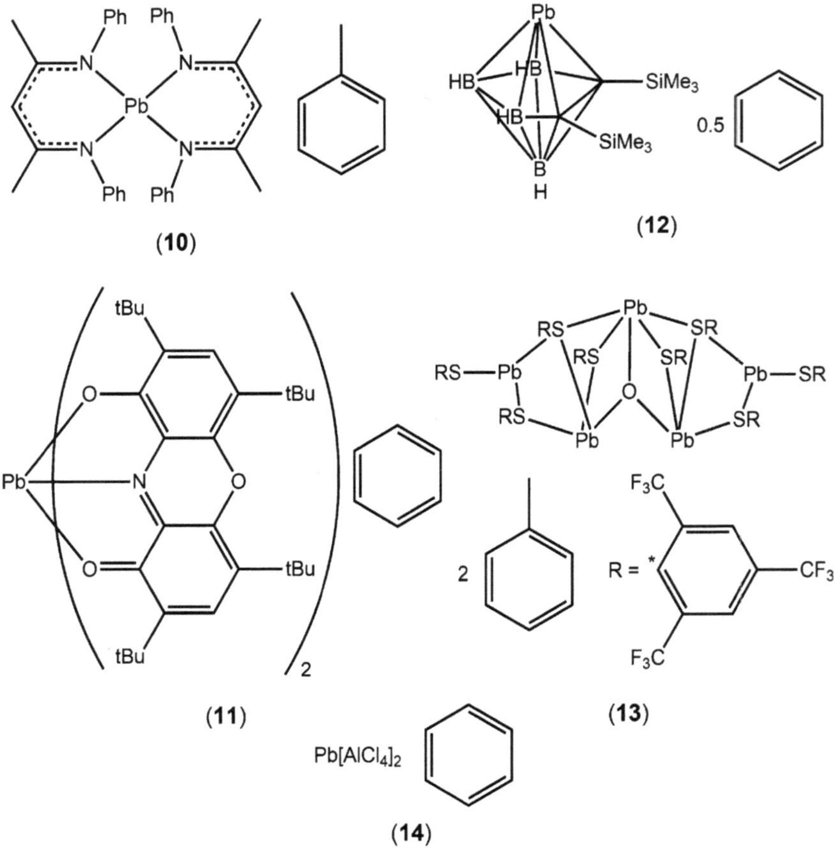

Fig. 14.6: Chemical diagrams for (10)–(14).

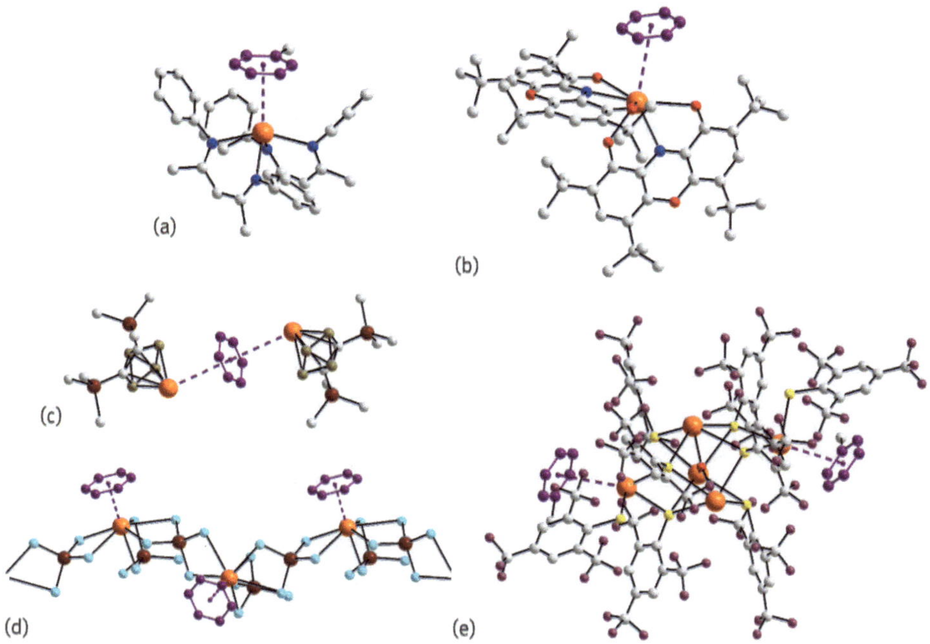

Fig. 14.7: Supramolecular aggregation via Pb(lp)...π(arene) interactions in the crystal structures of (10)–(14) shown in (a)–(e), respectively. Additional colour code: aluminium, *brown*.

(12) [30] is isostructural with (8) (Figure 14.5b) [26]. Thus, the lead(II) centre is coordinated by a B_3C_2 donor set, but the atoms comprising the cluster lie to one side of the molecule, enabling the formation of a Pb(lp)···π(benzene) contact [d = 3.74 Å and θ = 16.9°] with a centrosymmetric benzene molecule, which results in a binuclear, three-molecule aggregate (Figure 14.7c).

Two of the five lead(II) atoms in pentanuclear bis(μ_3-2,4,6-tris-(trifluoromethyl) phenylthiolato)-(μ_3-oxo)-tetrakis(μ_2-2,4,6-tris(trifluoromethyl)phenylthiolato)-bis(2,4,6-tris(trifluoromethyl)phenylthiolato)-penta-lead(II), isolated as a 1:2 toluene solvate (13) [31] forms Pb(lp)···π(arene) interactions [d = 3.55 Å and θ = 19.1°; d = 3.70 Å and θ = 17.8°] (Figure 14.7d). The common feature of the two interacting lead(II) atoms is that they exist in pyramidal coordination geometries, defined by S_3 donor sets, exposing the lone-pair to the rings. By contrast, the remaining lead atoms have higher coordination numbers, i.e. 4 and 5.

[Lead(II) bis(tetrachloridoaluminium(IV))]$_n$ (14) [32] is a coordination polymer that was isolated as a benzene solvate, with the ratio between lead atoms and solvent molecules being 1:1, enabling each lead atom to form a Pb(lp)···π(arene) contact [d = 2.78 Å and θ = 1.1°]. As seen from Figure 14.7e, each lead atom, despite being coordinated by six chlorido atoms, is in a distorted pentagonal-bipyramidal coordination geometry with the benzene ring lying to one side of the pentagonal plane.

14.3.3 Group V

14.3.3.1 Arsenic(III)

Three arsenic(III) solvates, (15)–(17), feature As(lp)···π(arene) interactions in their molecular packing. Chemical diagrams for these are shown in Figure 14.8.

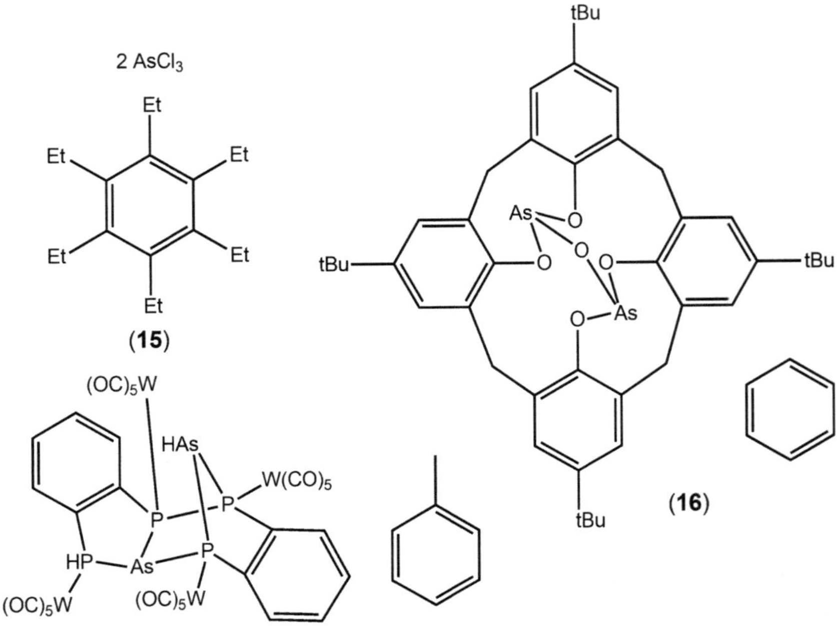

(17)

Fig. 14.8: Chemical diagrams for (15)–(17).

Each of the trichloroarsane arsenic(III) centres in bis[trichloroarsane(III)] hexaethylbenzene (15) [33] associates with the hexaethylbenzene ring via a As(lp)···π(arene) interaction [d = 3.14 Å and θ = 0.0°], with the entire three-molecule aggregate (Figure 14.9a) having crystallographic 3-fold symmetry. In (16), a binuclear compound, (μ_2-5,11,17,23-tetra-t-butylcalix(4)arene-25,26,27,28-tetraolato)-(μ_2-oxo)-di-arsenic(III) benzene solvate [34], both arsenic(III) atoms have three-coordinate, pyramidal geometries defined by an O_3 donor set, but only one centre forms a As(lp)···π(arene) interaction [d = 3.57 Å and θ = 3.6°] (Figure 14.9b).

In the heterometallic complex, (μ_4-7H-5,13-arsano(1,3,2)benzodiphospharsolo (1,2-b)(1,2,4,3)benzotriphospharinine)-icosacarbonyl-tetra-tungsten toluene solvate (17) [35], two distinct coordination environments are noted for arsenic(III). One is pyramidal and based on a P_3 donor set, whereas the other is also pyramidal, but based

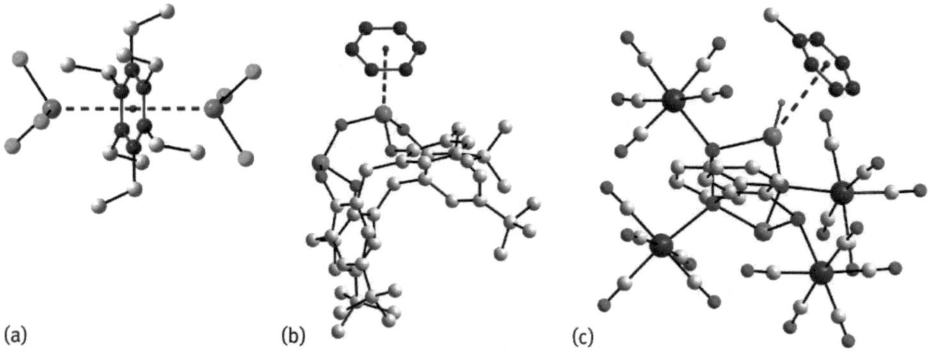

(a) (b) (c)

Fig. 14.9: Supramolecular aggregation via As(lp)...π(arene) interactions in the crystal structures of (15)–(17) shown in (a)–(c), respectively. Additional colour code: tungsten, *indigo*; phosphorous, *pink*; hydrogen, *green*.

on a HP_2 donor set. It is the latter that forms the As(lp)···π(arene) interaction leading to a two-molecule, zero-dimensional aggregate [d = 3.72 Å and θ = 10.5°].

14.3.3.2 Antimony(III)

The first four structures in this category, (18)–(21), are examples of classic Menschutkin complexes [36, 37]. Chemical diagrams for these and for two further structures featuring Sb(lp)···π(arene) interactions in their crystal structures, (22) and (23), are shown in Figure 14.10. Although the common feature of (18)–(21) is the presence of mononuclear species, in trichlorido-antimony toluene (18) [38] and tribromidoantimony(III) mesitylene (19) [39] solvate a single interaction occurs between antimony(III) and the solvent molecule. In (18), there are two independent complex:solvent pairs and each of these forms a Sb(lp)···π(arene) interaction [d = 3.10 Å and θ = 7.7°; d = 3.16 Å and θ = 9.0°]. The structure of (19) is even more remarkable in that there are four independent pairs of molecules in the asymmetric unit, and each of these forms a Sb(lp)···π(arene) interaction [d = 3.22 Å and θ = 8.8°; d = 3.22 Å and θ = 9.4°; d = 3.30 Å and θ = 5.3°; d = 3.33 Å and θ = 5.0°]. One of these complex : solvent pairs is shown in Figure 14.11a. In each of bis(trichlorido-antimony) benzene solvate (20) [40] and bis[trichloridoantimony(III)] p-xylene solvate (21) [41], each solvent molecule forms two Sb(lp)···π(arene) interactions. The absence of symmetry in (20) (Figure 14.11b) implies two independent set of parameters [d = 3.25 Å and θ = 7.6°; d = 3.33 Å and θ = 7.4°] whereas the solvent in (21) lies about a centre of inversion [d = 3.14 Å and θ = 11.5°].

The hexa-antimony(III) cluster complex constructed about a $\{SbO\}_6$ ladder, bis(μ-diphenylsilanediolato)-bis(μ-hydroxo)-hexakis(μ-oxo)-octaphenyl-hexa-antimony(III), crystallises as a 1 : 1 toluene solvate (22) [42]. From the formulation and as illustrated in Figure 14.11c, only one Sb(lp)···π(arene) interaction is formed [d = 3.67 Å and θ = 9.6°].

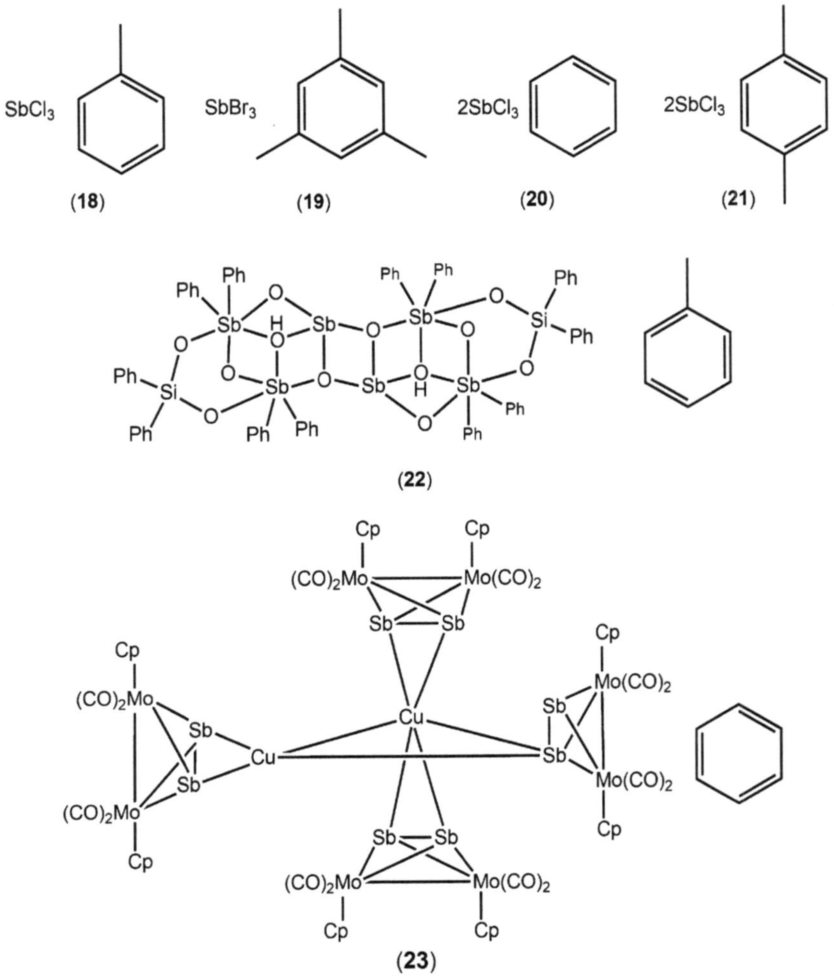

Fig. 14.10: Chemical diagrams for (18)–(23).

The final antimony(III) complex to be described is that of heterometallic and organometallic tris[μ_4-di-antimony(III)]-[μ_3-di-antimony(III)]-tetrakis(tetracarbonyl-bis(η^5-cyclopentadienyl))-di-copper(I)-octa-molybdenum(0) bis[tetra-chlorido-gallium(III)] benzene solvate (23) [43]. The antimony(III) atom that formally does not form an interaction to a copper(I) atom forms the Sb(lp)···π(arene) contact with the solvent benzene molecule [d = 3.47 Å and θ = 1.6°] (Figure 14.11d).

14.3.3.3 Bismuth(III)

With nine structures featuring Bi(lp)···π(arene) interactions, bismuth(III) compounds are the most frequently represented in the present survey. The nuclearity of the species

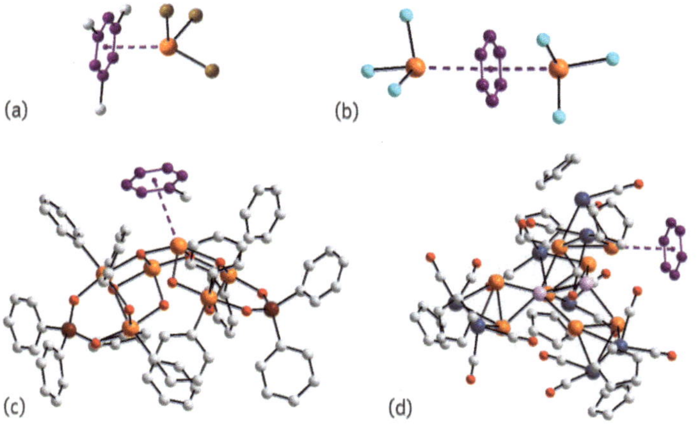

Fig. 14.11: Supramolecular aggregation via Sb(lp)...π(arene) interactions in the crystal structures of (19), (20), (22) and (23) shown in (a)–(d) respectively. Additional colour codes: molybdenum, *indigo*; copper, *mauve*.

ranges from 1 up to 14; therefore, a large number of distinct structural motifs are observed. The first three structures to be discussed, (24)–(27) (Figure 14.12) are formally mononuclear $BiCl_3$ and are typical Menschutkin complexes. Despite these similarities, three quite distinct structural motifs are observed in their crystal structures. In trichlorido-bismuth(III) toluene (24) [44] (Figure 14.13a), two independent molecules comprise the crystallographic asymmetric unit and each bismuth(III) centre forms a single Bi(lp)···π(arene) interaction to a toluene molecule [d = 3.04 Å and θ = 3.6°; d = 3.09 Å and θ = 7.9°]. In the other two structures, significant supramolecular association occurs, owing to strong Bi···Cl secondary bonding interactions. Thus, in catena[(bis(μ_2-chlorido))-chlorido-bismuth(III) (η^6-m-xylene)] (25) [45], a coordination polymer is observed (Figure 14.13b), decorated with m-xylene molecules held in place by Bi(lp)···π(arene) interactions [d = 2.99 Å and θ = 5.8°]. By contrast, an oligomeric aggregate is observed for bis(hexamethylbenzene) tetrakis(μ_2-chlorido)-octachloro-tetra-bismuth(III) (26) [36] (Figure 14.13c). The oligomer features four bismuth(III) centres and is disposed about a crystallographic four-fold inversion centre. The independent bismuth(III) atom associates with a hexamethylbenzene molecule [d = 3.07 Å and θ = 2.6°].

The next four structures to be described, (27)–(30) (Figure 14.12), are tetranuclear bismuth(III) compounds. In bis(μ_3-oxo)-tetrakis(μ_2-1,1,1,3,3,3-hexafluoro-2-propoxo)-tetrakis(1,1,1,3,3,3-hexafluoro-2-propoxy)-tetra-bismuth(III) toluene solvate (27) [46], the cluster is situated about a centre of inversion. There are two distinct coordination environments for bismuth(III): the endocyclic atoms based on a O_5 donor set and the exocyclic bismuth(III) atoms, which are coordinated within a O_4 donor set. Are the exocyclic bismuth(III) atoms that form the Bi(lp)···π(arene) interactions [d = 3.56 Å and θ = 15.5°] leading to distorted trigonal bipyramidal geo-

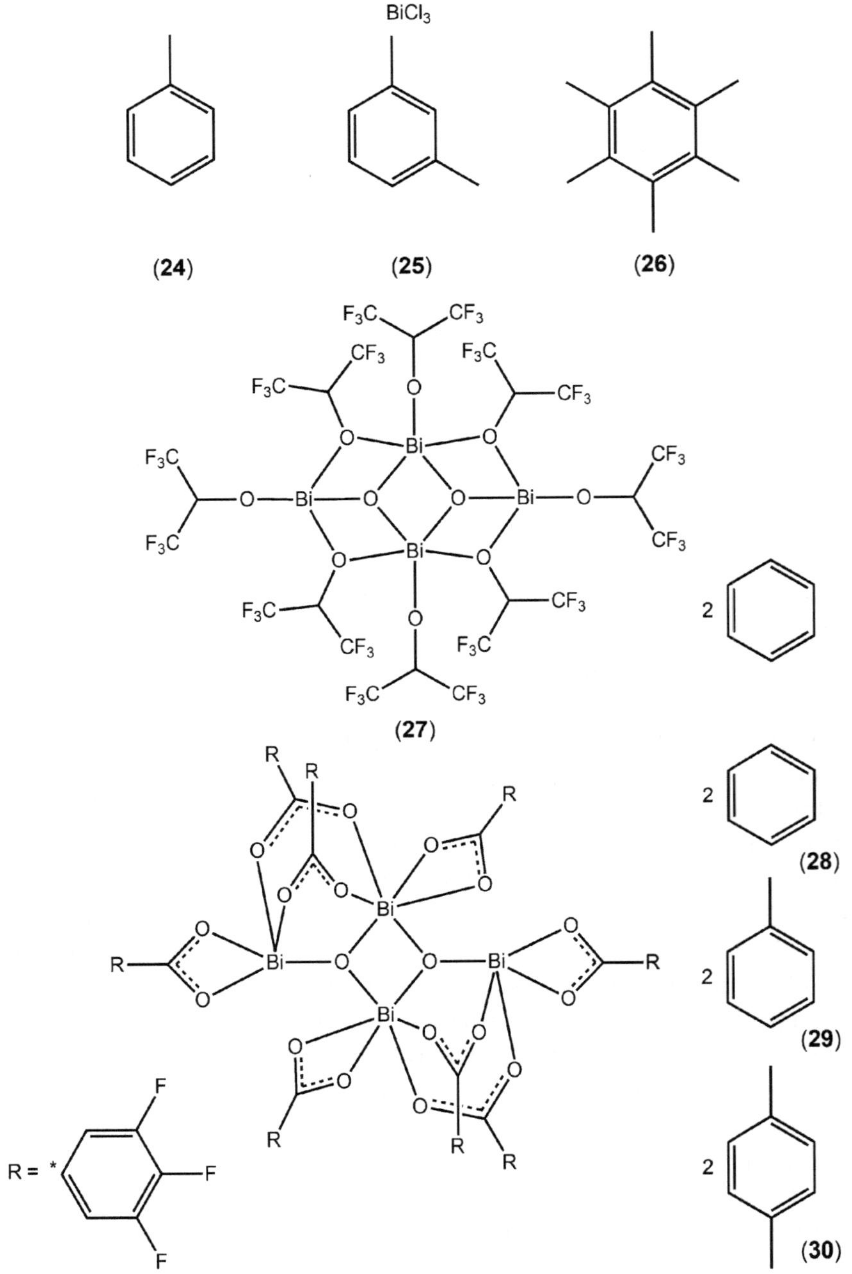

Fig. 14.12: Chemical diagrams for (24)–(30).

metries (Figure 14.13d). There are three tetranuclear structures that differ only in the pattern of fluoride substitution in the benzoate anions. These are exemplified in Figure 14.13e by bis(μ_3-oxo)-octakis(μ_2-3,4,5-trifluorobenzoato)-tetra-bismuth(III),

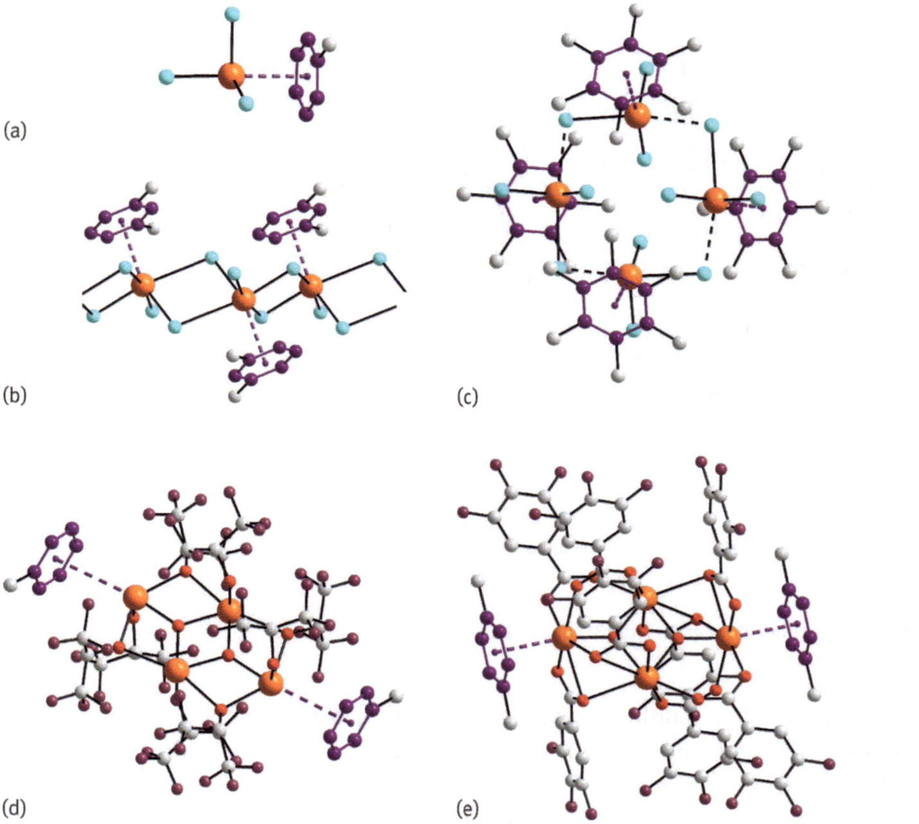

Fig. 14.13: Supramolecular aggregation via Bi(lp)...π(arene) interactions in the crystal structures of (24)–(28) shown in (a)–(e), respectively.

characterised as a 1,4-dimethylbenzene disolvate (28) [47]. The centrosymmetric molecular framework of the cluster may also be described as comprising endocyclic and exocyclic bismuth(III) atoms. The coordination numbers here, each based on donor sets provided by oxygen atoms, are 7 and 6 respectively. The exocyclic bismuth(III) centres form the Bi(lp)···π(arene) interactions [d = 3.13 Å and θ = 5.6°], leading to distorted pentagonal bipyramidal geometries. Very similar structures and modes of association with solvent are found in both (29), i.e. having benzene rather than 1,4-dimethylbenzene as the solvent [d = 3.02 Å and θ = 3.4°] [47], and (30) having toluene as the solvent [d = 3.13 Å and θ = 2.7°] [48].

The molecular structures of (31) and (32) contain eight and 14 bismuth(III) atoms, respectively (Figure 14.14). Compound (31) has composition bis(μ_4-5,11,17,23,29,35,41,47-octa-t-butyl-49,50,51,52,53,54,55,56-octaoxycalix(8)arene)-tetrakis(μ_4-oxo)-octa-bismuth(III) acetonitrile diethyl ether toluene solvate [49]. The cluster is disposed about a four-fold inversion centre. There are two independent bismuth(III) atoms and each

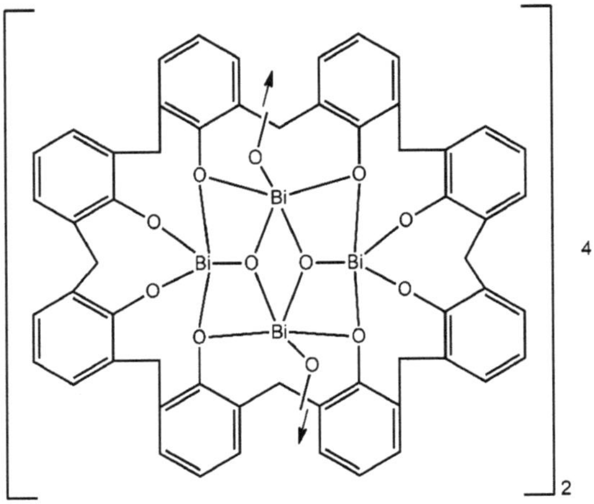

(31)

$[Bi_{14}(\mu\text{-}O)_{10}(\mu_4\text{-}O_2P(=O)tBu)_6(\mu_3\text{-}O_2P(=O)tBu)_4(\mu_4\text{-}O_2P(OH)tBu)_2].6C_6H_6$

(32)

Fig. 14.14: Chemical diagrams for (31) and (32).

exists within a five-coordinate geometry defined by oxygen atoms. One of these associates with a toluene molecule [d = 3.51 Å and θ = 6.4°], leading to a heavily distorted octahedral geometry (Figure 14.15a).

The final bismuth(III) structure revealed to possess Bi(lp)···π(arene) interactions is that of hexakis(μ_4-t-butylphosphonato)-bis(μ_4-t-butylhydrogenphosphonato)-decakis(μ_3-oxo)-tetrakis(μ_3-t-butylphosphonato)-tetradeca-bismuth(III) benzene solvate tetrahydrate (32) [50]. There is a high degree of crystallographic symmetry in the compound, and both the cluster and benzene molecules have *mmm* symmetry. Six of the 14 bismuth(III) centres associate with benzene molecules, because there are two independent interactions [d = 3.19 Å and θ = 0.0°; d = 3.21 Å and θ = 5.8°]. Four of the interactions are terminal, but two are bridging, leading to the formation of a linear supramolecular chain (Figure 14.15b).

14.3.4 Group VI

14.3.4.1 Selenium(II)

The two main oxidation states for selenium in their molecular compounds are +IV and, especially, +II leading to, one and two lone-pairs of electrons, respectively. The element in both oxidation states has been shown to form Se(lp)···π(arene) interactions in the crystal structures of its molecular compounds [7, 8]. When the accepting

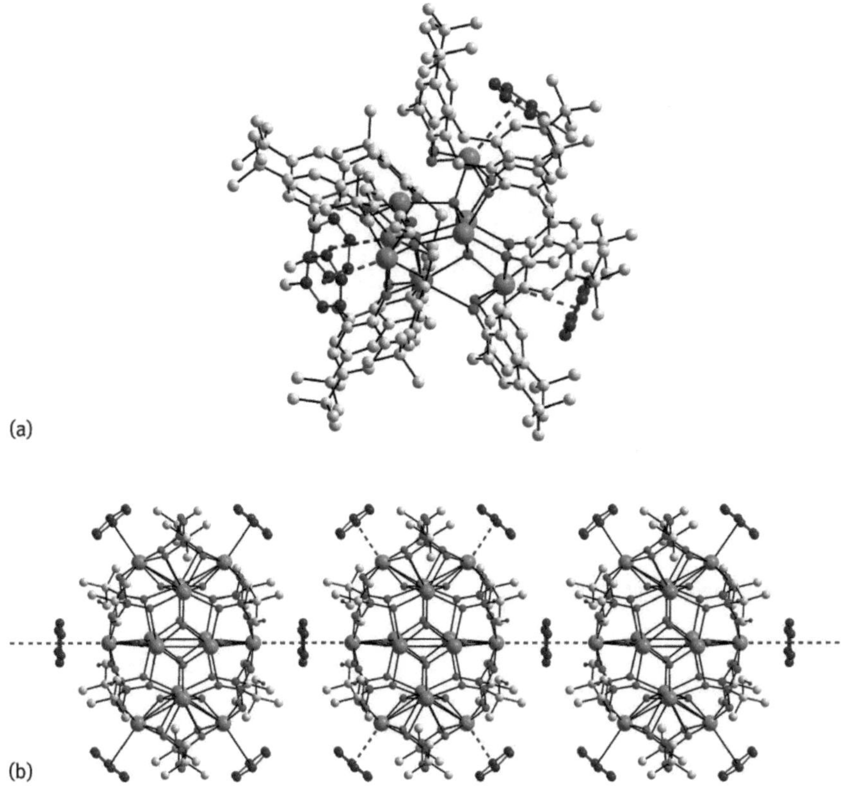

Fig. 14.15: Supramolecular aggregation via Bi(lp)...π(arene) interactions in the crystal structures of (31) and (32), leading to zero- and one-dimensional aggregation patterns in (a) and (b), respectively.

π-system is a solvent molecule, all four examples having Se(lp)···π(arene) interactions (33)–(36) (Figure 14.16) feature selenium in the +II oxidation state.

Compound (33), 3,4:7,8-bis(1,2-dicarba-closo-dodecaborano(1,2))-1,2,5,6-tetraselenacyclo-octane toluene solvate [51], is a zero-dimensional aggregate whereby the centrosymmetric molecule, with the eight-membered $[CSe_2C]_2$ core being based on a chair, is connected to two solvent toluene molecules [d = 3.30 Å and θ = 0.5°] to form a three-molecular aggregate, as shown in Figure 14.17a. By contrast, when the same compound is recrystallised from benzene, which results in a non-symmetric, 1 : 1 solvate, a one-dimensional supramolecular chain is formed [52]. As shown in Figure 14.17b, the solvent benzene molecules provide almost equivalent Se(lp)···π(arene) links between the selenium(II) centres [d = 3.45 Å and θ = 13.3°; d = 3.46 Å and θ = 13.4°] to generate the chain.

Compounds (35) and (36) (Figure 14.16) are heterometallic and feature four and eight selenium(II) centres respectively. The molecular structure of the complex in (35), (tris(t-butylimido)tetraselane)-dichlorido-palladium benzonitrile solv-

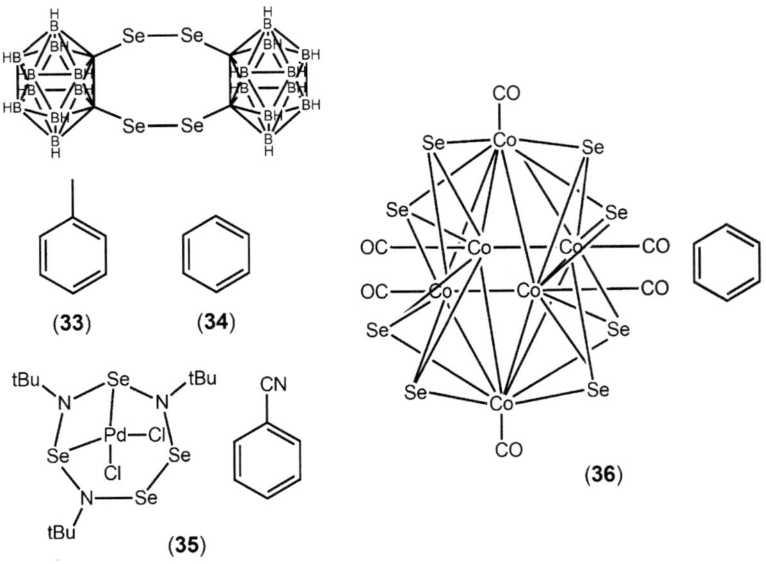

Fig. 14.16: Chemical diagrams for (33)–(36).

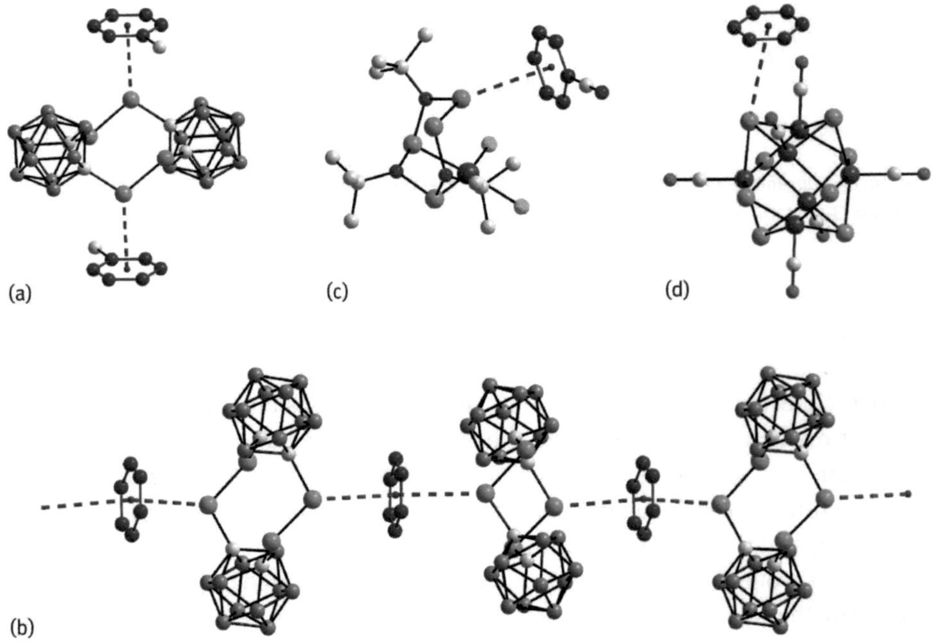

Fig. 14.17: Supramolecular aggregation via Se(lp)...π(arene) interactions in the crystal structures of (33)–(36) shown in (a)–(d), respectively.

ate [53], features a buckled, seven-membered [$SeNSe_2NSeN$] ring with two selenium(II) atoms coordinated to a palladium atom. One of the non-coordinating selenium(II) atoms forms a Se(lp)···π(arene) interaction with the benzonitrile ring [d = 3.57 Å and θ = 14.4°] (Figure 14.17c). In the second heterometallic complex, octakis(μ_3-selenido)-hexacarbonyl-hexa-cobalt benzene solvate (36) [54], each face of the central Co_6 octahedron is capped by a selenium(II) atom. One of these selenium atoms is connected to the benzene ring via a Se(lp)···π(arene) interaction [d = 3.94 Å and θ = 15.8°] (Figure 14.17d).

14.3.4.2 Tellurium(II)

There is a sole example of a tellurium compound with a Te(lp)···π(arene) interaction between tellurium and a solvent molecule, namely (μ_3-1,4-ditelluridopertelluracyclohexane)-tris[(η^5-pentamethylcyclopentadienyl)rhenium(III,VII)] benzene solvate (37) [55] (see Marsh and Clemente [56] for a space group revision). In (37) (Figure 14.18), there are eight potential tellurium acceptors. It is one of the tellurium(II) atoms directly bound to a rhenium atom and not part of the Te_6 six-membered ring that forms the interaction [d = 3.75 Å and θ = 8.7°] (Figure 14.19).

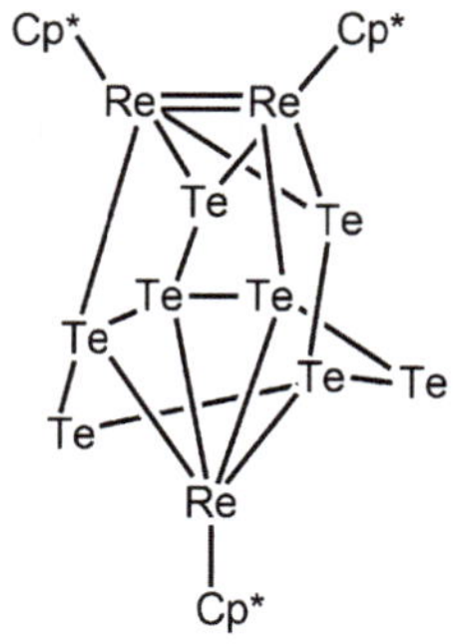

(37)

Fig. 14.18: Chemical diagrams for (37). Cp* represents the η^5-pentamethylcyclopentadienyl anion.

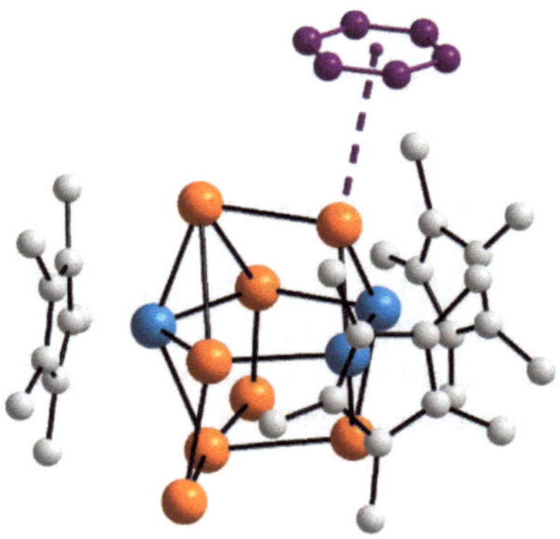

Fig. 14.19: Supramolecular aggregation via a Te(lp)...π(arene) interaction in the crystal structure of (37).

14.4 Conclusions

In the foregoing sections, examples of main group element(lp)···π(arene) interactions have been described for elements in oxidation states, consistent with the presence of a lone-pair of electrons. These interactions usually lead to zero-dimensional aggregates with a few rare examples where, owing to the presence of bridging solvent, supramolecular chains are observed. The bonding description for element(lp)···π(arene) interactions is akin to that accepted for halogen bonding. The energies of these interactions are not high, as was shown in a recent computational chemistry study [38]. This suggests that the (gas phase) energy of attraction in the series of Menschutkin complexes, $SbCl_3$.benzene, $SbCl_3$.toluene (18) and $SbCl_3$.hexamethylbenzene, might vary systematically with increasing methyl substitution, i.e. 7.7, 8.5 and 15.5 kcal mol^{-1} respectively [38]. With this information in mind, it is apposite to recall the series of $BiCl_3$ structures (24)–(26), i.e. co-crystallised with similar solvents such as benzene, m-xylene and hexamethylbenzene. Increasing the methyl content in the rings resulted in very different supramolecular architectures; a result that may correlate with the energies of attraction between rings and steric effects. In summary, the experimental and theoretical studies both indicate that the insight offered by systematic study within this fascinating area of supramolecular chemistry will be significant.

Acknowledgement: Sunway University is thanked for their support. The Brazilian agencies National Council for Scientific and Technological Development (305626/2013-2) and Brazilian Federal Foundation for Support and Evaluation of Graduate Education (CAPES) are acknowledged for their financial support.

Bibliography

[1] Zukerman-Schpector J, Haiduc I. Tellurium···π-aryl interactions: a new bonding motif for supramolecular self-assembly and crystal engineering. CrystEngComm, 2002, 4, 178–193.
[2] Haiduc I, Tiekink ERT, Zukerman-Schpector J. in Tin chemistry fundamentals, frontiers, and applications, ed. A. G. Davies, M. Gielen, K. H. Pannell and E. R. T. Tiekink, John Wiley & Sons Ltd., Chichester, 2008, p. 392–407.
[3] Tiekink ERT, Zukerman-Schpector J. Pb···π aryl interactions as supramolecular synthons. Aust. J. Chem. 2010, 63, 535–543.
[4] Caracelli I, Haiduc I, Zukerman-Schpector J, Tiekink ERT. M···π(arene) interactions for M = gallium, indium and thallium: influence upon supramolecular self-assembly and prevalence in some proteins. Coord. Chem. Rev. 2014, 281, 50–63.
[5] Zukerman-Schpector J, Otero-de la-Roza A, Luaña V, Tiekink ERT. Supramolecular architectures based on As(lone pair)···π(aryl) interactions. Chem. Commun. 2011, 47, 7608–7610.
[6] Caracelli I, Haiduc I, Zukerman-Schpector J, Tiekink ERT. Delocalised antimony(lone pair)- and bismuth-(lone pair)···π(arene) interactions: supramolecular assembly and other considerations. Coord. Chem. Rev. 2013, 257, 2863–2879.

[7] Caracelli I, Zukerman-Schpector J, Tiekink ERT. Supramolecular aggregation patterns based on the bio-inspired Se(lone pair)···π(aryl) synthon. Coord. Chem. Rev. 2012, 256, 412–438.

[8] Caracelli I, Haiduc I, Zukerman-Schpector J, Tiekink ERT, in The chemistry of organic selenium and tellurium compounds, ed. Z. Rappoport, John Wiley & Sons, Ltd, Chichester, 2013, vol. 4, p. 973–988.

[9] Caracelli I, Zukerman-Schpector J, Haiduc I, Tiekink ERT. Main group metal lone-pair···π(arene) interactions: a new bonding mode for supramolecular associations. CrystEngComm. 2016, 18, 6960–6978.

[10] Caracelli I, Haiduc I, Zukerman-Schpector J, Tiekink ERT, in Aromatic interactions: frontiers in knowledge and application, eds D. W. Johnson and F. Hof, Royal Society of Chemistry, Abingdon, 2016, p. 98–123.

[11] Egli M, Sarkhel S. Lone pair-aromatic interactions: to stabilize or not to stabilize. Acc. Chem. Res. 2007, 40, 197–205.

[12] Jain A, Ramanathan V, Sankararamakrishnan R. Lone pair···pi interactions between water oxygens and aromatic residues: quantum chemical studies based on high-resolution protein structures and model compounds. Protein Sci. 2009, 18, 595–605.

[13] Cavallo G, Metrangolo P, Milani R, Pilati T, Priimagi A, Resnati G, Terraneo G. The halogen bond. Chem. Rev. 2016, 116, 2478–2601.

[14] Groom CR, Bruno IJ, Lightfoot MP, Ward SC. The Cambridge Structural Database. Acta Crystallogr. Sect. B: Struct. Sci. Cryst. Eng. Mat. 2016, 72, 171–179.

[15] Bruno IJ, Cole JC, Edgington PR, Kessler M, Macrae CF, McCabe P, Pearson J, Taylor R. New software for searching the Cambridge Structural Database and visualising crystal structures. Acta Crystallogr. Sect. B: Struct. Sci. 2002, 58, 389–397.

[16] Schollmeyer D, Shishkin OV, Rühl T, Vysotsky MO. OH–π and halogen–π interactions as driving forces in the crystal organisations of tri-bromo and tri-iodo trityl alcohols. CrystEngComm, 2008, 10, 715–723.

[17] Spek AL. Single-crystal structure validation with the program PLATON. J. Appl. Crystallogr. 2003, 36, 7–13.

[18] DIAMOND, Visual crystal structure information system, version 3.1. CRYSTAL IMPACT, Bonn, Germany, 2006.

[19] Fillebeen T, Hascall T, Parkin G. Bis- and tris(pyrazolyl)hydroborato ligands with bulky triptycyl substituents: the synthesis and structural characterization of Tl[BpTrip] and Tl[TpTrip]. Inorg. Chem. 1997, 36, 3787–3790.

[20] Yap GPA, Jove F, Urbano J, Alvarez E, Trofimenko S, Diaz-Requejo MM, Perez PJ. Unusual polybrominated polypyrazolylborates and their copper(I) complexes: synthesis, characterization, and catalytic activity. Inorg. Chem. 2007, 46, 780–787.

[21] Yi R, Palmer JH, Parkin G. Benzannulated tris(2-mercapto-1-imidazolyl)hydroborato ligands: tetradentate κ^4-S_3H binding and access to monomeric monovalent thallium in an [S_3] coordination environment. Dalton Trans. 2014, 43, 1397–1407.

[22] Schmidbaur H, Bublak W, Riede J, Muller G. $[\{1,3,5\text{-}(CH_3)_3H_3C_6\}_6Tl_4][GaBr_4]_4$ – synthesis and structure of a mixed mono- and bis(arene)thallium complex. Angew. Chem. Int. Ed. 1985, 24, 414–415.

[23] Harvey S, Lappert MF, Raston CL, Skelton BW, Srivastava G, White AH. Conversion of the silicone poly(dimethylsiloxane) by thallium(I) ethoxide into the ladder polymer $[\{Tl_2(OSiMe_2)O\}_2]_n$; X-ray structure of the product and of thallium(I) triphenylsilanolate. Chem. Commun. 1988, 1216–1217.

[24] Hellmann KW, Gade LH, Fleischer R, Kottke T. Aggregation, reaggregation and degradation of a trifunctional thallium(I) amide induced by weak $Tl^{I}\cdots Tl^{I}$ attraction. Chem. Eur. J. 1997, 3, 1801–1806.

[25] Boyle TJ, Doan TQ, Steele LAM, Apblett C, Hoppe SM, Hawthorne K, Kalinich RM, Sigmund WM. Tin(II) amide/alkoxide coordination compounds for production of Sn-based nanowires for lithium ion battery anode materials. Dalton Trans. 2012, 41, 9349–9364.
[26] Hosmane NS, Zhang H, Maguire JA, Demissie T, Oki AR, Saxena A, Lipscomb WN. New stanna- and plumba-carboranes: synthetic and structural investigation. Main Group Met. Chem. 2001, 24, 589–596.
[27] Birchall T, Johnson JP. Crystal structure of di-μ3-oxo-octakis-μ-(trifluoroacetato)-ditin(II) ditin(IV)–benzene (1/1). J. Chem. Soc., Dalton Trans. 1981, 69–73.
[28] Tam ECY, Coles MP, Smith JD, Fulton JR. The steric influence of β-diketiminato ligands on the coordination chemistry of lead(II). Polyhedron. 2015, 85, 284–294.
[29] Ranis LG, Werellapatha K, Pietrini NJ, Bunker BA, Brown SN. Metal and ligand effects on bonding in Group 6 complexes of redox-active amidodiphenoxides. Inorg. Chem. 2014, 53, 10203–10216.
[30] Hosmane NS, Siriwardane U, Hong Zhu, Zhang G, Maguire JA. The first structural evidence for the insertion of an unsubstituted lead into a carborane cage. Organometallics, 1989, 8, 566–568.
[31] Edelmann FT, Buijink J-KF, Brooker SA, Herbst-Irmer R, Kilimann U, Bohnen FM. Formation and structure of the oxygen-centered lead thiolate cluster $Pb_5O(SR_F)_8 \cdot 2\,C_7H_8$ [R_F = 2,4,6-Tris(trifluoromethyl)phenyl]. Inorg. Chem. 2000, 39, 6134–6135.
[32] Gash AG, Rodesiler PF, Amma EL. Metal ion-aromatic complexes. XV. Synthesis, structure, and bonding of π-$C_6H_6Pb(AlC1_4)_2 \cdot C_6H$. Inorg. Chem. 1974, 13, 2429–2434.
[33] Schmidbaur H, Nowak R, Steigelmann O, Muller G. π-Complexes of p-block elements: synthesis and structures of adducts of arsenic and antimony halides with alkylated benzenes. Chem. Ber. 1990, 123, 1221–1226
[34] Shang S, Khasnis DV, Zhang H, Small AC, Fan M, Lattman M. Insertion of one and two Arsenic atoms into calix[4]arenes. Inorg. Chem. 1995, 34, 3610–3615.
[35] Stubenhofer M, Kuntz C, Balazs G, Zabel M, Scheer M. 1,2,3-Triphosphole derivatives as reactive intermediates. Chem. Commun. 2009, 1745–1747.
[36] Schier A, Wallis JM, Müller G, Schmidbaur H. $[C_6H_3(CH_3)_3][BiCl_3]$ and $[C_6(CH_3)_6][BiCl_3]_2$, arene complexes of bismuth with half-sandwich and "inverted" sandwich structures. Angew. Chem., Int. Ed. Engl., 1986, 25, 757–759.
[37] Schmidbaur H, Schier A. π-Complexation of post-transition metals by neutral aromatic hydrocarbons: the road from observations in the 19th century to new aspects of supramolecular chemistry. Organometallics 2008, 27, 2361–2395.
[38] Lo R, Svec P, Ruzickova Z, Ruzicka A, Hobza P. On the nature of the stabilisation of the E···π pnicogen bond in the $SbCl_3$···toluene complex. Chem. Commun. 2016, 52, 3500–3503.
[39] Schmidbaur H, Wallis JM, Nowak R, Huber B, Muller G. Aromaten-Komplexe mit Halbsandwich-Struktur: Die 1 : 1-komplexe des mesitylens mit $SbCl_3$, $SbBr_3$, $BiCl_3$ und $BiBr_3$. Chem. Ber. 1987, 120, 1837–1843.
[40] Mootz D, Handler V. Crystal structure of the Menshutkin complex benzene · 2 $SbCl_3$. Z. Anorg. Allg. Chem. 1986, 533, 23–29.
[41] Hulme R, Mullen DJE. Crystal structure of the 2 : 1 antimony trichloride: p-xylene intermolecular compound at –110 °C. J. Chem. Soc. Dalton Trans. 1976, 802–804.
[42] Kishore PVVN, Baskar V. Hexa- and trinuclear organoantimony oxo clusters stabilized by organosilanols. Inorg. Chem. 2014, 53, 6737–6742.
[43] Ly HV, Parvez M, Roesler R. An unusual ligand in copper chemistry: coordination oligomers and polymers containing the $[\{CpMo(CO)_2\}_2(\mu,\eta^2\text{-}Sb_2)]$ cluster. Inorg. Chem. 2006, 45, 345–351.
[44] Breunig HJ, Lork E, Rat C. The complex $BiCl_3 \cdot CH_3C_6H_5$. Z. Naturforsch., B: Chem. Sci. 2007, 62, 1224–1226.

[45] Muller-Becker S, Frank W, Schneider J. Schwermetall-π-Komplexe. IX. Die kettenpolymere $[(1,2\text{-}(CH_3)_2C_6H_4BiCl_3)_2]$, $[(1,3)(CH_3)_2C_6H_4BiCl_3)_2]$ und $[(1,4\text{-}(CH_3)_2C_6H_4BiCl_3)_2]$. Z. Anorg. Allg. Chem. 1993, 619, 1073–1082.

[46] Andrews PC, Junk PC, Nuzhnaya I, Spiccia L. Fluorinated bismuth alkoxides: from monomers to polymers and oxo-clusters. Dalton Trans. 2008, 2557–2568.

[47] Sharutin VV, Egorova IV, Sharutina OK, Ivanenko TK, Adonin NY, Starichenko VF, Pushilin MA, Gerasimenko AV. Tetranuclear bismuth complexes $Bi_4(O)_2(O_2CC_6H_2F_3\text{-}3,4,5)_8.2\,C_6H_6$ and $Bi_4(O)_2(O_2CC_6H_2F_3\text{-}3,4,5)_8.2\,C_6H_4Me_2\text{-}1,4$: synthesis and structures. Russ. J. Coord. Chem. 2005, 31, 2–8.

[48] Sharutin VV, Egorova IV, Sharutina OK, Ivanenko TK, Adonin NY, Starichenko VF, Pushilin MA, Gerasimenko AV. Tetranuclear bismuth complex $Bi_4(O)_2(O_2CC_6H_2F_3\text{-}3,4,5,)_8 \cdot 2\eta^6\text{-}C_6H_5Me$: synthesis and structure. Russ. J. Coord. Chem. 2003, 29, 838–844.

[49] Liu L, Zakharov LN, Rheingold AL, Hanna TA. Synthesis and X-ray crystal structures of the first antimony and bismuth calixarene complexes. Chem. Commun. 2004, 1472–1473.

[50] Mehring M, Schurmann M. The first bismuth phosphonate cluster. X-ray single crystal structure of $[(Bu^tPO_3)_{10}(Bu^tPO_3H)_2Bi_{14}O_{10} \cdot 3\,C_6H_6 \cdot 4\,H_2O]$. Chem. Commun. 2001, 2354–2355.

[51] Wrackmeyer B, Hernandez ZG, Kempe R, Herberhold M. Novel 1,2-Dicarba-closo-dodecaborane(12) derivatives of selenium. Eur. J. Inorg. Chem. 2007, 239–246.

[52] Garcia-Hernandez Z, Wrackmeyer B, Herberhold M. Irrgang T, Kempe R. Crystal structure of 3,4,7,8-bis(l,2-dicarba-closo-dodecaborano[l,2])-1,2,5,6-tetraselenacyclooctane benzene solvate, $[(C_2Bi_{10}H_{10})Se_2]_2.C_6D_6$. Z. Kristallogr. New Cryst. Struct. 2006, 221, 419–420.

[53] Risto M, Eironen A, Mannisto E, Oilunkaniemi R, Laitinen RS, Chivers T. Palladium complexes containing novel cyclic selenium imides. Dalton Trans. 2009, 8473–8475.

[54] Gervasio G, Kettle SFA, Musso F, Rossetti R, Stanghellini PL. Synthesis and characterization of three novel selenidohexacobalt carbonyl clusters. Crystal structures of $[(\mu_3\text{-}Se)Co_3(CO)_7]_2\mu_4\text{-}(Se_2)$ and $Co_6(\mu_3\text{-}Se)_8(Co)_6 \cdot 2\,C_6H_6$. Inorg. Chem. 1995, 34, 298–305.

[55] Jin G-X, Arikawa Y, Tatsumi K. Spontaneous formation of a diamond-crown structure of Re_8 polyselenide and a cage structure of Re_3 polytelluride. J. Am. Chem. Soc. 2001, 123, 735–736.

[56] Marsh RE, Clemente DA. A survey of crystal structures published in the Journal of the American Chemical Society. Inorg. Chim. Acta. 2007, 360, 4017–4024.

Index

https://doi.org/10.1515/9783110464955-203

Bei Fragen zur Produktsicherheit wenden Sie sich bitte an:
If you have any questions regarding product safety,
please contact:

Walter de Gruyter GmbH
Genthiner Straße 13
10785 Berlin
productsafety@degruyterbrill.com